Polar Lipids

Biology, Chemistry, and Technology

Polar Lipids

Biology, Chemistry, and Technology

Editors

Moghis U. Ahmad

Xuebing Xu

AOCS PRESS

Urbana, Illinois

ISBN 978-1-630670-44-3 (print)
ISBN 978-1-630670-45-0 (.epub)
ISBN 978-1-630670-46-7 (.mobi)

Library of Congress Cataloging-in-Publication Data
Polar lipids : biology, chemistry, and technology / editors, Moghis Ahmad, Xuebing Xu.
pages cm
ISBN 978-1-63067-044-3 (print : alk. paper)—ISBN 978-1-63067-045-0 (epub)—ISBN 978-1-63067-046-7 (mobi) 1. Lipids. 2. Oils and fats—Analysis. 3. Lecithin. 4. Chemistry, Organic. I. Ahmad, Moghis, editor. II. Xu, Xuebing, 1962– editor.
QP751.P635 2015
572'.57— dc23

2015007162

Printed in the United States of America
19 18 17 16 15 5 4 3 2 1

The paper used in this book is acid-free, and falls within the guidelines established to ensure permanence and durability.

www.aocs.org

Contents

Preface

Polar lipids from natural sources are the main constituents of natural membranes, occurring in all living organisms. Polar lipids isolated from vegetable (lecithin) are known for their emulsifying and structural-improvement properties in food matrices, whereas polar lipids derived from animal sources have great importance because they are rich in sphingolipids (in the case of dairy polar lipids) that have profound effects on cell metabolism and regulation. Sphingolipids are shown to be highly active compounds upon ingestion. Phospholipids and sphingolipids (polar lipids) are amphiphilic molecules with a hydrophobic tail and a hydrophilic head group. The glycerophospholipids consist of a glycerol backbone on which two fatty acids are esterified on positions *sn*-1 and *sn*-2. On the third hydroxyl, a phosphate residue with different organic groups such as choline, serine, ethanolamine, and others may be linked. Generally, the fatty acid chain on the *sn*-1 position is more saturated compared to *sn*-2 position. Lysophospholipids contain only one acyl group, predominantly at the *sn*-1 position of glycerol moiety. In sphingolipids, the characteristic structural unit includes a sphingoid base and a long-chain (C12–C22 carbon chain) aliphatic amine containing two or three hydroxyl groups. Sphingosine is the most prevalent sphingoid base in mammalian sphingolipids, containing an unsaturated C18 carbon chain and two hydroxyl groups. A ceramide is formed when the amino group of sphingoid base is linked with (usually) a saturated fatty acid. On this ceramide unit, an organophosphate group can be bound to form a sphingophospholipid or a saccharide to form a sphingoglycolipid (glycosylceramide). Monoglycosylceramides, such as glucosylceramide or galactosylceramide, are often called cerebrosides, and tri- and tetraglycosylceramides with a terminal galactosamine residue are called globosides. Similarly, gangliosides are highly complex oligoglycosyl ceramides containing one or more sialic acid groups in addition to glucose, galactose, and galactosamine. The basic chemical information of the polar lipids described has been studied in a number of fields, including occurrence, chemistry and chemical synthesis, enzymatic synthesis, analytical chemistry, food science and nutrition, physical properties, oxidation reaction, and health implications. Artificial polar lipids, including those that have similar surface properties to the ones previously mentioned , have received attention for different applications, such as sugar esters, phenolic lipids, and others. These compounds, although they do not exist in nature, have a lot of good properties for food and other applications. The potentials of those polar lipids have received significant attention.

A wealth of information about polar lipids is widely scattered in the literature, but this information was never compiled in one place to be used as a reference resource. Some of the polar lipids, such as soy lecithin, rice lecithin, sunflower lecithin, palm lecithin, serine phospholipids, glycolipids, milk and dairy polar lipids, sugar-based lipids, phenolic lipids, glycerolipids, lysophospholipids, and their spectral analyses, have been compiled in this book. In this book, authors who are recognized international authorities in their fields have addressed various domains of polar lipids such as chemistry and chemical synthesis, biosynthesis and biological effects, functional and nutritional properties, applications, processing technologies, and spectral analysis of different types of polar lipids. Each chapter contains the latest references, major technological advances, and important discoveries.

This book will be a valuable reference resource providing thorough and comprehensive coverage of different types of polar lipids known to lipid science and industry today. This book covers important applications and utilization of polar lipids in the areas of food, nutrition, health, and disease. We hope that the availability of so much information and the latest references in a single volume will help the next generation of scientists make further progress in this challenging field of lipid science.

Moghis U. Ahmad
Xuebing Xu

List of Abbreviations

1-LPs	1-Lysophospholipids
2-LPA	2-Lysophosphatidic acid
2-LPC	2-Lysophosphatidylcholine
31P-NMR	31P-Nucleic Magnetic Resonance
7-DHC	7-Dehydrocholesterol
8-OHdG	8-Hydroxy-2'-deoxyguanosine
AAMI	Age-associated memory impairment
AAPH	2,2'-azpbis(amidinopropane)
ACR	Acrolein
AD	Attention deficiency
ADHD	Attention-deficit hyperactivity disorder
AEP	Alkyl ether phospholipid
AFM	Atomic force microscopy
AI	Acetone insoluble
AICAR	5-aminoimidazole-4-carboxamide ribonucleoside
ALD	Alcoholic liver disease
ALDH2	Aldehyde dehydrogenase
ALP	Alkyl-lysophospholipid
AMPK	AMP protein kinase
AOCS	American Oil Chemists' Society
APCI	Atmospheric pressure chemical ionization
APHDL	Acute phase HDL
AQKR	Aldo-keto reductase
ARA	Arachidonic acid
AST	Aspartate aminotransferase
ATL	Antitumor lipid
ATX	Autotaxin
AV	Anisidine value
Aβ	Amyloid β-peptide
BC-PS	Bovine cortex
BEP-CE	Bicyclic endoperoxide-cholesteryl ester
BHT	Butylated hydroxyl toluene

CALB	Candida antarctica lipase B
CAP	Carboxyalkylpyrrole
CCP	Carboxypropylpropylpyrrole
CD	Cluster differentiation
CE	Cholesteryl ester
CE-16:0	Cholesteryl palmitate
CE-18:2	Cholesteryl linoleate
CHP	Carboxyheptylpyrrole
CID	Collision induced dissociation
CL	Cardiolipin
CLA	Conjugated linoleic acid
CLEAR	Canadian low erucic acid rapeseed
CMC	Critical micelle concentration
CML	Carboxymethyllysine
CNS	Central nervous system
COX	Cyclooxygenase
CPP	Carboxypentylpyrrole
CRP	C-reactive protein
CSLM	Confocal scanning laser microscopy
CVD	Cardiovascular disease
CW	Continuous wave
CYP	Cytochrome p450
DATA	Diacetyl tartaric acid
DF	Diafiltration
DGF	German Fat Society
DGDG	Digalactosyl diacylglycerol
DHA	Docosahexaenoic acid
DHET	Dihydroxyeicosatrienoic acid
DMAP	Dimethylaminopyridine
DME	Dimethyl ether
DMF	Dimethylformamide
DMPS	Dimyristoylphosphatidylserine
DMSO	Dimethylsulfoxide
DNPH	Dinitrophenylhydrazine
DOPS	Dioleoylphosphatidylserine
DPG	Diphosphatidylglycerol
DSHEA	Dietary Supplement and Health and Education Act

DSPC	Distearoyl phosphatidylcholine
DSPG	Distearoyl phosphatidylglycerol
EC	Electron capture
EDP	Epoxydocosapentaenoic acid
EDTA	Ethylenediaminetetra acetic acid
EEG	Electro-encephalography
EET	Epoxyeicosatrienoic acid
EFSA	European Food Safety Authority
EGF	Epidermal growth factor
EI	Emulsification indexes
ELISA	Enzyme linked immunosorbent assay
ELSD	Evaporating light scattering detector
EPA	Eicosapentaenoic acid
ESI	Electrospray ionization
Ether-PLs	Ether phospholipids
F2-IsoPs	Ring f-isoprostanes
FA	Fatty acid
FAB-MS	Fast atom bombardment MS
FCC III	Food Chemical Codex III
FDA	Food and Drug Administration (U.S.)
FFA	Free fatty acid
FFMP	Fat-filled milk powder
FG	Fat globule
FMOC	Fluorenylmethylcarbonate
FT	Fourier transformation
GA	Glucoalbumin
GC	Gas chromatographic
GLC	Gas liquid chromatography
GLP	Glycerol lacto palmitate
GL	Glycolipid
GM	Genetically modified
GMO	Genetically modified organism
GMP	Good manufacturing practice
GPA	Glycerol-3-phosphate
GPAA	Glycerophosphoacetic acid
GPC	Glycerol phosphocholine

GPCR	G protein-coupled receptor
GroPCho	Glycerophosphocholine
GSH	Gluthathione
H,P-TOXY	H-detected H,P total correlation spectroscopy
HA	Hyaluronic acid
HACCP	Hazard analysis and critical control points
HAEC	Human aortic epithelial cell
HAS	Human serum albumin
HAS2	Hyaluronan synthase 2
HBD	Hydrogen bond donor
HD	Hyperactivity disorder
HDL	High-density lipoprotein
HETE	Hydroxyeicosatetraenoic acid product
HHE	Hydroxyhexynal
HI	Hexane insoluble
HLB	Hydrophilic lipophilic balance
HM	Heat method
HMQC	Heteronuclear multiple-quantum correlation
HNE	4-Hydroxy-nonenal,4-hydroxy-2-nonenal
HODA	9-Hydroxy-12-oxo-10-dodecenoic acid
HODA-PC	9-Hydroxy-[12-oxo]-10-dodecenoyl-GroPCho
HODE	Hydroxyoctadecadienoic acid
HOOA-PC	Hydroperoxy arachidonoyl GroPCho
HPL	Hog pancreas lipase
HPLC	High performance liquid chromatography
HPLC-LSD	High performance liquid chromatography with a light scattering detector
HPODE	Hydroperoxyoctadecadienoic acid
HP	Carboxyethoxypyrrole
HPTLC	High performance thin-layer chromatography
HSE	Higher substitution esters
ICCO	International Cocoa Organization
IGF-1	Insulin-like growth factor 1
IL	Ionic liquids
ILPS	International Lecithin and Phospholipid Society
IP	Identity preserved

Iso-Fs	Isofurans
IsoLGs	Isolevuglandins
Iso-Ps	Isoprostanes
LAB	Lactic acid
LC	Liquid chromatrography
LC/ESI-MS	Liquid chromatograpy/electrospray-mass spectrometry
LCP	Liquid crystal phospholipid
LDH	Lactate dehydrogenase
LDL	Low density lipoprotein
LG	Levuglandins
LML	Lysine-MD-lysine
LOX	Lipoxygenase
LPA	Lysophosphatidic acid
LPATs	Lysophospholipid acyltransferases
LPC	Lysophosphatidylcholine
LPE	Lysophosphatidylethanolamine
LPI	Lysophosphatidylinositol
LPL	Lysophospholipid
LPS	Lysophosphatidylserine
LP	Lysophospholipid
LSE	Lower substitution ester
Lyso-PC	Lysophosphatidylcholine
Lyso-PLD	Lysophospholipase D
MCT	Medium chain triglycerides
MDA	Malonyl dialdehyde
MFGM	Milk fat globule membrane
MFG	Milk fat globule
MGDG	Monogalactosyl diacylglycerol
mmLDL	Minimally modified LDL
MMP-I	Matrix metalloproteinase I
MRP	Maillard reaction product
MS	Mass spectrometry
MS/MS	Tandem mass spectrometry
MWCO	Molecular weight cut-off
NHDL	Normal HDL
NIST	National Institute of Standards

NL	Neutral lipid
NMR	Nuclear magnetic resonance
NOE	Nuclear overhauser effect
NP	Neuroprostane
O/W	Oil-in-water
Oxo-FAs	Oxo-fatty acids
Oxo-HDL	Oxidized-HDL
Oxo-LDL	Oxidized LDL
Oxo-P/A-PC	Oxo-palmitoyl/arachidonoyl PC
Oxo-PAPE-N-biotin	1-Palmitoyl-2-[5-oxo]valeroyl GroPtn-N-biotin
Oxo-PC	Oxo-phosphatidylcholine
Oxo-PL	Oxo-phospholipids
P,H-COSY	P-detected P,H correlation spectroscopy
PA	Phosphatidic acid
PAF	Platelet activating factor
PC	Phosphatidylcholine
PCB	Polychlorinated biphenyl
PC-HAS	PtdCho γ-hydroxyalkenals
PD	Parkinson's disease
PE	Phosphatidylethanolamine
PEG	Polyethylene glycol
PES	Polyethersulphone
PFAD	Palm fatty acids distillates
PG	Phosphatidylglycerol
$PGF_{2\alpha}$	Prostaglandin $F_{2\alpha}$
PGME	Propylene glycol monomethyl ester
PGPC	Palmitoyl/glutarylGroPCho
PI	Phosphatidylinositol
PIPS	PS complex enriched with PI
PLA	Phospholipase A
PLC	Phospholipase C
PLD	Phospholipase D
PL	Polar lipid
PMSF	Phenylmethane sufonyl fluoride
POV	Peroxide value
PPAR	Peroxisome proliferator activator receptor

PPTS	P-toluensolfonate
PS	Phosphatidylserine
PSD	Particle size distribution
PtdCho	Phosdphatidylcholinc
PtdEtn	Phosphatidylethanolamine
PtdGro	Phosphatidylglycerol
PtdIns	Phosphatidyl inositol
SDS-PAGE	Sodium dodecyl sulfate polyacrylamide gel electrophoresis
SRA I/II	Scavenger receptor class A types I/II
SR-B1	Scavenger receptor B1
SRM	Selected ion monitoring
SR-RI	Scavenger receptor RI
SYK	Spleen tyrosine kinase
TAG	Triacylglycerol
TBARS	Thiobarbituric acid reactive substances
TEM	Transmission electron microscopy
TFA	Trifluroacetic acid
TG	Triglycerides
TGOOH	Triacylglycerol peroxide
TI	Toluene insoluble
TLC	Thin-layer chromatography
TLL	Thermomyces lanuginosus lipase
TLR	Toll-like receptor
TMS	Trimethylsilyl
TOF	Time of flight
T-OF-MS	Time of flight mass spectrometry
TPP	Triphenylphosphate
TTT	Thymol turbidity test
TXA2	Thromboxane receptor A2
UF	Ultrafiltrate
USP	U.S. Pharmacopeia
UV	Ultraviolet
VEGF	Vascular endothelial growth factor
VSMC	Vascular smooth muscle cell

W/O	Water-in-oil
WMP	Whole milk powder
XRT	X-ray micro-tomography

1

Soybean Lecithin: Food, Industrial Uses, and Other Applications

G.R. List ■ *USDA, Retired*

Introduction

Lecithin has long been an important component of a myriad of both food and nonfood products and is one of the most versatile and valuable byproducts of the oilseed industry. In foods, lecithin provides about a dozen functions, including as an emulsifier, as a wetting agent, for viscosity reduction, as release agents, and for crystallization control. Lecithin also provides functions in numerous industrial applications as well (Hilty, 1948; Sipos, 1989). Much of the lecithin technology was developed in Germany by Bollman and Rewald who held a number of patents describing the process of recovering phosphatides from oilseeds (Bollmann, 1923, 1928, 1929). Bollman also patented a process for improved salad oil manufacture (Bollman, 1926). By 1940, the U.S. lecithin industry was well established (Wendel, 2000). The emulsifying and other functional properties of lecithin were well-recognized and incorporated in numerous foods including margarine, shortenings, baked goods, chocolates, candy, macaroni and noodles, ice cream, instant cocoa beverages, powdered milk, and as crystal inhibitors and antioxidants in fats and oils (Eichberg, 1939; Wiesehahn, 1937). Moreover, a number of nonfood uses were described, including petroleum lubricants, textiles, leather, coatings, rubber, resins, and soaps. The benefits of lecithin were recognized by the cosmetic and pharmaceutical industries as well (Wittcoff, 1951). Phosphatides may undergo numerous chemical reactions including autoxidation, complexing with metal ions, and reactions within the fatty acids or the phosphatide groups (Ghyczy, 1989; Pryde, 1985). However, only a few have been exploited by the industry.

Much of the work on chemical modification of phosphatides, including sulfonation (Scharf, 1953), halogenation (Denham, 1939), hydroxylation (Wittcoff, 1948, 1949), hydrogenation (Arveson, 1942), acetylation (Davis, 1970, 1971; Hayes and Wolff, 1957), alcohol fractionated (Julian and Iveson, 1958), acid/base hydrolyzed (Davis, 1970, 1971), blending (Wiesehahn, 1940), and oil free lecithin (Thurman, 1947), have been detailed in the patent literature. Enzymatic modification was patented by Pardun (1972) and reviewed by Schneider (2008). Pardun was an early pioneer in lecithin usage and held a number of patents (Pardun, 1967, 1969, 1970, 1972a, 1972b, 1972c). His bio can be found on the Lipid Library website (www.lipidlibrary.org). Buer (1913) was an early worker in the lecithin field, especially in the nutritional arena. An excellent review of the chemistry of phosphatides can be found in Deuel (1951). Jordan (1932, 1935, 1949) was also a pioneer in lecithin applications.

Bonekamp (2008) wrote an excellent review of the chemical modification of lecithins and Doig and Diks (2003) have reviewed methods for the modification of the lecithin headgroup. Only three major chemical reactions have seen industrial adaptation. These include acetylation by the reaction of acetic anhydride with the amino group of phosphatidylethanolamine, hydroxylation by reaction of hydrogen peroxide with double bonds of unsaturated phospholipids, and chemical/enzymatic modification cleaving the acyl group on the fatty acid chains with the formation of lysophosphatides (Arrigo and Servi, 2010).

A number of reviews have covered the food, industrial, and manufacturing uses of lecithin (Brian, 1976; Dashiell, 1989; Flider, 1985; Gunstone, 2009; Hernandez and Quezada, 2008; List, 1989; Orthoefer and List, 2006; Prosise, 1985; Schmidt and Orthoefer, 1985; Schipinov, 2002; Schneider, 2008; Sipos and Szuhaj, 1996; Stanley, 1951; Szuhaj, 1983, 2005; van Nieuwenhyzen, 1976, 1981, 2008; van Nieuwenhuyzen and Tomas, 2008; Wendel, 1995). The reader is directed to an early discussion of phosphatides containing nearly 200 references up to 1944 (Markley and Goss, 1944).

Liposomes in drug delivery systems were reviewed by Sharma and Sharma (1997), Immordino et al. (2006), and Mufamadi (2011). By 1991, well over 100 refereed publications on liposomes had been published. The classification, preparation, and applications of liposomes were recently reviewed by Akbarzadeh et al. (2013).

The lecithin industry is a mature one, but several factors have affected it. Although historically soybean has been the major source of lecithin worldwide, others are being sought because of increased demands for non-GMO (genetically modified organism) lecithin, including canola and sunflower (van Nieuwenhuyzen, 2014). Although lecithin from GMO soybeans has been shown to be equivalent to non-GMO lines (List et al., 1999), the European market prefers non-GMO lecithin. Over the past few decades, lecithin has become more important as a neutraceutical and food supplement ingredient. Moreover, the discovery of liposomes has provided a new and more efficient means for drug delivery. This chapter will review the lecithin industry, manufacture and properties of commercial products, their quality control and modification, and food and nonfood uses.

An Overview of the U.S. Lecithin Industry

The American lecithin industry is composed of over 30 companies who manufacture and/or sell a wide variety of products, including fluid bleached/unbleached, acetylated, hydroxylated, water dispersible (chemically or enzymatically modified), fluid identity preserved, fluid non-GMO, and granular and powdered deoiled products.

Virtually all food-grade lecithin produced in the United States is derived from soybeans and oil. A U.S. company in Florida offers soy, sunflower, and canola lecithin. U.S. production has remained fairly constant over the period from 1994 to

2010 (the last year for which statistics are available). Production has varied from a low of about 83,000 short tons in 1994 to a high of 112,000 short tons in 2003. The 17 year average is slightly over 98,000 short tons and amounts to an annual production of nearly 197 million pounds.

Manufacturing and distribution of specialized phospholipids and lecithin is global in scope, with a number of plants located in China, India, Croatia, Peru, Egypt, Sri Lanka, and Japan. China and India have 20 plants combined. In 2010, well over 100 countries imported lecithin/phospholipids. The top five include the Netherlands ($124.4 million), the United States ($65.7 million), Germany ($54.7 million), China ($45.3 million), and Italy ($38.8 million).

Worldwide, prices of lecithin vary widely depending in part of the degree of refining, source, and purity. High purity phosphatidylcholine (PC) from vegetable oils may sell for $5–10 /kg up to over $1/gr. PC derived from eggs may sell for $15–100/kg up to $950–1000/kg. PC from soy extracts sell for around $20–30/kg. Very pure lyso PC may cost $740/gr or $800/500 gr. Soy-based lecithin sold in the United States is in the forms of fluid, granular, or modified. Although prices are not available, they vary with the degree of refining (in order of increasing costs): fluid, fluid single bleached, fluid double bleached, deoiled (95% AI), deoiled (97% AI), modified (acetylated, hydroxylated, chemically/enzymatically modified).

Soy-based fluid lecithin sold on the international market varies in price from $700–1200/ton in large lots. Imported prices range from $0.90–1/kg in Australia, $1/kg in South America, $2.07/kg in North America, and $3.50/kg in Europe and Japan.

Commercial Lecithins

Lecithins can be classified as fluid, oil-free, and modified. Gums from the treatment of crude oils (soybean, canola, sunflower) contain about 70% phospholipids and 30% oil and water. The lecithin is recovered by drying. Fluidization is accomplished by the addition of free fatty acids, metal salts, and/or oil. Specifications for fluid lecithin usually indicate a minimum of 62% acetone insolubles. Where color is important, lecithin may be bleached (Bollmann, 1929; Schwieger, 1934) via addition of hydrogen peroxide and benzoyl peroxide to the degumming step. Oil-free lecithin is obtained by treating the gums with acetone, resulting in 95–97% phospholipids. Commercial lecithins cover a wide range of hydrophobic lipophilic balance (HLB) values (Griffin, 1949), which serves as a guide to the emulsification required for various applications, depending whether oil-in-water (O/W) or water-in-oil (W/O) are needed. Lecithin serves to reduce surface tension at the oil–water interface and facilitates disruption of bulk phases into small droplets. HLB value serves as a guide for the type of emulsion required. For example, W/O emulsions require HLB values from 3–6, O/W from 8–18, and both types fall in the 6–8 range. HLB values are usually measured against other reference emulsifiers: W/O require mono/digylcerides, O/W or W/O require sorbitan esters, and O/W' require ethoxylated mono/digylcerides. Wetting agents require HLB values from 7–9, whereas detergents and solubilizing applications require 13–14 and 15–18, respectively (Archer Daniels Midland, 2014; Flider, 1985; van Nieuwenhuyzen, 2008).

Although phosphatides can be modified chemically, only acetylated (Hayes and Wolff, 1957), hydroxylated (Wittcoff, 1948, 1949), and enzymatically modified lecithins (Pardun, 1972) are commercially available and have been patented by Schmitt (2014). Chemically modified (acid, base, neutral pH) lecithins are described in the patent literature (Davis, 1970, 1971). Modification of lecithin by fractionation with alcohol to enrich individual phospholipids has been reported in the patent literature (Klenk et al., 1962; Myer et al., 1960). Treatment of oil-free lecithin with 90% ethanol yields a soluble fraction containing primarily PC and phosphatidyethanolamine (PE), while the insoluble fraction contains enriched phosphatidylinositol (PI). Oil-free lecithin favors both O/W and W/O emulsions. The soluble fraction is more suitable for O/W, whereas the insoluble fraction favors W/O emulsions (Flider, 1985). The PC/PE ratio of the soluble fraction is over 8 (van Nieuwenhuyzen, 1981). The alcohol soluble fraction can be further fractionated on alumina to yield very high contents of PC and to be virtually free of PE and PI (Betzing, 1980).

Manufacture of Soybean Lecithins

The manufacture of lecithin has been reviewed by a number of authors (Brian, 1976; Flider, 1985; Joshi et al., 2006; List, 1989; Markley and Goss, 1944; Sipos and Szuhaj, 1996; Stanley 1951; Szuhaj, 1983, 2005; van Nieuwenhuyzen, 1976, 1981; Wendel, 1995). The steps involved include: crude oil preparation, hydration of the gums, isolation of the crude gums by centrifugation, drying of the gums, and fluidization and/or further processing, which may include deoiling, addition of other surface active ingredients, or modification with chemicals or enzymes. Although alcohol fractionation is an excellent method to enrich phospholipids, the author is unaware of any U.S. commercial products based on alcohol fractionation.

The quality of soybeans influences the lecithin quality. Soil type, splits, and green immature or older beans may be sources of poor quality. Pretreatment of beans, including dehulling and removal of dust, are essential for high quality lecithin. The extraction step requires the proper flake depth and temperature within the extraction vessel. The degumming step requires control of water dosage (and metal-free water), keeping the temperature at a controlled 70–80 °C range, and sufficient agitation. The hydration of gums occurs within 15–30 minutes (List et al., 1981). Laboratory studies indicate bleaching occurs relatively fast (within 15 minutes) but may take longer in commercial practice.

Manufacture of lecithin begins with the crude oil, which is normally filtered to remove insoluble materials (hexane insolubles) such as dirt, seed fragments, and so forth. Degumming may be done with water, organic acids/anhydrides, or enzymes. Water degumming is the technology of choice in the United States, whereas acid or enzymatic degumming is favored in Europe. Water degumming will remove variable amounts of phosphatides ranging from about 80–95% (List et al., 1978). In the United States, crude soybean oils normally contain low levels of nonhydratable phosphatides (salts of phosphatidic acid). Oils high in nonhydratable phosphatides may require pretreatment with phosphoric acid, especially if the oil is to be physically refined. However, the oil should be degummed first with water then treated with acid. Direct treatment of crude oil with phosphoric acid produces dark-colored lecithin. Normally the addition of about 2% water (or an amount near the gum content of the oil) to the crude oil at a temperature of 70–80 °C in a stirred agitated tank for 15–30 minutes is sufficient to hydrate the gums to the point that they form gels and precipitate from the degummed oil. If bleached products are desired, the bleaching agent (30% hydrogen peroxide) is added during the gum hydration step. The crude gums are then recovered via a centrifuge. The crude gums consist of both water and phosphatides, which must be removed by drying under a vacuum and heating with steam. Although batch dryers have been used, continuous dryers in which residence time and temperature can be closely controlled are preferable. The conditions for batch (Bollman) and continuous (horizontal film evaporator) drying have been described by van Nieuwenhuyzen (1976). It is assumed that the starting material is 50/50 water and crude gums and the final moisture content is less than 1%. Batch dryers operate at 60–80 °C for 3–4 hours under a vacuum of 20–60 tor, whereas the continuous dryer is operated at 80–95 °C with a very short residence time (1–2 minutes) and a vacuum of 50–300 tor. The advantage of batch bleaching allows in situ bleaching, whereas continuous systems show less color degradation. The viscosity of lecithin is closely related to the moisture content and it is crucial to producing quality lecithin (van Nieuwenhuyzen, 1976). Overdrying causes dark-colored lecithin, whereas insufficient drying will produce a cloudy product. Color specifications for natural fluid lecithin, bleached lecithin, and double-bleached lecithins are 10, 7, and 4 on the Gardner Color Scale, respectively, with a viscosity of 150 P at 77 °F max. Acetone insoluble (AI) contents in the fluidized products are 62% or higher after fluidization with fatty acids and/or oil (Werly, 1957). Plastic grades have similar specs, but they have an acetone insoluble content of 65% or higher.

Deoiling of lecithin requires a high quality crude lecithin as a feedstock. Typically the ratio of acetone to lecithin should be 10–20:1 in order to reach 95% AI in the deoiled product. In batch extraction the process may be extracted a number of times. A complete description of deoiling is found in Flider (1985).

Modified lecithins available in commerce include hydroxylated, acetylated, and chemical/enzymatic modified. Hydroxylation is achieved by treating lecithin with hydrogen peroxide and lactic acid. The phosphatide group is not changed but the hydroxyl groups are formed by addition to the double bonds in the two fatty acid groups. This has been confirmed by a drop in iodine value with increased hydroxylation (Julian, 1953). The presence of hydroxyl groups in the phospholipid increases the HBL value to about 10, which favors O/W emulsions.

Acetylation occurs in the ethanolamine group in phosphatidyl ethanolamine and is catalyzed by the addition of acetic anhydride during degumming or on neat soy lecithin preparations. Acetylated lecithins also show an increased HLB value (9–10) and are useful in O/W emulsion systems.

An enzymatic process for preparation of acetylated phospholipids has been reported in the patent literature (Marellapudi et al., 2002). Enzymatically modified lecithins react with the fatty acids in the 1 position of the phospholipid molecule, leading to the formation of lysophosphatides, which show more surface activity than the parent phospholipid. Phospholipase A1 is the preferred enzyme, although a process was developed with A2 in the early 1990s. Since then, a number of enzymes have been developed and commercialized having different specificities. Lectase and lysomax A1 enzymes catalyze the reaction with all common phosphatides, whereas purifine is specific for PC and PE.

The preparation of lysophosphatides via chemical means was described in the early patent literature (Davis, 1970, 1971) in both acidic and basic media. The chemistry of lysophosphatides was recently reviewed by Arrigo and Servi (2010).

Commercial Soy-Based Lecithins

The following represents a typical line of soy oil-based lecithins produced in the United States (1–10) by major suppliers (Archer Daniels Midland, 2014).

1. Dry deoiled lecithin; light tan in color; minimum 97% AI; no carriers, diluents; very bland flavor and odor; oil soluble and water dispersible (HLB7).

   Control of fat in cooked ground meat products (chili, Mexican food fillings), and frostings in combination with polysorbates gives emulsification and air entrapment in baked goods, improves dough machinability. This is useful in products for which a dry form or a low flavor profile is required. It serves as O/W emulsifier, release agent, viscosity modifier, or wetting agent labeled as "soy lecithin" or "lecithin."

   It is packaged in 44 1-lb packages, 18 boxes/pallet, 794 lbs. It can be stored for 2 years in the original unopened container.
2. Same as above, with silicon dioxide added to prevent caking. Useful in hot, humid conditions. Labeled "soy lecithin with silicon dioxide (to prevent caking)." Packaged in 44 1-lb bags.
3. Dry deoiled lecithin granules. Labeled "soy lecithin" or "lecithin." Packaged in 44 1-lb bags per box.

4. Dry, deoiled lecithin granules. Packaged in vacuum-sealed pull-top canisters especially for sale in health food stores; 14 oz/can, 12 cans/case.
5. Hydroxylated soy lecithin, dry, deoiled, low pH applications (below 4). Highly effective wetting agent, O/W emulsifier; improves rollability, extensibility, and shelf life of flour tortillas; improves self-rising pizza crusts; controls hydration of fatty hydrophobic powders; solubilizes and disperses oil-soluble colors and flavors in aqueous systems; retards separation and improves emulsion stability in pourable salad dressings; oil soluble and water dispersible. Labeled "hydroxylated lecithin," 44 1-lb bags/box.
6. Heat-resistant, water-dispersible, essentially oil-free lecithin powder. Wetting agent; O/W emulsifier; specially processed to offer maximum heat resistance for release formulations; does not develop burned odor, color, or taste. Labeled as "soy lecithin."
7. Enzyme-modified soy lecithin, dry, deoiled. Extends shelf life of baked goods; provides smoothness in starch-based sauces and puddings; controls stickiness during dehydration of potatoes; oil soluble and water dispersible. Labeled "enzyme-modified soy lecithin," 44 1-lb bags/box.
8. Fluid, food-grade, light amber, medium viscosity blend of lecithin, ethoxylated mono- and diglycerides, and propylene glycol. Surfactants added to be a highly water dispersible O/W emulsifier (HLB 12). Wetting agent for high-fat powders and belt release. Labeled "soy lecithin with ethoxylated mono/diglycerides and propylene glycol," 450-lb drum.
9. Food-grade, fluid, bland, highly filtered clear and brilliant, low viscosity, refined lecithin. Sprayable wetting and release agent for flavor sensitive foods; instantizing agent controls hydration rate of hydrophilic powders; as a release agent, provides good solubility in pump and aerosol formulations; easily blended in both cold and warm oil; effective W/O emulsifier (HLB 4).
10. Food-grade, yellow, highly filtered fluid lecithin for release and emulsification. Maintains clear yellow color and resists darkening at temperatures up to 350 °F; no development of burned odor, color, or taste associated with traditional release agents; as an emulsifier, water dispersible (HLB 7–8), superior to standard-grade lecithins. Packaged in a 420-lb drum.

Other Sources of Lecithin

Traditionally soybeans have been the primary source of commercial lecithin. U.S. production figures from 1994–2010 average about 98,000 short tons annually. At various times, other oilseed-based lecithins have been commercially available, including corn and cottonseed (Cherry, 1985; Hougen et al., 1985; Weber, 1985). As mentioned previously, the demand for non-GMO lecithin has renewed interest in rapeseed and sunflower as commercial sources of lecithin. A number of European suppliers sell non-GMO rapeseed and sunflower lecithins. Other suppliers include China and India.

Rapeseed Lecithin

The composition and emulsifying properties of rapeseed lecithin were reported by Persmark (1968), Hougen (1985), and Sosada et al. (1992, 2003). Rapeseed lecithin has a phosphatide profile similar to soybean with respect to the three major ones (PC, PE, PI). Rapeseed lecithin contains 38–46% PC, 27–43% PE, and 18–33% PI compared to soybean (43% PC, 16% PE, and 41% PI). The total phospholipid content is dependent on the whether solvent extraction or expellers are used to produce crude oil. Solvent extracted oils contain 1–5.3%, whereas expeller oil contains 0.6–2.4%. Nonpolar solvents such as hexane yield crude oils with lower phosphatide contents compared to more polar solvents such as alcohols. Pressed oils generally have lower phosphatide content than solvent extracted oils.

An early assessment of the rapeseed industry is worth reviewing. In the 1980s, crushing plants in western Canada added the gums from processing to the meal intended for animal and poultry. Hougen et al. (1985) pointed out that in order for rapeseed to compete with soybean lecithin, only the highest quality of seed could be used, along with suitable refining treatments. Industrial acceptance of rape lecithin has been hampered because of poor color, flavor, taste, and appearance. However, rapeseed lecithin has been successfully used in chocolates and margarine in Europe. Important factors leading to high quality rapeseed lecithin include control of seed cooking conditions; avoidance of high temperatures during miscella stripping; control of the amount of water during hydration; control of temperature, pressure, and residence time in the dryer; and use of sound, undamaged seeds. Rapeseed lecithin prepared in this fashion had a light color, good consistency, and a pleasant smell and taste (Hougen, 1985).

The global expansion of the rapeseed lecithin industry has begun, with nine companies in India. A company in Florida markets rapeseed lecithin. It would appear that many of the early problems with rapeseed lecithin processing have been overcome.

Sunflower Lecithin

Morrison (1981, 1985) reviewed the early work on sunflower lecithin. Sunflowers are a potential source of lecithin in countries where large amounts are grown. Major growing regions include Argentina, France, Hungary, the Ukraine, and Russia. However, major markets are restricted to neutraceuticals and food supplements. The surface activity of sunflower lecithin is similar to soybean and its composition is similar, but phosphatidylcholine is higher in sunflower. Hollo et al. (1993) reported the alcohol fractionation and enzymatic modification of sunflower lecithin. Alcohol fractionation increased the phosphatidylcholine content from 41% to 65%. Lecithins modified by phospholipase A2 acylation showed excellent emulsification properties compared to a commercially modified soybean lecithin. Sunflower lecithin tends to be pasty and more difficult to handle than soybean lecithin. Fluidization with fatty acids and oil yields a fluid product (60% AI) with suitable viscosity (100 poise max). Hollo et al. (1985) reported that sunflower lecithin can be added to the meal or as an additive for poultry and animal feed. Poultry fed diets containing 2% sunflower lecithin showed improvements in weight, a shorter breeding period, and higher egg yields. Although cattle were fed, no nutritional data were available. Owing to the emulsifying properties and viscosity-reducing effects, sunflower lecithin is used in chocolate production at levels of 0.5%.

A sunflower lecithin plant opened in Hungary in 1993. Over the past 20 years, the sunflower lecithin industry has seen rapid expansion in Europe, Africa, the Middle East, and Japan. Today nearly 1200 companies supply sunflower lecithin, with the majority located in China and India. Sunflower lecithin reached GRAS status in the United States in 2009, and is not an allergen. As a result, Cargill entered the U.S. market with a crushing plant in the Ukraine and lecithin processing in Italy.

Production of sunflower lecithin requires dedicated crushing plants because the seeds are smaller than soybeans. Typically, expellers are used to reduce the oil content from 40% to 15%. The press cake is solvent extracted and combined with the expelled oil prior to degumming and lecithin processing. Lecithin yields are about 0.3% based on the seed weight. Intensive filtration of the crude nondegummed oil is critical in obtaining lecithin with low impurities. Processing of rapeseed lecithin is very similar to sunflower. However rapeseed/canola lecithin may contain chlorophyll and take on a greenish tone. In food applications in which levels of 0.5% are used, the green color is masked.

Sunflower lecithin is preferred in Europe as an alternative to soybean because of a favorable fatty acid composition, a pleasant nutty flavor, and good emulsifying properties (van Nieuwenhuyzen, 2014).

The emulsifying properties of modified sunflower lecithins were recently reported in O/W emulsions. The stability was evaluated by backscattering evolution, particle size distribution, and mean particle diameters (Cabezas et al., 2012). PC enriched fractions and alcohol fractionated and enzymatically modified lecithins had the best emulsifying properties during the main creaming and coalescence destabilization processes.

The emulsifying properties of canola, soybean, and sunflower lecithins degummed with different reagents has been reported (Guiotto et al., 2013; Smilles et al., 1989). Sixty percent O/W emulsions prepared with water degummed lecithin showed little difference with regard to oil type, but the authors believed soy lecithin to be slightly better than canola or sunflower. Organic acids including citric, phosphoric, and oxalic, along with acetic and maleic anhydrides, were all inferior to water degummed oil.

Organic Lecithins

Organic lecithin is produced from non-GMO soybean, rapeseed/canola, or sunflower oils without the use of traditional hexane extraction solvents. Instead, the seeds are cold pressed with expellers to recover the crude oil. Processing into food-grade lecithin follows the usual crude oil filtration, hydration with water, centrifugation, drying, and fluidization. Oil-free organic lecithin is not readily available because acetone extraction as normally done would not allow an organic claim. However, a dry-blended product is available containing 50% organic fluid lecithin blended with organic rice maltodextrin as a carrier. Furthermore, some suppliers avoid the use of alcohols or critical fluids to process their products. The organic lecithin industry is rather new and was developed in response to GMO issues in Europe (van Nieuwenhuyzen, 2014). Consumers there demanded lecithins from non-GMO sources. Because U.S.-produced lecithin may come from GMO soybeans, the organic lecithin industry has grown over the past decade.

Soybean organic lecithin is available in the United States through the use of identity preservation. Non-GMO beans are grown and kept separate through the supply chain. Organic soy lecithin is recovered as previously described. A typical line of organic soy lecithins includes standardized fluid products (bleached and unbleached) as well as a deoiled certified organic powder (96%). The fluid lecithins are packaged in 10, 50, and 450 lb lots, and the blended and powdered products are sold in smaller lots (20–100 lb or 5–20 kg).

Applications include chocolate, baked goods and bread, confections, ice cream, powdered mixes, infant formula, dietary foods, animal feeds, nutritional supplements, personal care items, and various industrial uses.

Quality Control of Lecithin Products

Several excellent reviews have been given for the industrial analysis of lecithin and phospholipid products (Lantz, 1989; Marmer, 1985). The latter reference discusses total phosphorus analysis and importance at length. A novel method for total phosphorus in oils based on turbidity measurements in acetone has been reported (Sinram, 1986; AOCS Method Ca 19-86).

Methods for analysis of lecithin are given in *Official Methods and Recommended Practices of the American Oil Chemists' Society* (AOCS). In Europe, methods approved by the German Fat Society (DGF) are used but they are quite similar to the AOCS Methods (and many are identical). The International Lecithin and Phospholipid Society (ILPS) also endorses AOCS Methods. Although 15 AOCS methods are available, the most important are for moisture, acetone insolubility, peroxide value, color, viscosity, and acid value. Commercial lecithin specifications in the U.S. contain this information. A typical unbleached fluid lecithin should contain 1% or less moisture, a minimum of 62% acetone insolubles, acid value of 32 max, a Gardner color of 10, and viscosity 150 P at 77 °F max.

Granular, powdered oil-free lecithins require high quality fluid lecithin as a starting material. A minimum of 65% acetone insolubles, a Gardner color of 12, and an acid value of 25 max is desirable. Dark-colored starting material will carry over into the de-oiled lecithin. Excess acid and divalent metal ions lead to tacky granules and granulation problems. High peroxide values tend to decrease shelf life and poor organoleptic properties. Acetone insolubles below 65% lower extraction efficiency. High moisture may cause sticky granules, poor flow, and inconsistent granule rigidity (List, 1989).

Methods for preparation of custom-blended products, including refined, de-oiled, and alcohol fractionated lecithins, are given in several reviews (Flider, 1985; List, 1989).

Lecithin Modification in the Patent Literature

As mentioned in the introduction, the chemical modification of lecithin is well documented in the early patent literature. However, research directed toward improvement of lecithin as a functional ingredient is ongoing. Orthoefer (1980) describes lecithins with improved cold water–dispersible properties. Lecithin, in combination with non-ionic emulsifiers, readily disperses in cold water forming stable emulsions. Lecithin treated with ethyl lactate, propylene glycol, or carbitols/cellosolves was described by Jordan (1939), who suggested that water-dispersible lecithins have many industrial uses including insecticides, dyeing, and leather tanning. Kass (1979) reported that water-soluble lecithin can be prepared by mixing with tertiary amine oxides, alkyl sulfate salts, and other sulfur-containing compounds. The solubility in water is infinite, and it shows limited solubility in aqueous ethanol. Possible uses include water-based cosmetics or toiletries.

Compressed lecithin preparations have been described (Orthoefer, 2001). Granular (oil-free) lecithin is compressed and extruded under at least 100 psig to form a solid mass called a liquid crystal phospholipid (LCP). Possible applications include the oral administration of neutraceutical and pharmaceutical drugs and to moisturize and protect the skin. The solid nature of LCP allows formulation into pills or tablets.

Hydrogenation of commercial soybean lecithin (containing 65% phosphatides and 35 % oil) is difficult at high pressures and temperatures because any moisture will promote deterioration of the phospholipids. Jacini (1959) and Cole (1959) reported methods to circumvent the problem; hydrogenation at 75–80 °C at 100–150 atmospheres in the presence of a flaked nickel catalyst (3–4%) yields a product having an iodine value of 10–30. The phospholipids are extracted with ethyl acetate. The hydrogenated oil is chilled and separated by filtration with recovery of the catalyst. Other variations include pretreating the oil/lecithin mixture with peracetic acid prior to hydrogenation with nickel/platinum catalysts as previously described.

Similarly, the preparation of iodized lecithin is a simple one. However, the products are unstable and decompose readily into free iodine, thus rendering them unsuitable sources of dietary iodine. To solve this problem, deoiled lecithin was dissolved in chloroform or carbon tetrachloride and treated with a periodide of iron or antimony and iodine and heated with stirring for 30 hours. The reaction products, washed with sodium thiosulfate and dried to yield a light brown powder, are nonhydroscopic, have good stability, can be preserved for long periods, and have no harmful effects (Hayashi, 1963).

The oxyalkylation of lecithin was reported by DeGroote and Keiser (1943), who identified a number of compounds suitable for the reaction. These authors pointed out that oxyalklated lecithin may be further modified by sulfonation to produce demulsifying agents for breaking oil field emulsions (DeGroote, 1937).

The reaction of gylcidol (oxyranyl methanol) with soybean lecithin yields oxyalkylated products having improved water dispersibility (Chang, 1963).

Hydroxylated Lecithin

In the presence of hydrogen peroxide and lactic acid, hydroxylation of lecithin occurs at the double bonds within the fatty acids. Laboratory work indicates that the reaction is very slow. A drop of 35 IV units at 70–75 °C requires 18 hours. However, carrying out the reaction under microwave-assisted conditions (70–75 °C, 600 W power) resulted in a reaction time of 1 hour (IV drop 37 units) (Karuna et al., 2011).

Hydroxylation of lecithin was thoroughly investigated by Julian et al. (1953), who showed that the reaction produces dihydroxystearic acid and increases the acetyl values, indicating that double bonds are hydroxylated. Julian's (1953) results indicate that only modest reductions (5–25%) in IV are required to improve the functional properties of hydroxylated lecithin. A more recent report describes a process for the preparation of hydroxylated lecithin for crude soybean oil (Vandana et al., 2003). These authors also describe a simple process to enrich the phospholipid content of commercial soybean lecithin (Vandana et al., 2001).

Acetylated Lecithin

Szuhaj and Yaste (1976) employed a combination of acetylation, hydroxylation, and neutralization to improve the water dispersibility of soybean lecithin. These authors recommend the acetylation reaction be continued until the free amino nitrogen value is reduced at least 10% and the acylated hydroxylated lecithin should have a reduction of 5–20% in iodine value. The reaction is carried out under acidic conditions. Control of the acylation/hydroxylation reactions is critical because higher yields in one step will lower yields in the other. To achieve both hydration and water dispersibility, the degree of either acetylation or hydroxylation has to be high. For example, IV reductions of 5–10% and free amino nitrogen of 60–70% gave good hydration and water dispersibility. However, IV reductions of 3–4% and free amino nitrogen of 30–60% failed to give the desired dispersibility. Other anhydrides including maleic, succinic, phthalic, valeric, and benzoic are suitable as acylation reagents. Lecithin modified by a combination of both acetylation and hydroxylation has been described by Szuhaj (1975) for pan release applications.

Eichberg (1967) describes acetylation of wet gums coming from the degumming centrifuge. Typically crude gums consist of about 35% water. Treatment with 1–3% acetic anhydride, drying (1% moisture) under vacuum at 160 °F for several hours yielded lecithin with improved water dispersibility.

Aneja (1971) describes a method for separation of acetylated PE from mixtures containing other phosphatides. It is claimed that when PE and PC are present together, the emulsifying properties are impaired. In addition, parenterally administered emulsions containing PE may have hypertensive reactions. Mixtures of PE and PC are acetylated with acetic anhydride in the presence of a nonacylatable tertiary amine and the pH is adjusted to at least 8.5. The mixture is then extracted (4 times) with acetone or methyl acetate. The product was essentially free of PC. Soy phosphatides (100 gr) yielded 42.1 gr of a straw-colored solid (3.2% phosphorus, 32% PC, 1% oil, and 1–2% acetyl PE).

Methods to separate oil, PE, and PC have been reported by Ginther (1984). Crude soybean phosphatides were extracted with 95% ethanol at 35 °C (1:2.5), yielding a solid (43% PC, 12% PE, and 21% oil). Column chromatography on silicic acid (heated column, 70 °C) yielded PC of 92% purity. Similar results were obtained with deoiled soybean lecithin. Others have purified soy lecithin by extraction with methylene chloride or mixtures with 5% ethanol as slurries with aluminum oxide. Pure PC with small amounts of PE was obtained after removal of solvents by distillation. Oily, highly purified phosphatidylcholines from soybean and sunflower were reported by Betzing (Betzing, 1980; Betzing and Eikerman, 1970). The crude lecithins are extracted with short-chain alcohols (1–3 carbons) and the alcohol phase treated with aluminum oxide followed by column chromatography with ethanol as the eluting solvent.

The acetylation of lecithin with enzymes has been reported (Marellapudi et al., 2002). The reaction is carried out with a 1–3 specific enzyme from mucor miehei in the presence of vinyl acetate. The acetylated products are separated from the enzyme by filtration and the excess vinyl acetate removed by distillation. The acetylation was carried out on soybean and rapeseed lecithin as well as pure PE and egg lecithin, and examples are given. The reaction is specific for PE and does not acetylate hydroxyl groups in phosphatidyl inositol present in vegetable lecithins.

Food Uses

Several reviews for food uses of lecithin have appeared (Dashiell, 1989; Prosise, 1985), as well have targeted applications for baking (Knightly, 1989; Pomeranz, 1985), beverages (Sander, 1985), and confections (Appl, 1989). According to Prosise (1989), 10 food groups utilize lecithin in some form to provide emulsification, crystallization control, wetting of powders, and as release agents, dietary supplements, browning agents, and anti-spattering agents in margarines and spreads (Houben and Jonker, 1966; Weiske et al., 1972; Vermas and Sonneveid, 1992; Duin, 1963; Mattikow, 1953). Lecithin also functions in food processing equipment as release agents and lubricators (Szuhaj, 1983), as well as a preservative coating for foods (Allingham, 1946).

Lecithin has the unique property of being both hydrophilic (water loving) and lipophilic (oil loving). The hydorophilic lipophilic balance (HLB) index can be used to predict the preference of an emulsifier for oil or water and is based on a scale of 1 through 12. The larger the HLB, the more hydrophilic the lecithin becomes, and a lower HLB indicates the lecithin becomes more lipophilic. Water-in-oil emulsions have HLB values of 1 through 6, whereas oil-in-water emulsions range from 6 through 8. Commercially available lecithins cover a wide range of HLB values. Usage levels for W/O emulsions vary from 1% to 5% while O/W emulsions require 5–10% (fat basis).

The functions of soybean lecithin in emulsions were reported by Rydhag and Wilton (1981). This study reports factors responsible for the stability and properties of O/W and W/O emulsions.

Major uses of lecithin in foods include pan release agents, belt release agents, product separation aids, and heat resistant applications (Dashiell and Prosise, 1984; Doumani, 1979; Liu, 2006; Szuhaj and Yaste, 1976). Pan release lecithins are designed to form fluid lipid barriers to ensure quick, clean separations from the baking surface and may be dispensed as aerosols (Clapp and Torrey, 1994) or as brush-on products. In commercial bakeries, bread pan release agents are needed to ensure quick separation of the loaves from the pan. Release agents are also required in cookie and cake baking as well as for other high moisture foods. Nonfoaming lecithin emulsions as cookware lubricants have been patented (Vollmer, 1975). Typically, lecithins designed for spray applications have low viscosity and acetone insoluble contents ranging from 32% to 50%. Other applications include instant beverage mixes, instant foods, agglomerated powders, and aqueous release agents.

Pan release lecithins are useful in continuous cooking and baking processes using belts or conveyors and are included in oil or melted shortening. Other benefits include water dispersibility in dip tanks, providing release and assistance in cleanup. Pan release agents also prevent sticking in finished products and allow separation of slices from high moisture pasteurized cheese.

Enzymatically modified lecithins are water dispersible and heat resistant up to 350 °F. Uses include baked goods, instant beverages/foods, confections, and release agents. Where prolonged heating is needed and resistance to darkening is desirable, these lecithins perform well in mold release, pan sprays, and griddle and pan frying. Liquid seasonings containing lecithin have been described (Todd and Haley, 1981). Lecithin in combination with monogylcerides has been reported as an improved dough additive (Gregerson, 1979).

Standard fluid lecithins contain 58–62% acetone insolubles and find applications in baked goods, cheese products, confections, icings, frostings, instant beverage mixes, instant foods, margarine, and release agents. Deoiled lecithin contains 97% acetone insolubles and finds uses in baked goods, dairy products, ice cream, icings, frostings, instant beverages/foods, and meat in sauces and gravies.

Bakery Applications

The literature on the uses of lecithin in baking has been reviewed by Pomeranz (1985) and Knightly (1989). Lecithin is useful from a rheological aspect. Doughs formulated with lecithin show improvement at make-up, are more stable and elastic, are drier, show less dusting flour requirements, and form a thin skin in molding and are less tender. Lecithin has no effect on adsorption, fermentation, proof, and bake time. Lecithin at 0.3% is most effective in dough mixing. Lecithin is thought to interact with dough components by physical enrichment on the surface of flour particles, combination with the surface particles, and formation of interaction products (lipoproteins), all of which improve water binding capacity, rheological dough handling, and overall bread quality. Lecithin dosage at 0.2–0.5% improves dough consistency by softening it and making it more pliable. Stiff, viscous ("bucky") doughs require higher levels.

The baking industry requires a versatile multifunctional ingredient, and lecithin is ideal. The primary function is that of an emulsifier, which ensures the even blending of components, increased moisture retention, and better release characteristics. Lecithin also has the ability to blend dissimilar ingredients evenly and thoroughly, which improves the functionality of baked goods.

Lecithin functions in baking by acting as an emulsifier (alone or in conjunction with other emulsifiers), as a wetting agent to provide instant wetting of powders (to decrease mixing time), as a parting agent providing cleaner faster release from molds, and as an antioxidant to stabilize vegetable and animal fats.

Reduced-Fat Baked Goods

The growing trend toward reduced-fat bakery products has posed a number of problems. These include lowered lubricity due to decreased availability of fat throughout the mix, doughs that are difficult to machine because of sticking to the equipment, moisture migration leading to dry texture and mouth feel, poor aeration, and weak cell structure. Lecithin can restore some of the desired properties in low-fat baked goods.

Bread

In bread baking, lecithin aids in increased loaf volumes and increased shelf life. Lecithin is most effective when used in conjunction with other emulsifiers such as monogylcerides and SSL. In addition to pan release applications, lecithin allows for a broader range of flours and improved sheeting of yeast-raised doughs. Inclusion rates range from 0.2–0.7% based on flour weight. Lecithin in bread does not affect volume and color but does yield tender crusts, a finer grain, smoother texture, more symmetrical appearance, and longer keeping quality.

Lecithin may reduce shortening requirements in bread. Bread with 0.3% lecithin and 1% shortening had keeping qualities equal to loaves baked with 4% shortening and no added lecithin. Bread with 0.25% lecithin and 3% shortening exhibited superior keeping qualities. Optimum levels of lecithin were 0.25–0.5 % for shortening improvement (Pratt, 1945, 1946). Walrod (1947) found that 1–5 oz of lecithin per 100 lbs flour permitted reduced mixing time and slightly increased adsorption. The doughs were drier, had improved machinability, greater elasticity/smoothness, and nearly complete absence of buckiness.

Cookies and Crackers

Lecithin functions in cookies to improve mixing of ingredients, improve moisture retention and fat sparing properties, and act as an internal release in rotary die cookies. In crackers, lecithin improves mixing and reduces dough stickiness. Dough handling and machining properties improve with the addition of lecithin. Lecithin in conjunction with oil in top dressings aids in the dispersion of spices. Usage levels are 0.2–0.5% based on flour weight.

Cakes and Donuts

In cake baking, lecithin improves volume, crumb, and freshness. Usage level is 1–3% based on the shortening. Lecithin is an important component of cake donut mixes. Without lecithin, the donuts do not achieve proper fat absorption and are dry and unsavory. In addition, adhesion of sugar to the donut will be difficult.

Typically, lecithin will be a part of an emulsifier system containing monogylcerides and propylene glycol monomethyl ester (PGME). Usage levels are 0.2–0.5%

of formula weight or 1–3% based on shortening weight. Hydroxylated soy lecithin performs well in prepared cake mixes (Elsesser and Bogyo, 1960; Obenauf and Tutter, 1962). Other workers have employed shortening systems based on liquid oils, lecithin, and monogylcerides for prepackaged cake mixes (Weiss et al., 1965).

Icings and Fillings

Lecithin in icing applications imparts a smoother texture and prevents drying. Increased creaminess and stiffness of the icing are also observed. Usage levels are 0.2–1% based on total formula weight.

Pizza, Pie Crusts, Tortillas, and Flat Breads

Deoiled lecithin acts as a natural release agent and, as such, dough stickiness is reduced and the dough is easier to process, saving time. Usage is 0.2–0.5% based on flour weight. Studies conducted at the American Institute of Baking showed that tortilla doughs containing 0.5% deoiled lecithin performed better than controls over 1 to 10 days. Distance to break tests showed that lecithin-treated tortillas gave superior stretching performance (elasticity) and moistness.

Lecithin finds applications in noodle manufacture as a dough-handling aid and for improved machining properties. Deoiled lecithin at 0.2–0.4% has found usage in production of Asian pre-fried instant noodles.

Instantizing

Formulation of instant foods faces the problem of incorporation of difficult materials (fat, proteins, fibers, vitamins/minerals) into a product that can be readily dissolved in cold water, milk, or other aqueous solutions. High-fat powders are difficult to wet and disperse because they are hydrophobic and repel water. On the other hand, high protein ingredients are hydrophilic (attracted to water) and can hydrate too quickly, forming lumps that do not hydrate internally. Lecithin solves these problems. Usage ranges from 0.2–10% and depends on fat and protein levels, particle sizes, and the desired degree of wetting. Foods in which instantizing is needed include beverage powders, meal replacement shakes, soups and gravies, high protein nutrition beverages, powdered dairy/nondairy products, and dry instant formulas. Instantization is usually accomplished by spray coating onto the powder's surface. Deoiled filtered lecithin is preferred in hydrophilic instantizing applications because it has excellent emulsifying properties and the absence of trigylcerides yields no off flavors or aromas, thereby maintaining product quality and integrity. Complexed lecithins are low viscosity and sprayable at ambient temperatures and are recommended for lipophillic instantizing applications.

Lecithins for instantizing milk and calf milk replacers require O/W emulsifiers and are designed for a high HLB. Standardized fluid lecithins (62% AI) provide moisture retention and emulsification in high viscosity applications.

Confections

Confections are the most widely known and firmly established lecithin applications. Lecithin performs as a natural emulsifier, an instantizer, an antioxidant, and a flavor protector for chocolates, hard candy, and related confections. For chocolates and compound coating, lecithin functions by lowering viscosity of molten chocolate (Julian, 1945), which allows high speed production. Other advantages include reduction of cocoa butter requirements and reduction of fat and sugar bloom. Production of caramels and fudge benefit from lecithin by facilitation of cutting, minimizing oiling out, and preventing or reducing sticking to wrappers. Clumping of caramel corn is reduced. Toffee and brittles tend to stick together and oil out; lecithin decreases or minimizes these problems. Lecithin improves the softness and tackiness of chewing gum and improves moisture retention. Lecithin functions in syrups and toppings by prevention of separation and viscosity reduction. Cocoa powders and mixes are instantized by lecithin. Typical usage levels for chocolate, compound coatings, caramels/fudge, and toffee/brittles range from 0.2% to 0.6%. Chewing gum and cocoa powder require 0.5–5% and syrups and toppings require 0.2–0.6%.

Appl (1985) reviewed the use of lecithin in confections. Lecithin functions as an emulsifier, an anti-sticking agent, and a viscosity modifier (Julian, 1945). A number of confections are made by dissolving and dispersing the ingredients in water and removal of the water to the desired level. These include hard candies, caramels, toffees, creams, gums/jellies, fudge, and marshmallows, all of which tend to stick together. Lecithin alleviates this problem. Chocolate may be made without lecithin. However, viscosity must be achieved by increasing the amount of cocoa butter. When lecithin is added in amounts ranging from 0.1% to 0.7%, the amount of cocoa butter needed to produce equal viscosities is substantially reduced. The effects of lecithin as an emulsifier for caramels are shown by the addition of 0.2% to the formula. Without lecithin, the product immediately shows phase separation after removal from the cooking kettle. In addition, lecithin prevents stickiness when the candy is cooled and cut by wire or knives. Because lecithin is a natural antioxidant, some positive effects on shelf life are possible.

Beverages

The use of lecithin in beverages has been reviewed by Sander (1989). Lecithinization of dry powders can be done in several ways. Lecithin is a natural surfactant that can be applied on individual ingredients or in a complex formula that may consist of proteins, cocoa powders, trigylcerides, and carbohydrates. Direct incorporation of lecithin can be accomplished by high shear mixing, coblending with spraying onto a powder, or cospraying a total emulsion of ingredients. Agglomeration may be done by continuous batch or fluid bed systems. A formula for a reduced-lactose drink consists of 2% lecithin whey protein concentrate, fructose, maltodextrin, vitamins, and flavors.

Miscellaneous Food Uses

Lecithin can be used as an egg yolk extender (Chess, 1980) as well as in meat curing compositions (Chandler et al., 1984; Haggerty and Corbin, 1984). The use of lecithin in meats, sauces, gravies, and fillings has been reported by Chandler et al. (1984). Pan and grill frying shortenings containing lecithin have been patented by Reid (1969). Lecithin usage in chewing gum dates to 1940 (Connor, 1940). Further work has been reported by Grey (1995) and Patel (2001). Egg yolk replacements/extenders have been described as employing lecithin/lipoprotein emulsions (Tan et al., 1982). These products reportedly perform well in baking applications. Reduced-fat shortenings employing lecithin as an emulsifier were patented by Desai and Bodor (1995). Nonaerosol vegetable oil compositions containing lecithin and ethyl alcohol are disclosed by Sejpal (1979). Schwied et al. (1985) describe lecithin-based emulsification systems for creamy products. Lecithin may serve as a nutrient for mushroom production (Holtz, 1983).

Industrial Uses of Lecithin

Schmidt and Orthoefer (1985) list 26 uses of lecithin in nonfood applications. These include adhesives, adsorbents, agricultural processing, corrosion inhibitors, lubricants (Hendry, 1940; Nickoloff, 1999; Nickoloff and Pelter, 1988), release agents (Olson et al., 1999), paints and inks, as catalysts, in ceramics and glass, coatings, detergents, explosives, leather, and for dust control. Others include masonry and asphalt, metal processing, pesticides, polymers, release agents, textiles, and water treatment/pollution control. At least 33 functions of lecithin in these applications were identified. In many applications, lecithin provides more than one function. For example, in paints and coatings lecithin acts as an antioxidant, color intensifier, dispersing agent, promoter, spreading aid, stabilizer, and synergist. Lecithin in animal feeds, cosmetics, pesticides, polymers, textiles, and leather also provides multiple functional properties. Most applications require dosage levels of 0.05–1.5%, but some cosmetics, paints and coatings, and release agents may require higher dosages. Space does not permit a detailed review of all nonfood applications, but some of the more important ones will be discussed.

Although lecithin is often used as a release agent, it can also be useful in adhesives as a plasticizer and as a dispersant in tape and ceiling tiles. Magnetic tape coatings may contain a lecithin-based adhesive layer. Lecithin is particularly suited as an adhesive for laminated materials such as paper for high-speed passage through computers. Lecithin finds uses in adsorbents and flocculents useful for cleanup of petroleum from sea water. Alumina coated with phosphatidylcholine not only adsorbs the petroleum, but the mixture also sinks to the bottom without release of oil. The primary use of lecithin in agriculture is in animal feed. Lecithin not isolated for commercial use is added to the feed. Lecithin functions as an emulsifier for fats in the feed, thus improving digestion and weight gain.

Lecithin is well recognized as an oxygen scavenger, and in combination with tocopherols synergistic effects may be observed. Corrosion is a form of oxidation, and lecithin added to turbine or spindle engines forms an anticorrosive lubricant. Oil-free and hydroxylated lecithins are useful for inhibition of the rusting of steel. Lecithin shows catalytic activity in the vulcanization of rubber as an initiator.

Cosmetics

Lecithin has a long history of inclusion in cosmetic formulations falling into a number of product categories. Among these are cleansers, moisturizing liquid makeup, beauty lotions, lipstick, skin penetration enhancers (Fazwi, 1988), and performing cremes. The functions of lecithin in cosmetics include improvement in the ease of application and emolliency of lipsticks, modification of films produced by face masks through increased elasticity, protection of hair in heat waving compositions, addition to creams to maintain elasticity of nails, production of bubble lathers in shaving creams, as a synergist to phenolic antioxidants, and as an emulsifier in denture pastes (Baker, 1989; Sagarin, 1957; Wilkinson and Moore, 1982). Inclusion of lecithin in most products fall into the 0.5–1.0% range.

Coatings

Sipos (1989) presents a thorough review of lecithin in coatings and paint applications and lists 14 functions, some of which are closely related depending on the nature of the phases with which the lecithin interacts. Paints consist of pigments and a carrier that is usually water based. At one time oil-based paints were used extensively. Oil-based paints require surfactants to disperse and stabilize, for which lecithin usage is 0.1–0.2%. Lecithin has traditionally been used as a grinding aid that facilitates rapid pigment wetting and dispersion, shortens grinding and mixing time, stabilizes by preventing pigment sedimentation, and aids pigment redispersion after storage.

A number of surfactants are available for manufacture of water-based paints. Ordinary grades of lecithin are unsuitable in water-based paints and special formulations are required. Water dispersibility is a prime requirement. Hydroxylated, fractionated, and refined grades are suitable as low-cost emulsifiers, stabilizers, thickening agents, and spreading aids. Although high-solid paints (80%) are difficult to formulate, products in the 70% range are available for a variety of applications. High-solid paints, putty, and caulking compounds can be formulated with commercially available compound lecithins, which provide thorough and complete wetting of pigments, prevention of separation, and reduced mixing time. Putty requires dosages of 0.1–0.25% and caulking compounds require 1%. Sincroft and Sipos (1961) reported coating compositions comprising lecithin.

Miscellaneous Uses of Lecithin

Many uses of lecithin can be found in the patent literature. These include as an egg yolk extender (Chess, 1980) as a flow agent (Ehrhardt, 1993), in culinary mixes (Elsesser and Bogyo, 1960), as a corrosion inhibitor (Eichberg and MacDonald, 1992), in softening compositions (Jacques and Pirotton, 1967), in solvent-based rubber gaskets (Cahill and Arons, 1989), as mold release agents (Meyer, 1985), as polyurethane curing agents (Caruso, 1981), as concrete cleaner and stripper (Bove, 1974), for cleaning soils from hardware (Bernardino, 1981), as an initiator for production of polymethylenes (Setterquist, 1968), to suspend pigment mixtures (Kronstein and Eichberg, 1977), as a scratch filler and primer in aerosol form (Kendall, 1983), for the drying of honey and molasses (Bateson et al., 1990), as steam condensate corrosion inhibitor (Whitekettle and Crovetto, 1992), for reduction of friction and flow properties (Brois et al., 1992), and as a leather cleaner and conditioner (Vasblomi, 1995). Lecithin has been used in the manufacture of bombs (Plauson, 1976), nail enamel (Socci et al., 1981), and as a hair conditioner (Spitzer et al., 1982). Lecithin finds use in gasoline as a stabilizer (Rathbun, 1940) and as an additive for the prevention of gum formation (Sollman, 1932). Lecithin was shown to clarify and prevent turbidity of turpentine (Eichberg, 1944). Phospholipids have found uses in plant protection (Ghyczy et al., 1985, 1987) and as foliage fertilizer (Bauer et al., 1986). The preparation of lecithin based food grade inks for printing on edible substrates was recently patented (Baydo et al., 2010). Glabe (1948) describes lecithin compositions containing dihydroxyphosphatides and amylacous materials. Stabilization of organic oils with lecithin has been reported (Lau and Schrier, 1996). Lecithin has found uses in cream icing (Howard and Koren, 1965) and in fluid shortenings (Handschmaker, 1961; Widlak, 2001).

Liposomes

Since their discovery in 1961 (Bangham and Home, 1964), liposomes have attracted much attention as drug delivery systems and as nutrients. Liposomes are composite structures made from phospholipids, but they may contain small amounts of other molecules. Liposomes may vary in size from one micrometer to tens of micrometers (Aitcheson, 1989). Unilamellar liposomes are generally in the lower size range with various ligands attached to their surface, allowing for their surface attachment and in pathological areas for treatment of diseases. A liposome encapsulates a region of aqueous solution inside a hydrophobic membrane, and when dissolved in hydrophilic solutes it cannot pass through the lipids. Hydrophobic chemicals can be dissolved into the membrane and as such can carry both hydrophobic and hydrophilic molecules. Delivery to the sites of action involves fusion of the bilayer to other bilayers, such as cell membranes. Liposomes serve as models for artificial cells and can be designed to deliver drugs in other ways. Liposomes containing high or low pH can be made such that dissolved aqueous drugs will be charged in the solution (the pH is outside the drug's range). As the pH naturally neutralizes within the liposome, the drug will be neutralized, allowing free passage through the membrane. In effect, liposomes delver drugs through diffusion rather than by cell fusion. Liposomes can be utilized in biodetoxification by injection of an empty liposome with a transmembrane pH gradient. The vesicles act as scavengers for drugs in the circulation system and prevent any toxic effect. Lecithin organogels have been reported for the delivery of bioactive agents for treatment of skin aging (Raut et al., 2012).

Methods of Preparation

Excellent reviews of liposomes and nanoliposomes were recently presented by Mazafari et al. (2008; Mazafari, 2005). Basically, four methods have been developed to manufacture liposomes: drying down of lipids from organic solvents, dispersion of the lipids in aqueous media, purification of the resultant liposomes, and analysis of the final product. Organic solvents (chloroform, ether, and methanol) are major disadvantages because of potential toxicity problems if not removed during preparation. Other methods not employing solvents have been reported, including the polyol dilution method (Kikuchi et al., 1994), the bubble method (Talsma et al., 1994), and the heating method (Mozafari et al., 2002a, 2002b, 2004). Various methods have been reported to remove solvents and detergents, including gel filtration, vacuum, and dialysis, with only partial success (Allen, 1984; Weder and Zumbuel, 1984). Sterilization is an important step in the manufacture of liposomes for human and animal use. Lipid vesicles can only be sterilized by filtration, and any other method (chemical or physical) would destroy the liposomal structure with release of the encapsulated material. Kikuchi et al. (1994) described the sterilization of liposomes (at 121 °C for 20 minutes), which maintain their structural integrity and high encapsulation efficiency.

Mozafari et al. (2008) developed a method for preparation of liposomes without the use of any organic solvents. The process consists of hydration of the components in aqueous media followed by heating in the presence of glycerol (3%) at 120 °C. Thus, the need for organic solvents and a separate sterilization step are eliminated. Liposomes prepared by this method are known as heat method (HM) liposomes. Incorporation of drugs into HM liposomes can be achieved by adding the drug to the reaction media along with liposomal ingredients and glycerol; adding the drug to the reaction media when the temperature is not lower than the transition temperature of the lipids; or adding the drug to the HM liposomes after they are prepared at room temperature. Alving et al. (1989) describe methods for encapsulating liposomes for drug delivery or as vaccines.

A very recent review covers new techniques for liposome preparations (Huang et al., 2014). These include supercritical fluid technology, dual asymmetric centrifugation, membrane contactor technology, cross-flow filtration detergent depletion, and freeze drying double emulsion. Each has advantages and drawbacks.

As of 2012, 12 drugs based on liposomes have been clinically approved, with 5 still pending.

Until recently, liposomes were directed toward drug delivery. However, their use in dietary and nutritional supplements is fairly new. The low bioavailability and absorption of oral supplements is well known (Williamson and Manach, 2005). Liposome technology provides a method for bypassing the destructive gastric system and allows the encapsulated nutrients to reach the cells and tissues (Brey and Liang, 2003).

As the term implies, *nanoliposomes* are exceedingly small molecules measuring in nanometers, compared to ordinary liposomes ranging in size from 15 to 1000 nm. Applications in the food industry include altering the texture of food ingredients, encapsulation of food components or additives, development of new sensory attributes, flavor release, and increasing the bioavailability of nutrients (Chaudhry et al., 2008; Farhang et al., 2012). Nanoliposomes have been employed in cheese manufacture to encapsulate food preservatives. Washed curd cheeses are highly susceptible to the growth of pathogenic and spoilage microorganisms and are commonly controlled by the addition of nitrate to the milk during processing. A replacement for nitrate has been proposed based on an antimicrobial enzyme derived from eggs. Lysozyme, however, binds to the casein in milk, reducing potency and effectiveness. Liposome-entrapped lysozyme has the potential to prevent binding and to target regions in the cheese matrix where bacteria are present (Mozafari et al., 2008).

The so-called stealth liposomes are made by the inclusion of polyethylene glycol (PEG) on the surface of the liposome (Immordino et al., 2006). These preparations have been shown to extend blood circulation times. A large number of liposomal preparations having high target efficacy and activity are possible. Furthermore, through modification of the terminal end of PEG, stealth liposomes can be actively targeted with monoclonal antibodies or ligands. Relatively few stealth liposomes have been approved for clinical use. Those approved have found application in Kaposi's sarcoma and head, neck, and lung cancer, along with several other cancer types.

Composite liposomal technologies for specialized drug delivery are the subject of a current review (Mufamadi et al., 2011). The modification of liposomes with hydrophilic polymers has been discussed, but a new approach involves integration of preencapsulated drug-loaded liposomes within depot polymer-based systems (Stenekes et al., 2000). Suitable natural depot polymers include collagen, gelatin, chitosan, fibrin, dextran, and aliginate. Synthetic polymers include carbopol, a hydrogel, and polyvinyl alcohol. Both techniques suffer from poor release of the drugs. Liposomes have been used to encapsulate antiviral drugs (Bergeron and Desormeaux, 1998), for gene delivery to the lungs (Mozafari et al., 2002), and for treatment of periodontal disease (Kurtz, 2001).

Dietary and Nutritional Benefits of Phospholipids

The reader is referred a number of reviews on choline, phospholipids, and health and disease (Orthoefer, 1998; Orthoefer and List, 2006; Szuhaj, 1989; Zeisel, 1985; Zeisel and Szuhaj, 1998). Phosphatidyl choline drug derivatives have received considerable attention in the patent literature (Chasalow, 1997, 1999, 2000; Duttaroy, 2008). Oil in water phospholipid emulsions for parenteral administration have been reported (Chang and Lindmark, 1986). A review of emulsions for the delivery of nutraceutical drugs can be found in Moottoosingh and Rousseau (2006). Davis and Ullum have reviewed the general use of phospholipids in drug delivery systems (Davis and Ullum, 1993).

Various attributes of lecithin as dietary supplements include: cardiovascular health, liver and cell function, fat transport and metabolism, reproduction and child development, physical performance and muscle endurance, cell communication, memory, learning and reaction time, arthritis, skin and hair health, and treatment of gallstones (Orthoefer and List, 2006).

Choline is essential as a precursor of acetylcholine, which controls muscle functions. Choline prevents fats from accumulating in the liver, and a deficiency may disturb lecithin synthesis, which is needed to export trigylcerides as part of lipoproteins. Phosphatidylcholine is involved in the hepatic export of very-low-density lipoproteins. Choline deficiency compromises kidney functions.

Lecithin reportedly improves pet animal health. After 2–3 weeks, dogs and cats of all ages showed glossier healthier hair coats, reduced shedding, and increased vigor and activity. Older dogs and cats showed less stiffness, better appetite, and return of learned habits. Working dogs showed increased alertness and quicker recovery after stress. Dogs with a tendency to suffer seizures showed improvement (Kullenberg, 1989).

Conclusion

The lecithin industry is global in scope as evidenced by a large number of internationally based companies offering non-GMO lecithin. The time-honored soybean lecithin, some of which is processed in the United States from GMO, glyphosate-tolerant beans, is being challenged by other sources, including rapeseed/canola and sunflower. As a result, the U.S.-based lecithin industry faces competition. However, U.S.-processed soybean lecithin is relatively cheap, in good supply, and is extremely functional. As such, a wide variety of products are offered for virtually every food or industrial application. The rapeseed and sunflower lecithins must be technically equivalent to soybean and price competitive as well. Both canola and sunflower oils contain smaller amounts of lecithin compared to soybean, and their performance has been evaluated mainly in the laboratory. A Florida-based company offers a line of non-GMO canola, sunflower, and soybean lecithins intended for the natural food and nutritional supplements markets. Liposomes as drug delivery systems will continue to attract attention from researchers in the medical and health professions. Lecithin, whether GMO or non-GMO, will continue to be a valuable and versatile component in foods, industrial products, and in the pharmaceutical industry.

References

Aitcheson, D.; Tenzel, R. Preparation of Uniform Size Liposomes and Other Lipid Structures. WO 1989011335, 1989.

Akbarzadeh, A.; Sadabady, R. R.; Davaran, S.; Joo, S.; Zarhami, N.; Hanifehpour, Y.; Samiei, M.; Kouhi, M.; Koshki, K. Liposome: Classification, Preparation and Applications. *Nanoscale Res. Letters* **2013,** *8,* 102–110.

Allen, T. *Removal of Detergent and Solvent Traces from Liposomes in Liposome Technology;* Gregoriadis, G., Ed.; CRC Press: Boca Raton, FL, 1984; Vol. 1, pp 109–122.

Allingham, W. Preservative Coatings for Foods. U.S. Patent 2,470,281, 1946.

Allingham, W. Lecithin Composition. U.S. Patent 2,447,726, 1948.

Alving, C.; Owens, R.; Waseff, N. Process for Making Liposome Preparation. U.S. Patent 6,007,838, 1999.

Appl, R. C. Lecithin in Confection Applications. In *Lecithins: Sources, Manufacture and Uses*; Szuhaj, B. F., Ed.; AOCS Press: Urbana, IL, 1989; pp 207–212.

Archer Daniels Midland. Lecithin Overview, 2014. http://www.adm.com/en-US/products/food/lecithin/Pages/default.aspx.

Arrigo, P.; Servi, S. Synthesis of Lysophospholipids, *Molecules* **2012,** *15,* 1354–1377.

Arveson, M. H. Lubricating Compounds and Modified Addition Agents for Same. U.S. Patent 2,295,192, 1942.

Baker, C. Lecithins in Cosmetics. In *Lecithins: Sources, Manufacture and Uses*; Szuhaj, B. F., Ed.; AOCS Press: Urbana, IL, 1989; pp 253–260.

Bangham, A. D.; Home, R. W. Negative Staining of Phospholipids and Their Structural Modification by Surface Active Agents as Observed in the Electron Microscope. *J. Mol. Biol.* **1964,** *8,* 660–668.

Bateson, G.; Morris, C.; Heuer, G. Methods for Drying Honey and Molasses. U.S. Patent 4,419,956, 1990.

Bauer, K.; Ghyczy, M.; Osthoff, H. Foliage Fertilizers. U.S. Patent 4,576,626, 1986.

Baydo, R.; Fabian, B.; Fathollahi, Z.; Graff, D.; Lee, C.; Martin, P. Food Grade Ink Jet Inks for Printing on Edible Substrates. U.S. Patent 7,842,320, 2010.

Bergeron, M.; Desormeaux, A. Liposomes for Encapsulating Antiviral Drugs. U.S. Patent 5,773,027, 1998.

Bernardino, L. Process for Removing Hard to Remove Soils from Hardware. U.S. Patent 4,297,251, 1981.

Betzing, H. Process to Obtain Oily Highly Purified Phosphatidylcholines. U.S. Patent 4,235,793, 1980.

Betzing, H.; Eikermann, H. Process for Obtaining Highly Purified Phosphatidylcholine and the Product of this Process, U.S. Patent 3,544,605, 1970.

Bollmann, H. Process of Obtaining Lecithin from Vegetable Raw Materials. U.S. Patent 1,464,557, 1923.

Bollmann, H. Process of Increasing the Durability of Pure Salad or Sweet Oils. U.S. Patent 1,575,529, 1926.

Bollmann, H. Process of Purifying Phosphatides from Oilseeds and the Like. U.S. Patent 1,667,767, 1928.

Bollmann, H. Improvements in and Relating to the Production and Purification of Phosphatides. British Patent 356b384, 1929.

Bonekamp, A. Chemical Modification. In *Phospholipids, Technology and Applications;* Gunstone, F. D., Ed.; Oily Press: Bridgewater, UK, 2008; pp 141–152.

Bove, F. Composition for Cleaning Adhering Hardened Concrete and for Stripping Concrete and Plaster. U.S. Patent 3,819,523, 1974.

Brey, R.; Liang, L. Polymerizable Fatty Acids, Phospholipids and Polymerized Liposomes Therefrom. U.S. Patent 6,511,677, 2003.

Brian, R. Soybean Lecithin Processing Unit Operations. *J. Am. Oil Chem. Soc.* **1976,** *53,* 27–29.

Brois, S.; Ryer, J.; Kearney, F.; Deen, H.; Guiterrez, A. Hydrogenated Lecithin for Friction and Flow Properties. U.S. Patent 5,135,669, 1992.

Buer, H. Process for Obtaining Tasteless and Inodorous Lecithin. U.S. Patent 1,055,514, 1913.

Cabezas, D.; Madoery, R.; Diehl, B.; Tomas, M. Emulsifying Properties of Different Modified Sunflower Lecithins. *J. Am. Oil. Chem. Soc.* **2012,** *89,* 355–361.

Cahill, G.; Arons, I. Gasket Forming Solvent Based Rubber Compositions Containing Lecithin. U.S. Patent 3,444,099, 1967.

Caruso, P. Polyurethane Curing Agent Dispersion, Process and Product. U.S. Patent 4,282,344, 1981.

Chandler, W.; Wilkens, W.; Heiss, J. Meat Curing Composition. U.S. Patent 4,434,187, 1984.

Chang, S.; Lindmark, L. Oil in Water Emulsion for Parenteral Administration. U.S. Patent 4,563,354, 1986.

Chang, S. S. Oxyalkylated Lecithin. U.S. Patent 3,085,100, 1963.
Chasalow, F. Phosphatidylcholine Drug Derivatives. U.S. Patents 5,703,063; 5,888,990; 6,127,349, 1997, 1999, 2000.
Cherry, J.; Gray, M. S.; Jones, L. A. A Review of Lecithin Chemistry and Glandless Cottonseed as a Potential Commercial Source. *J. Am. Oil Chem. Soc.* **1981,** *58,* 903–913.
Chess, W. Egg Yolk Extender. U.S. Patent 4,182,779, 1980.
Clapp, C.; Torrey, G. Dispersible Foodstuff Parting Compositions. U.S. Patent 5,296,021 (also 5,431,719; 5,567,456), 1994.
Cole, R. Hydrogenated Lecithin and Process for Making Same. U.S. Patent 2,907,777, 1959.
Connor, H. W. Chewing Gum. U.S. Patent 2,197,718, 1940.
Dashiell, G.; Prosise, W. Method of Preparing Heat Resistant Lecithin Release Agent. U.S. Patent 4,479,977, 1984.
Dashiell, G. L. Lecithins in Food Processing Operations. In *Lecithins: Sources, Manufacture and Uses*; Szuhaj, B. F., Ed.; AOCS Press: Urbana, IL, 1989; pp 213–224.
Davis, P. Alkaline Hydrolyzed Phosphatides. U.S. Patent 3,499,017, 1970.
Davis, P. Acid Hydrolyzed Phosphatides. U.S. Patent 3,576,831, 1971.
Davis, S.; Ullum, L. The Use of Phospholipids in Drug Delivery. In *Phospholipids, Chacterization, and Novel Biological Applications;* Geve. G., Paltauf, F., Eds.; AOCS Press: Urbana, IL, 1993; pp 67–79.
DeGroote, M. Process for Breaking Petroleum Emulsions. U.S. Patent 2,086,217, 1937.
DeGroote, M.; Keiser, B. Oxyalkylated Lecithin and Method of Making Same. U.S. Patent 2,310,679, 1942.
Denham, R. E. Untitled. U.S. Patent 2,166,286, 1939.
Desai, G.; J. Bodor, J. Reduced Fat Shortening Substitute for Bakery Products. European Patent 06732001, 1995.
Deuel, H. Chemistry of the Phosphatides and Cerebrosides. In *Lipids: Their Chemistry and Biochemistry;* Interscience Publishers: New York, 1951; Vol. 1, pp 405–505.
Doig, S.; Diks, R. Toolbox for Modification of the Lecithin Headgroup. *Euro. J. Lipid Technol.* **2003,** *105,* 359–376.
Doumani, C. Mineral Oil Modified Lecithin Cookware Spray Composition. U.S. Patent 4,155,770, 1979.
Duin, H. J. Phosphatide Anti Spattering Agents for Margarine. British Patent 1,113,241, 1963.
Duttaroy, A. Clinical and Nutritional Properties of Phospholipids. In *Phospholipids Technology and Applications;* Gunstone, F. D., Ed.; Oily Press: Bridgewater, UK; 2008; pp 153–167.
Ehrhardt, G. Flow Agent–Dispersant Composition. U.S. Patent 5,188,765, 1993.
Eichberg, J. Lecithin—Its Manufacture and Use in the Fats and Oils Industry. *Oil and Soap* **1939**, *16,* 51–53.
Eichberg, J. Turpentine Composition and Method of Making. U.S. Patent 2,355,061, 1944.
Eichberg, J. Lecithin Product and Method. U.S. Patent 3,359,201, 1967.
Eichberg, J.; MacDonald, K. Lecithin Corrosion Inhibitor. U.S. Patent 5,120,357, 1992.
Elsesser, C.; Bogyo, S. Culinary Mix. U.S. Patent 2,954,297, 1960.
Farhang, B.; Kakuda, Y.; Corredig, M. Encapsulation of Ascorbic Acid in Liposomes Prepared with Milk Fat Globule Membrane Derived Phospholipid. *Dairy Sci. and Techno.* **2012**, *92,* 353–366.

Fazwi, M.; Iyer, U.; Mahjour, M. Use of Commercial Lecithin as a Skin Penetration Enhancer. U.S. Patent 4,783,450, 1988.

Flider, F. The Manufacture of Soybean Lecithins. In *Lecithins: Sources, Manufacture and Uses*; Szuhaj, B. F., Ed.; AOCS Press: Urbana, IL, 1985; pp 21–37.

Ghyczy, M. Synthesis and Modification of Phospholipids. In *Lecithins: Sources, Manufacture and Uses;* Szuhaj, B. F., Ed.; AOCS Press: Urbana, IL, 1989; pp 131–144.

Ghyczy, M.; Imberge, P.; Wendel, A. Process for the Spray Applications of Plant Mixtures and Packing Units for Concentrates. U.S. Patent 4,506,831, 1985.

Ghyczy, M.; Imberge, P.; Wendel, A. Phospholipid Compositions and Their Use in Plant Protection. U.S. Patent 4,681,617, 1987.

Ginther, B. Process for the Separation of Oil and Phosphatidylcholine from Alcohol Soluble Phosphatidylcholine Products Containing the Same. U.S. Patent 4,425,276, 1984.

Glabe, E. Compositions Containing Amylaceous Materials and a Dihydroxyphospholipid. U.S. Patent 2,513,638, 1948.

Gregerson, J. Method of Preparing a Monogylceride Dough Additive. U.S. Patent 4,178,393, 1979.

Grey, R.; Patel, M.; Dubina, E.; Myers, M. Chewing Gum Containing Lecithin/Glycerol Triacetate Blend. U.S. Patent 5,474,787, 1995.

Griffin, W. Classification of Surface Active Agents by HLB. *J. Soc. Cosmet. Chem.* **1949,** *1,* 311–326.

Guiotto, E., Cabezas, D.; Diehl, B.; Tomas, M. Characterization and Emulsifying Properties of Different Sunflower Phosphatidylcholine Enriched Fractions. *Euro. J. Lipid Sci. Tech.* **2013,** *115,* 865–873.

Gunstone, F. D., Ed. *Phospholipid Technology and Applications;* The Oily Press: Bridgewater, UK, 2008.

Haggerty, J.; Corbin, D. Method for Preparing Meat in Sauce, Meat in Gravy, and Meat Filling. U.S. Patent 4,472,448, 1984.

Handschumaker, E. Fluid Shortening and Method of Making Same.U.S. Patent 2,999,755, 1961.

Hayashi, S. Process for Preparation of Iodized Lecithin. U.S. Patent 3,072,689, 1963.

Hayes, L.; Wolff, H. Refining Vegetable Oils. U.S. Patent 2,782,216, 1957.

Hernandez, E.; Quezada, N. Uses of Phospholipids as Functional Ingredients. In *Phospholipids, Technology and Applications;* Gunstone, F. D., Ed.; Oily Press: Bridgewater, UK, 2008; pp 83–94.

Hilty, W. Review of the Technical Applications of Soybean Lecithin. *J. Am. Oil Chem. Soc.* **1948,** *24,* 186–188.

Hollo, J.; Peredi, J.; Ruzics, A.; Jeranek, M.; Erdely, A. Sunflower Lecithin and Possibilities for Utilization. *J. Am. Oil Chem. Soc.* **1993,** *70,* 997–1001.

Holtz, R. B. Nutrient for Mushroom Growth and Process for Producing Same. U.S. Patent 4,370,159, 1983.

Hougen, F. W.; Thompson, J. K.; Daun, J. K. Rapeseed Lecithin. In *Lecithins: Sources, Manufacture and Uses*; Szuhaj, B. F., Ed.; AOCS Press: Urbana, IL, 1989; pp 79–95.

Howard, N.; Koren, P. Fluid Shortening for Cream Icings. U.S. Patent 3,208,857, 1965.

Houben, G.; Jonker, E. Low Spattering Margarine. U.S. Patent 3,248,230, 1966.

Huang, Z.; Li, X.; Zhang, T.; Song, Y.; She, Z.; Li, J.; Deng, Y. Progress Involving New Techniques for Liposome Preparation. *Asian J. Pharmaceutical Sci.* **2014,** *9,* 1–7.

Immordino, M.; Dosio, F.; Cattel, L. Stealth Liposomes: A Review of the Basic Science, Rationale and Clinical Applications, Existing and Potential. *Intl. J. Nanomedicine* **2006,** *3,* 297–315.

Jacini, G. Hydrogenation of Phosphatides. U.S. Patent 2,870,179, 1959.

Jacques, A.; Pirotton, P. Stable Aqueous Fabric Softener Compositions Based on Lecithin, Saponin and Sorbic Acid and Methods for Making and Using Same. U.S. Patent 4,816,170, 1989.

Jordan, S. Food Product and Method for Producing Same. U.S. Patent 1,859,240, 1932.

Jordan, S. Flavoring Material and Method of Using Same. U.S. Patent 2,019,494, 1935.

Jordan, S. Water Dispersible Lecithin. U.S. Patent 2,296,933, 1939.

Jordan, S. Water Dispersible Lecithin. U.S. Patent 2,193,873; 1949.

Joshi, A., Partatkar, S.; Thorat, B. Modification of Lecithin by Physical, Chemical and Enzymatic Methods. *Euro. J. Lipid Sci. Tech.* **2006,** *108,* 363–373.

Julian, P.; Iveson, H. T. Method of Preparing Dispersions of Vegetal Phosphatide Fractions. U.S. Patent 2,849,318, 1958.

Julian, P.; Iveson, P. T.; McCelland, M. The Hydroxylation of Phospholipids. U.S. Patent 2,629,662, 1953.

Julian, P.; Meyer, E.; Iveson, H. T. Phosphatide Product and Method of Making. U.S. Patent 2,373,686, 1945.

Karuna, M.; Vandana, V.; Prasad, P.; Lakshmi, P.; Prasad, B. Rapid Hydroxylation of Soybean Lecithin Under Microwave Assisted Conditions. *J. Am. Oil Chem. Soc.* **2011,** *88,* 1081–1082.

Kass, G. Water Soluble Lecithin. U.S. Patent 4,174,296, 1979.

Kendall, S. Combination Scratch Filler and Primer in Aerosol Form. U.S. Patent 4,372,991, 1983.

Kikuchi, M.; Matsumoto, H.; Yamada, T.; Koyama, Y.; Takakuda, K.; Tanaka, J. Glutaraldehyde Crosslinked Hydroxyapatite/Collagen Self-organized Nanocomposites. *Biomaterials* **1994,** *24,* 63–69.

Klenk, E.; Eikeraann, H.; Reuter, G. Process for Production of Natural Phospholipids and Substances Produced Thereby. U.S. Patent 3,031,478, 1962.

Knightly, W. H. Lecithin in Baking Applications. In *Lecithins: Sources, Manufacture and Uses*; Szuhaj, B. F., Ed.; AOCS Press: Urbana, IL, 1989; pp 174–196.

Kronstein, M.; Eichberg, J. Joint Suspensions of Mixed Pigmentations Achieved by Active Effects of Plant Phosphatides. U.S. Patent 4,056,494, 1977.

Kullenberg, F. W. Lecithin in Animal Health and Nutrition. In *Lecithins: Sources, Manufacture and Uses*; Szuhaj, B. F., Ed.; AOCS Press: Urbana, IL, 1989; pp 237–252.

Kurtz, S. Method for Treating Periodontal Disease. U.S. Patent 6,361,597, 2001.

Lantz, R. Industrial Methods of Analysis. In *Lecithins: Sources, Manufacture and Uses*; Szuhaj, B. F., Ed.; AOCS Press: Urbana, IL, 1989; pp 162–173.

Lau, J.; Schrier, B. Organic Oil Stabilization Techniques. U.S. Patent 5,492,648, 1996.

List, G. R. Commercial Manufacture of Lecithin. In *Lecithins: Sources, Manufacture and Uses*; Szuhaj, B. F., Ed.; AOCS Press: Urbana, IL, 1989; pp 145–161.

List, G. R.; Evans, C.; Black, L. T.; Mounts, T. L. Removal of Phosphorus and Iron by Commercial Degumming of Soybean Oil. *J. Am. Oil Chem. Soc.* **1978,** *55,* 275–276.

List, G. R.; Avellenada, J.; Mounts, T. L. Effect of Degumming Parameters on Lecithin Removal and Quality. *J. Am. Oil Chem. Soc.* **1981,** *58,* 892–898.

List, G .R.; Orthoefer, F. T.; Taylor, N.; Nelsen, T.; Abidi, S. Characterization of Phospholipids from Glyphosate Tolerant Soybeans. *J. Am. Oil Chem. Soc.* **1999,** *76,* 57–60.

Liu, L. Sprayable Cookware Release Composition with Reduced Heat Induced Browning. U.S. Patent 7,078,069, 2006.

Marellapudi, S.; Vemulapalli, V.; Penumarthy, V.; Narayana, B.; Rachipudi, P. Enzymatic Process for the Preparation of an Acetylated Phospholipid. U.S. Patent 6,403,344, 2002.

Markley, K.; Goss, W. *Soybean Chemistry and Technology;* Chemical Publishing, Inc.: New York, 1944; pp 100–133.

Marmer, W. N. Traditional and Novel Approaches to the Analysis of Plant Phospholipids. In *Lecithins: Sources, Manufacture and Uses*; Szuhaj, B. F., Ed.; AOCS Press: Urbana, IL, 1989; pp 247–288.

Mattikow, M. Antispattering Margarine. U.S. Patent 2,640,780, 1953.

Meyer, L. Compositions Containing Mold Release Agents. U.S. Patent 4,500,442, 1985.

Morrison, H. Sunflower Lecithin. In *Lecithins: Sources, Manufacture and Uses*; Szuhaj, B. F., Ed.; AOCS Press: Urbana, IL, 1989; pp 97–103.

Morrison, W. H., III. Sunflower Lecithin. *J. Am. Oil Chem. Soc.* **1981,** *58,* 902.

Mozafari, M. Liposomes: An Overview of Manufacturing Techniques. *Cell. Mol. Bio. Lett.* **2005,** *10,* 711–719.

Mozafari, M.; Johnson, C.; Hatziantoniou, S.; Demetzos, C. Nanoliposomes and Their Applications in Food Nantechnology. *J. Liposome Res.* **2008,** *18,* 309–327.

Mozafari, M.; Reed, C.; Rostron, C. Development of Non-toxic Liposomal Formulations for Gene and Drug Delivery to the Lung. *Technol. Health Care,* **2002,** *10,* 342–344.

Mufamadi, M.; Pillay, V.; Choonara, Y.; DuTolt, L.; Modi, G.; Naidoo, D.; Ndesendo, M. A Review of Composite Liposomal Technologies for Specialized Drug Delivery. *J. Drug Delivery* **2011,** *2011,* 1–19.

Obenauf, C.; Tutter, C. Lecithinated Product. U.S. Patent 3,060,030, 1962.

Orthoefer, F. Cold Water Dispersible Lecithin Concentrates. U.S. Patent 4,200,551, 1980.

Orthoefer, F. Compressed Lecithin Preparations. U.S. Patent 6,312,703, 2001.

Orthoefer, F. T.; List, G. R. Phopholipids/Lecithin: A Class of Neutraceutical Lipids. In *Neutraceutical and Specialty Lipids and Their Co-Products;* Shahidi, F., Ed.; CRC/Taylor and Francis: Boca Raton, FL, 2006; pp 509–530.

Pardun, H. Process for Extraction of Vegetable Phosphatides. German Patent 1,692,568, 1967.

Pardun, H. Phosphatide Emulsifier. British Patent 1,287,201, 1969.

Pardun, H. Phosphatides and Their Method for Preparation. U.S. Patent 3,305,074, 1970.

Pardun, H. Phosphatide Emulsifiers. U.S. Patent 3,361,795, 1972a.

Pardun, H. Phosphatide Extraction. U.S. Patent 3,661,946, 1972b.

Pardun, H. Preparation of Phosphatides. U.S. Patent 3,652,397, 1972c.

Patel, M. Method of Adding Lecithin to Chewing Gum. U.S. Patent 5,041,293, 2001.
Persmark, U. Main Constituents of Rapeseed Lecithin. *J. Am. Oil Chem. Soc.* **1968,** *45,* 742–743.
Plauson, R. Method for Bomb Manufacture. U.S. Patent 3,998,676, 1976.
Pomeranz, Y. Lecithins in Baking. In *Lecithins: Sources, Manufacture and Uses*; Szuhaj, B. F., Ed.; AOCS Press: Urbana, IL, 1989; pp 289–322.
Pratt, D. *Bakers Helper,* **1945,** *84,* 1053–1054.
Pratt, D. *Food Ind.* **1946,** *18,* 16.
Prosise, W. E. Commercial Lecithin Products: Food Use of Lecithin Products. In *Lecithins: Sources, Manufacture and Uses*; Szuhaj, B. F., Ed.; AOCS Press: Urbana, IL, 1985; pp 163–182.
Pryde, E. H. Chemical Reaction of Phosphatatides. In *Lecithins: Sources, Manufacture and Uses*; Szuhaj, B. F., Ed.; AOCS Press: Urbana, IL, 1985; pp 213–246.
Rathbun, R. Stabilization of Light Hydrocarbon Distillates. U.S. Patent 2,208,105, 1940.
Raut, S., Bhadoriya, S.; Uplanchiwar, V.; Mishra, V.; Ganhane, A.; Jain, S. Lecithin Organogel: A Unique Micellar System for the Delivery of Bioactive Agents in the Treatment of Skin Aging. *Acta Pharmaceutica Sinica* **2012,** *20,* 8–15.
Reid, E. Pan and Grill Fry Shortening. U.S. Patent 3,443,966, 1969.
Rydhag, L.; Wilton, I. The Function of Phospholipids of Soybean Lecithin in Emulsions *J. Am. Oil Chem. Soc.* **1981,** *58,* 830–837.
Sagarin, E. *Cosmetics Science and Technology.* Interscience: New York, 1957.
Sander, E. H. Lecithin in Beverage Applications. In *Lecithins: Sources, Manufacture and Uses*; Szuhaj, B. F., Ed.; AOCS Press: Urbana, IL, 1989; pp 197–206.
Scharf, A. Lecithin Compositions. U.S. Patent 2,632,705, 1953.
Schipunov, Y. Lecithin. In *Encyclopedia of Surface and Colloid Science;* Marcel Dekker: New York, 2002; pp 297–3017.
Schmidt, J.; Orthoefer, F. T. Nonfood Uses of Lecithin. In *Lecithins: Sources, Manufacture and Uses*; Szuhaj, B. F., Ed.; AOCS Press: Urbana, IL, 1985; pp 183–202.
Schmitt, H.; Falk, M.; Schneider, M. Enzymatic Production of Hydrolyzed Lecithin Products. European Patent 17407008, 2014.
Schneider, M. Fractionation and Purification of Phosphatides. In *Lecithins: Sources, Manufacture and Uses;* Szuhaj, B. F., Ed., AOCS Press: Urbana, IL, 1989; pp 109–130.
Schneider, M. Major Sources, Composition, and Processing. In *Phospholipids Technology and Applications;* Gunstone, F. D., Ed.; Oily Press: Bridgewater, UK, 2008; pp 21–40.
Schweid, J., Cohee, A.; Dee, A. Emulsification System for Creamy Food Products. U.S. Patent 4,539,215, 1985.
Schwieger. Bleaching of Phosphatides with Di Benzoylperoxide. German Patent 603,933, 1934.
Scocoa, P. Utilization of Lecithin. *J. Am. Oil Chem. Soc.* **1976,** *53,* 428–429.
Sejpal, V. Non-aerosol Vegetable Oil Compositions Containing Lecithin and Pure Ethyl Alcohol. U.S. Patent 4,142,003.
Setterquist, R. Process for Production of Polymethylenes in the Presence of Lecithin as an Initiator. U.S. Patent 3,394,092, 1968.

Sharma, A.; Sharma, U. Liposomes in Drug Delivery: Progress and Limitations. *Intl. J. Pharmaceutics* **1997,** *154,* 123–140.
Sincroft, D.; Sipos, E. Treating Compound and Method. U.S. Patent 2,987,527, 1961.
Sinram, R. Nephelometric Determination of Phosphorus in Soybean and Corn Oil Processing. *J. Am. Oil Chem. Soc.* **1986,** *63,* 667–670.
Sipos, E. Industrial Coating Applications for Lecithin. In *Lecithins: Sources, Manufacture and Uses*; Szuhaj, B. F., Ed.; AOCS Press: Urbana, IL, 1989; pp 261–276.
Sipos, E.; Szuhaj, B. F. Lecithins. In *Baileys Industrial Oil and Fat Products,* 5th ed.; Hue, Y. H., Ed.; Wiley-Interscience: New York, 1996; Vol. 1, pp 311–395.
Smilles, A.; Kakuda, B. E.; MacDonald, B. Effect of Degumming Reagents on the Composition and Emulsifying Properties of Canola, Soybean and Sunflower Acetone Insoluble. *J. Am. Oil Chem. Soc.* **1989,** *66,* 348–352.
Socci, R.; Gunderman, A.; Fottiu, E.; Kabacoff, B. Nail Enamel. U.S. Patent 4,302,442, 1981.
Sollman, E. Nondegumming Gasoline and Mode of Preparing Same. U.S. Patent 1,884,899, 1932.
Sosoda M.; Pasker, B.; Bogocz, M. Improving the O/W Emulsifying Properties of Rapeseed Lecithin Insoluble Fraction by Acetylation. *Drug Tech.* **2003,** *60,* 303–308.
Spitzer, J.; Marra, D.; Osipow, L.; Claffey, K. Process for Conditioning Hair. U.S. Patent 4,314,573, 1982.
Stanley, J. Production and Utilization of Lecithin. In *Soybeans and Soybean Products;* Markley, K., Ed.; Interscience: New York, 1951; Vol. 2, pp 593–647.
Szuhaj, B. Pan Release Product and Process. U.S. Patent 3,928,056, 1975.
Szuhaj, B. *Lecithins: Sources, Manufacture and Uses.* AOCS Press: Urbana, IL, 1989.
Szuhaj, B. Lecithin. In *Baileys Industrial Oil and Fat Products,* 6th ed.; Shahidi, F., Ed.; Wiley Interscience: New York, 2005; Vol. 3, pp 361–456.
Szuhaj, B; Yaste, J. R. Phosphatide Preparation Process. U.S. Patent 3,962,292, 1976.
Szuhaj, B. F. Lecithin Production and Utilization. *J. Am. Oil. Chem. Soc.* **1983,** *60,* 306–309.
Tan, C. T.; Howard, G.; Turner, E. Lipoproteins for Food Use and Methods for Preparing Same. U.S. Patent 4,360,537, 1982.
Thurman, B. Recovery of Valuable Fractions from Glyceride Oils. U.S. Patent 2,415,313, 1947.
Todd, P. H.; Haley, H. E. Liquid Seasoning Composition. U.S. Patent 4,283,429, 1981.
Vandana, V.; Karuna, M.; Vijayalaskshmi, P.; Prasad, R. A Simple Process to Enrich Phospholipid Content of Commercial Soybean Lecithin. *J. Am. Oil Chem. Soc.* **2001,** *78,* 555–556.
Vandana, V.; Karuna, M.; Prassad, P.; Prassad, R. Process for the Preparation of Hydroxylated Lecithin from Crude Soybean Oil. U.S. Patent 6,638,544, 2003.
van Nieuwenhuyzen, W. Lecithin Production and Properties. *J. Am. Oil Chem. Soc.* **1976,** *53,* 425–427.
van Nieuwenhuyzen, W. The Industrial Uses of Special Lecithins: A Review. *J. Am. Oil Chem. Soc.* **1981,** *58,* 886–888.
van Nieuwenhuyzen, W. Industrial Production of Lecithin and Its Derivates. Presented at Oresund Food Network, AOCS Phospholipid Seminar, Copenhagen, 2008.

van Nieuwenhuyzen, W. The Changing World of Lecithins. *INFORM* **2014,** *25,* 254–259.
van Nieuwenhuyzen, W.; Tomas, M. Update on Lecithin and Phospholipid Technologies. *Euro. J. Lipid Technol.* **1981,** *110,* 472–486.
Vermaas, L.; Jan, B.; Sonneveid, P. Anti-Spattering Agent and Spreads Comprising Same. European Patent Application 0,532,082 A2, 1992.
Vlasblom, J. Leather Cleaner and Conditioner. U.S. Patent 5,415,789, 1995.
Vollmer, D. Nonfoaming Lecithin Emulsion Cookware Lubricant. U.S. Patent 3,986,975, 1975.
Walrod, F. E. Proceedings 23rd Annual Meeting American Society of Baking Engineers, 1947, p 76.
Weber, E. J. Corn Lecithin. In *Lecithins: Sources, Manufacture and Uses*; Szuhaj, B. F., Ed.; AOCS Press: Urbana, IL, 1989; pp 39–55.
Weiss, R.; Sinner, J. M.; Bloch, W. Culinary Mix Utilizing Liquid Oil Shortening and Process for Preparing Same. U.S. Patent 3,222,184, 1965.
Wendel, A. Lecithin. In *Kirk-Othmer Encyclopedia of Chemical Technology;* Grant, M., Ed.; Wiley Interscience: New York, 1995; pp 192–210.
Wendel, A. Lecithin: The First 150 Years, Part 1, From Discovery to Early Commercialization; Part 2, Evolution for a Global Pharmaceutical Industry. *INFORM* **2000,** *11,* 885–892, 992–997.
Werly, E. Fluidizing Lecithin. U.S. Patent 2,777,817, 1957.
Whitekettle, W.; Crovetto, R. Steam Condensate Corrosion Inhibitor Compositions and Methods. U.S. Patent 7,407,623, 2008.
Widlak, N. Fluid Emulsified Shortening Composition. U.S. Patent 6,387,433, 2002.
Williamson, G.; Manach, C. Bioavailabilty and Bioefficacy of Polyphenols in Humans II. Review of 93 Intervention Studies. *Am. J. Clin. Nutrit.* **2005,** *81,* 243S–255S.
Wiesehahn, G. A. Soybean Phosphatides and Their Uses: A Review. *Oil and Soap* **1937,** *14,* 119–122.
Wiesehahn, G. A. Soft Lecithin Preparation. U.S. Patent 2,194,842, 1940.
Wittcoff, H. Hydroxyphosphatides. U.S. Patents 2,445,948; 2,483,748, 1948, 1949.
Wittcoff, H. *The Phosphatides, ACS Monograph No. 112;* Reinhold: New York, 1951.
Zeisel, S. Lecithin in Human Health and Nutrition. In *Lecithins: Sources, Manufacture and Uses*; Szuhaj, B. F., Ed.; AOCS Press: Urbana, IL, 1989; pp 225–236.
Zeisel, S. H. Lecithins in Health and Disease. In *Lecithins: Sources, Manufacture and Uses*; Szuhaj, B. F., Ed.; AOCS Press: Urbana, IL, 1985; pp 323–345.
Zeisel, S. H.; Szuhaj, B. F., Eds. *Choline, Phospholipids, Health and Disease;* AOCS Press: Urbana, IL, 1998.

2

Rice Bran Lecithin: Compositional, Nutritional, and Functional Characteristics

Ram Chandra Reddy Jala and R.B.N. Prasad ▪ *Centre for Lipid Research, CSIR-Indian Institute of Chemical Technology, Hyderabad, India*

Introduction

The word *lecithin* is derived from the Greek term *lekithos,* meaning "egg yolk." In 1846 Gobley isolated lecithin from egg yolk and in 1850 gave it its present name. Lecithin is a natural complex mixture of phosphatides that varies in color from light tan to dark reddish brown and in consistency from a fluid to a plastic solid. The most acceptable definition for lecithin is "a mixture of phospholipids (PLs) extracted from foods of vegetable or animal origin with a minimum of 60% acetone insoluble substances" (Schneider, 2006). Lecithin is the gummy material contained in crude vegetable oils that is removed by degumming. Globally, lecithins are being used as emulsifiers, wetting and instantizing agents, viscosity modifiers, releasing agents, separating agents, and anti-dusting agents. They are also utilized in nutritional supplements and to extend the shelf life of food products. Further, highly purified phospholipids can be used as raw materials for the preparation of lipid vesicles, liposomes, which are the vehicles of choice in some drug delivery systems and are also being explored in the development of artificial lung surfactants (Yukihiro Namba, 1993). Presently soybean, rapeseed, sunflower, corn, and other lecithins are exploited for several applications. However, soybean lecithin is the most commonly used lecithin for various applications. Worldwide soybean production forecasts for 2012–2013 were 271 million metric tons (USDA, May, 2011). Although the lecithin market grows steadily, only an estimated 15–20% of the potential gums are processed into standard quality lecithin and specialties with dedicated technological and nutritional functions in high value pharmaceutical, food, and feed segments (van Nieuwenhuygen, 2014). The widespread planting of genetically modified (GM) soybeans, which began in 1995, has caused disruptions in supply streams of commercial food-grade lecithin from traditional soybean varieties that previously flowed from the United States and Latin America to Europe. As the availability of traditional non-GM soybeans from the United States and Latin America has decreased, the sourcing of identity-preserved soybean lecithin from other regions and lecithin from sunflower and rapeseed has grown (van Nieuwenhuygen, 2014). Hence, rice bran lecithin (RBL) may be an attractive source for those who prefer non-GM-based lecithins.

Even though RBL has high potential for food and industrial applications, this has not enough attention. Therefore, our objective in preparing this chapter is to enhance the knowledge on RBL with respect to its composition of major and minor constituents and its nutritional properties.

Advantages of Rice Bran Lecithin

Rice bran lecithin is superior to soybean lecithin due to lower content of polyunsaturated fatty acids and the presence of natural antioxidants like phytosteryl esters, oryzanol, and tocotrienols. The active portion of RBL is a complex mixture of phosphatidylcholine (PC), phosphatidylethanolamine (PE), and phosphatidylinositol (PI) as major PLs (Figure 2.1), along with glycolipids (GLs, Figure 2.2), triglycerides, free fatty acid, oryzanol, tocols, sterols (Figure 2.3, p. 38), and waxes as minor components (Adhikari and Adhikari, 1986; Moazzami et al., 2011).

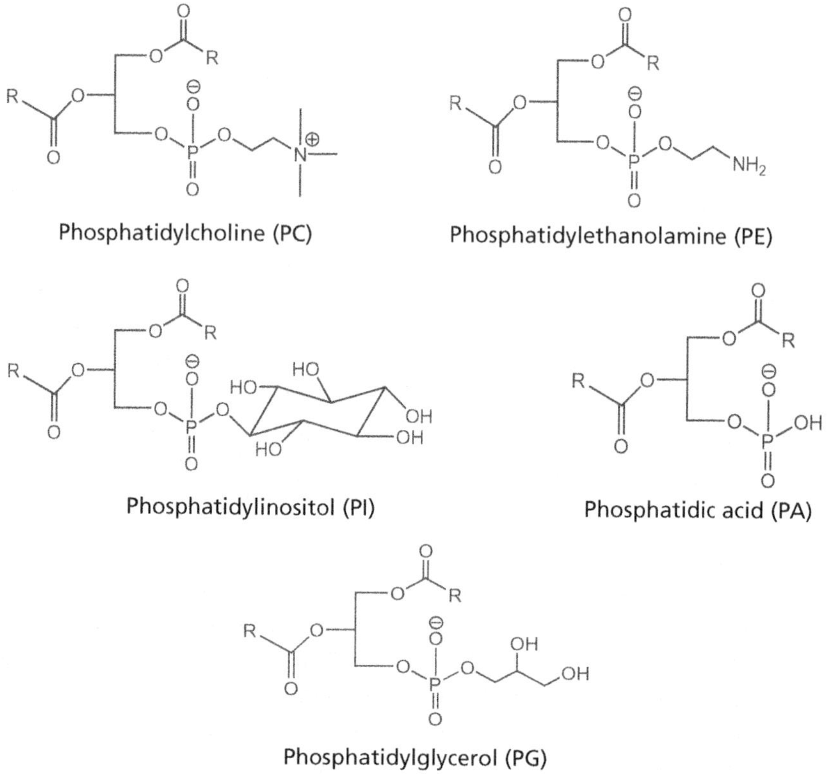

Figure 2.1 Phospholipids present in rice bran lecithin.

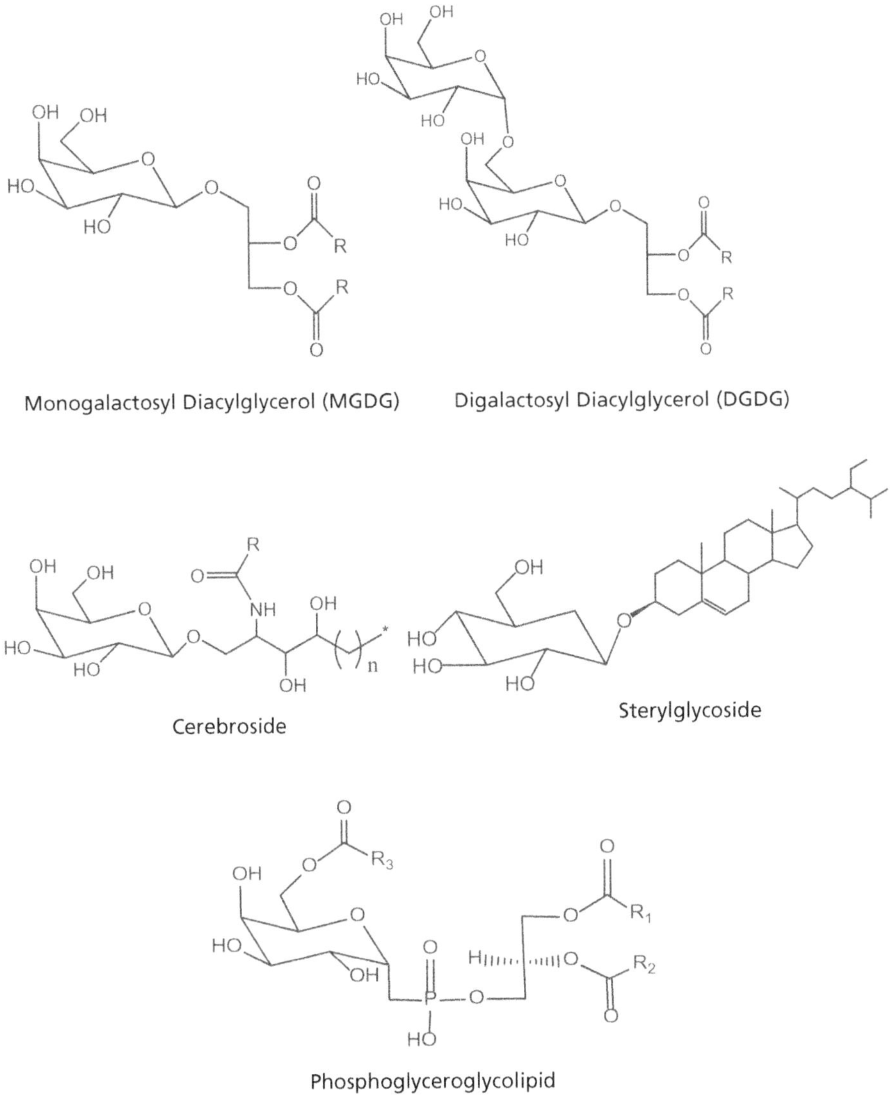

Figure 2.2 Commonly observed glycolipids in rice bran lecithin.

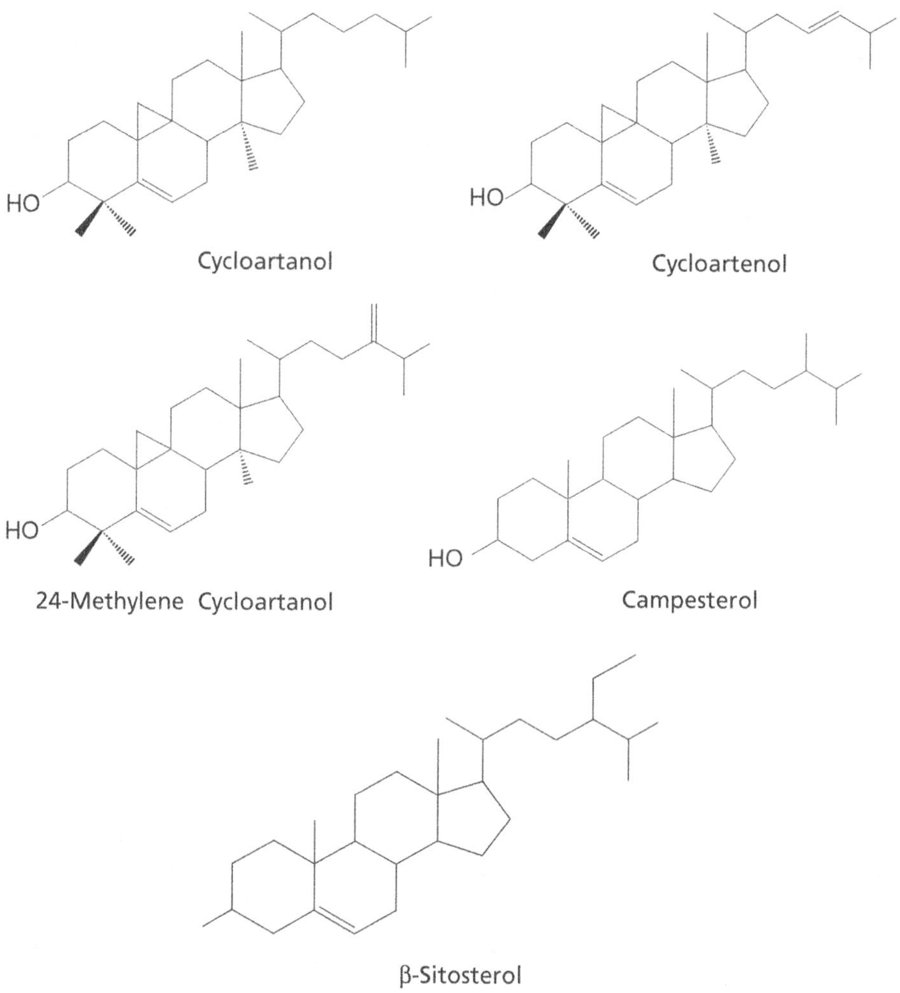

Figure 2.3 Sterols and tocopherols present in unsaponifiable matter of rice bran lecithin.

Global Scenario of Rice and Rice Bran Lecithin

According to the FAO, global rice production is expected to increase to a record 497.6 million tons (milled basis) in 2013–2014, and most of the increase in rice production can be seen in Asia (FAO, Rice Market Monitor, 2013). The recovery of rice bran oil (RBO) from rice bran is usually 15 to 20%. The RBL amounts to roughly 1 to 2% of oil weight depending on the drying conditions. As global production of RBO fluctuates between 1 million and 1.4 million tons, depending on the extraction of RBO, the global RBL production is expected to fluctuate between 10,000 and 20,000 tons.

Composition and Significance of Rice and Phospholipids

Paddy rice, a monocotyledon plant, is known as *Oryza*. The genus *Oryza* is composed of two cultivated species, namely *Oryza sativa* and *Oryza glaberrima*, plus 21 wild species (Khush, 1997). Rice (*Oryza sativa*) is one of the cereal crops, as well as staple food for most of the world's population, especially Asian countries (Bird et al., 2000). Frequently, rice is eaten in cooked form by humans to obtain various nutrients, as well as to supplement their caloric intake (Kim et al., 2011). Paddy rice milling yields 70% of rice (endosperm) as the major product, and byproducts consist of rice husk (20%), rice bran (8%), and rice germ (2%) (Norhaizan et al., 2013). Rice grains contain a much smaller proportion of lipids than starch; however, these lipids make a significant contribution to processing and nutritional properties (Moazzami et al., 2011). For instance, RBO is a popular cooking oil in several Asian countries, which has a direct impact on human nutrition and health (Ghosh, 2007). PLs consisting of a covalently bound phosphate and lipid is a major class of lipid in rice, comprising up to 10% of total grain lipid content. Lecithin contains a minimum of 60% PLs and GLs and a maximum of 40% neutral lipids (NLs), depending on the degumming conditions. PLs are amphipathic in nature, that is, each molecule consists of a hydrophilic portion and a hydrophobic portion. Natural PLs can be classified into two major categories, glycerophospholipids (GPLs) and sphingophospholipids. The most abundant types of naturally occurring GPLs are PC, PE, phosphatidyl serine (PS), PI, phosphatidyl glycerol (PG), and cardiolipin (CL). The structural diversity within each type of phosphoglyceride is due to the variability of the head group, variability of the chain length, and degree of saturation of the fatty acid ester groups. GPLs consist of fatty acids esterified to a glycerol backbone, a phosphate group, and a hydrophilic head group. PLs are a major component of lipids in the oil-rich rice embryo and bran (Yoshida et al., 2011). Lysophospholipids (LPLs) are an important subcategory of GPLs with a free alcohol in the *sn*-2 position (Choudhury and Juliano, 1980). The PLs in rice bran and endosperm make a significant contribution to the quality of rice, affecting properties such as the rancidity of paddy or brown rice (Aibara et al., 1986) and the physicochemical properties of starch (Pérez and Bertoft, 2010). Even though PLs are only minor components compared to starch and protein, they have both nutritional and functional significance. Dietary PLs have a positive impact on several human diseases and reduce the side effects of some drugs. Rice has long been consumed as a staple food in many Asian countries, and rice PLs may have significant health benefits for those populations. The amphiphilic lipids found in plant and animal cell membranes are critically important to all cellular organisms. Dietary PLs have beneficial effects on a range of human health conditions such as coronary heart disease, cancer, and inflammation (Kullenberg et al., 2012). Because rice is the single most important staple food in the world, it is likely to represent a significant source of dietary PLs for a large proportion of the world's population. In addition to GPLs, 1,2-diacyl-3-*O*-phospho-*O*-[6-*O*-acyl-(α-D-galacto-pyranosyl)]-*sn*-glycerol and 1,2-diacyl-3-*O*-phospho-*O*-[6-*O*-acyl-(β-D-galacto-pyranosyl)]-*sn*-glycerol were also isolated (Sarode Manjula et al., 2009; Shaik Ramjan et al., 2004) from RBO.

Isolation and Compositional Studies on Rice Bran Lipids

Initially, Sakata et al. (1974) investigated the classification and fatty acid composition of PLs in rice grain belonging to Japan. The PLs in rice grain were isolated by silica gel column chromatography and further separated into neutral and acidic PLs by DEAE-cellulose column chromatography. Individual PLs of each group were isolated and purified either by silica gel column chromatography or thin layer chromatography and characterized by IR spectroscopy. The PLs were cleaved with 5% methanolic HCl and the fatty acid methyl esters produced were analyzed by gas chromatography. Among the eight PLs in rice grain identified by them, PC, PE, and PI were the principle components. Palmitic, oleic, and linoleic acids were the main fatty acids present in these three PLs. The fatty acid composition of PC and PE were similar but slightly different from that of PI. Yasuo et al. (1999) investigated the variation in fatty acid composition of three Japonica and two Indica rices grown in Japan. Kirara397, Yukihikari, and Koganenishiki are the Japonica varieties, and IR64 and RD19 are the Indica varieties. Unsaturated fatty acid contents in the PL classes were largely different from each other among Japonica rices harvested in separate districts (Mano et al., 1999). In Indica varieties, oleic acid, linoleic acid, and palmitic acids were found to be major factors in PC, PE, and PI, respectively. Similarly, in Japonica varieties, no specific trend was observed in cases of PC and PI; however, linoleic acid was found to be significant in PE (see Table 2.A). Yoshida et al. (2011a, 2011b) investigated the fatty acid compositions of lipid extracts from rice brans belonging to Japan. The lipids of these rice brans comprised mainly triglycerides (84.9–86.0 %), free fatty acid (4.2–4.6 wt %), and PLs (6.5–6.7%), and other components were detected in minor proportions (0.2–2.1%). Among the PLs, PC (43.3–46.8%), PE (25.0–27.3%), and PI (20.2–23.2%) were identified as major components. Comparison of the different cultivars showed, with a few exceptions, no substantial difference in fatty acid distribution. Fatty acid distribution of triglycerides among the five cultivars was characterized. They found unsaturated and saturated fatty acids predominantly concentrated at the *sn*-2 and *sn*-1 (or *sn*-3) position, respectively.

According to Ahn et al. (1984), total lipid contents in South Korean-based rice bran from for Poong-San (Tongil) and Dong-Jin (Japonica) varieties were 16.1 and

Table 2.A Fatty Acid Composition (%)[a] from Brown Rice Phospholipid Classes

Phospholipid	Rice Variety	Fatty Acid Composition (wt%)									
		14:0	16:0	18:0	18:1	18:2	18:3	20:0	20:1	16/18[b]	18:1/Σ(18:2 + 18:3)[c]
PC	Kirara397	0.5	21.1	1.0	37.4	38.8	0.9	<0.1	0.3	0.27	0.94
	Yukihikari	ND[d]	20.0	<1.0	36.0	43.0	1.0	ND[d]	ND[d]	0.25	0.82
	Koganenishiki	0.8	19.7	1.0	44.4	32.3	1.4	0.1	0.3	0.25	1.32
	IR64	1.0	19.6	1.1	48.7	28.7	0.6	<0.1	0.3	0.25	1.66
	RD19	1.0	19.9	1.7	49.3	26.9	0.6	0.2	0.4	0.25	1.79
PE	Kirara397	0.8	23.3	1.7	25.0	46.7	2.1	0.1	0.3	0.31	0.51
	Yukihikari	ND[d]	23.0	<1.0	22.0	52.0	3.0	ND[d]	ND[d]	0.29	0.41
	Koganenishiki	1.3	22.9	1.5	29.2	43.0	1.8	0.3	<0.1	0.30	0.65
	IR64	1.3	23.1	2.2	34.2	38.0	0.8	0.2	0.2	0.31	0.88
	RD19	1.2	20.9	1.8	36.8	37.9	0.8	0.3	0.3	0.27	0.95
PI[e]	Kirara397	0.6	33.5	3.1	25.8	35.6	1.0	0.3	0.1	0.51	0.70
	Koganenishiki	0.6	36.5	2.5	29.1	29.8	1.4	0.1	0.1	0.58	0.93
	IR64	0.7	37.6	2.3	28.0	30.5	0.7	0.1	0.2	0.61	0.90
	RD19	0.7	38.9	2.5	27.3	29.2	0.9	0.4	0.1	0.65	0.91

[a]Values represent the average of two or three analyses.

[b]The ratio of 16:0 to the sum of C18 acids.

[c]The ratio of 18:1 to Σ(18:2 + 18:3).

[d]Not detected.

[e]Yukihikari sample was not sufficient to examine the fatty acid composition.

Source: Adapted from Mano et al. (1999) with kind permission of Taylor & Francis Group.

17.0%, respectively. NLs for Tongil (75.2%) were slightly higher than those for Japonica (73.7%), whereas contents of GL for Tongil (16.7%) were lower than those for Japonica (22.80%). Contents of PL in Tongil (8.1%) were much higher than those in Japonica (3.5%). Acid value, peroxide value, and thiobarbituric acid values of total lipids extracted from rice bran of Tongil were slightly lower than those of Japonica, whereas the iodine value exhibited the reverse trend. The NLs were fractionated and identified as hydrocarbon, sterol ester, free sterol, triglyceride, free fatty acid, diglyceride, and monoglyceride. Triglyceride contents were less than those of common edible oils, but diglyceride and monoglyceride contents were higher. Among the GLs, sterylglycoside ester (11.5%) was the most abundant, and in case of PLs, PE, PI, and PC were the major components. The main fatty acids present in the entire lipid components were oleic, linoleic, and palmitic acids. The fatty acid composition of the NLs was similar to that of the total lipids. In GLs, the content of linoleic acid was higher than that of oleic acid, and palmitic acid was predominant in the fatty acid composition of steryl glycoside ester. Shin and Yang (1986) performed the comparative studies on the composition of GLs and PLs of milled rice grains grown in Korea. Main classes of the GLs identified were steryl glycoside ester, steryl glycoside, monogalactosyl diglyceride, digalactosyl diglyceride, and cerebroside in both varieties. Among the PLs, LPC, PC, PE, PI, and PS were the major components (>85%) and diphosphatidyl glycerols and phosphatidyl glycerols were the minor components. In both the varieties, the major fatty acids of GL and PL fractions were palmitic, linoleic, and oleic acids. However, the content of palmitic acid in GL fractions and stearic, oleic, and linoleic acids in PLs showed differences between two varieties. Later, Kwon et al. (1996) conducted comparative studies on the composition of polar lipids in rice grains of Japonica and Indica varieties grown in Korea. They found that the ratio of GLs and PLs was 4.1:6.5 in Japonica RBO and 2.6:3.7 in Indica RBO. Polar lipid content was significantly higher in Japonica RBO. PLs in RBOs consisted of diphosphatidyl glycerol, PE, PI, PS, PC, and LPC. Major fatty acids of the GL and PL fractions were oleic, linoleic, and palmitic in Japonica and Indica RBOs. According to these authors, the main components of GLs were steryl glycoside ester, steryl glycoside, monogalactosyl diglyceride, digalactosyl diglyceride, and cerebroside. The content of steryl glycoside ester was the highest (48.8–52.1% of total GLs).

Nasirullah and Nagaraja (1987) found that NLs, GLs, PLs, ceramides, and wax lipids were present in rice bran of Indian origin. NLs were very much similar to groundnut lipids as indicated by fatty acid composition. As in the NLs, the major fatty acids in rice bran PLs were palmitic, oleic, and linoleic. Ceramide monohexoside, ceramide dihexoside, ceramide trihexoside, and wax lipids were also present in appreciable amounts. According to them, GLs of rice bran deserve special mention because they contained sterylglycosides along with glyceroglycolipids. Sterylglycosides are generally not found in other oilseeds. Moreover, many other GLs, such as tetraglycosyl diglyceride, monoglycosyl monoglyceride, diglycosyl diglyceride, and triglycosyl diglyceride were also present, and these contribute to the sweetness of the rice. Hemavathy and Prabhakar (1987) reported the composition of lipids of rice bran lipids from three varieties of rice belong to India. Lipids extracted amounted to 21.9–23.0% of the bran (on dry weight basis) and consisted of 88.1–89.2% NLs, 6.3–7.0% GLs, and 4.5–4.9% PLs. NLs consisted mostly of triglycerides (83.0–85.5%), monoglycerides (5.9–6.8%), and small amounts of diglycerides, sterols, and free fatty acids. Three GLs and eight PLs were separated and characterized. The major PLs were PC, PE, and PI, and minor PLs were PA, PG, LPC, LPE, and acyl PE. These researchers reported that acylated steryl glucoside and digalactosyl diglyceride were the main GLs and monogalactosyl monoglyceride was present in small amounts.

El-Shattory et al. (1979) conducted a study on isolation and purification of Egypt-based rice bran PLs (1.2–1.94% of local solvent-extracted RBOs). The seven RBO varieties were extracted with chloroform and methanol (2:1, v/v). The mixed total PLs of the seven samples were isolated by using silicic acid. Thin layer chromatographic investigations of the individual PLs were carried out with chloroform, methanol, and water (65:25:4, v/v/v). The isolated PLs were fractionated by using silicic acid columns. The columns were eluted with chloroform and methanol mixtures (93:7, 80:20, 50:50, 30:70, v/v/v/v) and purity of fractions was examined by TLC. They found that the major fatty acids in the PLs were oleic, palmitic, and linoleic, and other fatty acids were myristic, stearic, and palmitoleic acids.

Production of Rice Bran Lecithin and Lyso Lecithin

PLs are generally removed along with soap stock in case of chemical refining of crude RBO and the recovery of PLs or lecithin is not commercially feasible. In cases of phosphoric acid degumming, lecithin cannot be used for food applications because it becomes contaminated with the phosphoric acid.

Pragasam et al. (2002) reported the preparation of RBL from crude RBO (20 kg/batch) employing degumming at 70 °C by adding water (800 ml) under stirring (250 rpm) for 30 minutes. The wet gum sludge (2, 5 kg) was subjected to drying using the roller drum drier. However, the resulting lecithin was dark in color due to oxidation during the drying operation for both phosphoric acid degummed and water degummed wet gum sludges (Indira et al., 2000; Pragasam et al., 2002). Dark-colored lecithin is generally obtained after cooling the dried gums. Carotenoids and chlorophyll are the major pigments present in RBL, along with porphyrin and brown pigments (Scholfield and Dutton, 1954). The other sources affecting color are solvent extraction and degumming conditions. Vandana et al. (2003) carried out a detailed study to remove the color of RBL using different bleaching agents namely, hydrogen peroxide, benzoyl peroxide, and sodium chlorite. The authors also reported a simple method to enrich PL content to a required extent in commercial RBL by adding wet acetone insolubles.

Lysolecithin is obtained as a byproduct during enzymatic degumming process. The phospholipases A_1 or A_2 are the most suitable enzymes for the enzymatic degumming process of vegetable oils. During this process, both hydratable and non-hydratable PLs are converted to lyso gums that can be easily removed by centrifugation (Chakrabarti et al., 2006; Roy et al., 2002). Reddy et al. (2008) adopted hexane fractionation method for removing the insoluble impurities by taking the lyso gums in hexane (1:10, w/v).

Physicochemical Characterization of Rice Bran Lecithin and Lyso Lecithin

Pragasam et al. (2002) reported several analytical characteristics of RBL, such as moisture content (0.7%), acetone insoluble matter (43.5%), benzene insoluble matter (0.1%), acid value (20.9), PL content (42.5%), wax content (1%), peroxide value (zero), and color (5R+Y, 1 cm cell, 60.4 Lovibond Units). Vandana et al. (2003) determined the physicochemical characteristics of RBL such as acid value (37–38), iodine value (69–72), peroxide value (15–21), and color (12–18+). The physicochemical properties (Reddy et al., 2008) of lyso lecithin obtained from enzymatic degumming process (Chakrabarti et al., 2006; Roy et al., 2002) were reported as follows: FFA, 23.5; peroxide value, 0 ppm; iodine value, 61.4 g/100g; calorific value, 7.7 kcal/g; acetone insoluble matter, 71.5; acetone solubles, 28.5; hexane insolubles, 7.5; unsaponifiable matter, 6.1; oryzanol, 1.5; cupper, cadmium, lead elements, below detection levels; and moisture, 7.7 %. The calorific value of lysolecithin (7.7 kcal/kg) and its polar lipid fraction (6.03 kcal/kg) and non-polar lipids fractions (8.9 kcal/kg) were also evaluated (see Table 2.B).

Identification and Isolation of Components in Rice Bran Lecithin and Lyso Lecithin

Adhikari and Adhikari (1986) conducted a detailed study on isolation, identification, and purification of the PLs from RBL. The crude RBL was separated into individual lipid classes by preparative thin layer chromatography using hexane, diethyl ether, and acetic acid (80:20:1, v/v/v). The components were identified by comparing R_f values against standard samples and by using various spray reagents. PLs were isolated from crude lecithin using acetone fractionation method. The isolated PLs were resolved by silica gel thin layer chromatography (TLC) of the mixture using either chloroform, methanol, and water (65:25:4, v/v/v); chloroform, methanol, and 28% ammonia (65:25:5, v/v/v); or chloroform, methanol, acetone, acetic acid, and wa-

Table 2.B Physicochemical Characterization of Rice Bran Lecithin

Characteristic	Value	Reference
Moisture (%)	0.3	Adhikari and Adhikari (1986)
	0.75	Pragasam et al. (2002)
Acetone Insolubles (%)	63	Adhikari and Adhikari (1986)
	43.5	Pragasam et al. (2002)
	40.0	Vandana et al. (2003)
Phospholipids Content (%)	42.5	Pragasam et al. (2002)
Benzene Insoluble (%)	0.1	Adhikari and Adhikari (1986)
	0.11	Pragasam et al. (2002)
Acid Value	20.8	Adhikari and Adhikari (1986)
	20.9	Pragasam et al. (2002)
	37.0	Vandana et al. (2003)
Peroxide Value (ppm)	0.0	Pragasam et al. (2002)
	15.0	Vandana et al. (2003)
Wax Content (%)	1	Pragasam et al. (2002)
Viscosity, P at 25 °C	150 (max)	Adhikari and Adhikari (1986)
Color (on Gardner scale)	10 (max)	
Color (5R+Y) 1 cm Cell (Lovibond Units)	60.4	Pragasam et al. (2002)
Color (5% solution) 1 cm Cell (Lovibond Units)	24.4	
Color (on Gardner scale)	18+	Vandana et al. (2003)

ter (6:2:8:2:1, v/v/v/v/v). TLC spots were visualized using iodine vapour (revealing mainly unsaturated lipid materials), ammonium molybdate-perchloric acid reagent (specific for PLs), ninhydrin (specific for primary and secondary amino groups), and Dragendorff's reagent (specific for choline lipids). Different PL classes were subjected to strong acid hydrolysis for the liberation of bases and glycerol or inositol, so as to have further confirmation on the identity of the PLs. The hydrolytic products were identified using TLC and compared with literature R_f values. PLs were separated into individual classes by preparative TLC (silica gel G, 0.5 mm; chloroform, methanol, and water [65:25:4, v/v/v]) and PL components were determined gravimetrically

Table 2.C Composition of Crude RBO and Lecithin

Composition of Crude Rice Bran Oil

	TG	DG	MG	FFA	Wax	GL	PL	Unsap	Reference
Crude RBO	81–84	2–3	1–2	2–6	3–4	0.8	1–2	4	Mahua (2007)
Crude RBO	83–86	3–4	6–7	2–4	6–7	6–7	4–5	2–4	Orthoefer (2011)

Composition of Rice Bran Lecithin

	PC	PE	PI	PA, LysoPC, Lyso PE	TG	Wax	Carbohydrates, Sterols, FFA	Moisture	Reference
From Crude Oil	20.4	17.8	5.8	9.4	39.2	3.1	4.0	0.3	Adhikari and Adhikari (1986)
From Dewaxed Oil	23.1	20.2	6.6	10.8	35.2	—	3.8	0.3	

(Table 2.C). PL composition of RBL prepared from crude RBO was reported by them as follows: PC, 20.4; PE, 17.8; PI, 5.8; other phosphatides, 9.4; triglycerides, 39.2; wax, 3.1; carbohydrates, sterols, FFA, 4; and moisture, 0.3%. Further, these authors reported the PL composition of RBL prepared from dewaxed RBO as follows: PC, 23.1; PE, 20.2; PI, 6.6; other phosphatides, 10.8; triglycerides, 35.2; carbohydrates, sterols, and FFA, 3.8; and moisture, 0.3% (Table 2.C). The data of PL compositions of different deoiled lecithins are depicted in Table 2.D. The authors reported the fatty acid composition of commercial RBL as follows: palmitic, 18.1; stearic, 4.0; oleic, 42.8; linoleic, 33.6; and linolenic, 1.5%. Palmitic, oleic, and linoleic acids were the major fatty acids present in all the PL classes. They found that palmitic acid was high in PC and LPC, followed by oleic acid. The linoleic acid contents of PE, PI, and PG were in the range of 47.9–55.0%, and oleic acid contents were in the range of 17–20.8%. PA contained less palmitic acid than the other fractions, about 40% each of oleic and linoleic acid and 4.4% of linolenic acid, a considerably higher amount than in the other fractions (0.3–1.3%).

According to Vandana et al. (2003), the fatty acids present in RBL were myristic (0.2%), palmitic (19.1%), stearic (1.2%), oleic (43.2%), linoleic (35.4%), and linolenic (0.9%). In general, RBL has higher proportion of saturated fatty acid compared to soybean lecithin. The other major differences in fatty acid composition were the presence of higher oleic and lower linoleic acids in RBL compared to soybean, sunflower,

Table 2.D Phospholipid Composition of Deoiled Lecithins

	Rice Bran		Soybean			Sunflower	Rapeseed	Corn
PC	38.0	30.0	22.0	28.1	23.0	25.0	25.0	30.0
PE	33.2	24.0	14.0	21.9	20.0	11.0	22.0	3.0
PI	10.9	—	12.0	14.1	14.0	19.0	15.0	16.0
PS	—	—	—	—	—	1.0	—	1.0
PG/DPG	8.6	—	2.0	—	—	1.0	—	1.0
PA	3.6	14.0	6.0	7.8	8.0	3.0	—	9.0
NAPE	—	—	3.0	—	—	1.0	—	3.0
Lyso PLs	5.7	23.0	2.0	—	—	—	19.0	5.0
Minor PLs	—	—	—	3.1	8.0	—	—	—
Unidentified PLs	—	5.0	—	—	—	—	—	—
Glycolipids, Complex Sugars, Traces of NLs	—	—	4.0	25.0	27.0	—	19.0	—
Reference	Adhikari and Adhikari (1986)	Pragasam et al. (2002)	Schneider (2008)	Central Soya (2002)	Cherry and Kramer (1989)	Grewal et al. (1978)	Appleqvist and Ohlson (1972)	Weber (1981)

rapeseed, and corn lecithins (Table 2.E). RBL contains lower linolenic acid compared to soybean and rapeseed lecithins. In principle, the lower level of linoleic and linolenic acids gives RBL greater resistance to autoxidation and development of off flavors than soybean lecithin, and therefore, RBL exhibits better oxidative stability compared to other lecithins (Table 2.E; Adhikari and Adhikari, 1986; Grewal et al., 1978; Hougen et al., 1985; Reddy et al., 2008b; Vandana et al., 2003; Weber, 1981).

Reddy et al. (2008) performed isolation of polar lipids from RB lysolecithin produced during enzymatic degumming of RBO (Chakrabarti et al., 2006; Roy et al., 2002). RB lysolecithin, which is free from hexane insolubles (25 g), was taken in a centrifuge tube, and to this chilled acetone was added (1:5, w/v). The contents were thoroughly mixed and kept at 0–5 °C for a period of 15 minutes, with occasional mixing. At the end of this period the contents were centrifuged in a refrigerated centrifuge for about 10 minutes. The acetone solubles were separated and the residue was washed twice with chilled acetone to completely remove the oil. The residue was dried under reduced pressure to obtain acetone insolubles (17.875 g, 71.5%) containing polar lipids. The acetone solubles were pooled and concentrated to recover NLs (7.125 g, 28.5%). Separation of traces of NLs, GLs, and PLs from polar lipids was achieved by silica gel column chromatography by eluting with chloroform, acetone, and methanol, respectively. The separation pattern was monitored by a micro-TLC technique using a solvent system of chloroform, methanol, and water (65:25:4, v/v/v). The PL mixture was further separated to individual PLs using silica gel column chromatography employing gradient elution of chloroform along with methanol in the ratios starting from 85:15 to 60:40 (v/v). The separation pattern was monitored by a micro-TLC technique using characteristic spray reagents and standard samples. The individual PLs were identified by spraying with characteristic spray reagents. GL fraction was analyzed by TLC using reference compounds and α-naphthol spraying reagent. The solvent system employed was chloroform, methanol, acetic acid, and water (170:24:25:4, v/v/v/v).

Similar to RBL, the major fatty acids present in rice bran lysolecithin were oleic (43.2%), linoleic (34.1%), and palmitic (19.8%); however, lysophosphatidylcholine (LPC) and lysophosphatidylethanolamine (LPE) appeared to be rich sources of oleic and linoleic acids. The major fatty acids present in the triglyceride (TG) and NLs isolated from lysolecithin were oleic (41.0% and 42.4%), linoleic (34.9% and 31.4%), and palmitic (20.2% and 22.6%). The polar lipid fraction, PC, LPC, PE, LPE, GLs contain oleic (45.6, 47.3, 60.0, 40.2, 60.5, 25.3%), linoleic (35.9, 26.0, 30.1, 29.3, 33.6, 35.6%), and palmitic (16.1, 22.8, 8.3, 25.7, 4.9, 25.7%) as major fatty acids, respectively (Reddy et al., 2008). In this study, PLs fraction was found to contain PC 8.6, PE 6.5, PA and PG 9.3, LPE 17.8, and LPC and LPI mixture 30.8%. The authors claimed that RB lysolecithin was a rich source of GLs (20.1% based on polar lipid fraction). The GL fraction of lysolecithin was qualitatively characterized by TLC and it was found that cerebrosides, monogalactosyl diglyceride, and

Table 2.E Fatty Acid Composition of Deoiled Lecithins

	Rice Bran		Soybean			Sunflower	Rapeseed	Corn
Myristic	—	0.2	—	—	—	—	—	—
Palmitic	18.1	19.1	10.8	21.7	20.0	15.0	18.0	22.0
Stearic	4.0	1.2	3.9	3.0	4.0	3.0	1.0	2.0
Oleic	42.8	43.2	24.2	16.7	12.0	13.0	21.0	27.0
Linoleic	33.6	35.4	54.8	53.2	57.0	69.0	48.0	48.0
Linolenic	1.5	0.9	6.3	5.4	7.0	—	7.0	1.0
Other	—	—	—	—	—	—	5.0	—
Reference	Adhikari and Adhikari (1986)	Vandana et al. (2003)	Adhikari and Adhikari (1986)	Reddy et al. (2008b)	Weber (1981)	Grewal et al. (1978)	Hougen et al. (1985)	Weber (1981)

digalactosyl diglyceride were the major components and phosphoglycolipids were the minor components. The authors mentioned that campesterol (15.8%) stigmasterol (9.9%), β-sitosterol (35.0%), fucosterol (6.9%), stigmasta-7-en-3-ol (1.8%), cycloartenol (8.9%), 24-methylene cycloartenol (20.9%), and squalene (0.8%) were the unsaponifiables present in rice bran lysolecithin.

Significance of Nutrients for RBL's Nutritional and Oxidative Stability Properties

In comparison with soybean and sunflower lecithins, RBL has a qualitatively different composition of bioactive minor components, such as γ-oryzanol, tocotrienols, and phytosteryl esters. γ-Oryzanol covers the whole group of ferulic acid esters of triterpene alcohols and phytosterols. The unsaponifiable constituents present in rice bran lysolecithin were campesterol, stigmasterol, β-sitosterol, fucosterol, stigmasta-7-en-3ol, cycloartenol, 24-methylene cycloartenol, and squalene (Reddy et al., 2008). In general, vegetable oils are regarded as a rich natural source of dietary plant sterols. The majority of crude vegetable oils contain 1–5 g/kg^{-1} of phytosterols, but RBO contains up to 30 g/kg^{-1} of phytosterols. The level of tocotrienols in RBO is also very high compared with other vegetable oils (Ghosh, 2007). Because of the high level of active components, RBO is considered to promote good health. For example, γ-oryzanol and phytosterols have the capacity to lower blood cholesterol and decrease cholesterol absorption (Ghosh, 2007). Tocotrienols and γ-oryzanol are known as powerful antioxidants, and this is associated with the prevention of cardiovascular diseases and some cancers. Being an important by-product obtained during degumming of RBO, RBL possesses these additional benefits. Further, because the stability of oil is a significant factor with regard to the nutritional quality of oil, the oxidative stability of lecithin is also a significant factor with regard to the quality of lecithin. Antioxidants are compounds that either prevent the autoxidation or shorten the radical generating paths in lipids. Many naturally occurring compounds, such as thymol, carvacrol, tocopherol, tocotrienols, and others, were reported as antioxidants. Among them, tocopherols and tocotrienols were the most active. Apart from tocols, γ-oryzanol present in RBO as well as RBL is reported to have antioxidant properties due to the ferulic acid group present in steryl ferulates. In RBO and RBL, both vitamin E and γ-oryzanol reduce the formation of oxidized byproducts. Perhaps this is because the structures of the ferulic acid derivatives (γ-oryzanol) closely resemble the common antioxidants such as BHA and TBHQ.

Raju et al. (2011) conducted a study for evaluating rice bran lysolecithin as an energy source in broiler chicken diets. They found that the body weight, food consumption, and food conversion efficiency were unaffected by feeding rice bran lysolecithin. Fat digestibility was increased, and liver and meat fat content were not affected. Due to the balanced fatty acid profile and nutritionally beneficial minor components such as oryzanol, tocopherols, tocotrienols, squalene, and others, RBL is likely to exhibit several positive effects through consumption.

Functional and Biological Applications/Properties of Rice Bran Lipids

Shujiro (2011) reported the health functions of RBL and rice bran due to the presence of nutrients such as vitamin B1, dietary fiber, phytic acid, ferulic acid, γ-oryzanol, and phytosterols. Liu et al. (2013) discussed rice bran PLs and their significance in grain quality and health benefits. Although PLs are only a minor component compared to starch and protein, they have both nutritional and functional significance (Liu et al., 2013). Reddy et al. (2008a) conducted a study on the polar lipids fraction of rice bran lysolecithin for surfactant properties, namely, surface tension, CMC, emulsifying power, calcium tolerance, foaming power, and wetting ability. According to them, the polar lipids fraction of rice bran lysolecithin was found to have superior emulsifying, foaming, wetting, and calcium tolerance properties compared to polar lipids of soybean lecithin and RBL. Sayantani et al. (2013) conducted a study on a formulation in which GLs and PLs isolated from RBL were used in conjunction with gene-carrying lipids to test its efficacy in selectively delivering genes to cancer cells. This formulation did not show any negative effect on noncancerous cells. Therefore, they have claimed potential use of the formulation to deliver anticancer therapeutics to cancer cells without eliciting treatment-related toxicity to normal cells. Dull (2002) described rice bran extract as a novel ingredient for use in personal care products. The nutrient rich, water-soluble portion of the bran was captured in a spray-dried powder after enzymatic stabilization to prevent rancidity. These nutrients include protein, PLs, GLs, tocopherols, tocotrienols, B vitamins, inositol, γ-oryzanol, and eight minerals. He claimed that these nutrients have the potential for improving skin and also have hypoallergenic properties. The rice bran GLs help in providing emulsification capacity to whichever system incorporates the ingredient. The significant amounts of antioxidants provide enhanced product stability as well as beneficial effects on skin. Tessmann (2007a, 2008b, 2007c) developed three processes for natural fungicides based on essential oils, namely, eucalyptus oil and mentha piperita oil. In this study, W/O emulsions were prepared for crop protection applications using RBL as a natural emulsifier.

Future Trends and Concluding Remarks

RBL exhibits an excellent combination of a balanced fatty acid profile, micronutrients, and natural antioxidants. Therefore, RBL is an outstanding product and can act as an alternative to soybean and sunflower lecithins for food and industrial applications. The realization of the proper utilization of RBL may lead to the development of several nutritionally-important products. RBL is rich in unsaponifiable matter and, therefore, it may be used for cosmetic applications.

Isolation protocols, identification techniques, and physicochemical, nutritional, and functional properties of rice bran lipids are described in this chapter. Further, PL and fatty acid compositions of lecithins of various sources such as soybean, sunflower, and rapeseed are compared with RBL. Mostly compositional studies were reported from countries such as Japan, India, and Korea. PC, PE, and PI were the major PLs found, and cerebrosides, monogalactosyl diacylglycerol, digalactosyl diacylglycerol, and steryl glycoside were the commonly observed GLs in RBL. Palmitic, oleic, and linoleic were the major fatty acids present in all the rice bran lipids. Evaluation of polar lipid fraction of lysolecithin for surfactant properties (surface tension, CMC, emulsifying power, calcium tolerance, foaming power, and wetting ability), as an energy source in broiler chicken diet, in formulation of a gene delivery system for anticancer applications, as a natural emulsifier in preparation of a natural fungicide based on essential oils, and in personal care products were the notable studies performed for projecting RBL for a variety of applications. To date, no RBL-based commercialized technology is available for food and industrial applications. The demand for neutraceutical and personal care products is continuously increasing. Further, because of exponential growth in global rice production, food chemists are seeing RBL as an alternative. After 30 years of idea accumulation on RBL production and utilization, in the near future a fast growth in the RBL utilization for food and industrial applications can be expected.

References

Adhikari, S.; Adhikari, J. Indian Rice Bran Lecithin. *J. Am. Oil Chem. Soc.* **1986,** *63*, 1367–1369.

Ahn, T. H.; Rhee, C. O.; Kim, D. Y. Lipid Components of Rice Bran of Tongil and Japonica Type Varieties [in Korean]. *Han'guk Sikp'um Kwahakhoechi* **1984**, *16*, 192–200.

Aibara, S.; Ismail, I. A.; Yamashita, H.; Ohta, H.; Sekiyama, F.; Morita, Y. Changes in Rice Bran Lipids and Free Amino Acids During Storage. *Agric. Biol. Chem.* **1986**, *50*, 665–673.

Appleqvist, L. A.; Ohlson, R., Eds. *Rapeseed: Cultivation, Composition, Processing and Utilization;* Elsevier: Amsterdam, 1972.

Bird, A. R.; Hayakawa, T.; Marsono, Y.; Gooden, J. M.; Record, I. R.; Correll, R. L.; Topping D. L. Coarse Brown Rice Increases Fecal and Large Bowel Short-Chain Fatty Acids and Starch but Lowers Calcium in the Large Bowel of Pigs. *J. Nutr.* **2000**, *130*, 1780–1787.

Central Soya. http://www.centralsoya.com (accessed May 2002).

Chakrabarti, P. P.; Roy, S. K.; Rao, B. V. S. K.; Kale, V.; Prabhavathi Devi, B. L. A.; Prasanna Rani, K. N.; Prasad, R. B. N. An Improved Process for the Pretreatment of Vegetable Oils for Physical Refining. Indian Patent 202379, 2006.

Cherry, J. P.; Kramer, W. H. Plant Sources of Lecithin. In *Lecithins: Sources, Manufacture & Uses*; Szuhaj, B. F., Ed.; AOCS Press: Urbana, IL, 1989; pp 16–31.

Choudhury, N. H.; Juliano, B. O. Effect of Amylose Content on the Lipids of Mature Rice Grain. *Phytochemistry* **1980**, *19*, 1385–1389.

Dull, B. J. Stabilized Rice Bran Extract for Personal Care Products. *NutraCos.* **2002**, *1*, 23–26.

El-Shattory, Y.; Fadel L. H.; El-Khalafy, H. M. Studies on Phospholipids of Local Rice Bran Oils. *Grasas Aceites* **1979**, *30*, 309–313.

FAO. *Rice Market Monitor*, 2013. http://www.fao.org/economic/est/publications/rice-publications/rice-market-monitor-rmm/en/ (accessed Feb. 2015).

Ghosh, M. Review on Recent Trends in Rice Bran Oil Processing. *J. Am. Oil Chem. Soc.* **2007**, *84*, 315–324.

Grewal, S. S.; Sukhija, P. S.; Bhatia, L. S. Polar Lipid Composition During Sunflower *Helianthus annuus* Seed Development. *Biochemie, Physiologie der Pflanzen* **1978,** *173*, 11.

Hemavathy, J.; Prabhakar, J. V. Lipid Composition of Rice (*Oryza sativa L.*) Bran. *J. Am. Oil Chem. Soc.* **1987**, *64*, 1016–1019.

Hougen, F. W.; Thompson, V. J.; Daun, J. K. Rapeseed Lecithin. In *Lecithins*; Szuhaj, B. F., List, G. R., Eds.; AOCS Press: Urbana, IL, 1985; pp 79–96.

Indira, T. N.; Hemavathy, J.; Khatoon, S.; Gopala Krishna, A. G.; Bhattacharya, S. Water Degumming of Rice Bran Oil: A Response Surface Approach. *J. Food Eng.* **2000**, *43*, 83.

Khush, G. S. Origin, Dispersal, Cultivation and Variation of Rice. *Plant Mol. Biol.* **1997**, *35*, 25–34.

Kim, S. P.; Yang, J. Y.; Kang, M. Y.; Park, J. C.; Nam, S. H.; Friedman, M. Composition of Liquid Rice Hull Smoke and Anti-inflammatory Effects in Mice. *J. Agric Food Chem.* **2011,** *59*, 4570–4581.

Kullenberg, D.; Taylor, L. A.; Schneider, M.; Massing, U. Health Effects of Dietary Phospholipids. *Lipids in Health and Disease* **2012**, *11*, 3.

Kwon, K. S.; Choi, K. S.; Kim, H. K. Comparative Studies on the Composition of Polar Lipids in Japonica and Indica Rice Bran Oils [in Korean]. *Han'guk Sikp'um Yongyang Kwahakhoechi* **1996**, *25*, 735–740.

Liu, L.; Waters, D. L. E.; Rose, T. J.; Bao, J.; King, G. J. Phospholipids in Rice: Significance in Grain Quality and Health Benefits: A Review. *Food Chem.* **2013**, *139*, 1133–1145.

Mahua, G. Review on Recent Trends in Rice Bran Oil Processing, *J. Am. Oil Chem. Soc.* **2007**, *84*, 315–324.

Mano, Y.; Kawaminami, K.; Kojima, M.; Ohnishi, M.; Ito, S. Comparative Composition of Brown Rice Lipids (Lipid Fractions) of Indica and Japonica Rices. *Biosci. Biotechnol. Biochem.* **1999**, *63*, 619–626.

Moazzami, A. A.; Lampi, A. M.; Kamal-Eldin, A. Bioactive Lipids in Cereals and Cereal Products. In *Fruit and Cereal Bioactives, Sources, Chemistry, and Applications;* Hall, C. I., Tokusoglu, O., Eds.; CRC Press: Boca Raton, FL, 2011; pp 229–249.

Nasirullah; Nagaraja, K. V. Rice Bran Lipids: A Review. *J. Oil Tech. Ass. India.* **1987** (Jan–Mar), 2–6.

Norhaizan, M. E.; Tan, B. L.; Loh, S. P. Byproducts of Rice Processing: An Overview of Health Benefits and Applications. *J. Rice Research* **2013**, *1*, 1–11

Orthoefer, F. T. In *Proceedings of the World Conference on Oil Seed Processing and Utilization*; Wilson R. F., Ed.; AOCS Press: Urbana, IL, 2011.

Perez, S.; Bertoft, E. The Molecular Structures of Starch Components and Their Contribution to the Architecture of Starch Granules: A Comprehensive Review. *Starch* **2010**, *62*, 389–420.

Pragasam, A.; Indira, T. N.; Gopala Krishna, A. G. Preparation and Physico-Chemical Characteristics Evaluation of Rice Bran Lecithin in Relation to Soya Lecithin. *Beverage and Food World*, **2002**, *29*, 19–22.

Raju, M. V. L. N.; Rama Rao, S. V.; Chakrabarti, P. P.; Rao, B. V. S. K.; Panda, A. K.; Prabhavathi Devi, B. L. A.; Sujatha, V.; Reddy, J. R. C.; Shyam Sunder, G.; Prasad, R. B. N. Rice Bran Lyso Lecithin as a Source of Energy in Broiler Chicken Diet. *British Poult Sci.* **2011**, *52*, 769–774.

Reddy, J. R. C.; Rao, B. V. S. K.; Chakrabarti, P. P.; Karuna, M. S. L.; Prabhavathi Devi, B. L. A.; Prasad, R. B. N. Characterization of Rice Bran Lyso Lecithin and Evaluation of Its Polar Lipids for Surfactant Properties. *J. Lipid Sci. Technol.* **2008a**, *40*, 10–15.

Reddy, J. R. C.; Rao, B. V. S. K.; Karuna, M. S. L.; Sagar Rao, K.; Prasad, R. B. N. Lipase-catalyzed Preparation of Stearic Acid-rich Phospholipids. *J. Lipid Sci. Technol.* **2008b**, *40*, 124–128.

Roy, S. K.; Rao, B. V. S. K.; Prasad, R. B. N. Enzymatic Degumming of Rice Bran Oil. *J. Am. Oil Chem. Soc.* **2002**, *79*, 845–846.

Sakata, S.; Fujino, Y.; Mano, Y. Classification and Fatty Acid Composition of Phospholipids in Rice Grain [in Japanese]. *Obihiro Chikusan Daigaku Gakujutsu Kenkyu Hokoku, Dai-1-bu* **1974**, *8*, 719–723.

Sarode, M.; Soundar, D.; Rangaswamy, S. Evaluation of Phosphoglycolipid Elimination from Rice Bran Oil by a Nonporous Membrane Using NMR Spectroscopy. *Eur. J. Lipid Sci. Technol.* **2009**, *111*, 1020–1026.

Sayantani, R.; Banerjee, R.; Chakrabarti, P. P.; Prasad, R. B. N. Rice Bran Lipids Based Formulation and Process for Preparation Thereof for Selective Delivery of Genes to Cancer Cells. PCT Int. Appl., WO 2013186793 A1 20131219, 2013.

Schneider, M. Phospholipids—What's New? *Lipid Technology* **2006**, *18*, 249–255.

Schneider, M. Major Sources, Composition and Processing. In *Phospholipid Technology and Applications*; Gunstone, F. D., Ed.; The Oily Press: Bridgwater, England, 2008; pp 21–40.

Scholfield, C. R.; Dutton, H. J. Sources of Color in Soybean Lecithin. *J. Am. Oil Chem. Soc.* **1954**, *31*, 258–261.

Shaik, R. V.; Chakrabarti, P. P.; Kaimal, T. N. B. Characterization of a Novel Phosphoglycolipid from Rice Bran Oil. *J. Oil Tech. Ass. India* **2004**, *36*, 3–14.

Shin, H. S.; Yang, J. H. Comparative Studies on the Polar Lipids Composition in Nonglutinous and Glutinous Rice [in Korean]. *Han'guk Sikp'um Kwahakhoechi* **1986**, *18*, 143–148.

Shujiro, E. Current and New Technologies Available for Use of Rice Bran [in Japanese]. *Gekkan Fudo Kemikaru* **2011**, *27*, 54–62.

Tessmann, W. S. Natural Fungicide Based on Essential Oils [in Portuguese]. *Braz. Pedido PI.* BR 2006003005 A 20070116, 2007a.

Tessmann, W. S. Natural Fungicide Based on Eucalyptus Oils [in Portuguese]. *Braz. Pedido PI.* BR 2006003004 A 20070116, 2007b.

Tessmann, W. S. Natural Fungicide Based on Mentha Piperita Essential Oils [in Portuguese]. *Braz. Pedido PI.* BR 2006003001 A 20070116, 2007c.

USDA. http://www.fas.usda.gov/data/oilseeds-world-markets-and-trade (accessed May 2011).

Van Niewenhuygen, W. The Changing World of Lecithins. *Inform* **2014**, *25*, 254–259.

Vandana, V.; Karuna, M. S. L.; Reddy, J. R. C.; Vijeeta, T.; Rao, B. V. S. K.; Prabhavathi Devi, B. L. A.; Prasad, R. B. N. Bleaching and Enrichment of Phospholipid Content in Commercial Rice Bran Lecithin. *J. Oil Tech. Ass. India.* **2003**, *35*, 145–150.

Weber, E. J. Composition of Commercial Corn and Soybean Lecithins. *J. Am. Oil Chem. Soc.* **1981**, *58*, 898.

Yasuo, M.; Kumi, K.; Michiyuki, K.; Masao, O.; Seisuke, I. Comparative Composition of Brown Rice Lipids (Lipid Fractions) of Indica and Japonica Rices. *Biosci. Biotechnol. Biochem.* **1999**, *63*, 619–626.

Yoshida, H.; Tanigawa, T.; Kuriyama, I.; Yoshida, N.; Tomiyama, Y.; Mizushina, Y. Variation in Fatty Acid Distribution of Different Acyl Lipids in Rice (*Oryza Sativa* L.,) Brans. *Nutrients* **2011a**, *3*, 505–514.

Yoshida, H.; Tanigawa, T.; Yoshida, N.; Kuriyama, I.; Tomiyama, Y.; Mizushina, Y. Lipid Components, Fatty Acid Distributions of Triacyl Glycerols and Phospholipids in Rice Brans. *Food Chem.* **2011b**, *129*, 479–484.

Yukihiro, N. Medical Applications of Phospholipids. In *Phospholipids Handbook*; Cevc, G., Ed.; Marcel Dekker: New York, 1993; pp 879–894.

3

Sunflower Lecithin

Estefania N. Guiotto and Mabel C. Tomás ▪ *Centro de Investigación y Desarrollo en Criotecnología de Alimentos, Universidad Nacional de La Plata, La Plata, Argentina*

Bernd W.K. Diehl ▪ *Spectral Service GmbH Laboratoriumfür Auftragsanalytik, Cologne, Germany*

Introduction

Vegetable lecithin was considered a byproduct of the oil refining process, in which substances containing phosphorus have to be removed to give the oil better stability and/or facilitate further refining. However, vegetable lecithin is now considered a valuable coproduct and is used as a natural emulsifier in a wide range of industrial applications: dietetic, pharmaceutical, food, cosmetics, and so on (Nguyen et al., 2013). This coproduct of the oil industry represents a multifunctional additive for the manufacture of chocolate, margarine, mayonnaise, and bakery and instant products due to the characteristics of its phospholipids (Cabezas et al., 2013). Vegetable lecithin is usually defined as a mixture of acetone insoluble polar lipids and triglyceride oil with other minor components produced by water-degumming crude vegetable oils and separating and drying the hydrated gums (Szuhaj, 2005). Vegetable lecithins mainly contain phosphatidylcholine (PC), phosphatidylethanolamine (PE), phosphatidylinositol (PI), minor components such as phosphatidic acid (PA), and other substances (e.g., triglycerides, glycolipids, sterols, fatty acids, carbohydrates and sphingolipids) (Schneider, 1989). Its composition does not only depend on the processing parameters and type of seeds, but also on the weather conditions and geographical characteristics. As a result, vegetable lecithin is not a uniform, standard material, but it is a complex mixture of polar and nonpolar components that contribute to its overall emulsifying performance (Nguyen et al., 2013).

Production of Vegetable Lecithins

Phospholipids are found in all living cells, whether of animal or plant origin. In humans and animals, the phospholipids are concentrated in the vital organs, such as the brain, liver, and kidneys; in vegetables, they are highest in the oilseeds (sunflower, soybean, canola), nuts, and grains (corn) (Szuhaj, 2005).

In order to stabilize vegetable oils against sedimentation and also to enable further refining steps, phospholipids and glycolipids must be removed. Oils extracted by solvent generally have higher phospholipid content than those obtained by pressing (Grompone, 2005). During the so-called degumming process, the residual phosphorus content in the oil is minimized. The crude oil is heated at about 70 °C, mixed with 2% water, and subjected to stirring for about 30 minutes to 1 hour. The addition of water to the oil hydrates the polar lipids in the oil, making them insoluble. The resulting lecithin sludge is then separated by centrifugation. This wet sludge is made up of water, phospholipids and glycolipids, some triglycerides, carbohydrates, traces of sterols, free fatty acids, and carotenoids. The crude vegetable lecithin is then obtained by careful drying (Bueschelberger, 2004). The phospholipid (lecithin) content of crude sunflower oil ranges from 0.5% to 1.2% (Grompone, 2005).

Sunflower lecithin is not produced in considerable amounts worldwide. This fact is mainly because of the low lecithin content of crude sunflower oil as compared with 2.9% for soybean, 1.9% for rapeseed, 2.4% for cottonseed, and 2.0–2.7% for corn oil (normalized at 70% of insolubles in acetone) (Grompone, 2005). In Argentina, the production of sunflower oil is of utmost importance from an economic point of view. In this country, sunflower lecithin could represent an alternative to soybean lecithin because it is considered a non-GMO (non-genetically modified organism) product, which is currently preferred by certain consumers (Cabezas et al., 2012a).

Sunflower seeds usually have black hulls that tend to stick to the kernels. Dehulling efficiency may influence the natural wax content of 0–1% in lecithin. Hence, additional expelling (pressing) is often used to squeeze out oil and reduce the original oil content of the seed (~40%) down to 15% in the press cake; the press cake is then hexane-extracted to reduce the oil to under 0.5%. Both press and extraction oil streams are combined for degumming. In addition, intensive filtration of the non-degummed oil is important for obtaining lecithin with low-impurity contents (van Nieuwenhuyzen, 2014).

Table 3.A gives the phosphatide composition of vegetable lecithins obtained from different oils. Distribution of the main phospholipid components of sunflower lecithin appears to be rather similar to that of soybean lecithin. Moreover, the fatty acid composition of the phosphatides reflects the fatty acid composition of the oil in

Table 3.A Composition of Phosphatides of Various Lecithins by ^{31}P NMR

Phospholipid (%)	Sunflower	Soybean	Rapeseed
PC	16	15	17
PE	6	11	9
PI	17	10	10
PA	2	4	4
Others	1	7	6
All phospholipids	42	47	46

Source: Adapted from van Nieuwenhuyzen and Tomás (2008).

which these phosphatides occur, but it tends to have a higher palmitic acid content and a lower oleic acid content than the oil, as illustrated by Table 3.B (Dijkstra, 2011).

Functional and Nutritional Properties

Phospholipids, along with proteins, are the major structural components of membranes of all kinds of cells. Phospholipids play an important role in biological functions such as maintaining cell membrane integrity, prevention of neurological diseases, and regulation of basic biological processes such as cell to cell signalling, which is essential for life (Guo et al., 2005). The key of phospholipids' biological role is the ability of the phosphate group to combine with water and hydrophilic molecules such as proteins, sugars, and oxygen, while the fatty acids interact with hydrophobic substances such as hormones and nonesterified fatty acids (Quezada, 2007). Beneficial effects of dietary phospholipids (PLs) have been mentioned since the early 1900s in relation to different illnesses and symptoms, such as coronary heart disease, inflammation, and cancer (Küllenberg et al., 2012). Thus, phospholipids have been used as dietary supplements.

Vegetable lecithin is regarded as a well-tolerated and nontoxic surfactant approved for human consumption by the United States Food and Drug Administration (FDA). Small quantities of phospholipids occur naturally in a wide variety of foods, and food technologists frequently use them as emulsifiers in processed products such as margarine, chocolate, and bakery goods. For these applications, nutritional properties of phospholipids do not play any role because the quantities used in food products normally only range between 0.3% and 1% (Duttaroy, 2008).

The entire metabolism of phospholipids has not been completely investigated, and many questions remain to be answered. More research is needed to understand

Table 3.B Fatty Acid Compositions of Vegetable Lecithins and Oils

	Sunflower		Soybean		Rapeseed	
Fatty Acid	Lecithin	Oil	Lecithin	Oil	Lecithin	Oil
16:0	11	7	16	11	7	4
18:0	4	5	4	4	1	2
18:1	18	29	17	23	56	61
18:2	63	58	55	54	25	22
18:3	0	0	7	8	6	10
Others	4	1	1	0	5	1

Source: Dijkstra (2011).

the role of phospholipids after their ingestion and how they interact with the various processes of our complex metabolisms (Küllenberg et al., 2012).

Modification Processes

The properties of crude lecithins can be improved by modifying the phospholipid structure or composition (van Nieuwenhuyzen and Tomás, 2008). The most common modification processes used in sunflower lecithin are fractionation by deoiling to separate oil from phospholipids, introduction of enzymatic changes in phospholipid molecules to obtain lysolecithins, and fractionation with ethanol or ethanol and water systems to produce enriched fractions in specific phospholipids (Cabezas et al., 2012a).

Deoiling

Fluid lecithins contain about 30–40% neutral lipids, mainly triglycerides. To improve dispersability properties and the processing characteristics of these highly viscous materials, the industry makes use of the fact that polar lipids (phospholipids and glycolipids) are almost insoluble in acetone, whereas neutral lipids dissolve in it. Acetone extraction leads to powdered or granulated products that contain 2–3% residual neutral lipids (van Nieuwenhuyzen and Tomás, 2008). The key result of the deoiling process is that the phospholipids, as the components that provide functionality, have now been concentrated and purified. This results in significantly lower dosage requirements and higher functionality. Moreover, deoiled products have a more neutral taste than the corresponding liquid products (Bueschelberger, 2004).

Enzymatic Modification

Enzymes can be used to modify the structure of phospholipids in a wide variety of ways. There are various known hydrolytic enzyme systems (phospholipases) that differ in their point of attack on the phospholipid molecule (Figure 3.1). The phospholipases A_1, A_2, C, and D can be easily distinguished. Phospholipase A_2 (PLA_2) catalyzes the specific hydrolysis of fatty acid ester bonds in the second position, forming lysophospholipids and free fatty acids. In contrast, Phospholipase A_1 splits the fatty acid in the first position. After hydrolysis, however, a fatty acid may also migrate from the C_2 to the C_1 position. One method that is of growing interest is enzymatic hydrolysis with phospholipase D. This reaction increases the phosphatidic acid (PA) content in the lecithin (Bueschelberger, 2004).

Enzymatic hydrolyzed lecithins may present technological and commercial advantages over native lecithins; consequently, the demand for lysolecithins has been increasing in recent years (Cabezas et al., 2011a). PLA_2 is the most employed family of enzymes for the enzymatic modification of lecithin. Porcine pancreatic PLA_2 has been

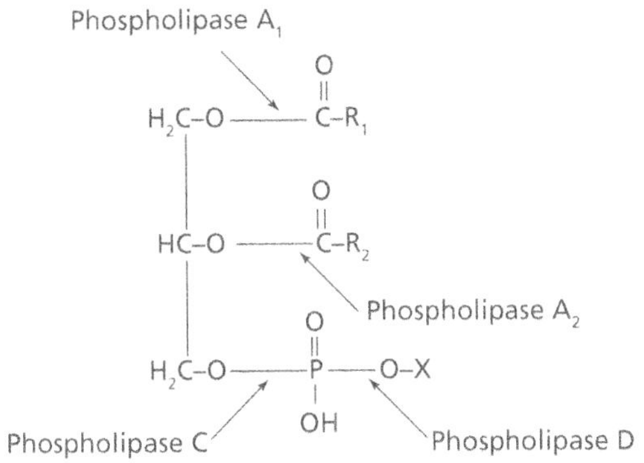

Figure 3.1 Scheme of the action sites of various types of phospholipases.

used on an industrial scale for decades, for example, in the production of food products such as mayonnaise and degumming of edible oils. Advances in biotechnology and certain requirements of consumers (kosher or *halal* foods) have influenced the development of the production of microbial enzymes (bacteria, fungi, yeasts) (Cabezas et al., 2011b). This fact has led to the production of new phospholipases and the development of novel industrial applications (De María et al., 2007; Wilton, 2005).

Cabezas et al. (2012b) carried out the enzymatic hydrolysis of sunflower lecithins using a porcine pancreatic and a microbial PLA_2 (*Streptomyces violaceoruber*). The hydrolyzed lecithins presented a high concentration of major lysophospholipids compared to the native sunflower lecithin. In particular, the pancreatic PLA_2 produced a higher hydrolysis degree of the main phospholipids in comparison with the microbial phospholipase.

Fractionation

The industrial production of vegetable lecithin fractions mainly uses ethanol or ethanol and water systems. The principle of the fractionation process is that PC is preferentially soluble in ethanol, whereas PE distributes more or less evenly between the soluble and the insoluble phase. The more acidic phospholipids, such as PI and PA, remain practically insoluble (Schneider, 2008). Therefore, the phospholipid mixture in crude sunflower lecithin can be fractionated into the alcohol-soluble and alcohol-insoluble fractions. Various types of alcohol and concentrations can be used to obtain specific extraction yields and selectivity of the PC:PE ratio (van Nieuwenhuyzen and Tomás, 2008).

The phospholipid distribution of each enriched fraction has been shown to be quite strongly dependent on the pH of the solvent extraction. Moreover, the solubility of the phosphatides is affected differently: PC and PE are zwitterionic at neutral pH because of the presence of the amino group in the molecule, whereas PI has a net negative charge (anionic) at this pH (Dijkstra and De Kock, 1993; Penci, 2009). The various head groups of phospholipids, furthermore, give them different polarities and emulsification properties (Hernández and Quezada, 2008) that are, in turn, directly associated with the net charge of the molecules (Schneider, 1989).

According to Cabezas et al. (2009b), in the fractionation process of sunflower lecithin with absolute ethanol, the temperature has the highest effect on the extraction of PC from the original material to the PC-enriched fraction followed by the incubation time. Also, these authors found that the presence of water in the solvent extraction and the solvent:lecithin ratio are important factors to be considered in the development and application of the fractionation process for optimizing the PC-enriched fraction yield (Cabezas et al., 2009a; Guiotto et al., 2013).

The fractionation process was also carried out on native sunflower lecithin with the addition of absolute ethanol and water mixtures (96:4, 90:10 v/v) at different pH levels (3.3, 7.5, and 10.0) at 65 °C for 90 minutes with moderate agitation, and then centrifuged at 1880 g and 10 °C for 10 minutes. Afterward, the corresponding ethanolic extracts and residues were obtained and ethanol was eliminated by evaporation under vacuum (Figure 3.2).

The PC- and PI-enriched fractions' yield percentages obtained by this fractionation and the subsequent deoiling process are presented in Figure 3.3. When

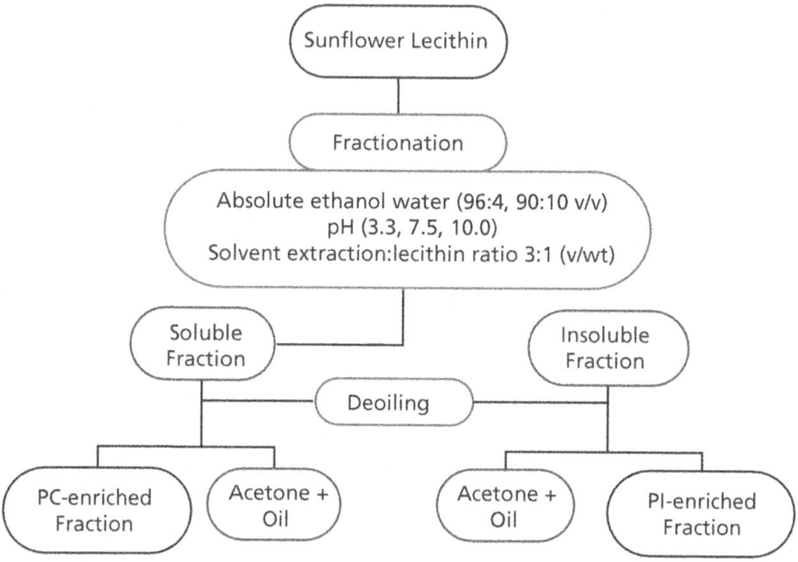

Figure 3.2 Flow diagram of the fractionation process of sunflower lecithin.

considering only the PC-enriched fraction's yield percentage, a significant increase ($p < 0.05$) was recorded as a function of the increase of absolute ethanol in the solvent extraction. In contrast, an opposite behavior was observed in the PI-enriched fractions. This fact could be because PLs exhibit a poor solubility in water and aqueous solutions (Carlsson, 2008; Dijkstra and De Kock, 1993). These results evidence the efficiency of the fractionation process. No significant differences ($p > 0.05$), however, were detected for the two different fraction yields as a function of the pH for each of the absolute ethanol and water mixtures assayed.

Phospholipid Composition (^{31}P NMR)

The phospholipid composition of PC- and PI-enriched fractions obtained under different conditions of the fractionation process was usually determined by ^{31}P NMR analysis. Phospholipid composition was expressed in terms of molar concentration (mol/100 mol lecithin), thus the phospholipid content (% PC, % PI, and % PE) of each fraction was obtained.

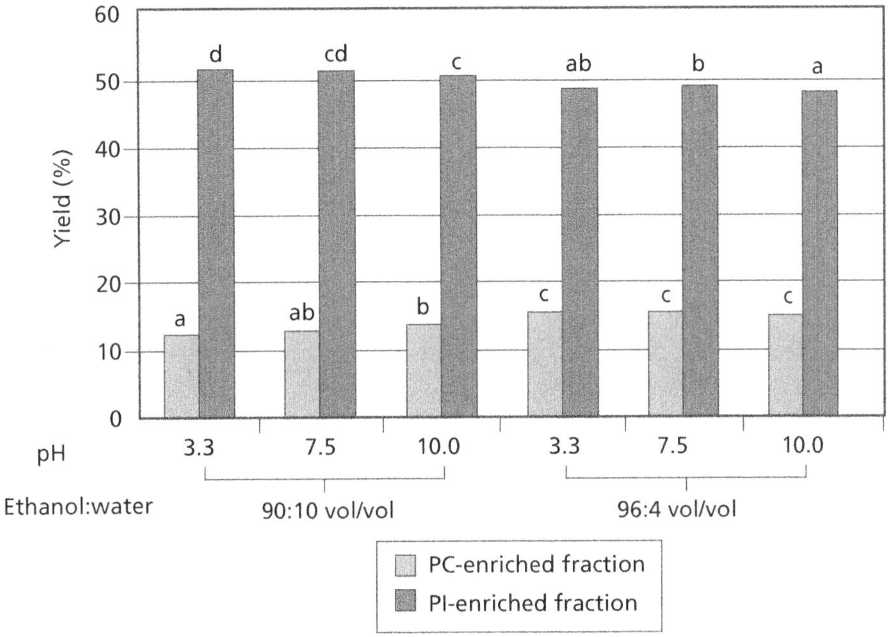

Figure 3.3. PC- and PI-enriched fractions' yields percentages obtained by fractionation with different absolute ethanol and water mixtures and pHs. Average values $n = 2$. Bars of the same color with different letters are significantly different ($p < 0.05$).

The ^{31}P NMR analyses of the different PC- and PI-enriched fractions showed that PC-enriched fractions obtained by fractionation with the absolute ethanol and water mixtures (96:4 and 90:10 v/v) at different pHs contained high levels of PC but only low amounts of PI (Figure 3.4). These enriched fractions from the absolute ethanol and water 90:10 (v/v) extracts exhibited a greatly enhanced concentration of PC (≥45), but the percentage of PI was also significantly increased ($p < 0.05$) compared to its concentration in the fraction extracted with absolute ethanol and water 96:4 (v/v).

In contrast, the PI-enriched fractions presented significant differences ($p < 0.05$) in phospholipid composition, with the proportion of PI increasing as a function of absolute-ethanol concentration and that of PC and PE decreasing. These results are in agreement with those previously obtained with different concentrations of absolute ethanol in the fractionation of sunflower lecithin (Cabezas et al., 2009a, 2009b; Guiotto et al., 2013).

The PC:PE ratio for the PC- and PI-enriched fractions obtained as a function of the composition and pH of different absolute ethanol and water mixtures are shown in Figure 3.4. The ethanol-soluble fractions exhibited the marked differences in phospholipid composition between the enriched fractions and the native sunflower lecithin (PC:PE ratio 2.6). The PC-enriched fraction obtained by fractionation at pH 7.5 with an absolute ethanol and water mixture of 96:4 (v/v) resulted in the highest PC:PE ratio (8.6); whereas the insoluble fraction had an extremely low PC:PE ratio (≤1.6).

Taking into account the enriched fraction yield percentage values and phospholipid composition, Cabezas et al. (2009a, 2009b) determined the percentage extraction coefficient ($\%E_{PC}$, $\%E_{PE}$, $\%E_{PI}$) values as a function of the different processing conditions for each fraction. These values represent the percent contribution of each phospholipid in these enriched fractions according to the following equation:

$$\%E_{PL}\,(\text{PC enriched fraction}) = \frac{m_{PL}\,(\text{PC enriched fraction})}{m_{PL}\,(\text{PC enriched fraction}) + m_{PL}\,(\text{PI enriched fraction})} .100 \qquad (1)$$

where PL: PC, PE, or PI

$$m_{PL}\ (\text{PC enriched fraction}) = \text{PC enriched fraction yield}\ \% \bullet \%\ \text{PL}\ (\text{PC enriched fraction}) \qquad (2)$$

$$m_{PL}\ (\text{PI enriched fraction}) = \text{PI enriched fraction yield}\ \% \bullet \%\ \text{PL}\ (\text{PI enriched fraction}) \qquad (3)$$

The following expression must be considered for calculations:

$$\%E_{PL}\ (\text{PC enriched fraction}) + \%E_{PL}\ (\text{PI enriched fraction}) = 100\% \qquad (4)$$

The efficiency of sunflower lecithin fractionation was carried out analyzing the $\%E_{PL}$ extraction coefficients ($\%E_{PC}$, $\%E_{PI}$, and $\%E_{PE}$) values based on the solubility in aqueous ethanol of the different PLs (Cabezas et al., 2009a). The $\%E_{PL}$ under

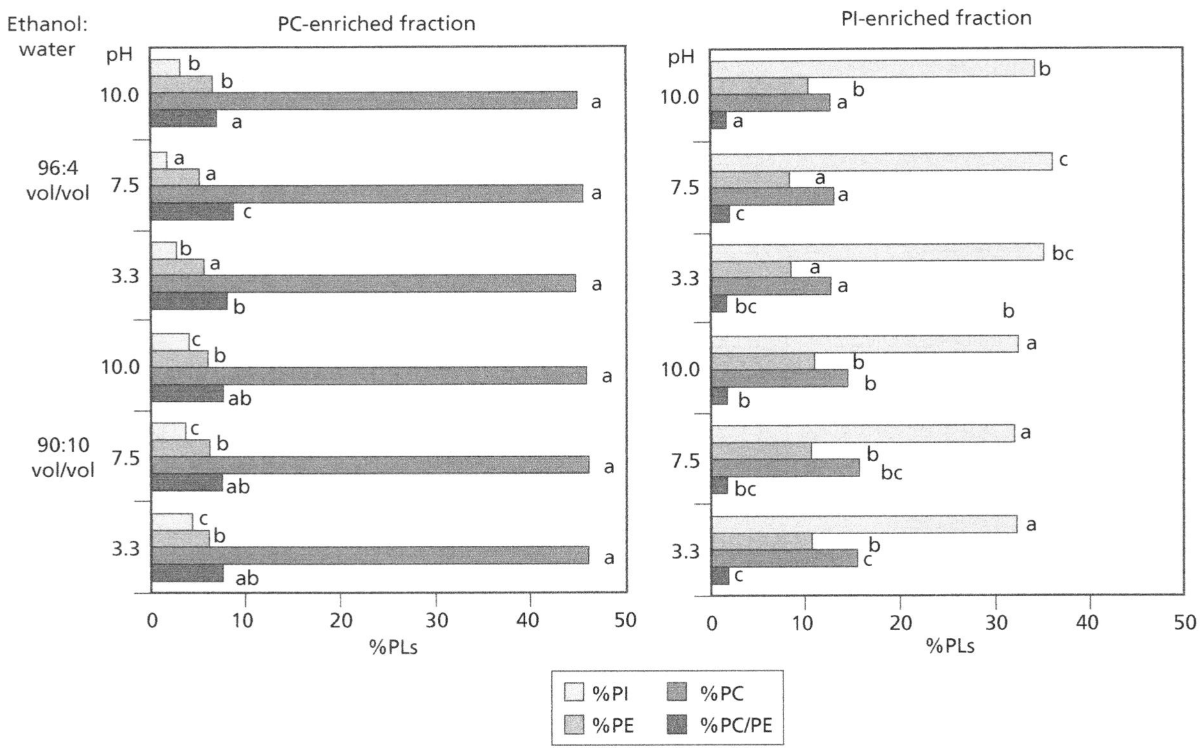

Figure 3.4. Phospholipid composition of PC- and PI-enriched fractions (percent molar concentration relative to the total concentration of the fractions) determined by ^{31}P NMR. Average values $n = 2$. Bars of the same color with different letters are significantly different ($p < 0.05$).

different processes are shown in Figure 3.5. The solubilities of PC and PE were significantly higher ($p < 0.05$) when the fractionation process was carried out with lower water concentration in the solvent extraction. These results are according to Cabezas et al. (2009a), who found that the E_{PC} values were considerably lower for the fractionation process with aqueous ethanol mixtures than those obtained with absolute alcohol. These results could be explained in terms of the extraction kinetics of PC, which is favored by the pure solvent. E_{PI} presented low values (<4.0%) under all the conditions assayed, indicating a very low solubility of PI. Extraction coefficients of the PI-enriched fractions should be considered, taking into account Equation 4.

Applications

The main application of lecithin in the food industry is associated with its role as an emulsifier agent for dispersions or emulsions (Hernández and Quezada, 2008). Sunflower lecithin may be used as a food additive in view of its high phosphatidylcholine and essential fatty acid content. Its use in the manufacture of foods and cosmetics

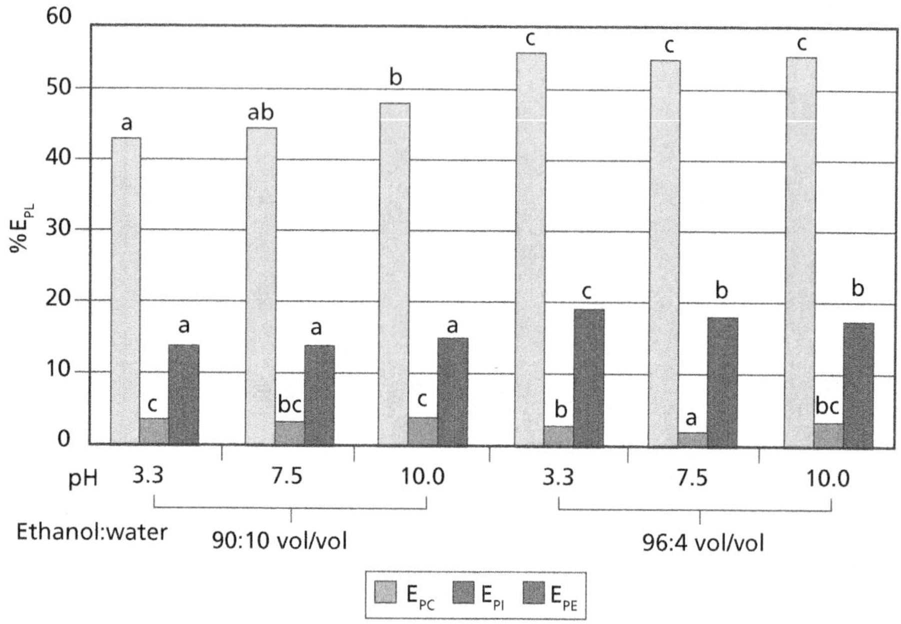

Figure 3.5. Extraction coefficients (%E_{PL}) of PC-enriched fractions obtained after fractionation of sunflower lecithin at different ethanol:water ratios and pH of the partitioning solvent. Bars of the same color with different letters are significantly different ($p < 0.05$).

can be increased by refining and fractionation and/or modifications (Szuhaj, 2005). Commercial interest in modified lecithins arises from the difference in their technological functionality. Regarding this, Cabezas et al. (2012a, 2012b) reported positive results of deoiling, fractionation process, and enzymatic modification of sunflower lecithin for improving the emulsifying properties.

Phospholipids are nature's principal surface-active agents and are polar lipids with a hydrophilic and a lipophilic part. These compounds concentrate at the interface between oil and water and reduce the interfacial tension. This property facilitates the formation of emulsions (van Nieuwenhuyzen, 2014). Emulsions are thermodynamically unstable systems from a physicochemical point of view, separating into two immiscible phases according to the kinetic stability. Physical destabilization mechanisms of emulsions include oil droplet size variation processes, such as flocculation and coalescence, and particle migration phenomena, such as sedimentation and creaming (McClements, 1999).

PC-enriched fraction, due to its high PC:PE ratio and the lamellar phase structure of the PC at the interface between oil and water, is recognized to be a good O/W emulsifier, whereas the PI-enriched fraction is good for W/O emulsions (van Nieuwenhuyzen and Tomás, 2008).

Oil-in-Water Emulsions

In a previous study, Cabezas et al. (2012a) evaluated the emulsifying properties of five modified sunflower lecithins that were obtained by deoiling (deoiled lecithin), fractionation with absolute ethanol (PC- and PI-enriched fractions), and enzymatic hydrolysis with phospholipase A_2 from pancreatic porcine and microbial sources (hydrolyzed lecithins). According to these authors, PC-enriched fractions and both hydrolyzed lecithins presented the best emulsifying properties against the main destabilization processes (creaming and coalescence) for O/W emulsions (30:70 wt/wt). These modified lecithins represent a good alternative for the production of new bioactive agents.

Characterization of Oil-in-Water Emulsions with Sunflower Lecithin

The physical stability of the O/W emulsions (30:70 wt/wt) immediately after the homogenization step were followed by measuring the evolution of backscattering (% BS) with a QuickScan Vertical Scan Analyzer (Coulter Corp., Miami, FL). The backscattering of monochromatic light (λ = 850 nm) of the emulsions was determined as a function of the height of the sample tube (65 mm) in order to quantify the rate of the different destabilization processes during the first 60 minutes. This methodology allowed for discrimination between particle migration (sedimentation, creaming) and particle size variation (flocculation, coalescence) processes (Pan et al., 2002). The basis of the vertical scan analyzer profiles has been exhaustively studied by Mengual et al. (1999).

The destabilization of the O/W emulsions was followed by monitoring the sequential profiles obtained in the bottom zone I (10–15 mm) of the typical QuickScan profiles in Figure 3.6, an area in which a clarification of the emulsion is produced by the migration of the emulsified droplets from the bottom toward the top of the tube (i.e., creaming). The upper zone II (50–55 mm) is the location of the potential destabilization of the cream phase of the emulsions through coalescence as a function of time.

The O/W emulsions in Figure 3.6 include the addition of 1.0% of the PC-enriched fractions obtained with absolute ethanol and water mixtures (A) 96:4 (v/v), and (B) 90:10 (v/v) at pH 7.5, respectively. The % BS values in zone I decreased versus time due to a process of destabilization akin to creaming produced by the migration of the oil droplets from the bottom to the top of the tube in both O/W emulsions. Figure 3.6B shows a fast separation of the creamed phase followed by a destabilization by coalescence at the top of the tube (zone II), with both creaming and coalescence occurring simultaneously (Guiotto et al., 2013; McClements, 1999).

The destabilization kinetics of the different O/W emulsions (30:70 wt/wt) with a range of 0.1% to 2.0% of PC-enriched fraction extracted with an absolute ethanol and water mixture of 90:10 (v/v) at the different pH levels are shown in Figure 3.7 (p. 70). A rapid decrease in the % BS values at the bottom of the tube was evidenced for all conditions assayed (Figure 3.7A). These destabilizations result from the ascent of oil particles to the uppermost part of the tube. Figure 3.7B illustrates a decrease in the % BS values at concentrations of 0.1% to 0.5% (v/v) of the emulsifying agent, showing in the coalescence destabilization of the cream phase. When the concentration of the emulsifier was increased from 1.0% to 2.0% (v/v), however, a stable, coalescence-resistant cream phase was formed at all pHs (Figure 3.7B). The lowest cream-phase stability was observed at pH 3.3 with these PC-enriched fractions.

Figure 3.8 (p. 71) shows the destabilization kinetics of O/W emulsions containing different PC-enriched fraction concentrations extracted with an absolute ethanol and water mixture of 96:4 (v/v) at the different pH levels assayed. In particular, the addition of PC-enriched fraction obtained at pH 7.5 of the solvent extraction generated a higher stability in O/W emulsions against the creaming process for all concentrations studied. In contrast, the addition of PC-enriched fractions obtained at pH 3.3 and pH 10.0 at this high ethanol:water ratio showed a low emulsifying capacity under all the conditions studied (zone I, Figure 3.8A). Figure 3.8B shows the QuickScan profiles corresponding to zone II. A coalescence destabilization of the cream phase occurred at low concentrations (0.1% and 0.5% v/v) of emulsifier. Similar results were obtained for emulsions with PC-enriched fractions from absolute

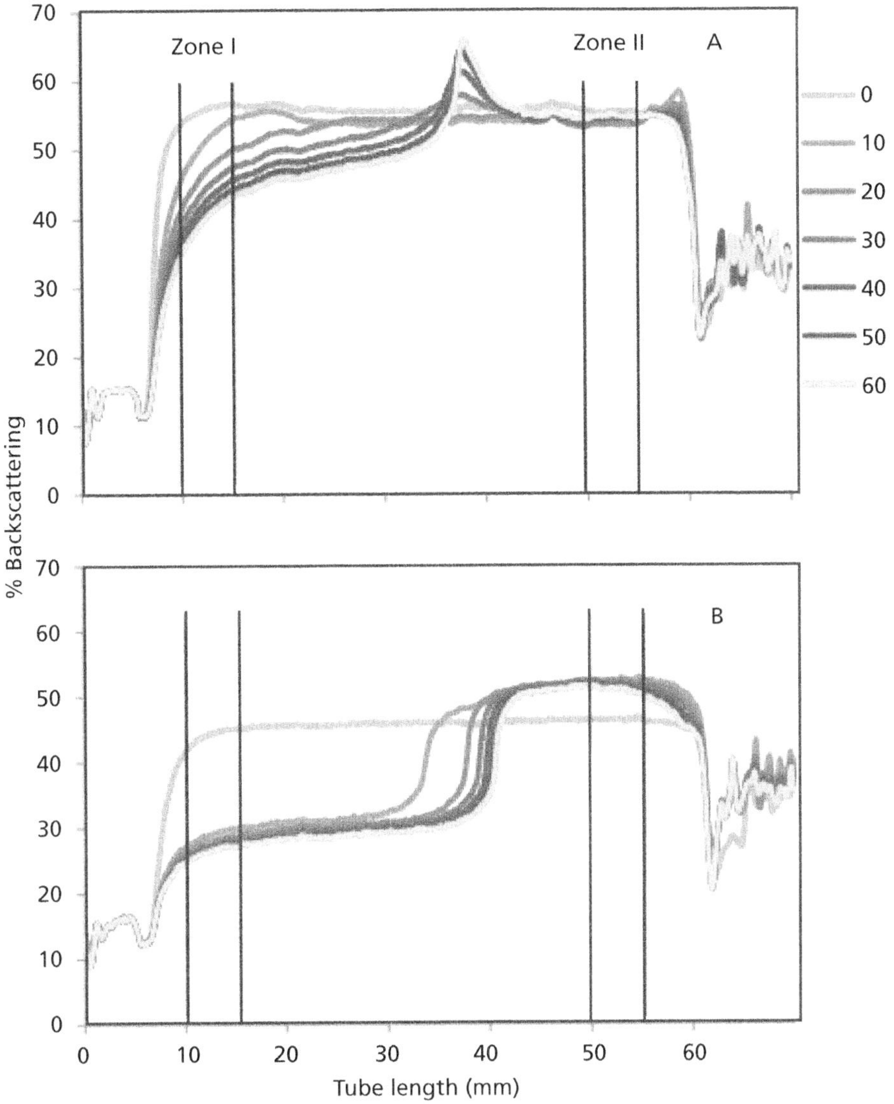

Figure 3.6. Backscattering (% BS) profiles of O/W emulsions (30:70 wt/wt) containing 1.0% of PC-enriched fraction obtained by extraction with an absolute ethanol:water mixture at pH 7.5 (A) 96:4 (v/v) and (B) 90:10 (v/v).

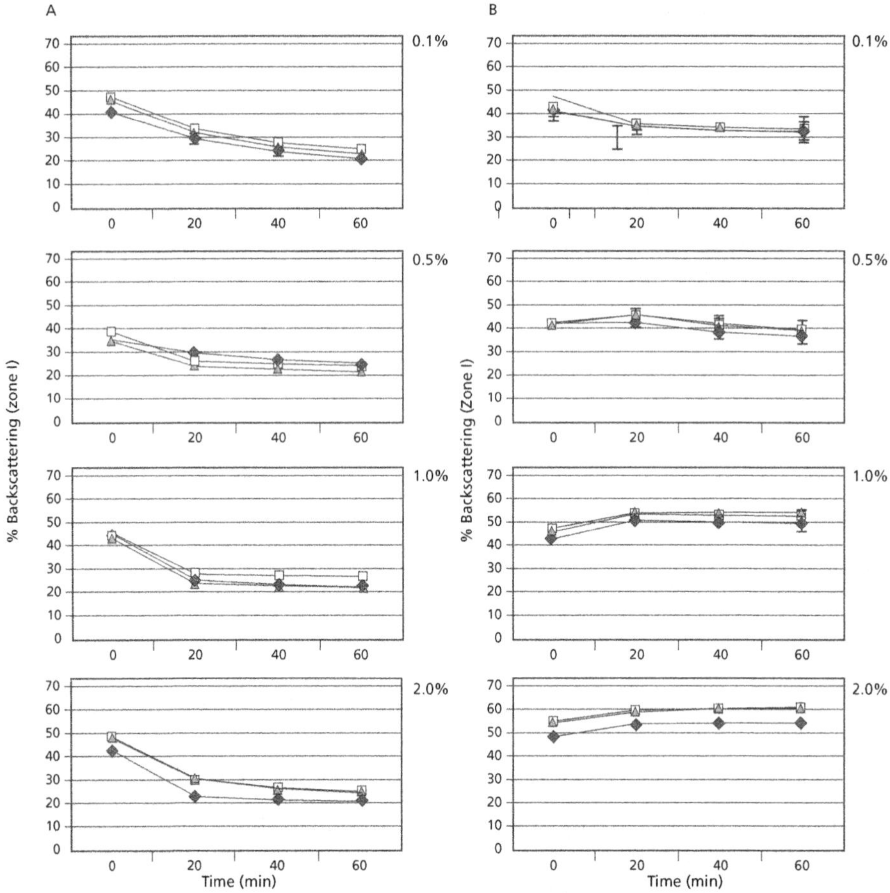

Figure 3.7. Backscattering (%) versus time of O/W emulsions (30:70 wt/wt) containing the PC-enriched fraction obtained by extraction with an absolute ethanol:water mixture of 90:10 (v/v) ranging from 0.1% to 2.0% (wt/wt) at different pH levels: 3.3; 7.5, 10.0. (A) Zone I (10–15 mm), (B) zone II (50–55 mm). Mean values ± SD (*n* = 2).

ethanol:water ratio of 90:10 (v/v) (see Figure 3.7B). However, at 1.0% and 2.0% of the emulsifying agent, high percentages of BS were recorded and the emulsions with the PC-enriched fractions extracted at pH 3.3 and 7.5 formed a stable cream phase at both concentrations. This fact could be related to the formation of a cream phase with a low proportion of continuous phase that, as a result of Stokes's law, produces a slow creaming process (Palazolo, 2006).

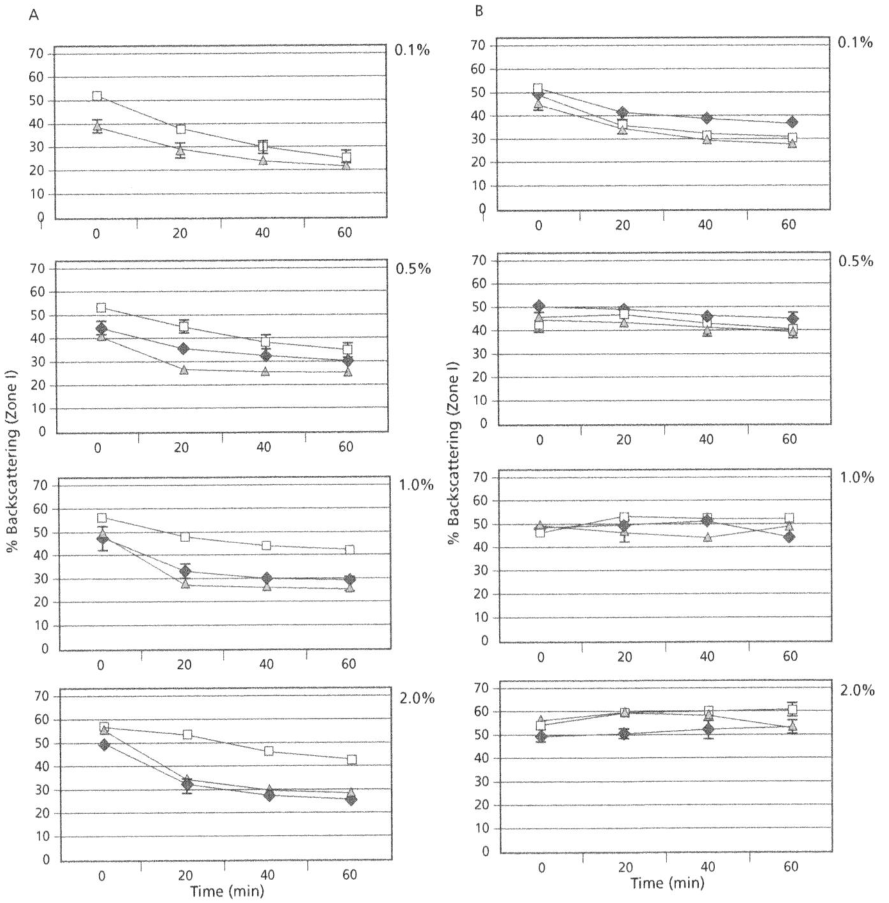

Figure 3.8. Backscattering (%) versus time of O/W emulsions (30:70 wt/wt) containing the PC-enriched fraction obtained by extraction with an absolute ethanol:water mixture of 96:4 (v/v) ranging from 0.1% to 2.0% (wt/wt) at different pH levels: 3.3; 7.5, 10.0. (A) Zone I (10–15 mm), (B) zone II (50–55 mm). Mean values ± SD (n = 2).

The structure of the different phospholipids at an O/W interface influences emulsion formation and their stability (van Nieuwenhuyzen and Tomás, 2008). PC forms a lamellar structure at such an interface with well-ordered monolayers and bilayers, whereas PE assumes a reverse-hexagonal arrangement at the interface that is more difficult to attain (van Nieuwenhuyzen and Szuhaj, 1998). Taking into account the high PC:PE ratio and the consequent interface arrangement of the PC-enriched fraction obtained by sunflower lecithin fractionation with an ethanol:water 96:4 ratio at pH 7.5, this fraction could be considered a good O/W emulsifier.

Particle size of O/W emulsions formulated with PC-enriched fractions obtained by fractionation with absolute ethanol and water mixtures (96:4, 90:10 v/v) at different pHs was measured just after emulsification ($t = 0$) by laser scattering using a Malvern Mastersizer 2000E (Malvern Instruments Ltd, Worcestershire, UK) according to Cabezas et al. (2012a). De Brouker mean diameter (D [4,3]) is a sensitive parameter for analyzing oil-droplet aggregation (through coalescence and/or flocculation) (Relkin and Sourdet, 2005).

The emulsion containing 0.1% (v/v) of the PC-enriched fractions tended to coalesce quickly because the amount of emulsifying agent was not enough to completely cover the surface of the oil droplets (data not shown). The D [4,3] mean diameter decreased progressively and significantly ($p < 0.05$) with increased concentrations of PC-enriched fractions above 0.5% (v/v), so the concentration of 2.0% (v/v) resulted in the

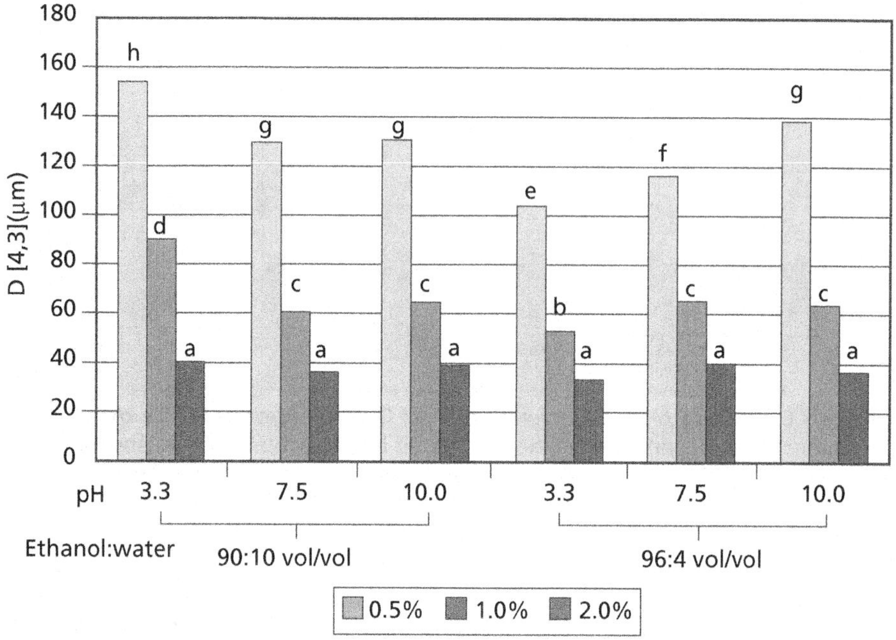

Figure 3.9. De Brouker D [4,3] mean diameter of O/W emulsions obtained with PC-enriched fractions as a function of composition and pH of solvent extraction (absolute ethanol:water v/v) at 0.5%, dark-gray bars; 1.0%, light gray; and 2.0%, black bars. Mean values ± SD ($n = 2$). Bars of the same color with different letters are significantly different ($p < 0.05$).

lowest droplet diameters for the different pH levels in the emulsions made with both absolute ethanol:water extracts (Figure 3.9). Emulsions with the higher percentages of the PC-enriched fractions (2.0% v/v) did not exhibit significant differences ($p > 0.05$) in mean diameters D [4,3], whether those fractions had been obtained at either of the absolute ethanol and water ratios or at any of the pH levels used (Figure 3.9).

Conclusion

Sunflower lecithin is a promising alternative to soybean lecithin because it is the product of a non-GMO. Lecithin modification under industrial conditions with adequate techniques of analysis may be useful for evaluating the potential applications of these sunflower byproducts to the production of new emulsifiers. Regarding fractionation of sunflower lecithin at two absolute ethanol and water ratios and different pH levels, the highest ethanol concentration in the solvent extraction resulted in an enhancement of the PC fraction yield in the extract obtained at the different pHs studied. The PC-enriched fraction extracted at an absolute ethanol:water ratio of 96:4 (v/v) and a pH 7.5 presented the highest PC:PE ratio. The relative proportion of these two phospholipids was associated with their optimal emulsification properties through their interaction with water and their geometry of assembly at the O/W interface.

Acknowledgments

This work was supported by grants from the Agencia Nacional de Promoción Científica y Tecnológica (ANPCyT), Argentina, PICT 1085 (ANPCyT); the Consejo Nacional de Investigaciones Científicas y Técnicas (CONICET), Argentina, PIP 1735 (CONICET); and the Universidad Nacional de La Plata (UNLP), Argentina, 11/X610 (UNLP).

References

Bueschelberger, H. G. Lecithins. In *Emulsifiers in Food Technology;* Whitehurst, J., Ed.; Blackwell Publishing: West Sussex, UK, 2004.

Cabezas, D. M.; Diehl, B. W. K.; Tomás, M. C. Effect of Processing Parameters on Sunflower PC Enriched Fractions Extracted with Aqueous-ethanol. *Eur. J. Lipid Sci. Technol.* **2009a,** *111,* 993–1002.

Cabezas, D. M.; Diehl, B. W. K.; Tomás, M. C. Sunflower Lecithin: Application of a Fractionation Process with Absolute Ethanol. *J. Am. Oil Chem. Soc.* **2009b,** *86,* 189–196.

Cabezas, D. M.; Madoery, R.; Diehl, B. W. K.; Tomás, M. C. Application of Enzymatic Hydrolysis on Sunflower Lecithin Using a Pancreatic PLA_2. *J. Am. Oil Chem. Soc.* **2011a,** *88,* 443–446.

Cabezas, D. M.; Madoery, R.; Diehl, B. W. K.; Tomás, M. C. Enzymatic Hydrolysis of Sunflower Lecithins Using a Microbial PLA_2. In *Sunflowers: Cultivation, Nutrition, and Biodiesel Uses;* Hughes, V. C., Ed.; Nova Science Publishers: New York, 2011b; pp 303–316.

Cabezas, D. M.; Madoery, R.; Diehl, B. W. K.; Tomás, M. C. Emulsifying Properties of Different Modified Sunflower Lecithins. *J. Am. Oil Chem. Soc.* **2012a,** *89,* 355–361.

Cabezas, D. M.; Madoery, R.; Diehl, B. W. K.; Tomás, M. C. Emulsifying Properties of Hydrolyzed Sunflower Lecithins by Phospholipases A_2 of Different Sources. In *Food Industrial Processes—Methods and Equipment, The Food Industry/Book 2*; Benjamin Valdez, B., Ed., InTech: Croatia, 2012b; pp 39–50.

Cabezas, D. M.; Guiotto, E. N.; Diehl, B. W. K.; Tomás, M. C.: Antioxidant and Emulsifying Properties of Modified Sunflower Lecithin by Fractionation with Ethanol-Water Mixtures. In *Food Industry*; Muzzalupo, I., Ed.; Intech: Croatia, 2013; pp 589–602.

Carlsson, A. Physical Properties of Phospholipids. In *Phospholipids Technology and Applications*; Gunstone, F. D., Ed.; The Oily Press: St. Andrews, Scotland, 2008; pp 95–137.

De María, L.; Vind, L.; Oxenbøll, K. M.; Svendsen, A. Phospholipases and Their Industrial Application. *Appl. Microbiol. Biotechnol.* **2007,** *74,* 290–300.

Dijkstra, A. J. Edible Oil Processing—Refining. Deguming, 2011. http://lipidlibrary.aocs.org/processing/degum-intro/index.htm (accessed March 2014).

Dijkstra, A. J.; De Kock, J. Process for Fractionating Phosphatide Mixtures. U.S. Patent 5,214,171, May 25, 1993.

Duttaroy, A. K. Clinical and Nutritional Properties of Phospholipids. In *Phospholipid Technology and Applications*; Gunstone, F. D., Ed.; The Oily Press: St. Andrews, Scotland, 2008; pp 153–164.

Grompone, M. Sunflower Oil. In *Bailey's Industrial Oil and Fat Products. Edible Oil and Fat Products: Edible Oils;* Shahidi, F., Ed.; Wiley: New Jersey, 2005; Vol. 2, pp 655–725.

Guiotto, E. N.; Cabezas, D. M.; Diehl, B. W.; Tomás, M. C. Characterization and Emulsifying Properties of Different Sunflower Phosphatidylcholine Enriched Fractions. *Eur. J. Lipid Sci. Technol,* **2013,** *115* (8), 865–873.

Guo, Z.; Vikbjerg, A.; Xu, X. Enzymatic Modification of Phospholipids for Functional Applications and Human Nutrition. *Biotechnol Adv.* **2005,** *23,* 203–259.

Hernández, E.; Quezada, N. Uses of Phospholipids as Functional Ingredients. In *Phospholipids Technology and Applications;* Gunstone, F. D., Ed.; The Oily Press: St. Andrews, Scotland, 2008; pp 83–94.

Küllenberg, D.; Taylor, L. A.; Schneider, M.; Massing, U. Health Effects of Dietary Phospholipids. *Lipids Health Dis.* **2012,** *11* (3), 1–16.

McClements, D. J. *Foods Emulsions: Principles, Practice and Techniques;* CRC Press: Boca Raton, FL, 1999.

Mengual, O.; Meunier, G.; Cayré, I.; Puech, K.; Snabre, P.; Turbiscan, M. A. 2000: Multiple Light Scattering Measurement for Concentrated Emulsion and Suspension Instability Analysis. *Talanta* **1999,** *50,* 445–456.

Nguyen, M. T.; Van de Walle, D.; Petit, C.; Beheydt, B.; Depypere, F.; Dewettinck, K. Mapping the Chemical Variability of Vegetable Lecithins. *J. Am. Oil Chem. Soc.* **2014,** *91,* 1093–1101.

Palazolo, G. G. Formación y Estabilidad de Emulsiones O/W Preparadas con Proteínas Nativas y Desnaturalizadas de Soja. Ph.D. Thesis, Universidad Nacional de La Plata, Argentina, 2006.

Pan, L. G.; Tomás, M. C.; Añón, M. C. Effect of Sunflower Lecithins on the Stability of Water in Oil (W/O) and Oil in Water (O/W) Emulsions. *J. Surfactants Deterg.* **2002,** *5,* 135–143.

Penci, M. C. Modificación Enzimática de Lecitinas. Ph.D. Thesis, Universidad Nacional del Sur, Bahía Blanca, Argentina, 2009.

Quezada, A. N. Synthesis of Structured Phospholipids with Conjugated Linolenic Acid, and Evaluation of Their Physical Properties. Master's Thesis, Texas A&M University, 2007. http://hdl.handle. net/1969.

Relkin, P.; Sourdet, S. Factors Affecting Fat Droplets Aggregation in Whipped Frozen Protein-stabilized Emulsions. *Food Hydrocolloid.* **2005,** *19,* 503–511.

Schneider, M. Fractionation and Purification of Lecithin. In *Lecithin: Sources, Manufacture & Uses*; Szuhaj, B. F., Ed.; AOCS Press: Urbana, IL, 1989; pp 109–130.

Schneider, M. Major Sources, Composition and Processing. In *Phospholipids Technology and Applications*; Gunstone, F. D., Ed.; The Oily Press: St. Andrews, Scotland, 2008; pp 21–39.

Szuhaj, B. F. Lecithins. In *Bailey's Industrial Oil and Fat Products,* 6th ed.; Shahidi, F., Ed.; Wiley: New York, 2005; Vol. 2, pp 361–456.

van Nieuwenhuyzen, W. The Changing World of Lecithins. *Inform* **2014,** *25* (4), 254–259.

van Nieuwenhuyzen, W.; Szuhaj, B. F. Effects of Lecithins and Proteins on the Stability of Emulsions. *Fett/Lipid.* **1998,** *100,* 282–291.

van Nieuwenhuyzen, W.; Tomás, M. C. Update on Vegetable Lecithin and Phospholipid Technologies. *Eur. J. Lipid Sci. Technol.* **2008,** *110,* 472–486.

Wilton, D. C. Phospholipases A_2: Structure and Function. *Eur. J. Lipid Sci. Technol.* **2005,** *107,* 193–205.

4

Palm Phospholipids

Worawan Panpipat and Manat Chaijan ■ *Division of Food Technology, Department of Agro-Industry, School of Agricultural Technology, Walailak University, Thasala, Nakhon Si Thammarat, Thailand*

Introduction

Crude palm oil contains about 1% of minor components including carotenoids, tocopherols, sterols, triterpene alcohols, phospholipids, glycolipids, and terpenic and paraffic hydrocarbons (Goh et al., 1985). Some of these low-abundance components constitute key intermediates in lipid biosynthesis (Cheong et al., 2014). For example, phospholipids such as phosphatidylcholine (PC) are important precursors in triacylglycerol biosynthesis (Bourgis et al., 2011; Lu et al., 2009). Phosphatidic acid can be dephosphorylated by phosphatidic acid phosphatases to yield diacylglycerol, which in turn is stored as triacylglycerol (Nakamura et al., 2007). In addition, phospholipids are crucial metabolites that act as signaling molecules and are involved in regulating plant growth and development (Xue et al., 2009). However, these minor lipids in palm oil affect the bleachability, stability, and nutritional value of the palm oil. Carotenes and tocopherols are nutritionally important components that can improve stability of the oil. Phospholipids, which are complex esters containing phosphorus, nitrogen bases, sugars, and long-chain fatty acids, are one of the undesired impurities, and they should be removed to avoid the darkening of the oil during deodorization at a high temperature. Phospholipids can be removed during degumming by coagulating the phosphatides contents with phosphoric acid (Zin, 2006). Generally, the maximum phosphorus content in an ideal quality target of crude palm oil is 15 ppm (Ai, 1990).

Phospholipids play two major roles in palm oil: as an antioxidant synergist and a surface active agent to disperse impurities in oil. Phospholipids are the polar lipids that constitute an important part of cellular membranes and possess unique structures containing both lipophilic and hydrophilic functions. These naturally occurring emulsifiers can bind oil molecules together, leading to increased viscosity and refining and/or flow losses. Phospholipids of palm oil are receiving considerable attention because of the suspected deleterious effect of phosphorus on oil quality (Goh et al., 1982, 1985), and they must be removed from the oils via a series of steps and processes that together are called *refining*.

Crude palm oil contains relatively low levels of phospholipids, ranging from 5 to 130 ppm, because the wet-milling process reduces the original phospholipid content (Choo et al., 2004). Approximately 4% of the phospholipids in the palm fruits remain in the crude palm oil after the wet-milling process (Zin, 2006). Palm-pressed fiber, a by-product of palm oil milling, contains a higher level of phospholipids than crude palm oil (Choo et al., 2004). Additionally, phospholipids can be found in oil droplets from the centrifuge sludge of a palm oil mill (Chow and Ho, 2002). The main phospholipids composed in crude palm oil and palm-pressed fiber are represented by phosphatidylcholine (PC) (Cheong et al., 2014; Goh et al., 1985; Kulkarni et al., 1991), whereas phosphatidylglycerol (PG) is the main phospholipid found in palm oil mill sludge (Chow and Ho, 2002). Compositional analysis of phospholipids in palm oil was carried out back in the 1980s using thin layer chromatography and colorimetric techniques whereby phospholipid levels were established by measuring phosphorus content (Cheong et al., 2014).

Phospholipids are vital in some essential biological functions in live organisms. They are important in maintaining cell membrane integrity and play a significant role in functions such as membrane permeability, membrane fluidity, membrane interactions (lipid–protein), and membrane deterioration (Bezrukov 2000; White et al., 2001). Phospholipids help regulate biological processes, such as signaling, and are involved in metabolic and neurologic diseases (Guo et al., 2005).

Phospholipids are high value-added products. Commercial lecithins are sold at 2 to 10 times the price of soybean oil (Choo et al., 2004). Phospholipids can be valuable byproducts that have potential as multifunctional additives for food, pharmaceutical, and industrial applications. Nowadays, phospholipids have been recognized as being one of the major groups of food emulsifiers (Chua et al., 2009). Thus, recovery of phospholipids from palm-based products such as crude palm oil, palm-pressed fiber, and palm oil mill sludge is very significant and challenging.

Occurrence of Palm Phospholipids

Phospholipids are normally byproducts from the refining process of plant oils, including palm oil. In crude palm oil, the phospholipid content is small because most of it is removed during milling (Goh et al., 1985). In the mill, most of the phospholipids of palm fruit mesocarp remain in the sludge water and the fiber waste (Goh et al., 1982). During processing, a certain amount of residual oil still remains in the palm-pressed fiber after the mesocarp oil is extracted (Majid et al., 2013). The residual oil is reported to contain some minor components including phospholipids. In sludge water, phospholipids are directly responsible for oil loss due to micellar and emulsifying properties. Fiber presumably contains membrane material with considerable phospholipid content (Goh et al., 1982). Palm-pressed fiber constitutes about 15% by weight of the fresh fruit bunches (Yee et al., 1984). Choo et al. (2004) estimated that palm-pressed fiber can provide large amounts of palm lecithin (about 21,645 tons in 2001) based on the present total world production of fresh fruit bunches; thus, this could be an alternative source of lecithin, which is normally obtained from soybeans. Goh et al. (1982) reported that stearin, which is the solid (precipitated) fraction of palm oil, contains larger amounts of phospholipids, presumably as a consequence of coprecipitation of phospholipid vesicles with particulate matter.

Goh et al. (1982) reported that Malaysian palm mesocarp oil previously solvent-extracted from the mesocarp of ripe palm oil fruits (*Elaeis guineensis*) after sterilization provides 1000–2000 ppm of phospholipids. Also, Sambanthamurthi et al. (2000) reported that the solvent-extracted mesocarp oil usually contained 1000–2000 ppm phospholipids, however it only presents at a level of 20–80 ppm in commercial crude palm oil. Thin layer chromatography revealed that the major components of the solvent-extracted Malaysian palm mesocarp oil are PC (36%), phosphatidylethanolamine (PE) (24%), phosphatidylinositol (PI) (22%), and PG (9%). Minor components are phosphatidic acid (PA) (3%), diphosphatidylglycerol (DPG) (4%), and lysophosphatidylethanolamine (LPE) (2%), and traces of lysophosphatidylcholine (LPC) and phosphatidylserine (PS) are detectable (Goh et al., 1982) (Table 4.A). Kulkarni et al. (1991) identified the phospholipid compositions of oil from two varieties of the oil palm in India: dura and pisifera. The oil was extracted from mesocarp with chloroform and methanol (2:1, v/v) and the total phospholipids were isolated from the oil by silicic acid column chromatography using chloroform, methanol, and acetone as the eluting solvents. The major components were found to be 34–35% PC, 22–26% PE, 21–25% PI, 7–8% cardiolipin (CL), 5–7% PG, and 4–6% unidentified substances. The predominant fatty acids in phospholipids from palm mesocarp oil were palmitic, stearic, oleic, and linoleic acids.

Oil droplets from the centrifuge sludge of a palm oil mill are an alternative source of phospholipids (Chow and Ho, 2002). The oil droplets were organic solvent (methanol and chloroform) extractable and the extract consists of 84 wt% of neutral lipids and 14 wt% of complex lipids. Phospholipids account for 10 wt% in the complex

Table 4.A Composition of Phospholipids of Palm Oil

Phospholipid	Mole %	Phospholipid	Mole %
Phosphatidylcholine (PC)	36	Phosphatidic acid (PA)	3
Phosphatidylethanolamine (PE)	24	Lysophosphatidylethanolamine	2
Phosphatidylinositol (PI)	22	Phosphatidylserine	trace
Phosphatidylglycerol	9	Lysophosphatidylcholine	trace
Disphosphatidylglycerol	4		

Source: Goh et al. (1982).

lipid fraction. Five major types of phospholipids were determined and identified in the composition of oil droplets from the centrifuge sludge of a palm oil mill: PE (21%), PG (37%), PC (17%), and PS together with PI (11%) (Chow and Ho, 2002). The main fatty acids associated with the phospholipids from the oil droplets were palmitic and oleic acids (Table 4.B). Normally, the oil droplets found in the sludge are formed in the milling process and stabilized by the surface active agents, especially phospholipids and glycolipids, covering the surface of the oil droplets. The diameter of these oil droplets varies between 1.2 and 12 microns and they are extremely stable because even when separated they would not coalesce to form a homogenous layer of oil (Chow and Ho, 2002). Thus, the oil droplet from the centrifuge sludge of a palm oil mill may have commercial potential as a biosurfactant or other use in innovative value-added products.

The major phospholipids of palm oil derived from ethanolic extract of palm-pressed fiber were similar to those of soybean, which consists of 35–46 wt% PC and 25–27% PE (Choo et al., 2004). Fatty acid compositions of palm phospholipids are more unsaturated with higher linoleic acid content, compared with triacylglycerol in crude palm oil (Choo et al., 2004). The high linoleic acid content indicates that phospholipids are formed at early stages of development, but as the fruit matures, enzymatic processes provide for triacylglycerol with more saturated fatty acid. Jacobsberg (1975) reported that higher degrees of unsaturation observed in phospholipids can also be due to differences arising from the statistical distribution of fatty acid in the *sn*-1,3 and *sn*-2 positions of the triacylglycerol. Position *sn*-2 of triacylglycerol is esterified mainly with unsaturated oleic and linoleic acids. Because position 3 of phospholipids has been esterified by the phosphoric acid ester, there will be a statistical reduction in the amount of saturated acids in the molecule (Choo et al., 2004). Fatty acids in phospholipids from palm-pressed fiber contain relatively little unsaturated fatty acid compared with soybean phospholipids, which comprise about 60% linoleic acid, 10% oleic acid, and 5% linolenic acid. Thus, owing to their similar composition to phospholipids in soybeans, palm-pressed fiber phospholipids may have greater antioxidant stability than phospholipids in soybeans owing to enhancement with a greater proportion of saturated fatty acid (Chigozie et al., 1997; Choo et al., 2004).

Chow and Ho (2002) reported that the fatty acid composition of the individual phospholipids may vary from different parts of the plants and also based on the types of organelles from which they are extracted, but generally palmitic, linoleic, and linolenic acids are the main fatty acids (Table 4.B). Data showed that the major fatty acid compositions of phospholipids from both crude palm oil and the oil droplets from the centrifuge sludge of a palm oil mill were similar (Chow and Ho, 2002; Goh et al., 1982).

Goh (1982) reported an anomalous component that constituted up to 45% of the total phospholipids from solvent-extracted mesocarp palm oil. This compo-

Table 4.B Fatty Acid Composition of Phospholipids from Palm Oil Mill Sludge

	Fatty Acid (%)						
Phospholipid	**C12:0**	**C14:0**	**C16:0**	**C18:0**	**C18:1**	**C18:2**	**Other**
Phosphatidylserine	0.5	1.2	45.3	5.0	37.3	0.4	1.0
Phosphatidylinositol	3.0	1.4	55.7	6.1	24.8	0.5	4.1
Phosphatidylcholine	1.3	1.1	41.1	2.7	42.9	0.4	1.9
Phosphatidylglycerol	0.6	1.1	53.1	5.9	34.1	0.4	2.3
Phosphatidylethanolamine	3.0	2.1	51.0	2.1	26.9	0.2	8.4

Source: Chow and Ho (2002).

nent is now found to be the artifact phosphatidylmethanol (PM), which arose from enzymatic transphosphatidylation of natural phospholipids during extraction of unsterilized fruits by methanolic solvents. This component is absent in commercial crude palm oil and is not formed if the fruits are sterilized at or above 100 °C prior to solvent extractions. Generally, commercial grades of palm oil actually contain relatively small amounts of phospholipids. Refining removes considerable amounts of phospholipids through alkali washing or phosphoric acid degumming, and bleaching earth finally eliminates most of them (Goh et al., 1982).

Phospholipids are only present at low levels (20–80 ppm) in commercial crude palm oil, and they usually account for a minor part of the total element phosphorus of the oil. It is desirable to have low levels of phosphorus in the oil to obtain better oxidative stability and bleaching properties (Goh et al., 1982). Inorganic phosphorus accompanies the pro-oxidant metal iron in correlating with the free fatty acid values of the oil (Goh et al., 1985). The presence of excessive amounts of phospholipids in palm oil can result in some refinery problems, such as loss of oil and loss of bleaching power of the bleaching earth (Goh et al., 1982).

Recovery of Palm Phospholipids

As a constituent of all cells, phospholipids are present at different concentrations in various palm raw materials. Different quantities of phospholipids had been reported in solvent-extracted palm mesocarp oil, crude palm oil, recovered oil from spent bleaching earth, centrifuged gummy residue (Goh et al., 1982), oil droplets from the centrifuge sludge of a palm oil mill (Chow and Ho, 2002), and palm-pressed fiber (Choo et al., 2004). Goh et al. (1982) noted that palm mesocarp oil, crude palm oil, recovered oil from spent bleaching earth, and centrifuged gummy residue provide 1000–2000 ppm, 20–80 ppm, 10,000 ppm, and 14,000 ppm of phospholipids, respectively. The recovery of phospholipids from different sources of palm oil consisted of two steps including extraction of oil and separation of phospholipids from the oil using different techniques, such as column chromatography (Goh et al., 1982) and membrane filtration (Lin et al., 1997; Majid et al., 2013; Ong et al., 1999). The extraction of phospholipids by using solvent was a time-intensive and solvent-consuming technique (Choo et al., 2004; Chua et al., 2009).

In crude palm oil, in which the major part of the phospholipids is not particularly soluble, most of the phospholipids are expected to remain in reverse micelles, vesicles, or emulsion droplets. During long storage, some phospholipid gums precipitate together with suspended insoluble materials. Similar precipitation was done by centrifugation in the laboratory, and the oily gums obtained were found to contain as much as 1.4% of phospholipids together with a relatively high amount of iron and copper, which are undesirable prooxidants in palm oils. The iron and copper contents can be attributed to the sequestering action of phospholipids. Micellar action would also cause water-soluble metal ions to be associated with phospholipids, whereas hydrated insoluble metal salts could possibly be dispersed by phospholipids in a similar action (Goh et al., 1982). Removal of phospholipid gums by phosphoric acid treatment also leads to the removal of substantial quantities of iron and copper with a resultant increase of oxidative stability of the oil (Goh et al., 1982).

Palm-pressed fiber, a byproduct of palm oil milling, can be used as a potential raw material for phospholipids extraction with hexane and 95% ethanol (Choo et al., 2004). Palm-pressed fiber oil was first purified by open-column chromatography over acid-treated Florisil. The isolated phospholipids fractions were further concentrated based on their different solubilities in hexane and ethanol. The ethanolic extract yielded 46,800 ppm of phospholipids, whereas only 1367 ppm of phospholipids and most of the neutral lipids were present in the hexane extract. The predominant phospholipids found in palm-pressed fiber are PC, PE, PG, and PA as analyzed by high performance liquid chromatography (HPLC) coupled with an evaporating light scattering detector (ELSD) (Choo et al., 2004). The phospholipid compositions in palm-pressed fiber are similar in crude palm oil (Choo et al., 2004; Goh et al., 1982). PI, which is another major component of phospholipids previously found in crude palm oil, was not observed in palm-pressed fiber (Choo et al., 2004).

Ultrasound is a new technique applied for phospholipid extraction from palm-pressed fiber (Chua et al., 2009). Application of ultrasound for oil separation and recovery of palm oil has been studied by Juliano et al. (2013). They suggested that the use of high frequency ultrasound for improving oil recovery is a significant advance for palm oil milling operations. Ultrasound enhances the mass transfer and the solvent's ease of access to the cell material of the fiber. Cavitations in ultrasound extraction produced high energies around the solvent molecules and caused cell disruption for better penetration of solvent into the cell. The asymmetric collapse of microbubbles near the surface of cellular material has been suggested to disrupt biological cell wall and ease the release of extractable compounds into solvents (Maricela et al., 2001). Furthermore, ultrasound-assisted extraction has been proven to improve the extraction efficiency of the targeted compound by increasing the yield and shortening extraction time, as compared to conventional extraction methods in plant materials (Chua et al., 2009). They reported the optimum operating ultrasound condition for phospholipids extraction from palm-pressed fiber. Data showed that the maximum phospholipid extraction efficacy was achieved with 30 minutes sonication time, 20% of amplitude, and 0.2 (W/s) cycle with the solid:liquid ratio of 1:4 (g/ml). This optimum condition gave 25,213 ppm PC, 14,920 ppm PG, 12,870 ppm PE, and 328 ppm LPC. From this study, PC was found to be the most abundant phospholipid in palm-pressed fiber, as mentioned by Goh et al. (1985), Kulkarni et al. (1991), and Choo et al. (2004). The physical structure of palm-pressed fiber before and after ultrasonic treatment was visualized by a scanning electron microscope. From the scanning electron micrographs, ultrasound-treated palm-pressed fiber had more fractured sections on the surface structure compared to palm-pressed fiber without ultrasound treatment. This confirmed that ultrasound treatment caused the degradation of vegetal cell walls and facilitated the extraction of phospholipids from palm-pressed fiber (Chua et al., 2009). The normal solvent extraction of solid dried materials basically consists of two stages: first is steeping vegetal materials in solvent to ease the swelling and the hydration process; and second is the mass transfer of soluble compounds from the vegetal material to solvent by osmotic and diffusion processes (Vinatoru, 2001). Nevertheless, ultrasonic treatment enhances the swelling and softening process of the cell wall via the hydration of pectinous material from middle lamella, which might lead to the breakup of vegetal tissue during sonication (Maricela et al., 2001).

Recently, the separation of phospholipids from palm oil was be achieved with the aid of membrane filtration. However, only a few studies have been conducted to separate phospholipids from palm oil (Lin et al., 1997; Majid et al., 2013; Ong et al., 1999). The membrane-based crude oil degumming has been reported by Lin et al. (1997) and Ong et al. (1999). This process has potential for energy and cost savings and produces permeate- and retentate-containing triglyceride and phospholipids, respectively (Lin et al., 1997; Ong et al., 1999). The majority of the coloring materials and some of the free fatty acids and other impurities are included in phospholipid micelles and removed as well (Lin et al., 1997). Polyethersulfone membrane with a molecular weight cut off of 9000 effectively removed phospholipids (96.4%) in the degumming of crude palm oil in which less than 0.3 ppm of phosphorus was found in permeate. Majid et al. (2013) separated phospholipids from residual palm-pressed fiber oil/hexane miscella using polymeric membranes, including polyethersulphone (PES) (10 and 20 kDa) and polyvinylidene fluoride (PVDF) (30 kDa). Data noted that the PVDF 30 kDa membrane had the highest overall flux with low operating pressure, and the phospholipid retention was above 80%. PVDF 30 kDa membrane was found to be the most hydrophobic membrane with the largest molecular weight cut off compared to 10 and 20 kDa PES membranes. In addition, the permeate flux increases with the increase of operating pressure, temperature, and agitation speed. Increasing operating pressure from 2 to 6 bar with a temperature of 40 °C and speed of 300 rpm resulted in an increase in the retention of phospholipids from 81% to 95%. However, the regression model obtained from the study showed that the temperature had a negative effect on the retention of phospholipids (Majid et al., 2013).

The small difference in molecular weight of phospholipids (about 800 Da) and triacylglycerols (about 900 Da) makes the separation by membrane quite difficult. Gupta (1977) reported that in nonpolar media such as hexane or neutral oil, phospholipid molecules tend to form reverse micelles with an average molecular weight of 20 kDa or more. In the oil extraction process, hexane is commonly used as a solvent. Thus, the separation of phospholipids by using appropriate membranes is possible in nonaqueous streams such as vegetable oil/hexane miscella (Majid et al., 2013). The membrane separation technique, which can simultaneously concentrate, fractionate, and purify products, thus offers an alternative method to recover valuable components as well as reject unwanted impurities from residual palm-pressed fiber oil (Majid et al., 2013).

Functional and Nutritional Properties of Palm Phospholipids

Phospholipids have shown important biological functions and health benefits and are currently used in several food products (Schneider, 2001). Phospholipids are primarily bound in cellular membranes in both animal and plant tissues and they are very important emulsifiers in living tissues. They are bound in lipoproteins and help to transport nonpolar lipids in blood and other intercellular fluids (Pokorny, 2003). These compounds play an important role in biological functions such as maintaining cell membrane integrity, prevention of neurological diseases, and regulation of basic biological processes such as cell-to-cell signaling (Guo et al., 2005). Phospholipids have shown health benefits such as reducing blood cholesterol and triglycerides (Knuiman et al., 1989), liver detoxification and repair of damaged liver tissue (Lieber et al., 1990), and improvement of cognitive functions (Pepeu et al., 1996) and visual acuity (Koletzko et al., 1995). Additionally, phospholipids may be considered an effective and versatile delivery system for functional compounds because they may increase their bioavailability and chemopreventive effect (Hossen and Hernandez, 2005). Phospholipids are involved in metabolism-related and neurological diseases (Lee, 1998; Lohmeyer and Bittmann, 1994) and in the regulation of basic biological processes as signaling compounds (Hannun et al., 2001; Izumi and Shimizu 1995). Moreover, phospholipids act as emulsifiers in the bile digestive fluid, ensuring fine dispersion of fatty food molecules in the water phase, thus improving digestion and absorption (Arboleda, 2007). Phospholipids act also as surface active wetting agents in the lungs (Robertson et al., 1990). Overall, they are likely to have a positive effect on human health.

From the review of Arboleda (2007), phospholipids combined with bioactive substances can enhance their bioavailability such as in the case long-chain polyunsaturated fatty acids (Carnielli et al., 1998), n-3 polyunsaturated fatty acids (Cansell et al., 2003), and tocopherol (Nacka et al., 2001).

Phospholipids have been used as nutritional supplements because they have shown beneficial physiological effects. Soy phospholipids mixtures have been shown to reduce elevated blood cholesterol and triglycerides (Wilson et al., 1998). Soy PC helps in liver detoxification and repair of damaged liver tissue (Lieber et al., 1990). PS from soybean can improve cognitive functions (Pepeu et al., 1996). Egg phospholipids have been used in infant formula to improve visual acuity (Koletzko et al., 1995). As a consequence, palm phospholipids are also able to show such health promoting effects.

Due to their amphiphilic character, phospholipids can adopt various molecular assemblies when dispersed in water, such as bilayer vesicles or micelles, which give them unique interfacial properties and render them very attractive in terms of emulsion or foam stabilization (Patino et al., 2007; Pichot et al., 2013). In foods, phospholipids are often present at the interface of emulsions or they cooperate in forming films on the surface of the solid particles (Pokorny, 2003). Thus, the main positive functions of phospholipids in foods are their surface-active properties. They act as emulsifiers and stabilizers of emulsion (van Nieuwenhuyzen and Szuhaj, 1998), facilitate the dispersion of solid particles in water phases, and improve the texture of multiphase food materials. Phospholipids make the texture smooth, improve the mouthfeel by increasing the viscosity, and make the food taste full and homogenous (Pokorny, 2003). The amphiphilic character and endogenous nature of phospholipids have proven that they are important raw materials in agriculture, food, pharmaceutical, medical, and cosmetic products (Chua et al., 2009; Krawczyk, 1996; Kovach et al., 2014; van Nieuwenhuyzen, 1981; Wendel, 2000). Pokorny (2003) reported the application of phospholipids in several food industries. In the bakery industry, phospholipids improve the volume of bakery products, improve the fat dispersion, possess anti-staling properties, and improve nutritional value. For the effect of phospholipids on chocolate quality, lecithin increases the fluidity of the chocolate mass and prevents the crystallization of high-melting triacylglycerol. In margarine, phospholipids prevent spattering, improve spreadability, and promote browning during frying. Amino groups of phospholipids, in particular of PE or PS, can react with carbonyl moieties, for example, reducing sugars and other products of sugar degradation to form brown melanoidins via the Maillard reaction (Pokorny, 2003). For sauce-type emulsions, phospholipids can stabilize the structure by orienting themselves at the oil–water interface, lowering the interfacial tension between the two phases and reducing the pressure gradients required to disrupt the droplets during the emulsification process (Mezdour and Relkin, 2011). In the case of confectionary and vegetable oils, phospholipids function as antisticking agents and antioxidants, respectively (Arboleda, 2007). Phospholipids are also utilized for nonfood purposes such as emulsifiers, dispersion agents, adhesives, or lubricants. They can also be added to coatings such as paints, waxes, and polishes (Pokorny, 2003). Their uses in the cosmetics and pharmaceutical industries are also important. Numerous innovative phospholipids-based drug carriers continue to be developed by the pharmaceutical industry, thus providing a big market for phospholipids applications (Chua et al., 2009).

Phospholipids can stabilize colloidal dispersions and can act as natural antioxidants. Adding phospholipids to palm oil resulted in the increase in oxidative stability of the palm oil (Jacobsberg, 1975). Choe (2008) reported that PC decreased the oxidation of docosahexaenoic acid (DHA) at 25–30 °C in the dark. Peroxide formation in the autoxidation of soybean oil at 50 °C for 8 weeks was decreased by addition of PC, PE, and PS at 0.03–0.05%. In addition, phospholipids can cause reverse micelle, vesicle, or emulsion droplet formation. This means that pro-oxidant metal ions and their hydrophilic salts may be removed from the lipid phase so that they are rendered less effective in promoting autoxidation (Goh et al., 1985; Zin, 2006). The direct sequestering of metal ions by the polar lipids can be one of the modes of action in the antioxidant activity. Zin (2006) reported that antioxidant-synergistic effects of phospholipids can be attributed to the sequestering of soluble prooxidant metal ions to form inactive species. Hydratable insoluble metal ions could also dispersed by phospholipids through miscellar action. Khan and Shahidi (2001) showed a synergism between phospholipids and naturally occurring antioxidants such as α-tocopherol, flavonoids, and other phenolic antioxidants. King et al. (1992) found that nitrogen-containing phospholipids such as PC and PE were efficient antioxidants under most conditions. The nitrogen moieties can donate hydrogen or electron to antioxidant radicals such as tocopheroxy radical and regenerate the antioxidants. Reische et al. (2008) noted that possible antioxidative actions of phospholipids include regeneration of primary antioxidants, metal chelation, and decomposition of hydroperoxides. Moreover, PC, PE, and PS display antioxidant activity that is possibly linked to chelating ability.

Future Trends

Extraction of palm oil phospholipids from palm oil byproducts associated with the refining process is one of the important zero-waste strategies, which are related to the sustainable agriculture. Transformation of palm oil byproducts to value-added products would significantly increase the profit of a palm oil refinery, and consequently, palm oil would remain competitive in the industry (Haslenda and Jamaludin, 2011; Gobi and Vadivelu, 2013). Novel lipidomic tools for isolation and characterization of palm phospholipids from different palm raw materials can be of importance to give a better understanding of the properties of palm phospholipids. Recently, Cheong et al. (2014) applied multiple lipidomic approaches, including high-sensitivity and high-specificity multiple reaction monitoring, to comprehensively quantify individual lipid species including phospholipids in crude palm oil.

Membrane separation and ultrasound-assisted extraction of phospholipids from crude palm oil and palm residues can be applied to improve the yield of phospholipids (Chua et al., 2009; Majid et al., 2013). Utilization of isolated palm phospholipids as functional ingredients in food, pharmaceutical, and cosmetics products can be achieved to increase the value of palm oil.

References

Ai, T. Y. Analytical Techniques in Palm and Palm Kernel Oil Specifications. 10th Palm Oil Familiarization Programme (POFP), Bangi, 1990.

Arboleda, N. Q. Synthesis of Structured Phospholipids with Conjugated Linolenic Acid, and Evaluation of Their Physical Properties. Master's Thesis, Texas A&M University, August 2007.

Bezrukov, S. Functional Consequences of Lipid Packing Stress. *Curr. Opin. Coll. Inter. Sci.* **2000,** *5*, 237–243.

Bourgis, F.; Kilaru, A.; Cao, X.; Ngando-Ebongue, G.; Drira, N.; Ohlrogge, J. B.; Arondel, V. Comparative Transcriptome and Metabolite Analysis of Oil Palm and Date Palm Mesocarp That Differ Dramatically in Carbon Partitioning. *Proc. Natl. Acad. Sci. USA.* **2011**, *108*, 12527–12532.

Cansell, M.; Nacka, F.; Combe, N. Marine Lipid-based Liposomes Increase in Vivo Fatty Acid Bioavailability. *Lipids* **2003**, *38*, 551–559.

Carnielli, V.; Verlato, G.; Perderzini, F.; Luijendijk, I.; Boerlage, A.; Pedrotti, D.; Sauer, P. Intestinal Absorption of Long-chain Polyunsaturated Fatty Acids in Preterm Infants Fed Breast Milk or Formula. *Am. J. Clin. Nutr.* **1998**, *67*, 97–103.

Cheong, W. F.; Weng, M. R.; Shui, G. Comprehensive Analysis of Lipid Composition in Crude Palm Oil Using Multiple Lipidomic Approaches. *J. Genet. Genomics* **2014**, *41*, 293–304.

Chigozie, V. N.; Leon, C. B.; Brain, S. Effect of Fatty Acid Composition of Phospholipids on Their Antioxidant Properties and Activity Index. *J. Am. Oil Chem. Soc.* **1997**, *74*, 471–475.

Choe, E. Effects and Mechanisms of Minor Compounds in Oil on Lipid Oxidation. In *Food Lipids Chemistry, Nutrition, and Biotechnology*; Akoh, C. C., Min, D. B., Eds.; CRC Press: New York, 2008; pp 449–474.

Choo, Y. M.; Bong, S. C.; Ma, A. N.; Chuah, C. H. Phospholipids from Palm-pressed Fiber. *J. Am. Oil Chem. Soc.* **2004**, *81*, 471–475.

Chow, M. C.; Ho, C. C. Chemical Composition of Oil Droplets from Palm Oil Mill Sludge. *J. Oil Palm Res.* **2002**, *14*, 25–34.

Chua, S. C.; Tan, C. P.; Mirhosseini, H.; Lai, O. M.; Long, K.; Baharin, B. S. Optimization of Ultrasound Extraction Condition of Phospholipids from Palm-pressed Fiber. *J. Food Eng.* **2009**, *92*, 403–409.

Gobi, K.; Vadivelu, V. M. By-products of Palm Oil Mill Effluent Treatment Plant—A Step towards Sustainability. *Renew. Sust. Energ. Rev.* **2013**, *28*, 788–803.

Goh, S. H.; Khor, H. T.; Gee, P. T. Phospholipid of Palm Oil (*Elaeis guineensis*). *J. Am. Oil Chem. Soc.* **1982**, *59*, 296–299.

Goh, S. H.; Choo, Y. M.; Ong, S. H. Minor Constituents of Palm Oil. *J. Am. Oil Chem. Soc.* **1985**, *62*, 237–240.

Guo, Z.; Vikbjerg, A. F.; Xu, X. Enzymatic Modification of Phospholipids for Functional Applications and Human Nutrition. *Biotechnol. Adv.* **2005**, *23*, 203–259.

Gupta, A. K. S. Process for Refining Crude Glyceride Oils by Membrane Filtration. U.S. Patent 4,062,882, 1977.

Hannun, Y.; Luberto, C.; Argraves, K. Enzymes of Sphingolipid Metabolism: From Modular to Integrative Signaling. *Biochemistry* **2001**, *40*, 4893–4903.

Haslenda, H.; Jamaludin, M. Z. Industry to Industry By-products Exchange Network towards Zero Waste in Palm Oil Refining Processes. *Resour. Conserv. Recy.* **2011**, *55*, 713–718.

Hossen, M.; Hernandez, E. Enzyme-catalyzed Synthesis of Structured Phospholipids with Conjugated Linoleic Acid. *Eur. J. Lipid Sci. Technol.* **2005**, *107*, 730–736.

Izumi, T.; Shimizu, T. Platelet-activating Factor Receptor: Gene Expression and Signal Transduction. *Biochim. Biophys. Acta.* **1995**, *1259*, 317–333.

Jacobsberg, B. Characteristic of Malaysian Palm Oil. *Oleagineux* **1975**, *30*, 271–276.

Juliano, P.; Swiergon, P.; Mawson, R.; Knoerzer, K.; Augustin, M. A. Application of Ultrasound for Oil Separation and Recovery of Palm Oil. *J. Am. Oil Chem. Soc.* **2013**, *90*, 579–588.

Khan, M. A.; Shahidi, F. Tocopherols and Phospholipids Enhance the Oxidative Stability of Borage and Evening Primrose Triacylglycerols. *J. Food Lipids* **2001**, *7*, 143–150.

King, M. F.; Boyd, L. C.; Sheldon, B. W. Antioxidant Properties of Individual Phospholipids in a Salmon Oil Model System. *J. Am. Oil Chem. Soc.* **1992**, *69*, 545–551.

Knuiman, J.; Beynen, A.; Katan, M. Lecithin Intake and Serum Cholesterol. *Am. J. Clin. Nutr.* **1989**, *49*, 266–270.

Koletzko, B.; Edenhofer, S.; Lipowsky, G.; Reinhardt, D. Effects of a Low Birthweight Infant Formula Containing Human Milk Levels of Docosahexaenoic Acid and Arachidonic Acid. *J. Pediatr. Gastroenterol. Nutr.* **1995**, *21*, 200–208.

Kovach, I.; Koetz, J.; Friberg, S. E. Janus Emulsions Stabilized by Phospholipids. *Colloids Surf. A.* **2014**, *441*, 66–71.

Krawczyk, T. Lecithin: Consider the Possibilities. *Inform* **1996**, *7*, 1158–1167.

Kulkarni, A. S.; Khotpal, R. R.; Bhakare, H. A. Phospholipids and Glycolipids in the Oil from Some Varieties of *Elaeis guineensis* in India. *Elaeis* **1991**, *3*, 363–368.

Lee, T. Biosynthesis and Possible Biological Functions of Plasmogens. *Biochim. Biophys. Acta.* **1998**, *1394*, 129–145.

Lieber, C.; De Carli, K.; Mak, K.; Kim, C.; Leo, M. Attenuation of Alcohol Hepatic Fibrosis by Polyunsaturated Lecithin. *Heptology* **1990**, *12*, 1390–1398.

Lin, L.; Rhee, K. C.; Koseoglu, S. S. Bench-scale Membrane Degumming of Crude Vegetable Oil: Process Optimization. *J. Membrane Sci.* **1997**, *134*, 101–110.

Lohmeyer, M.; Bittmann, R. Antitumor Effect Lipids and Alkylphosphocholines. *Drug Fut.* **1994**, *19*, 1021–1037.

Lu, C.; Xin, Z.; Ren, Z.; Miquel, M.; Brown, J. An Enzyme Regulating Triacylglycerol Composition Is Encoded by the *ROD1* Gene of *Arabidopsis*. *Proc. Natl. Acad. Sci. USA*. **2009**, *106*, 18837–18842.

Majid, R. A.; Mohamad, A. W.; May, C. Y. Performance of Polymeric Membranes for Phospholipid Removal from Residual Palm Fiber Oil/Hexane Miscella. *J. Oil Palm Res.* **2013**, *25*, 253–264.

Maricela, T.; Vinatoru, M.; Paniwnyk, L.; Mason, T. J. Investigation of the Effects of Ultrasound on Vegetal Tissues during Solvent Extraction. *Ultrasonics Sonochem.* **2001**, *8*, 137–142.

Mezdour, S.; Relkin, D. P. Effect of Residual Phospholipids on Surface Properties of a Soft-refined Sunflower Oil: Application to Stabilization of Sauce-types' Emulsions. *Food Hydrocolloids* **2011**, *25*, 613–619.

Nacka, F.; Cansell, M.; Meleard, P.; Combe, N. Incorporation of α-Tocopherol in Marine Lipid-based Liposomes: In Vitro and in Vivo Studies. *Lipids* **2001**, *36*, 1313–1320.

Nakamura, Y.; Tsuchima, M.; Ohta, H. Plastidic Phosphatidic Acid Phosphatases Indentified in a Distinct Subfamily of Lipid Phosphate Phosphatases with Prokaryotic Origin. *J. Biol. Chem.* **2007**, *282*, 29013–29021.

Ong, K. K.; Fakhrul-Razi, A.; Baharin, B. S.; Hassan, M. A. Degumming of Crude Palm Oil by Membrane Filtration. *Artif. Cells Blood Substit. Immobil. Biotechnol.* **1999**, *27*, 381–385.

Patino, J. M. R.; Caro, A. L.; Nino, M. R. R.; Mackie, A. R.; Gunning, A. P.; Morris, V. J. Some Implications of Nanoscience in Food Dispersion Formulations Containing Phospholipids as Emulsifiers. *Food Chem.* **2007**, *102*, 532–541.

Pepeu, G.; Pepeu, L.; Amaducci, L. A Review of Phosphatidylserine Phamacological and Clinical Effects. Is Phosphatidylserine a Drug for the Ageing Brain? *Phamacol. Res.* **1996**, *33*, 73–80.

Pichot, R.; Watson, R. L.; Norton, I. T. Phospholipids at the Interface: Current Trends and Challenges. *Int. J. Mol. Sci.* **2013**, *14*, 11767–11794.

Pokorny, J. Phospholipids. In *Chemical and Functional Properties of Food Lipids;* Sikorski, Z. E., Kolakowska, A., Eds.; CRC Press: New York, 2003; pp 79–92.

Reische, D. W.; Lillard, D. A.; Eitenmiller, R. R. Antioxidants. In *Food Lipids Chemistry, Nutrition, and Biotechnology*; Akoh, C. C., Min, D. B., Eds.; CRC Press: New York, 2008; pp 409–434.

Robertson, B.; Curstedt, T.; Johansson, J.; Jornvall, H.; Kobayashi, T. Structural and Functional Characterization of Porcine Surfactant Isolated by Liquid-gel Chromatography. *Prog. Respir. Res.* **1990**, *25*, 237–246.

Sambanthamurthi, R.; Sundram, K.; Tan, Y. A. Chemistry and Biochemistry of Palm Oil. *Prog. Lipid Res.* **2000**, *39*, 507–558.

Schneider, M. Phospholipids for Functional Food. *Eur. J. Lipid Sci. Technol.* **2001**, *103*, 98–101.

van Nieuwenhuyzen, W. The Industrial Uses of Special Lecithin: A Review. *J. Am. Oil Chem. Soc.* **1981**, *58*, 885–888.

van Nieuwenhuyzen, W.; Szuhaj, B. Effects of Lecithins and Proteins on the Stability of Emulsions. *Lipid* **1998**, *100*, 282–291.

Vinatoru, M. An Overview of the Ultrasonically Assisted Extraction of Bioactive Principle from Herbs. *Ultrasonics Sonochem.* **2001**, *8*, 303–313.

Wendel, A. Lecithin: The First 150 Years, Part I: From Discovery to Early Commercialization. *Inform* **2000**, *11*, 885–894.

White, S.; Ladokhin, A.; Jayasinghe, S.; Hristova, K. How Membrane Shape Protein Structure? *J. Biol. Chem.* **2001**, *276*, 32395–32398.

Wilson, T.; Mescervey, C.; Nicolosi, R. Soy Lecithin Reduces Plasma Lipoprotein Cholesterol and Early Atherogenesis in Hypercholesterolemic Monkeys and Hamsters: Beyond Linoleate. *Atherosclerosis* **1998**, *140*, 147–153.

Xue, H. W.; Chen, X.; Mei, Y. Function and Regulation of Phospholipid Signaling in Plants. *Biochem. J.* **2009**, *421*, 145–156.

Yee, C. B.; Lim, K. C.; Ong, E. C.; Chan, K. W. The Effects of E.kamerunicus on Bunch Component of E. guineensis. *Proceedings of the Symposium on Impact of the Pollinating Weevils on The Oil Palm Industry,* Kuala Lumpur, Malaysia, 1984.

Zin, R. B. M. *Process Design in Degumming and Bleaching of Palm Oil*; Research Report for Universiti Teknologi Malaysia, November 2006.

5

Milk and Dairy Polar Lipids: Occurrence, Purification, and Nutritional and Technological Properties

Thien Trung Le ▪ *Faculty of Food Science and Technology, Ho Chi Minh City Nong Lam University, Vietnam*

Thi Thanh Que Phan, John Van Camp, and Koen Dewettinck ▪ *Department of Food Safety and Food Quality, Ghent University, Belgium*

Introduction

Cow's milk contains 12.8–40 mg/L of total polar lipids (PLs), which is equivalent to 0.32–1.0% of the total lipids of milk. A major part of PLs in milk the membrane surrounding fat globules, which is called the milk fat globule membrane (MFGM). In the dairy industry, processing techniques such as churning cream into butter or homogenizing (phase conversion) the concentrated cream into butter oil break the fat globules and the resulting MFGM fragments are preferentially distributed to aqueous phases such as buttermilk and butter serum. Technologies are available for isolation of MFGM fragments and for further purification to obtain PL concentrate from the byproducts. Recently, milk PLs have gained attention because of the beneficial health properties they possess. In addition, PLs are amphiphilic molecules that can be used as emulsifiers for food product development.

Occurrence, composition, and factors influencing composition of PLs in milk are discussed in the first part of this chapter. Distribution of MFGM and PLs during dairy processing is explained, and this helps determine which processing-derived streams/fractions can serve as a good sources for isolation of MFGM and purification of PLs. We present isolation and purification techniques before reviewing nutraceutical and technological properties of PLs, and the chapter closes with a discussion of possible applications of MFGM material and PL concentrate. This chapter elucidates the possibilities of obtaining MFGM material and PL concentrate from dairy processing–derived streams and the potential uses of such ingredients for development of functional foods and nutraceuticals.

Occurrence of Polar Lipids in Milk and Dairy Products

Origin of Milk Polar Lipids

Milk is defined as the secretion of the mammary glands of mammals. The primary natural function of milk is as a source of nutrition for the young. Unlike other animals, humans consume milk and dairy products even beyond their infancy, and the relation between consumption of milk and dairy products and health is of great consideration. In terms of colloidal chemistry, milk is not a solution of dissolved substances but rather a poly-dispersed system with particles of different sizes. Milks from different mammals have a similar structure to that of cow's milk, depicted in Figure 5.1.

Physicochemically, milk can be considered as an O/W emulsion with fat globules as the dispersed phase. The size of milk fat globules (MFGs) ranges from 0.1 to 15 μm in diameter. Milk plasma is milk without fat globules (FGs), and milk serum is milk plasma without casein micelles. Milk serum contains lipoprotein particles and soluble components such as whey proteins, lactose, minerals, and other minor components.

FGs consist of a core that is surrounded by a true biological member: the MFGM. MFGM is derived from secretory cell content and membrane. A depiction of an MFGM structure can be seen in Figure 5.2. According to this model, PLs are located in the single inner layer, connected directly to the triglyceride core, and in the

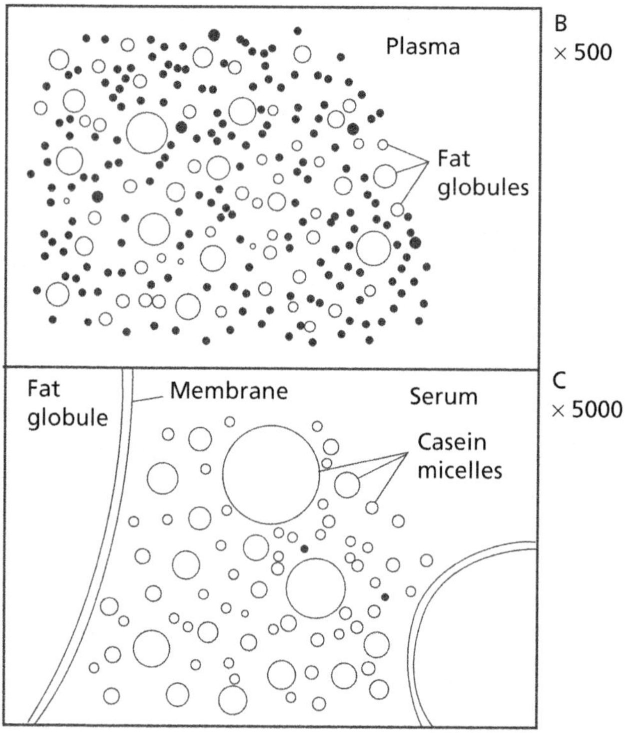

Figure 5.1 Illustration of milk structure at different magnifications (from Walstra et al., 2006).

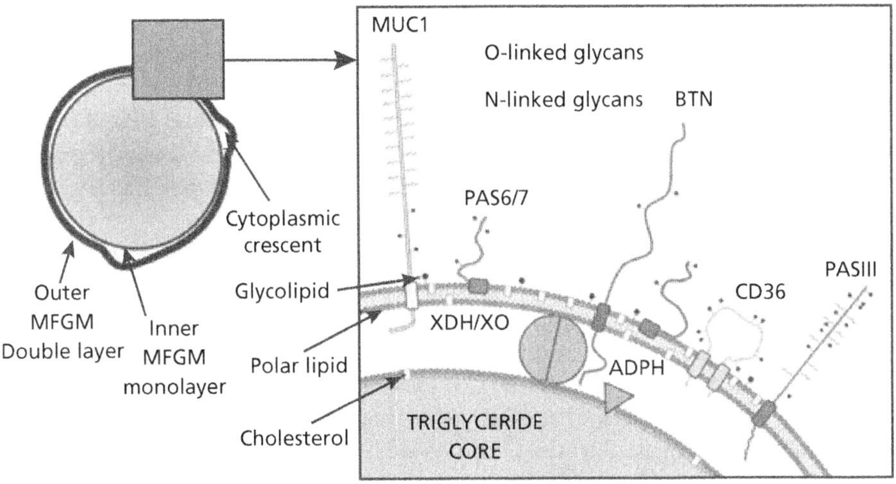

Figure 5.2 Schematic illustration of fat globules and fat globule membrane with arrangement of PLs, specific MFGM proteins, and cholesterol (from Dewettinck et al., 2008).

outer bilayer. MFGM is the first main source of PLs occurring in milk. Explanation of the intracellular origin and secretion of MFGs can be found in literature (Heid and Keenan, 2005). The physicochemical role of the MFGM is to stabilize MFGs against coalescence and protect the core fat against chemical oxidation and the attack of enzymes.

Milking, pumping, cooling, and agitating can cause mechanical impact on the surface of MFGs, and part of MFGM materials can be shredded to the serum phase. The shredded parts are called *MFGM fragments,* and they are the second source of PLs occurring in milk. Another source of PLs in milk is lipoprotein particles. These are dispersed in the serum phase and are believed to be remnants of mammary secretory cell membranes. Lipoprotein particles have an average diameter of 10 nm (Walstra et al., 2006). It is worth mentioning that somatic cells (e.g., leucocytes) also contribute to the total PL content in milk if they are not separated from milk during processing. Somatic cells are separated during centrifuging for cream separation.

Milk PL Composition

PLs consist of phospholipids and sphingolipids. Phospholipids consist of phosphatidylcholine (PC), phosphatidylethanolamine (PE), phosphatidylinositol (PI), and phosphatidylserine (PS), and sphingolipids consist of sphingomyelin (SM), lactosylceramide

(LacCer), glucosylceramide (GluCer), and trace amounts of galactosylceramide and gangliosides. The last four can be grouped as glycosphingolipids. Gangliosides are sialic acid–containing glycosphingolipids. Lysophosphatidylcholine (LPC) sometimes occurs in milk as a result of enzymatic hydrolysis of PC. Similarly, LPE and LPS also occur in very limited amounts. An example of high performance liquid chromatography–evaporating light scattering detector (HPLC–ELSD) separation of milk PLs is given in Figure 5.3.

PLs are amphiphilic molecules with a hydrophobic tail and a hydrophilic head group, which largely contribute to the emulsifying properties of the membrane (see more in the "Technological Properties of PLs" section). PL content and relative composition of phospholipids and sphingolipids of raw milk are given in Table 5.A.

Cow's milk contains 12.8–40 mg/L of total PLs, which is equivalent to 0.32–1.0 g/100 g total lipids, provided that milk has an average content of total lipids of 4.0%. These numbers are small, but they become very important when nutritional properties and technical functionalities of PLs are considered (Table 5.A).

There is a high variation in reported concentrations and composition of PLs in milk. This is due to the fact that the composition of milk in general and, as expected, composition of PLs in milk are influenced by many factors such as breed, feed and season, and lactation period. This will be discussed in detail in the next section, "Factors

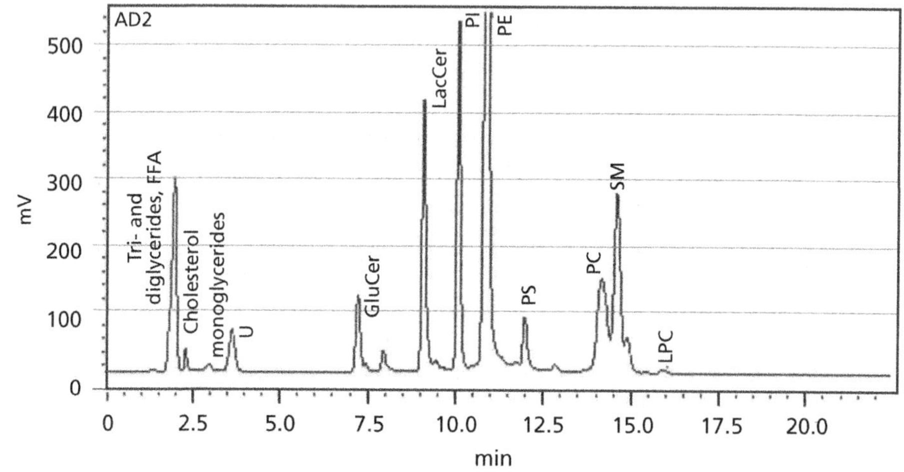

Figure 5.3 Separation of polar lipids from a milk polar lipids product using a normal phase HPLC-ELSD method (from Le et al., 2011a). FFA: free fatty acid; U: unknown; GluCer: glucosylceramide; LacCer: lactosylceramide; PI: phosphatidylinositol; PE: phosphatidylethanolamine; PS: phosphatidylserine; PC: phosphatidylcholine; SM: sphingomyelin; LPC: lysoPC.

Table 5.A Content of Polar Lipids and Their Relative Composition of Raw Cow's Milk

Polar Lipids (mg 100 g^{-1} product)	Relative Sphingolipid Content (g 100 g^{-1} polar lipid)			Relative Phospholipid Content (g 100 g^{-1} polar lipid)				Analysis Method	Reference
	GluCer	LacCer	SM	PE	PS	PI	PC		
23	5	2.9	23.6	34.2	2.8	6.2	25.4	HPLC-ELSD	Christie et al. (1987)
20.4	—	—	19.9	35.9[c]	11.2	3.6	28.7	P-31 NMR	Garcia et al. (2013)
29	2.7	6.7	—	42	6.7	4.8	19.2	HPLC-ELSD	Rombaut et al. (2005)
40	2.1	2.8	23.8	31.5	8.8	4.9	26	HPLC-ELSD	Rombaut et al. (2006b)
29.4	2.7	6.7	17.9	42.0	6.7	4.8	19.1	HPLC-ELSD	Rombaut et al. (2007b)
	—	—	29.2	38.6	—	—	32.2	HPLC-ELSD	Fagan and Wijesundera (2004)
25.1[a]	—	—	28.7	31.1	8.5	5.2	26.4	TLC-Densitometry	Bitman and Wood (1990)
12.8[b]	—	—	31.4	19.8	1.9	11.8	35.1	TLC-Densitometry	Bitman and Wood (1990)
24.2[a]	—	—	29.2	38.6	—	—	32.2	HPLC-ELSD	Fagan and Wijesundera (2004)

[a]At lactation day 42.

[b]At lactation day 180.

[c]Consists of PE (31.4%) and phosphatidylethanolamine plasmalogen.

— Indicates no data reported or not determined. Relative composition of PL species was calculated among the determined components.

Influencing Milk PL Composition." Other factors contributing to the variation are the extraction and analysis methods. As shown in Figure 5.2, PLs are strongly linked to other components and different structures in milk. Because PLs are amphiphilic, it is not easy to efficiently extract them for analysis if a suitable combination of organic solvents is not used. The expression of the results should also be considered. Some researchers injected standards with series of known concentrations to determine the composition, while others used detector output signals for the calculation. The latter case is subject to more errors in case the sensitivity of different PL species to a certain detector is not the same (Le et al., 2011a). The relative composition is also dependent on how many components can be detected and are considered in the calculation. It is accepted that the three most prominent PL components in milk are PE, PC, and SM.

Milk fat, in general, is characterized by short- and mid-chain fatty acids (C4–C14). However, these fatty acids (FAs) are virtually absent in the PL fraction of milk. PE is highly unsaturated, followed by PI and PS. PC is rather saturated compared to the other glycerophospholipids. The FA pattern of sphingomyelin (SM) is very uncommon. Although long-chain FAs occur, nearly all of them are saturated (~97%). In addition, the occurrence of C23:0 (>17%) is remarkable (Bitman and Wood, 1990; Jensen, 2002). This uncommon high degree of saturation gives SM the ability to form so-called lipid rafts with cholesterol, which are rigid domains in cellular membranes with a higher melting point and a higher degree of packing (liquid ordered state) in comparison with the glycerophospholipids and their packing (liquid-disordered state) (Gallier et al., 2010b). Similarly, the dietary intake of SM lowers cholesterol absorption in the intestines by lowering the membrane fluidity of the liposomes (see the section "Health-Beneficial Properties of PLs").

Classically, the total content of phospholipids in food can be determined by first analyzing the total phosphorous content in its lipid extract using spectrometry and relating this to the amount of PLs using a conversion factor that depends on the type of food analyzed (Bartlett, 1959). This method does not provide information about the composition of the phospholipids, and GluCer and LacCer are not included because they do not contain phosphorous. The prominent difference in composition between milk PLs and soy lecithin is that the latter does not contain SM and contains a very limited concentration of PS, whereas PLs fractionated from dairy sources contains about 24–32% SM and 3.2–12% PS (Burling and Graverholt, 2008; Miura et al., 2006; Thompson and Singh, 2006).

Factors Influencing Milk PL Composition

Effects of Natural and Environmental Factors

It is known that variation of milk composition, structure, and properties occurs according to the variability of natural factors such as breed and individuals (Graves et al., 2007), stage of lactation, feed, seasons (Puente et al., 1996), illness of the cow (e.g., mastitis) (Erwin and Randolph, 1975; Smolenski et al., 2007), and milking frequency (Wiking et al., 2006). Because of this, the content and composition of PLs are also expected to vary with those factors. This has been reviewed by several authors (Evers, 2004; Huppertz et al., 2006; McPherson and Kitchen, 1983; Singh, 2006). The discussion here is limited to the changes of MFGM according to three important factors: species, stage of lactation, and feed/diet.

Species

Milk from mammals is characterized by varying total fat content, PL content, and PL composition. Using P-31 NMR, a modern and sensitive analysis technique, Garcia et al. (2013) reported average total PL contents of cow, camel, horse, and human milk to be 204.0, 393.4, 77.8, and 250.3 µg/ml, respectively. This comparison is reliable because it came from the same analytical method carried out at the same period of time with a number of samples from each kind of milk. Horse milk had the lowest total PL content and was also the least varying among the analyzed samples. When considering percentage of PLs on total fat, it was interesting that horse milk had the highest value (Garcia et al., 2013). This is because horse milk has a low total lipid concentration, namely 0.66–1.23% (w/w) (Cagalj et al., 2014; Garcia et al., 2013) and relatively smaller FGs (Welsch et al., 1988). At the same volume, the surface of FGs increases with decreasing globule sizes. It should be noted that, when the fat globule sizes decrease, the total surface per unit fat volume will increase and the MFGM material, including PLs, compared to total fat will increase. PL content and composition of milk from cow (Fong et al., 2007; Garcia et al., 2013; Rodriguez-Alcala and Fontecha, 2010; Trenerry et al., 2013), buffalo (Menard et al., 2010), human (Benoit et al., 2010; Garcia et al., 2013; Giuffrida et al., 2013; Lopez and Menard, 2011; Zou et al., 2012), horse (Garcia et al., 2013), ewe (Rodriguez-Alcala and Fontecha, 2010; Zancada et al., 2013), camel (Garcia et al., 2013), and goat (Rodriguez-Alcala and Fontecha, 2010; Zancada et al., 2013) have been analyzed. Buffalo milk is the second most produced milk in the world. Buffalo milk had a higher fat content compared to cow's milk (7.3% vs. 4.1%) (Menard et al., 2010). Because FGs of buffalo milk were larger than those of cow's milk, the former contained significantly lower percentage of PLs expressed per gram of lipids (0.26% vs. 0.36%) but significantly higher percentage of PLs per liter of milk (as much as 26%) (Menard et al., 2010).

Comparisons on content and composition of PLs in milk from different mammals can be seen in in Lopez (2011) and Contarini and Povolo (2013). Generally, it is observed that human milk contains a higher relative concentration of SM in total PLs than milk of the other animals (Garcia et al., 2013; Lopez, 2011). In addition, human milk has about twice the gangliosides as cow's milk and goat's milk (Iwamori et al., 2008; Pan and Izumi, 2000).

Stage of Lactation

Within 2 to 3 days after parturition, milk is called *colostrum,* and during this short period the composition of milk changes markedly during transition into mature milk (Tsioulpas et al., 2007). After that, the lactation can be divided into early, middle, and late (final) stages.

Bitman and Wood (1990) analyzed milk from Holstein cows according to the progress of lactation and found that concentrations of PLs (% of total fat) were 0.72, 1.06, 1.11, and 0.56 at 3, 7, 42, and 180 days, respectively, after parturition. PL composition changed relatively. For example, the order of relative percentages of SM, PC, and PE changed, although these three still remained the major PL species. Changes in the fatty acid profiles of PL species were also noticed (Bittman and Wood, 1990).

The bovine MFG diameter is 4 μm on average in mature milk and increases with lactation time (Boersma et al., 1991; Michalski et al., 2005). In the same milk sample, no significant differences in the PL and sterol profiles were detected between MFGM extracted from small and from large milk fat globules (Fauquant et al., 2005, 2007). However, Mesilati-Stahy and Argov-Argaman (2014) analyzed bovine milk samples throughout the lactation period and found that from day 100 postpartum on, PE concentration was constant in the large MFG, but dropped two-fold in the small MFG. In simulated gastrointestinal tract conditions, it was found that small fat globules are digested better than large fat globules, and the free fatty acids released from digestion of small fat globules could be less atherogenic compared to large fat globules (Garcia et al., 2014). Results from simulated environments such as these need to be confirmed with studies in actual environments and within the context of complex meals and digestion duration (Garcia et al., 2014).

It is generally observed in milk from mammals that concentration of gangliosides is highest in colostrum and decreases when milk matures (Barello et al., 2008; Martin et al., 2001; Martin-Sosa et al., 2003; Nakamura et al., 2003; Rueda et al., 1995). It is also observed that in human milk, GD3 was the most abundant ganglioside in colostrum, whereas in mature milk it was GM3 (Rueda et al., 1995). Gangliosides contain sialic acids and these are believed to be very important for brain and immune system development of the newborn during the first months of life (see the "Health-Beneficial Properties of PLs" section in this chapter).

Colostrum of St. Lucian women contained higher cholesterol, vitamin E, and long-chain polyunsaturated FAs—the components derived from MFGM—compared to the mature milk (Boersma et al., 1991). Analyzing milk samples from 45 Danish mothers, Zou et al. (2012) observed that d_{32} (volume-surface average diameter) of fat globules decreased from 3.51 μm to 3.14 μm and increased again to 3.25 μm when lactation progressed from colustrum (first five days) to transitional milk (days 6–15) and mature milk (after 16 days). Correspondingly, concentrations of PLs expressed per total lipids increased from colostrum to transitional milk and decreased again after that. Michalski et al. (2005) observed the same trend: The mode diameter of fat globules decreased from 8.9 ± 1.0 μm less than 12 hours after parturition to 2.8 ± 0.3 μm at 96 hours postpartum and that the surface area of MFG increased from 1.1 ± 0.3 m^2/g to 5.4 ± 0.7 m^2/g between human colostrum and transitional milk. The diameter increased back to 4 μm on average in mature milk and continued to increase with advancing lactation (Michalski et al., 2005).

Feeding

Milk fat is characterized by a high content of lauric acid (C12:0) and palmitic acid (C16:0). Compared to unsaturated vegetable oils, milk fat is considered to be more hypercholesterolemic. The FA profile does not only affect the nutritional properties but also the technological functionalities of milk fat. Feeding affects the FA composition of milk fat. Because of that, investigation on the influence of modified feeding on FA composition of milk has been carried out quite extensively (Atwal et al., 1991; Chilliard et al., 2001; Denise, 1991; Ric, 1991).

MFGM in milk from cows fed with a diet rich in polyunsaturated FA (supplemented with extruded linseed) resulted in a higher amount of PLs, related to a smaller fat globule size of milk fat globules, an increase of 30% (w/w) of the concentration in sphingomyelin, and a higher content of unsaturated FAs (Lopez et al., 2008). Feeding dairy cows with a high corn diet or an infusion of soy oil increased the ratio of polyunsaturated:saturated FAs in MFGM, and the ratio of C18:2/C18:1 increased from 0.31 to 1.0 after infusing soy oil for 4 days (Palmquist and Schanbacher, 1991). The MFGM contained a higher concentration of unsaturated FAs (C18:1, C18:2, and C18:3) and very-long-chain FAs (C22:0, C23:0, C24:0, EPA, DHA) compared with total lipids extracted from milk (Lopez et al., 2008). All these results point out that MFGM PLs are good sources of essential FAs, especially when cows are fed with a diet high in unsaturated fat.

Dietary polyunsaturated fat supplements (linseed oil plus algae and sunflower oil plus algae) to German Holstein cows increased the proportion of phospholipids and decreased triglycerides in milk fat compared to rumen-stable fractionated palm fat (Angulo et al., 2013). These researchers also noticed that long-chain polyunsaturated fatty acids were preferentially deposited into phospholipids and the diet effect was more pronounced in triglycerides (increased the proportion of unsaturated fatty acids and decreased the proportion of saturated fatty acids) than in phospholipids (Angulo et al., 2013).

In a recent study on cow milk raised in French Brittany, it was found that the milks collected in spring (fresh pasture–based diet) contained a lower amount of total lipids: 39.7 ± 0.8 g/kg versus 41.7 ± 0.5 g/kg in winter (corn silage–based diet); a higher amount of polar lipids: 138 ± 11 mg/kg versus 112 ± 8 mg/kg milk equivalent to 3.5 ± 0.3 mg/kg versus 2.7 ± 0.4 mg/g total fat, which was related to a smaller size of fat globules; and a higher amount of sphingomyelin, 32 mg/kg milk versus 25 mg/kg milk in winter (Lopez et al., 2014). In addition, the polar lipids from the MFGM of spring milk contained a higher proportion of unsaturated fatty acids (Lopez et al., 2014). The increase in proportion of unsaturated fatty acids in the summer does not result in higher oxidative stability of milk fat because it was observed that milk fat from the period when cows were fed pasture had a higher ratio of fat-soluble vitamins to the content of unsaturated fatty acids (Rafalowski et al., 2014).

Effects of Processing Treatments

Several changes to milk occur after milk leaves the udder during milking and subsequent storage (McPherson and Kitchen, 1983). Physical changes can occur due to the inclusion of air during milking and agitation. Air bubbles cause damage of fat globule membranes (van Boekel and Walstra, 1989). Chemical changes can occur, such as oxidation caused by the presence of oxygen and light. Biochemical changes may occur under the activity of endogenous enzymes: lipases cause lipolysis; proteases cause proteolysis; and phosphatases cause hydrolysis of phosphoric ester acids. Microbial changes occur related to fermentation and hydrolysis caused by microbial growth (Walstra et al., 2006). In this section, effects of three important processing treatments—cooling, heat treatment, and homogenization—on MFGM will be discussed. The changes due to processing treatments are not necessarily the disappearance or destruction of some components, but rather that the MFGM components move from one phase to another (like from the fat globules to the serum phase).

Cooling

After milking, milk is normally brought to below 4 °C as soon as possible to prevent microbial and biological changes. Cooling of the milk slightly decreases the concentration of PLs in MFGM, that is, there is a release of PLs from the membrane to the serum phase (Baumruck and Keenan, 1973; Patton et al., 1980). Cooling of milk normally goes along with pumping and agitation, and these conditions may cause changes in composition and stability of MFGM. When part of the FG surface is damaged, the FGs will coalesce with each other, which results in loss of MFGM material to the serum and increases the analyzed diameter of FGs (Le et al., 2009). Effects of milk cooling on MFGM composition and structure have been reviewed (Evers, 2004; McPherson and Kitchen, 1983).

Heat Treatment

In dairy processing, heat treatment is employed as an operational unit to inactivate microorganisms and enzymes and hence to increase the stability of milk products. Thermal processes (e.g., drying, evaporation) are also applied to render products with certain moisture contents.

Several studies have been performed to investigate the changes in protein composition of MFGM as a result of heat treatment. Heating whole milk at temperatures above 60–65 °C caused adsorption of β-lactoglobulin and a lesser extent of α-lactalbumin to the fat globule surface (Corredig and Dalgleish, 1996; Dalgleish and Banks, 1991; Houlihan et al., 1992; Kim and Jimenez-Flores, 1995; Lee and Sherbon, 2002; Ye et al., 2004).

The involvement of other binding mechanisms in the interaction between whey proteins and MFGM cannot be ruled out. It is possible that membrane lipids could also be involved (Houlihan et al., 1992). β-Lactoglobulin can bind to both neutral lipids and PLs (Brown, 1984). Spector and Fletcher (1970) reported that β-lactoglobulin can bind tightly to long-chain FAs, which are present in substantial amounts in MFGM PLs (Kitchen, 1977). It is then logical to reason that MFGM lipids could complex with β-lactoglobulin, especially when the hydrophobic regions of the lipids are accessible, which could occur as membrane proteins are lost on heating (Houlihan et al., 1992). Using capillary electrophoresis, some whey proteins such as bovine serum albumin and β-lactoglobulin were shown to interact with PS and PC when they are incubated together (Bo and Pawliszyn, 2006; Hu et al., 2001).

Houlihan et al. (1992) reported that heating at 80 °C for 20 minutes caused significant losses of triacylglycerols but not PLs. Koops and Tarassuk (1959) heated milk for 15 minutes at 80 °C and for 15 seconds at 90 °C, followed by cooling, and found losses of 20% and 14% of PLs, respectively, to the serum phase. Results from Greenbank and Pallansch (1961) showed that the loss of phospholipids was both temperature and time dependent. Lee and Sherbon (2002) reported a loss of 25% of total lipids from the MFGM after heating the milk at 80 °C for 18 minutes. There should be a systematic investigation on this aspect in which the effect of heating is not confounded with the effects of pumping/shearing/agitation and inclusion of air. As noted by van Boekel and Walstra (1989), the effect of temperature depends on the presence or absence of air. Bandyopadhyay and Ganguli (1975) reported that cooling, heating, or pressurizing causes reduction of sialic acid content in the MFGM of buffalo milk to levels of 46–80%, depending on the treatment. This indicates losses of glycoproteins and/or glycosphingolipids from these treatments.

Homogenization

Homogenization causes disruption of fat globules into smaller ones. This process is applied to milk to counteract creaming (floating of fat globules to the top), to prevent partial coalescence of fat globules, and to render desirable rheological properties for processed products (Walstra et al., 2006). Because of the reduction in globule size, the surface area of the fat globules increases 5–10 times. Original MFGM materials become insufficient and because of that, casein micelles are adsorbed to the globule surface (Cano-Ruiz and Richter, 1997; Iametti et al., 1997; Lee and Sherbon, 2002).

Based on sodium dodecyl sulfate polyacrylamide gel electrophoresis (SDS-PAGE) and densitometric analysis, Cano-Ruiz and Richter (1997) observed that in heated and homogenized milk, caseins represented 70% of total newly formed MFGM proteins whereas native MFGM proteins represented 10% and the rest were whey proteins. Keenan et al. (1983) also found that about 10% of the surface of fat globules in homogenized milk was covered by natural MFGM. This indicates that the ratio of PLs in the newly formed MFGM after homogenization is reduced extensively. There has been no report on whether homogenization will release part of the PLs from original MFGM to the serum phase or the other way around.

Distribution of PLs during Dairy Processing

This section explains how PLs in milk are distributed to streams/fractions during milk and dairy processing. It provides information on PL content and its composition in milk and dairy products and helps evaluate which streams/fractions contain high concentrations of the material and/or which are possible sources for isolation of PL materials.

Following secretion, a fraction of the membrane surrounding the MFGs may be shed into the skim milk phase (Singh, 2006). In bovine milk, about 55–70% of the phospholipids are associated with fat globules and the rest are located in the skim milk (Gallier et al., 2010a; Mather and Keenan, 1998; Patton and Keenan, 1975). As mentioned, the presence of PLs in the skim milk phase is not from shed MFGM fragments alone but also from other sources, including secretory cell fragments (microvilli, cytoplasm, membrane particles) and leucocytes (Anderson et al., 1975; Plantz and Patton, 1973; Plantz et al., 1973). The compositions of PLs from skim milk membrane, MFGM, and the plasma membrane of the mammary secretory cell appear to be similar (Plantz and Patton, 1973). The distribution of PLs can be used to estimate the distribution of MFGM because it is assumed that most of the PLs are associated with the membrane, either on the MFG surface or in the skim milk fraction (Mather and Keenan, 1998).

Figure 5.4 summarizes the processes illustrating how cream, butter, and butter oil (or anhydrous milk fat [AMF]) and other related products are produced in industry. The decreaming step uses a cream separator (centrifuge) to separate bulk raw milk into cream and skim milk, and the MFGM that associates with fat globules is fractionated into cream. Heat treatment of cream causes certain changes in composition and properties of the fat globule surface.

During churning of the cream at reduced temperature and under the impact of agitation and inclusion of air, fat globules get disrupted and they are partially coalesced together to form granules of MFGs (butter granules) (Walstra et al., 2006). Part of the MFGM is broken into fragments that are fractionated to the aqueous

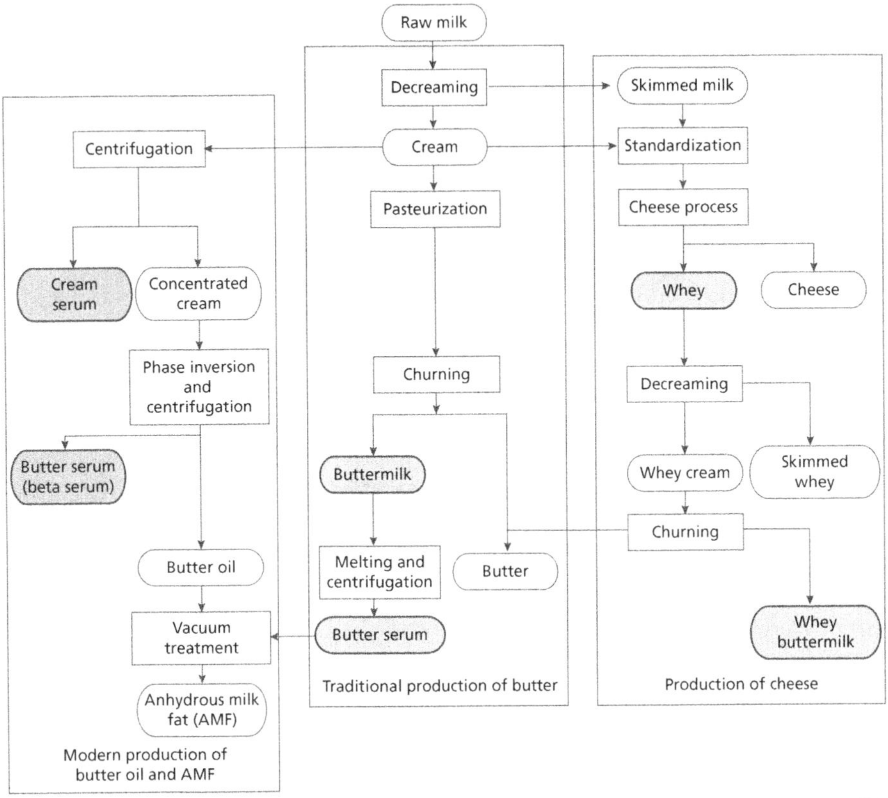

Figure 5.4 Outline of the processes for production of cream, butter, and butter oil. The shaded boxes indicate potential sources for isolation of PLs (adapted from Rombaut, 2006, and Vanderghem et al., 2010).

phase, the buttermilk (Keenan and Mather, 2006; Ward et al., 2006), and the rest are trapped in the butter. Buttermilk has a composition close to that of skim milk except that it contains more PLs (Gassi et al., 2008). Butter can be melted and centrifuged into butter oil and butter serum, an aqueous phase that recovers virtually all MFGM present in the previous butter. In buttermilk and butter serum, MFGM is present as fragments (Figure 5.5) or devoid of FGs (Le et al., 2011b).

The decreaming process can be applied to whey, derived from cheese production, to collect whey cream (Figure 5.4). Similar to cream from milk, the whey cream can be churned into butter and whey buttermilk. In a more modern routine of butter oil and AMF production, cream is first concentrated and then a phase inversion process (heating in combination with homogenization) is applied, which is followed

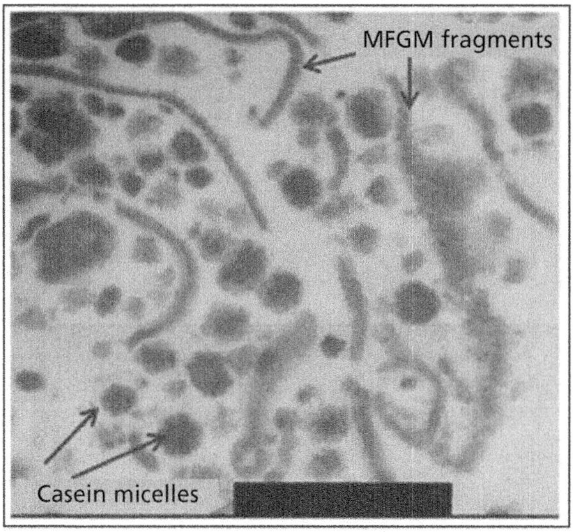

Figure 5.5 Electron microscopy of buttermilk, sedimented by centrifugation and fixed with glutaraldehyde and post-fixed with OsO4, showing MFGM fragments (from Ward et al., 2006). Bar = 0.7 μm.

by centrifugation to separate the suspension into butter oil and butter serum (which is also called beta serum or cream residue) (Figure 5.4).

In such a production line (Figure 5.4) it is seen that during dairy processing, MFGM is ruptured and distributed selectively into aqueous/serum phases such as buttermilk, butter serum, and whey buttermilk. Consequently, these byproducts are good sources for isolation of PLs. These byproducts do not contain a high content of lipids but have a high ratio of PLs:total lipids.

Table 5.B is an illustration of how PLs are distributed during a pilot dairy process. Butter serum contained about 11.5% total PLs in dry basis (Rombaut and Dewettinck, 2006) and it is the richest source of MFGM material among the mentioned byproducts. In a pilot dairy process, Rombaut et al. (2006b) found that butter serum represented 28.4% of milk PLs and only 0.9% of the initial milk mass. Govindasamy-Lucey et al. (2006) observed sweet cream buttermilk (buttermilk obtained from churning sweet cream, which is cream without lactic acid fermentation) over a year at a dairy factory and found that the total PLs of the product varied from 0.113% to 0.153%. The prominent difference in composition between buttermilk and whey buttermilk is that the later does not contain casein micelles because they have been coagulated and included into the cheese curd (Morin et al., 2006; Rombaut et al., 2007a).

Skim milk contains 40–45% PLs from the raw milk, but skim milk contains a low concentration of PLs (Table 5.B) because it contains a substantial amount of wa-

Table 5.B Polar Lipid Content of Various Dairy Products during Processing

Product	On Product (g/100g)	On Dry Matter (g/100g)
Raw milk	0.03–0.04	0.23–0.32
Skim milk	0.02	0.28
Cream	0.19	0.40
Pasteurized cream	0.14	0.31
Butter	0.14–0.23	0.17–0.26
Buttermilk	0.16	2.03
Butter serum	1.25	11.54
Fresh acid buttermilk quarg	0.31	1.86
Acid buttermilk whey	0.10	1.84
Cheddar cheese	0.15	0.25
Cheddar cheese whey	0.02	0.26

Note: The whey products contain a low concentration of PLs on dry matter basis because they contain a high amount of lactose and minerals. Acid buttermilk whey and acid buttermilk quarg are supernatant and coagulate, respectively, obtained from acidification of sweet cream buttermilk.

Source: From Rombaut and Dewettinck (2006) and Rombaut et al. (2007b).

ter and other dry matter materials of milk. Because of that, skim milk is not a potential source of PLs if it is not processed further into other products. For example, when skim milk is processed into cheese, a byproduct called whey is generated (as shown in Figure 5.4) which is a potential source for isolation of PLs. AMF is derived from the core fat of MFGs, so it contains virtually no PLs.

Potential sources for isolation of PLs are the ones that should contain a substantial amount of PLs over other dry matter components and preferably be of low economic value. In many cases, the proportion of phospholipids to total fat is higher in products with little fat (Park and Drake, 2014).

Isolation and Purification of PLs

Isolation of PLs from Raw Milk

Most of the research publications on isolation or extraction of PLs from raw milk are limited to the research purpose. Milk contains a low concentration of PLs and, therefore, it is not industrially realistic to extract the components directly from milk for commercial purposes. Extraction steps often produce waste and/or byproduct streams.

For the purpose of extraction for analysis, a mixture of hydrophobic and hydrophilic solvents is suitable to use. A hydrophobic solvent can be hexane or chloroform, and suitable hydrophilic solvents can be propanol or methanol (Bligh and

Dyer, 1959; Le et al., 2011a; Rombaut et al., 2005). While triglycerides are neutral lipids, which only need a hydrophobic solvent to extract, PLs are amphiphilic, thus a mixture of hydrophobic and hydrophilic solvents is more suitable to extract the components effectively from the milk matrix. When a phase separation is formed, PLs are dissolved to the hydrophobic phase. With some dairy products that have low mineral concentrations and are rich in protein (like microfiltered or ultrafiltered products), addition of $CaCl_2$ increases the ionic strength and facilitates the phase separation to extract PLs (Le et al., 2011a). The extract obtained contains both neutral lipids (triglycerides) and targeted PLs. Techniques to purify PLs are discussed in the "Purification and Fractionation of PLs," section of this chapter.

As previously discussed, MFGM is the first source of PLs in milk. A number of studies have been carried out to isolate MFGM from milk in small quantities for the purpose of research. Milk contains fat globules as its biggest natural particles. The remaining parts are casein micelles and milk serum, which consists of soluble whey proteins, nonprotein nitrogenous compounds, lactose, minerals, and other minor compounds. A requirement of MFGM isolation is to obtain a MFGM material with the least contamination of non-MFGM components. Depending on the investigation purposes, the procedures for isolation of MFGM material from milk may vary to some extent. An isolation procedure can be divided into four stages (Mather, 2000; Singh, 2006): fat separation, cream washing, release of MFGM from the FGs, and collection of the MFGM material. The isolation procedures in literature are summarized in Figure 5.6 and are explained in detail in a separate article (Le et al., 2014b). So far these isolation techniques have only been applied in laboratories (not on an industrial scale) for research purposes. The isolated MFGM material contained mainly PLs, MFGM-specific proteins, and triglycerides (Dewettinck et al., 2008). The triglyceride fraction is believed to be contaminants from the core fat during isolating steps. In most of the research publications, MFGM proteins were selected as a target to optimize or evaluate the isolation methods. Many proteins are loosely attached to the membrane and they are more subjected to loss during isolation (Anderson et al., 1974) (Figure 5.2). However, loss of PLs is also expected, especially when there are damaged fat globules and globules surface. A sign of damage of the MFG surface is the coalescence of FGs leading to increasing FG size, and sometimes small parts of butter granules are visible on the surface (Fong et al., 2007; Le et al., 2009; Walstra, 1985).

Production of PL Concentrate from Dairy Industrial Byproducts

Good sources of MFGM material, in general, or PLs specifically, such as buttermilk, butter serum, and whey, are still considered to be low value byproducts originating from dairy processing. Therefore, it is advantageous to use these sources for isolation of MFGM and purification of PLs. From that perspective, there have already

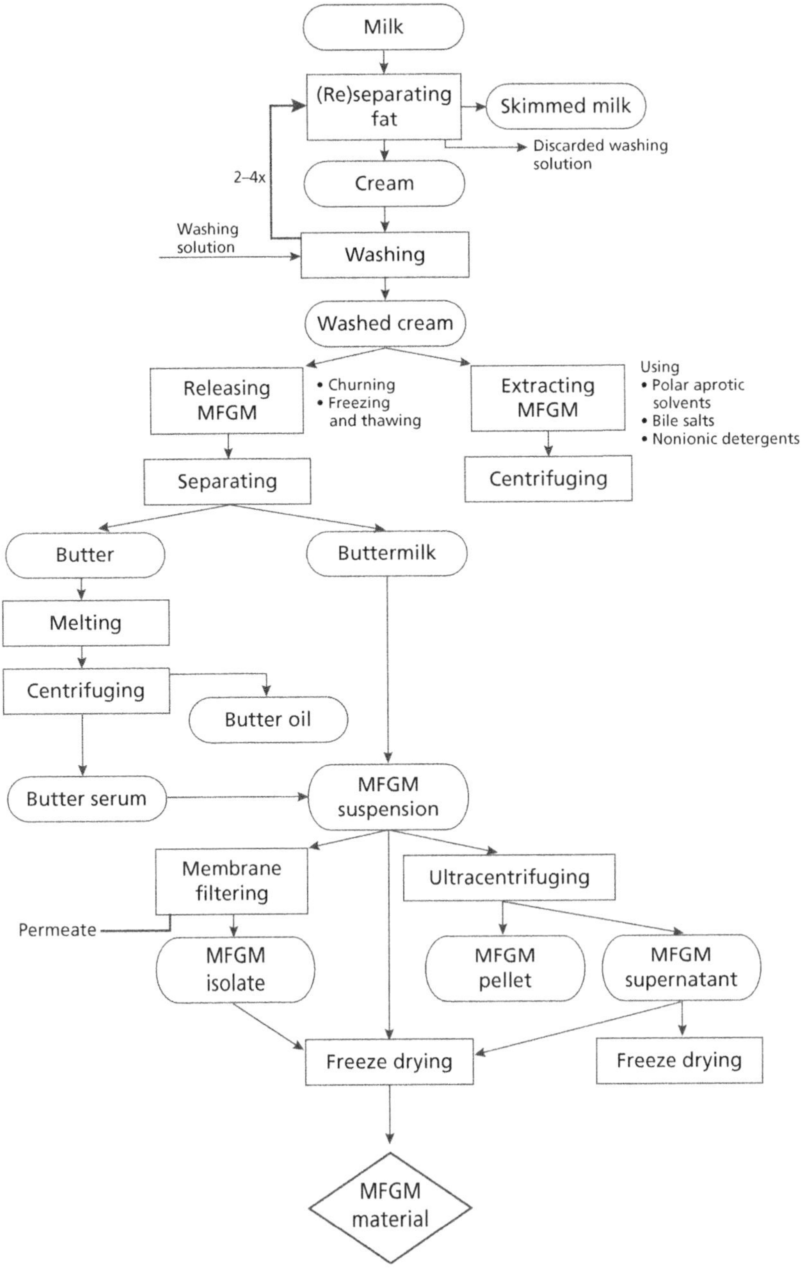

Figure 5.6 Summary of isolation methods to extract MFGM (PL enriched) materials from raw milk (reprinted from Le, 2012, with permission.)

been several techniques developed to isolate MFGM material from these byproducts. These are summarized in Figure 5.7 and explained in the following subsections.

A good production process must offer a high yield and high purity of the final MFGM material. In order to achieve that goal, there should be a selective removal of non-MFGM components such as casein, whey proteins, nonprotein nitrogen compounds, lactose, and minerals.

Application of Membrane Separation

The first technique is to use cross-flow membrane filtration. In this technique, a solution is circulated tangentially above a filtering membrane and pressure is applied perpendicular to the membrane. From this, a solution will be separated into two fractions: the retentate, the fraction that is retained by the filtering membrane, and the permeate or filtrate, the fraction that passes through the membrane. MFGM fragments are the target retentate. The two most important parameters for evaluation of performance of a membrane separation application are membrane selectivity (selective permeation) and permeate flux (permeate velocity of a membrane surface unit). These are, in turn, affected by properties of the feed, membrane characteristics (material, pore size, structure, configuration, etc.), and filtration conditions such as cross-flow (feed) velocity, transmembrane pressure, temperature, and pH (Dewettinck and Le, 2011). If buttermilk and butter serum are used as the source for microfiltration, the similarity in size of casein micelles and MFGM fragments is an issue that must be dealt with (Surel and Famelart, 1995).

Microfiltration of Buttermilk and Butter Serum

In a first approach, dissociation agents can be added to disintegrate casein micelles into molecules that then can permeate the filtering membrane. To do that, citrate (Udabage et al., 2000), alcohol (O'Connell et al., 2001; Rombaut et al., 2006a; Zadow, 1993), or Ethylenediaminetetraacetic acid (EDTA) (Udabage et al., 2000) can be used, among which the citrate is most often used in microfiltration for concentration of MFGM material (fragments). With this approach, microfiltration membranes with a pore size range of 0.1–0.2 μm are often used (Figure 5.7) (Corredig et al., 2003; Rombaut et al., 2006a).

The microfiltration is normally performed in combination with diafiltration (DF), which can be performed in either batch mode, adding a solvent to the retentate and reconcentrate it again, or in continuous mode, in which the rate of solvent addition is equal to the permeate rate. DF facilitates the separation of components and then increases the purity of the retentate (Corredig et al., 2003; Morin et al., 2006). Increasing the number of diafiltration steps with deionized water from two to six reduced the casein contamination in the retentate during microfiltration of buttermilk after dissociating casein micelles from 30 to 6% of total protein (Corre-

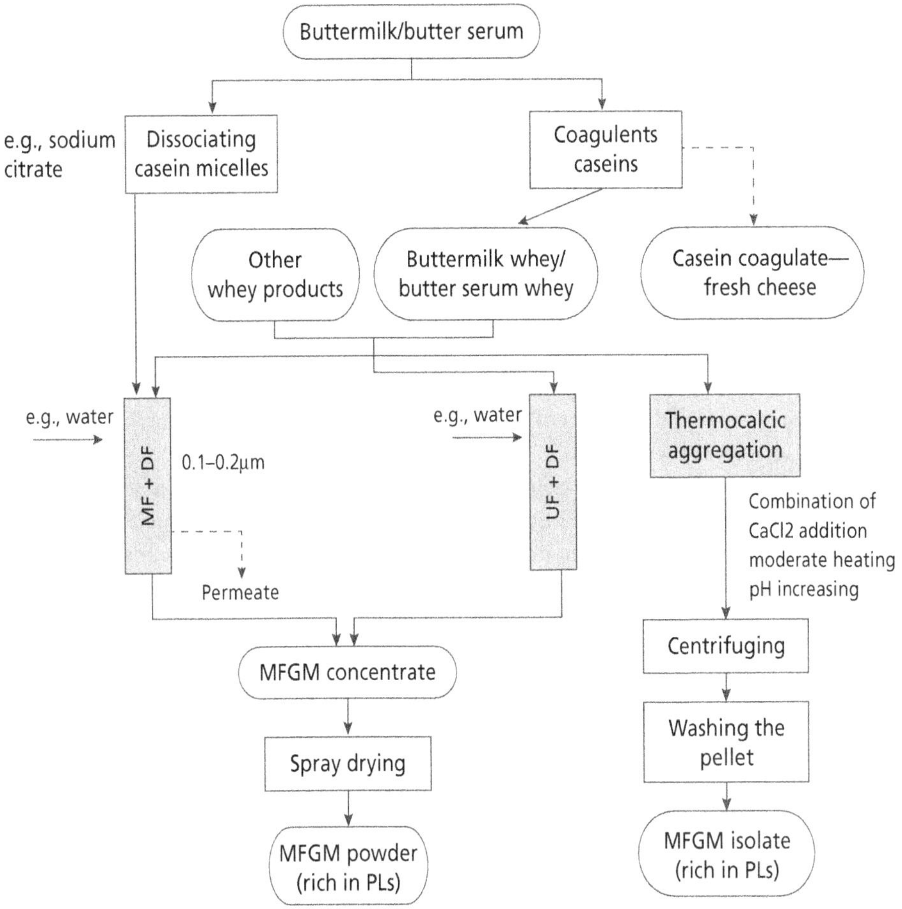

Figure 5.7 Various techniques and possible pathways for isolation of MFGM material, a PL concentrate, from dairy industrial byproducts (reprinted from Le, 2012, with permission). Highlighted boxes indicate key techniques behind the isolation. DF: diafiltration; MF: microfiltration; UF: ultrafiltration; LAB: lactic acid bacteria.

dig et al., 2003). However, it is logical that increasing DF steps are accompanied by increased loss of MFGM materials and of course PLs (permeated through the membrane) (Rombaut et al., 2007a; Sachdeva and Buchheim, 1997). Corredig et al. (2003) obtained 15–18% (w/w) MFGM material from the buttermilk using cross-flow microfiltration with hydrophilic polyethersulfone membrane of 0.1 μm pore size after four diafiltration steps. The isolated materials contained 35% total lipids.

The authors did not quantitatively analyze the PL content, but normally one-third of total lipids of MFGM fragments obtained by microfiltration of buttermilk was PLs (Le et al., 2010). The low yield of MFGM materials was obtained because the buttermilk contained high concentrations of lactose and minerals, which permeated the membrane during filtering. Le et al. (2010) employed a hydrophilized polyvinylidene difluoride (PVDF) of pore size 0.22 μm to concentrate MFGM fragments from reconstituted buttermilk and butter serum after dissociating casein micelles using citrate and with three steps of diafiltration. Total solid yields with the two starting materials were 19.96% and 24.58%, respectively. Total PLs increased from 3.36% to 8.43% and from 9.33% to 23.65% on dry basis after microfiltration of the buttermilk and butter serum, respectively. The PL recovery yields of microfiltration of buttermilk and butter serum were reported to be 50% and 62%, respectively. This suggests that a considerable part of the PLs was lost to the permeate. Further work can be carried out to reduce this loss.

In the second approach, casein micelles from buttermilk or butter serum can be coagulated and removed before the resulting whey is introduced to microfiltration to separate away whey proteins, nonprotein nitrogenous compounds, lactose, minerals, and water (Sachdeva and Buchheim, 1997). There are possible options to coagulate casein micelles, such as with addition of an acidulant, reduction in pH by means of fermentation with lactic acid (LAB), or rennet treatment. The best case is when only casein micelles are coagulated and the highest amount of PLs or actually MFGM material is still dispersed in the whey. However, MFGM material/fragments are always involved in the coagulation to a certain extent depending on the coagulation methods used and the history processing of the sourcing materials. Sachdeva and Buchheim (1997) obtained a recovery of 53%, 79%, and 83% of PLs from a reconstituted buttermilk in the resulting whey by coagulation of caseins with lactic acid bacteria, citric acid, and rennet treatment, respectively. In another study, Le et al. (2010) obtained a recovery of 32% PLs in the resulting whey from coagulation of reconstituted buttermilk with citric acid. This means that quite high amounts of MFGM materials coprecipitated with caseins upon lowering pH. In the second method of separation of caseins, it seems very possible to use the coagulated caseins for production of cheese or other caseinate ingredients.

In both cases of dealing with casein micelles, the obtained MFGM material is always contaminated with a considerable amount of caseins and whey proteins. This could be due to the selectivity of membrane separation, but mostly it is because these caseins already attached to the MFGM fragments before the isolation process during the history production of buttermilk or butter serum. For example, heat treatment of milk creates binding of skim milk proteins (caseins and whey proteins) to the MFG surface (see the "Effects of Processing Treatments" section). Morin et al. (2007) investigated another approach in which cream separated from whole milk was washed first with ultrafiltrate (UF) permeate of skim milk before the washed cream was churned into butter and buttermilk. After that, this buttermilk was subjected to microfiltration with a 0.5 μm ceramic membrane to concentrate the MFGM fragments. Compared to the microfiltration of regular buttermilk, the microfiltration of the washed cream buttermilk had a two-fold higher permeate flux. Although there was loss of PLs to the skim milk phase during the washing and reseparation of the cream, the final recovery yield of PLs after microfiltration of the washed cream buttermilk was higher and the obtained MFGM materials had lower contamination of caseins and whey proteins (Morin et al., 2007). This approach was also applied by Britten et al. (2008) where the reduction of total protein and increase in PLs in the obtained buttermilk from the washed cream were confirmed. Washing of cream (resuspension and reseparation) will lose some PLs from the former (Morin et al., 2007) and create a significant amount of waste water. This approach also requires considerable modification of the processing instruments before it can be integrated into the system.

Membrane Filtration of Whey

Whey products are considered more favorable than buttermilk and butter serum for concentration of MFGM materials because whey products supposedly contain no casein micelles. For example, whey buttermilk (Morin et al., 2006) (see Figure 5.7) and acid buttermilk cheese whey, the whey obtained from buttermilk after acidic coagulation and removal of casein micelles (Rombaut et al., 2007a), have been studied as sources for isolation of MFGM material. Morin et al. (2006) experimented with a ceramic tubular membrane of 0.45 μm and reported that the transmission/permeation of MFGM proteins through the membrane was lower when microfiltration of whey buttermilk compared to regular buttermilk was used as the feed. Upon the microfiltration with two-fold continuous DF (volume of added water was two times of initial volume of the feed), the losses of PLs were 9.9% and 39.1% for whey buttermilk and regular buttermilk, respectively (Morin et al., 2006). Le et al. (2010) found a recovery of up to 95% PLs in the retentate after microfiltration in combination with three diafiltration steps of a buttermilk whey, which was obtained by casein coagulation of the reconstituted buttermilk. Rombaut et al. (2007a) used a flat-sheet 0.15 μm cellulose acetate membrane and carried out the filtration of acid buttermilk cheese whey at 50 °C and pH 7.5 and found that the retention of PLs in the retentate decreased from 83.8% to 45.5% and 44.6% after 1, 2, and 3 diafiltration steps. Diafiltration, on the positive side, increases the purity of the retentate, but on the other side increases the permeate (loss) of PLs.

Membrane separations are widely applied in food processing. Concentration polarization (building up of high weight molecular components above the filtering membrane) and membrane fouling (deposition of material on and in the filtering membrane) are the two most important phenomena affecting the performance of a membrane separation process. The cause and mechanisms of these two phenomena, as well as techniques to improve membrane separation performance for applications in foods, are extensively discussed (Dewettinck and Le, 2011). As in membrane separation of other foods, membrane structure, materials and pore sizes, temperature, pH, and feed properties were found to influence the separation performance in cross-flow filtration of dairy byproducts for concentration of MFGM materials (Astaire et al., 2003; Corredig et al., 2003; Morin et al., 2004, 2006; Rombaut et al., 2007a).

Figure 5.7 also shows that, besides microfiltration, ultrafiltration can also be applied to concentrate MFGM material from whey products or buttermilk and butter serum after dissociating casein micelles. Ultrafiltration membranes are characterized with smaller pore sizes (< 0.1 μm), which are normally denoted as the molecular weight cut-off (MWCO) with the unit being kDa. So it should be noted that the use of ultrafiltration will result in the retention of other lower molecular weight components such as proteins, depending on the MWCO values, besides MFGM fragments. The use of ultrafiltration then becomes important if the obtained concentrate, which is rich in MFGM material and proteins (e.g., caseins and whey proteins) and low in lactose and minerals, would find its applications in food processing.

Aggregation of MFGM Fragments from Whey

Whey, in dairy processing, is the liquid byproduct from the manufacture of cheese and caseins. For example, during the manufacturing of cheese from milk (Figure 5.4), about 20% of total protein (e.g., whey proteins of milk) and 10% of total fat are distributed to the whey (Fox et al., 2000). Nowadays, whey is normally used to produce whey protein concentrates and isolates using ultrafiltration. The residue of fat in the whey, present in the form of small fat globules, lipoproteins, and MFGM fragments, reduces the performance of the ultrafiltration and also influences the purity, technical functionalities, and storage stability of the final products, so it is normally removed from the whey before it is further processed. The residue fat is normally aggregated and then removed either by sedimentation, centrifugation, or cross-flow microfiltration (Fox et al., 2000). A number of techniques have been developed for this so-called whey clarification or delipidation step. Among these, thermocalcic aggregation, the combination of the addition of a divalent cation (e.g., Ca^{2+}) with a pH adjustment to 7–7.5 and a moderate heat treatment of whey (Attebery, 1968; Fauquant et al., 1985; Maubois et al., 1987), seems to be the most often applied method. Nowadays, with PLs becoming a potential ingredient in food, pharmaceutical, and cosmetic processing, the residue fat in whey becomes a source to extract PLs. Rombaut and Dewettinck (2007) applied thermocalcic treatment to an acid buttermilk whey, and the experiments were designed to obtain the highest amount of PLs and the lowest amount of proteins in the aggregated sludge. The sludge was recovered by centrifugation. These authors suggested that washing the pellet with deionized water increases the purity of the MFGM material. As such, an isolate with 10.7% PLs on dry basis was obtained with a recovery of 76% PLs present in the starting whey. It is interesting to note that the obtained isolate contained a very high ash content, about 35% on dry basis (Rombaut and Dewettinck, 2007). This may affect the application possibilities of the material. Washing the pellet several times can reduce some minerals, but that will disperse the MFGM fragments into the washing liquid and cause loss of PLs. Using a combination of saponin and bile salts at reduced pH (Hwang and Damodaran, 1994) or zinc (Damodaran, 2010) has been recently evaluated for whey clarification, but the possible applications of the fat-rich fraction is questionable.

In conclusion, there are two methodological approaches to obtain MFGM isolates that are enriched in PLs: (1) membrane concentration of MFGM fragments from buttermilk, butter serum, or whey; and (2) aggregation of MFGM fragments from whey. Depending on the material sources and the techniques applied, the recovery yield and composition of the final material is expected to vary. Investment in any approaches requires consideration of other factors such as available instruments, required characteristics of the isolate, treatment of waste water, byproducts, and total cost of production.

Isolation of MFGM materials from raw milk using washing methods is at present for research purposes only, whereas membrane filtration/separation techniques for concentration of MFGM materials from dairy byproducts are industrially applicable. However, for any specific applications, separation conditions/parameters must be optimized further. There is not an optimal membrane filtration process for different feeds. For a successful/economical industrial application of membrane separation for concentrating MFGM material, whether or not it is possible to recover/utilize valuable components in the resulting permeate should also be considered. A possible direction is to use ultrafiltration to recover proteins in the microfiltration permeate and use nanofiltration to recover lactose in the ultrafiltration permeate (Dewettinck and Le, 2011; Le et al., 2014a; Pouliot, 2008). It can be concluded that it is technically possible to isolate MFGM from dairy byproducts on an industrial scale. Once MFGM isolate finds its applications, suitable isolation techniques to obtain the MFGM with desired functionalities for the applications can be optimized.

Purification and Fractionation of PLs

The MFGM material obtained using isolation techniques previously explained consists of two main fractions: lipids (neutral lipids and polar lipids) and proteins, including MFGM-specific proteins and contaminating proteins from skim milk. Several techniques have been developed to further purify or separate MFGM material into different fractions or components.

On a lab scale, the lipid moiety can be separated from the protein moiety using a mixture of chloroform and methanol (2:1). If a powder MFGM material is used, the protein will be precipitated while the lipids are dissolved in the solvent system.

Purified lipids can be obtained after removal of the solvents. The combination of solvents is a popular extraction procedure and based on extraction experiments of Folch et al. (1957) and Bligh and Dyer (1959). If a liquid MFGM isolate is used, the solvent mixture will cause a phase separation and the proteins will be dissolved in the aqueous phase (Le et al., 2011a). Chloroform and methanol, being toxic solvents, should not be used to extract PLs for human consumption. A combination of hexane and isopropanol can be considered for the extraction (Hara and Radin, 1978). These solvents are less toxic than chloroform and methanol. However, there is still a need to determine if the combination of hexane and isopropanol can separate lipids and proteins efficiently. The combination of hexane and isopropanol does not seem to dissolve gangliosides very well (Hara and Radin, 1978).

Acetone insolubility, a standard AOCS method (Ja 4–46, AOCS) for evaluation of soybean lecithin quality, can be used to separate milk PLs out of total lipids. Chilled acetone is used. Glycerides, fatty acids, and cholesterol are generally dissolved in acetone, whereas phospholipids and glycolipids are not soluble (Vandana et al., 2001). The yellow precipitated pellet obtained after centrifugation is the PLs fraction of milk. However, there is still a need to determine whether this technique can precipitate all milk PL species, including the minor ones. Boyd et al. (1999) applied alcohol fractionation to a whey powder and obtained an alcohol-insoluble fraction with an increased amount of PE and cerebrosides.

More fractions of lipids can be obtained with preparative solid phase extraction (SPE). Vaghela and Kilara (1995) used an aminopropyl silica column with a series of different solvents and solvent combinations to elute dairy lipids separately into free fatty acids, phospholipids, cholesterol ester, triacylglycerol, cholesterol, diacylglycerol, and monoacylglycerol. A review on SPE or column chromatography for separation of PLs components has been published by Ruiz-Gutierrez and Perez-Camino (2000) and in a more recent review by Rombaut and Dewettinck (2010). To separate species of PLs (e.g., for analysis or making standard components), the liquid chromatography technique can be used (Avalli and Contarini, 2005; Fagan and Wijesundera, 2004; Le et al., 2011a; Rombaut and Dewettinck, 2010; Rombaut et al., 2005; Vila et al., 2003).

Astaire et al. (2003) used microfiltration to obtain a MFGM-enriched material that was transformed into a powder from which neutral lipids were removed selectively by applying supercritical fluid extraction (SFE) with the use of supercritical carbon dioxide. The SFE decreased the concentration of neutral lipids from 21% to 4% and the PLs concentration increased from 9.6% to 19.7% on dry matter basis. They optimized the same approach (microfiltration and then SFE) on other materials, namely whey buttermilk and sweet cream buttermilk, and reported that up to 55% of neutral lipids could be removed while the total proteins did not change considerably (Spence et al., 2009). Using ultrafiltration and SFE on whey buttermilk, Costa et al. (2010) obtained a final powder consisting of 73% protein and 21% lipids, of which 61% were PLs.

Catchpole et al. (2008) used successfully a combination of SFE and near-critical dimethyl ether (DME) antisolvent fractionation process to produce a product of 70% PLs with depletion of neutral lipids, proteins, and other components from a proprietary dairy ingredient containing 19.7% total lipids and 7.9% total PLs. SFE could be applied either before or after DME. The SFE was meant to remove neutral lipids, whereas DME was used to selectively separate PLs from proteins and lactose (Catchpole et al., 2008). The protein fraction could also be recovered from the DME process. The application of DME on a liquid product caused denaturation of proteins, whereas this was not the case if it was applied on a powder. However, lactose of the powder needs to be reduced first to have an efficient extraction of PLs (Catchpole et al., 2008). Parameters influencing the DME process were also investigated (Catchpole et al., 2007, 2008). SFE can be applied only on powdered material (Astaire et al., 2003; Costa et al., 2010). However, it is more environmentally friendly because it does not employ toxic organic solvents.

Health-Beneficial Properties of PLs

Nutraceutical properties of PLs are summarized in Table 5.C. Other reviews on health effects of milk PLs can be found in Contarini and Povolo (2013) and in Kullenberg et al. (2012). Several PL components have been used already as therapeutics due to their medicinal properties.

Milk is a major source of exogenous SM in the human diet (Burling and Graverholt, 2008). Dietary SM was found to contribute to myelination of the central nervous system in developing rats, in which serine palmitoyltransferase, a rate-limiting enzyme for sphingolipid biosynthesis, is inhibited (Oshida et al., 2003). In addition, in experiments on rats, sphingolipids were found to inhibit colon carcinogenesis (Dillehay et al., 1994; Schmelz et al., 1996). This protective effect against colon carcinogenesis of dietary sphingolipids was confirmed by feeding weanling Fischer-344 Rats with MFGM (in a mixture form) (Snow et al., 2010). A possible mechanism of action is that exogenously supplied sphingolipids bypass a sphingolipid signaling defect that is important in cancer (for example, a loss of cellular SM turnover to produce ceramide and sphingosine) (Berra et al., 2002). The role of sphingolipids in relation to colon health has been recently reviewed (Kuchta et al., 2012).

Sphingolipids are also involved in the intestinal uptake of cholesterol. SM was found to be dose-dependent and to lower the intestinal absorption of cholesterol and fats in rats (Eckhardt et al., 2002; Noh and Koo, 2004; Nyberg et al., 2000). Sphingolipids, therefore, lower plasma cholesterol and triglycerides (TG) and protect the liver from fat- and cholesterol-induced steatosis (Duivenvoorden et al., 2006; Watanabe et al., 2011). However, the decreasing effect on plasma concentrations of these components was not evident in humans who were fed with an SM-enriched buttermilk formulation

Table 5.C Possible Health Effects of PL Species

PL Species and Their Nutraceutical Properties	References
Phosphatidylcholine (PC)	
Source of choline, which is an essential nutrient for brain development	Blusztajn (1998)
Improvement of athletic performance and recovery after exercise-induced stresses[a]	Jager et al. (2007b)
Support of recovery of the liver after toxic or chronic viral damage[a]	Niederau et al. (1998)
Expected protection of the gastrointestinal mucosa against the damage caused by anti-inflammatory drugs and chemicals[a]	Ehehalt et al. (2010); Lichtenberger et al. (2009); Stremmel et al. (2010b)
PC and LysoPC: anti-inflammatory effects in ulcerative colitis	Hartmann et al. (2009); Stremmel et al. (2010b); Tokes et al. (2010); Treede et al. (2007)
Sphingomyelin (SM)	
Source of choline, which is an essential nutrient for brain development	Blusztajn (1998)
Contribution to myelination of the central nervous system[b]	Oshida et al. (2003)
Inhibition of colon carcinogenesis/colorectal cancer prevention[b]	Berra et al. (2002); Dillehay et al. (1994); Schmelz et al. (1996); Snow et al. (2010);
Lowering the intestinal absorption of cholesterol and fats[b]	Eckhardt et al. (2002); Noh and Koo (2004); Nyberg et al. (2000)
Protection of liver from fat- and cholesterol-induced steatosis[b]	Duivenvoorden et al. (2006); Watanabe et al. (2011)
Digestion products of SM can protect against gastrointestinal infections[b]	Sprong et al. (2002)
Phosphatidylserine (PS)	
Theoretically, PS has regulatory and structural functions in the brain, lungs, heart, liver, and skeletal muscle	Pepeu et al. (1996); Starks et al. (2008)
Proposed function as an endogenous regulator of immune and (anti-) inflammatory responses	Carr et al. (1992); Gaitonde et al. (2011); Ponzin et al. (1989); Yamazaki et al. (1997)
Reduction of the risk of dementia and cognitive dysfunction and improvement of cognitive performance in the elderly[a]	Cenacchi et al. (1993); Crook et al. (1991); Engel et al. (1992); Hellhammer et al. (2010); Pepeu et al. (1996)

Table 5.C *Continued*

PL Species and Their Nutraceutical Properties	References
Potential for management of attention-deficit hyperactivity disorder in children[a]	Hirayama et al. (2006); Vaisman et al. (2008)
Improvement of athletic performance and recovery after exercise-induced stresses[a]	Jager et al. (2007a, 2007b); Starks et al. (2008)
Phosphatidylethanolamine (PE) and Phosphatidylinositol (PI)	
Source of unsaturated fatty acids	Fong et al. (2007); Lopez et al. (2008)
Lactosylceramide *(LacCer)*	
Pivotal role in the biosynthesis of nearly all the major glycosphingolipids in the body and involvement in regulating cellular function	Chatteriee and Pandey (2008)
Inhibition of/binding to pathogens	Sanchez-Juanes et al. (2009)
Glucosylceramide (GluCer)	
Anti-inflammatory properties[b]	Duan et al. (2011)
Gangliosides	
Important for development of the brain and positive effect for the learning ability of human infants	Carlson (2009); Iwamori et al. (2008); McJarrow et al. (2009); Schnaar et al. (2014); Wang (2009); Wang et al. (2001a)
Maturation process of the intestinal immune system[b]	Clandinin et al. (2005); Gil and Rueda (2002); Park et al. (2006); Vazquez et al. (2001)
Inhibition of *Escherichia coli* and *Vibrio cholerae* enterotoxins	Idota and Kawakami (1995); Otnaess et al. (1983); Rueda et al. (1998)
Prevent the adhesion of *Helicobacter pylori* to gastric epithelial cells[b]	Hata et al. (2004); Wada et al. (2010); Wang et al. (2001b)
Direct toxicity to and prevention of adhesion of parasites[b]	Suh et al. (2004)
Prebiotic effect	Rueda et al. (1998)

[a]Results from experiment with humans (i.e., clinical results).

[b]Results obtained with experiments on rats.

(Ohlsson et al., 2009). In this report, however, the authors noted that the SM-enriched formulation might counteract the increase in concentrations of blood lipids due to increased energy intake, as they observed that the equivalent control formulation (not rich in SM) but not the SM-supplemented formulation caused an increase in plasma concentrations of TG, LDL, and HDL cholesterol, and apolipoprotein B after four weeks (Ohlsson et al., 2009). Digestion products of sphingolipids were shown to have antibacterial activity to various food-borne pathogens and they can protect against food-borne gastroenteritis (Sprong et al., 2001, 2002). Physiological functions and clinical implications of sphingolipids in (and of) the gut have recently been reviewed (Duan, 2011).

PS is mostly concentrated in organs with high metabolic activity, such as the brain, lungs, heart, liver, and skeletal muscles. PS is located mainly in the internal layer of the cell membrane and has a variety of unique regulatory and structural functions, such as modulation of the activity of receptors, ion channels, enzymes, and signaling molecules and involvement in governing membrane fluidity (Pepeu et al., 1996; Starks et al., 2008). PS has been proposed to function as an endogenous regulator of immune and anti-inflammatory responses (Carr et al., 1992; Gaitonde et al., 2011; Ponzin et al., 1989; Yamazaki et al., 1997). Clinical studies have shown that consumption of PS may reduce the risk of dementia and cognitive dysfunction (of the Alzheimer's type) in the elderly (Cenacchi et al., 1993; Crook et al., 1991; Engel et al., 1992; Pepeu et al., 1996). Soy-derived PS is a safe nutritional supplement for older persons if taken in a dosage of up to 200 mg three times daily (Jorissen et al., 2002). Explanation of the pharmacological and clinical effects of PS on the central nervous system (CNS) has been reviewed (Pepeu et al., 1996). Interestingly, use of PS alone or in combination with omega-3 has a potential in the management of attention-deficit hyperactivity disorder (Hirayama et al., 2006; Vaisman et al., 2008; Veereman-Wauters et al., 2012).

Several randomized, double-blind, placebo-controlled studies have demonstrated that PC and PS may improve sport performance and can be an effective supplement for combating exercise-induced stress and preventing the physiological deterioration that can accompany too much exercise (Jager et al., 2007a, 2007b; Starks et al., 2008). Findings related to effects of PS supplementation on exercising humans, as well as the mechanisms of action, have been reviewed by Kingsley (2006).

PC and SM are sources of choline, which is an essential nutrient for humans; it is especially involved in brain development (Blusztajn, 1998). PC is effective in support of recovery of the liver after toxic or chronic viral damage, as shown in a multicenter, randomized, double-blind, placebo-controlled trial (Niederau et al., 1998). PC is an important constituent of the gastrointestinal tract. Oral administration of PC may help to maintain the defensive hydrophobic barrier and then protect the gastrointestinal mucosa against the damage caused by anti-inflammatory drugs and chemicals

(Anand et al., 1999; Bernhard et al., 1995, 1996; Dial et al., 2008; Dunjic et al., 1994; Ehehalt et al., 2010; Fittkau et al., 2002; Ghyczy et al., 2008; Lichtenberger et al., 2009; Stremmel et al., 2010a) and prevent bile salt toxicity to gastrointestinal epithelia and membranes (Dial et al., 2008). PC and LysoPC showed anti-inflammatory effects in ulcerative colitis, a chronic inflammatory disorder of the colon (Hartmann et al., 2009; Stremmel et al., 2010b; Tokes et al., 2010; Treede et al., 2007). Snow et al. (2011) showed that MFGM-enriched milk fat had a protective effect against gastrointestinal leakiness in mice treated with lipopolysaccharide. In double-blind trial conditions, PC demonstrated potentially lifesaving benefits against pharmaceutical and death cap mushroom poisoning, alcoholic liver damage, and the hepatitis B virus (Kidd, 2002a). PC is safe and well tolerated beyond several grams of daily intake and is cost effective for manufacture into functional foods (Kidd, 2002b). Pharmacological effects of PC along with its mechanisms of action, dosage, and side effects can be found in a monograph by Kidd (2002a). A monograph on PS is also available (Anonymous, 2008).

Gangliosides are sialic acid–containing glycosphingolipids. These are key regulatory components in the brain that contribute to proper development, maintenance, and health of the nervous system (Schnaar et al., 2014). In mammals, including humans, the brain contains the highest relative ganglioside content in the body, particularly in neuronal cell membranes concentrated in the area of the synaptic membrane. The human brain contains two- to four-fold concentrations of sialic acid compared to those in the brains of other mammals, including chimpanzees (Wang et al., 1998). Gangliosides are known to be essential nutrients for neuronal growth, migration and maturation, neuritogenesis, synaptogenesis, and myelination (Iwamori et al., 2008; McJarrow et al., 2009; Wang, 2009). Because the liver, where sialic acid can be synthesized from simple sugar precursors, is relatively immature in newborn infants, and due to the rapid growth and development of the brain, dietary sources of sialic acid may play a role in determining the final concentration of sialic acid in the brain and may possibly influence the learning ability of human infants (Wang et al., 2001a). There has been evidence from animal studies that sialic acid supplementation is associated with an increase of gangliosides in the brain and improved learning and memory (Carlson, 2009; Morgan and Winick, 1980; Wang et al., 2007).

Gangliosides are also believed to be essential for development of the gut immune system in infants (Gil and Rueda, 2002). From the results obtained from feeding experiments in mice, it is suggested that dietary gangliosides accelerate the maturation process of the intestinal immune system that takes place during weaning (Clandinin et al., 2005; Park et al., 2006 ; Vazquez et al., 2001). Human-milk gangliosides were found to be involved in the inhibition of *Escherichia coli* and *Vibrio cholerae* enterotoxins (Idota and Kawakami, 1995; Otnaess et al., 1983). Gangliosides were also found to prevent the adhesion of *Helicobacter pylori* to gastric epithelial cells (Hata et al., 2004; Wada et al., 2010; Wang et al., 2001b). All these results indicate that human milk gangliosides may play an important role in protecting infants against enterotoxin-induced diarrhea. The possible mechanism by which these compounds can prevent the infection is that dietary gangliosides can act as putative decoys, due to their sialic acid groups, that interfere with pathogenic binding in the intestine (Idota and Kawakami, 1995; Rueda, 2007; Yuyama et al., 1993). Suh et al. (2004) also found that feeding mice with gangliosides decreased the infection upon gastric incubation with *Giardia muris* (a parasite) compared with mice fed a control diet (Suh et al., 2004). In vitro tests in this work also indicated direct toxicity of gangliosides to parasites. In a clinical study in which preterm infants were fed with a ganglioside-supplemented formula, Rueda et al. (1998) found that gangliosides at concentrations present in human milk significantly modified the fecal flora, increasing the *Bifidobacteria* content (prebiotic effect) and lowering the content of *Escherichia coli*. Dietary gangliosides may also promote intestinal immunity development in the neonate and consequently prevent infections during early infancy (Rueda, 2007). PLs from MFGM showed dose-dependent inhibition against rotavirus in an in-vitro experiment (Fuller et al., 2013). However, the specific components possessing that effect are unknown and the inhibition mechanisms still remain to be discovered (Fuller et al., 2013). To learn more about potential uses of gangliosides as therapeutic interventions, readers are encouraged to read some of the reviews (Gagnon and Saragovi, 2002; McJarrow et al., 2009; Rueda, 2007). Methods for determination of sialic acid and gangliosides in biological samples and dairy products have been recently reviewed by Lacomba et al. (2010).

Other health-related effects of PLs and MFGM-specific proteins are described in the reviews of Contarini and Povolo (2013), Dewettinck et al. (2008), and Le et al. (2014b).

Technological Properties of PLs

Emulsifying Properties of MFGM Material

Phospholipids are amphiphilic, and their physicochemical role in milk is to stabilize the fat globules against coalescence (Walstra et al., 2006) and protect the core fat against lipolysis (Shimizu et al., 1982). Due to their dipolar nature, phospholipids in general or milk phospholipids more specificifically are considered to have good emulsifying properties.

MFGM isolated from raw milk (not heat treated) is comprised of 25–35% PLs depending on the isolation methods (Fong et al., 2007; Le et al., 2014b) and has a good emulsifying capacity. One percent of this MFGM material is enough to stabilize an emulsion made from 25% milk fat, and the emulsion stability increases with increasing concentrations of the MFGM (Kanno, 1989). The droplet size ranged from 0.9–17 μm in diameter (Kanno et al., 1991). The process conditions, such as emulsifying time, homogenization, temperature, pH, emulsifying properties, size, and surface of globules, were also investigated (Kanno et al., 1991). The emulsion had the highest viscosity and was least stable at around pH 5, which is close to the isoelectric point of MFGM, 4.9 (Kanno, 1989; Kanno et al., 1991).

Whey buttermilk resulting from churning of whey cream in the manufacture of whey butter (Figure 5.4) gave higher emulsifying properties and lower foaming ability compared to sweet or cultured buttermilk, probably due to a higher ratio of phospholipids to proteins (Sodini et al., 2006). However, commercial buttermilk was found to have an inferior emulsifying and stabilizing capacity compared to nonfat dried milk. It was also found to not contain MFGM (Wong and Kitts, 2003). Concentrated MFGM isolates, prepared by adding citrate followed by high-speed centrifugation to collect the membrane material/fragments, were reported to be inferior in emulsifying properties compared to industrial buttermilk (Corredig and Dalgleish, 1997). In a later work, Corredig and Dalgleish (1998b) discovered that the stability of O/W emulsions by MFGM material depends on the heat treatment of the previous cream and that MFGM material isolated from raw cream, which did not undergo any heat treatment, had a good emulsifying capacity. The emulsions stabilized by this isolate were stable, and the absorbed MFGM material at the interface could not be displaced by surfactants or caseins and β-lactoglobulin added to the emulsions. Heating the cream at temperatures higher than 65 °C resulted in loss of emulsifying capacity of the MFGM isolate, but pasteurization temperature had no effect on the emulsifying properties of the whole buttermilk (Corredig and Dalgleish, 1998a). These results suggest that it is not the MFGM components alone, but rather the ratio between the casein, whey, and MFGM content in buttermilk, that determines its functional properties (Wong and Kitts, 2003).

Sodini et al. (2006) reported that commercial sweet, sour, and whey buttermilks have better emulsifying properties and a lower foaming capacity compared to milk and whey. Furthermore, among the three, whey buttermilk was found to have the best emulsifying properties and the lowest foaming capacity, possibly due to a higher ratio of PLs to protein in whey buttermilk compared to the other two. Roesch et al. (2004) tested 10% soybean oil emulsions with 0.25% MFGM isolate (60% wt/wt proteins) from commercial buttermilk, obtained by citrate addition followed by microfiltration. These emulsions showed good stability to creaming and resulted in a small particle size distribution. Similar emulsions prepared with conventional buttermilk concentrate showed extensive flocculation (Roesch et al., 2004). Making emulsions with soybean oil, reported from our own laboratory, showed that MFGM-enriched materials isolated from buttermilk and butter serum using microfiltration after addition of sodium citrate or from buttermilk whey using microfiltration (Le et al., 2010) demonstrated superior emulsifying properties compared to the mother materials, buttermilk, butter serum, and sodium caseinate. The superior emulsifying properties of the MFGM-enriched materials were exhibited by the smaller and narrower distribution of droplet sizes, the absence of droplet clustering, and less separation during chilling storage (Miocinovic et al., 2014; Phan et al., 2013).

Different studies on the functionality of MFGM material may have inconsistent results because they depend on various factors, such as the dairy sources for MFGM isolation and the intensity and frequency of heat treatment (different processing for milk, cream, and butter). As discussed previously, dairy processing treatments (e.g., heating) cause interaction of caseins and serum proteins to the MFGM, and because of this the MFGM material isolated from industrial sources is not as purified as MFGM material isolated from untreated (raw) milk. This is considered to be an important reason for the not-always-superior emulsifying properties of MFGM from industrial sources (Corredig and Dalgleish, 1998b). MFGM materials isolated from different sources, namely buttermilk, butter serum, and whey buttermilk, using microfiltration were characterized with different compositions in PLs, caseins, whey proteins, and minerals. This, along with the interaction among the components, determines the emulsifying properties (Phan et al., 2014a). MFGM material isolated from industrial whey buttermilk was found inferior to that isolated from industrial buttermilk to stabilize soybean O/W emulsions, even though the former had a higher concentration of PLs. This could be due to a high content of mineral (calcium) in the former because this had been previously added during the production of cheese (Phan et al., 2014a). However, whey buttermilk MFGM isolate in combination with buttermilk powder was shown to improve whipping properties, representing high overrun, no serum loss, and reasonable firmness of reconstituted cream prepared from AMF (Phan et al., 2014b). It is important to note that all results described here did not come from the same emulsion preparation conditions. There are many parameters that can influence stability of an emulsion system, such as the composition of the interfacial components, composition of the dispersed phase, composition and properties of the continuous phase, as well as the preparation methods (Fredrick et al., 2010).

When talking about technological aspects of MFGM material, one refers directly to its emulsifying properties, which have been assumed to be due to the PL fraction. It is noteworthy that PLs in MFGM fragments strongly interact with other components, such as proteins and cholesterol (see Figure 5.2). Proteins also have an amphiphilic characteristic and can act as surface-active compounds (Singh, 2011). However, surface-active properties of glycoproteins and other proteins of MFGM have not been studied. Contribution of protein moiety to emulsifying properties of the MFGM isolates is under investigation within our group.

MFGM material isolated from raw milk is, so far, mainly used for research purposes; potential applications lie in the enriched MFGM material isolated from the industrial sources. As previously summarized, many researchers have been successful in isolation of MFGM fragments, which contain high concentrations of PLs and specific membrane proteins from byproducts of dairy industrial processing. This opens opportunities to make use of such cheap sources to manufacture added-value ingredients with special functionalities. However, the treatment during milk processing affects the composition of the obtained materials and, therefore, their technological functionalities. There has not been sufficient investigation on the effect of processing treatments (e.g., heat treatment) on PL moiety of MFGM material. Whether or not the changes on phospholipid composition, chemical structure, or alteration in interaction of phospholipids together with other components of the membrane or of the milk serum phase would change the technological properties is still to be explored. Corredig and Dalgleish (1998b) proposed that the denaturation of MFGM proteins, the complexation between MFGM proteins and lipid fractions, and the association of whey proteins to the MFGM, caused by heat treatment, may decrease the solubility of the MFGM isolate. It was reported that reconstituted cream made with a high-PL buttermilk was more stable, had higher droplet surface, and was more acid tolerant than that made with low-PL buttermilk (Ihara et al., 2011). These authors also demonstrated that the acid tolerance of cream was not due to phospholipids or lysophospholipids but could be partially due to the complexes of PLs and proteins. Further studies on emulsifying properties of MFGM material as a whole as well as its individual components are still needed.

Emulsifying Properties of Milk PL Concentrate

After a first step of isolation of MFGM fragments from industrial sources, several techniques can be applied to further purify PLs (see the "Purification and Fractionation of PLs" section). Several dairy functional ingredients that contain mainly PLs have been launched on the commercial market. Lacprodan-20 and Lacprodan-75 of Arla Foods Ingredients Amba (Denmark) contain 20% and 75% phospholipids, respectively, and Phospholac 600, produced by Fonterra Cooperative Group Ltd. (New Zealand), contains more than 70% phospholipids. One interesting result from the use of such enriched PL isolates has been reported by the research group of Snow Brand Milk Products (Japan), who found that bovine milk lecithin (85% PLs) stabilized soy lecithin (95% PLs) solidified cream reconstituted from butter oil (Miura et al., 2006). Figure 5.8 shows the difference of particle size distribution with the use of bovine or soy lecithin as emulsifying agents. Among the PL species, PC (but not PE or SM, regardless of their origin) was found to be the determinant

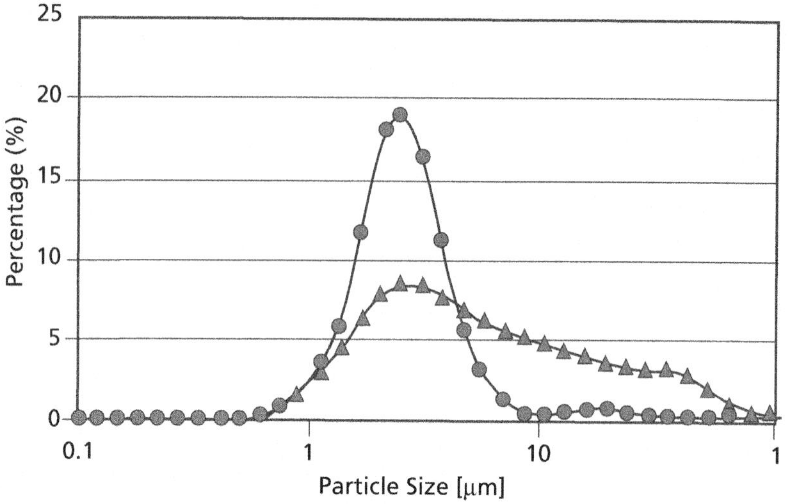

Figure 5.8 Particle size distribution of reconstituted cream with bovine milk PL (●) and soybean PL (▲) added to butter oil (from Miura et al., 2006).

of the emulsifying effect (Miura et al., 2004, 2006). PLs from milk differ from soy lecithin in both the classes of PLs as well as in FA composition (Boyd et al., 1999).

Zhu and Damodaran (2013) used ethanol to extract PLs from cheese-whey derived MFGM and obtained a dairy lecithin containing 31% total PLs. O/W emulsions made with less than 2% of this lecithin (relative to the total emulsion weight) were unstable; however, emulsions made with greater than 4% dairy lecithin were very stable for more than 60 days at room temperature (Zhu and Danodarn, 2013).

It is rather certain that PL concentrate (e.g., purified PLs) have good emulsifying properties while those properties of MFGM material are influenced by many factors such as composition, interaction among the components, and historical treatment of the MFGM (see the "Emulsifying Properties of MFGM Material" section). However, MFGM material is likely to have a lower cost of production, and MFGM-specific proteins also possess many significant health-beneficial properties, which are of interest for potential applications (Dewettinck et al., 2008; Jimenez-Flores and Brisson, 2008; Le et al., 2014b).

Applications of MFGM Material and PLs

Due to health-beneficial properties, PLs and PL-enriched materials seem very suitable for development of nutraceuticals and functional foods, especially for infants and children, the elderly, and athletes.

Due to good emulsifying properties, MFGM materials or PL-enriched materials have a high potential in production of emulsions, liposomes, and in emulsion-based foods such as reconstituted cream, mayonnaise, and salad sauces. Purified vegetable phospholipids (lecithins) have a long tradition of being employed in food products as well as other applications such as cosmetic products, animal feeds, medicinal and agrochemical products, paints, and lubricants. The use of PLs from dairy sources had been limited to pharmaceutical and cosmetic applications (Miura et al., 2006) but has gained more popularity in food development due to the availability of technology for extraction and the availability of the commercial PL ingredients. It is noticed that soy lecithin does not contain SM and contains a limited concentration of PS, whereas PLs isolated from dairy sources contain about 24–32% SM and 3.2–12% PS (Burling and Graverholt, 2008; Miura et al., 2006; Thompson and Singh, 2006). SM and PS possess several remarkable bioactivities that were previously reviewed.

Bezelgues et al. (2009) discovered that MFGM material (both proteins and lipids) was more efficient than milk proteins in stabilizing emulsions for delivering liposoluble nutrients (lycopene and α-tocopherol). Liu et al. (1995) studied the lymphatic absorption of vitamin D_3 in rats carried by a soybean oil emulsion stabilized with MFGM and reported that pancreatic lipases and bile salts are critical factors for absorption and for the promotive effect of MFGM that takes place in the lumen of the intestine rather than in the epithelial cells of the intestine. To protect fish oil emulsion, prepared at neutral pH against oxidation, protein (whey protein isolate or sodium caseinate) was found to be better than milk PLs and soy lecithin for use as an emulsifier (Horn et al., 2011). At low pH (4.5), Lacprodan PL-20 (containing 20% PLs and the rest mostly milk serum proteins) was better than Lacprodan PL-75 (containing 75% PLs) as an emulsifier to protect fish oil against oxidation (Horn et al., 2011). MFGM material isolated from industrial whey buttermilk was shown to improve whipping properties of reconstituted cream (Phan et al., 2014b). It should be noted again that milk PLs are more prone to oxidation than milk triglycerides because the former contain a higher proportion of unsaturated fatty acids. This challenge should be taken into account for future applications.

Liposomes for carrying and controlling release of various bioactive agents, including food ingredients and nutraceuticals, are normally prepared from polar lipids of soybean or eggs. Nowadays, preparation of liposomes from milk PLs get great attention due to the health-beneficial properties of MFGM and the commercial availability of milk PLs. Several forms of liposomes like unilamellar, multilamellar, and multivesicular can be created from Phospholac 600 using microfluidization, and these PL dispersions are quite stable toward oxidation (Thompson and Singh, 2006). The capacity to produce liposomal formulations gives rise to possibilities of using MFGM phospholipids to improve the moisturizing effect and to encapsulate and protect sensitive cosmetic ingredients or bioactive ingredients in functional foods.

More interestingly, MFGM PLs liposomes were found to be more stable than soy lecithin in a range of pH conditions, at a variety of storage and processing temperatures, and in the presence of mono- and divalent cations (Thompson et al., 2006). Thompson et al. (2007) compared three techniques—microfluidization, the traditional thin-film hydration, and the heating method—to prepare liposomes from Phospholipid Concentrate 700 isolated from buttermilk; they found that the liposomes prepared by both microfluidization and the heating method had high entrapment efficiencies. Liposomes produced using microfluidization were significantly smaller than those produced by the other methods, with a narrower size distribution and a higher proportion of unilamellar vesicles. Possible applications of PL-based liposomes are for nutrient delivery in food and feed, target delivery of drugs, and encapsulation of flavor, aroma, and natural coloring compounds in foods (Lu et al., 2011). Guelseren and Corredig (2013) used high-pressure homogenization to prepare nanoliposomes from milk PLs and soy phospholipids to encapsulate tea polyphenolic compounds and observed that the milk liposomes had smaller size, higher encapsulation efficiency, and lower extent of release of tea polyphenols during storage, compared to soy liposomes. The authors suggested that the tea catechins may be incorporated in the milk phospholipid bilayer more efficiently than in a soy phospholipid bilayer. Farhang et al. (2012) reported that 100 nm liposomes could be prepared from milk polar lipids derived from MFGM (NZMP Phospholipid Concentrate 700 containing 85% total lipids, Fonterra Cooperative Group, Palmerston North, New Zealand) using microfluidization. The incorporation efficiencies for ascorbic acid increased as the concentrations of phospholipid-rich powder increased from 5% to 10% (w/w) and reached a plateau value of 26%. With high levels of PLs (16%) and a high number of passes through the microfluidizer, liposome structure was changed from unilamellar to multivesicular/multilamellar vesicles (Farhang et al., 2012). The liposomes could protect ascorbic acid for at least 7 weeks at neutral pH and in refrigerated conditions (Farhang et al., 2012). A review on various aspects of nanoliposomes, including currently available preparation methods and their application in food technology, can be seen in Mozafari et al. (2008).

When a mixture of milk PLs was applied topically on rodents, the extent of hair cycle progression was comparable to that of animals treated with minoxidil, the most well-known reagent that initiates anagen. This led to speculation by Kumura et al. (2012) that milk PLs contain a dermal penetrative component that can regulate the hair cycle, and thus this preparation possesses potential for cosmetic use.

For the production of nonfat and low-fat yogurt, the reduction of total solids causes adverse effects on textural and sensory properties, such as severe syneresis and a lack of the typical flavor and mouthfeel. Several attempts have been carried out to improve these defects by using buttermilk powder and ultrafiltered buttermilk (Guinee et al., 1995; Kosikowski, 1979; Trachoo and Mistry, 1998). In an attempt to fortify a dairy product with milk PLs, Le et al. (2011b) replaced part of milk solids with MFGM material isolated from buttermilk using microfiltration and found that the PL-enriched yogurt had higher water-holding capacity and adhesiveness. There was no retardation of fermentation upon the fortification. It seemed that MFGM fragments weakened the casein network. However, this can be resolved by increasing casein concentration (Le et al., 2011b).

PLs have a polar nature and, as a result, they have high water-holding capacity. This explains why Emmental produced from native small MFGs (~3 μm) had 5.0% more moisture on a nonfat basis compared to the cheese made from large MFGs (~6 μm) after 52 days of ripening, and 2.2% more moisture in the case of camembert cheese after 40 days of ripening (Michalski et al., 2003, 2004). Addition of sweet buttermilk results in a significant decrease in free oil and an increase in cheese yield, but higher added levels could adversely affect the texture, melt, and sensory properties of pizza cheese (Govindasamy-Lucey et al., 2006; Lilbæk et al., 2006; Mistry et al., 1996). In experiments on the production of reduced-fat mozzarella cheese, Poduval and Mistry (1999) found that the addition of ultrafiltered buttermilk increased the moisture content and contributed to a further reduction of the free oil content during homogenization. When buttermilk was used for making cheese, the retention of MFGM in the rennet gel was higher if the buttermilk was obtained from heat-treated cream than when the buttermilk was from an untreated cream (Morin et al., 2008). This is because during heat treatment of the cream, caseins and whey proteins attach to the fat globule surface and as a result the MFGM fragments in the subsequent buttermilk contain a substantial amount of caseins. These caseins will coagulate and bring with them the MFGM into the rennet gel network. In conclusion, through the use of MFGM and PLs, new cheeses or yogurts with different functional and nutritional properties can be produced. This perspective could also bring economical profits by increasing the product yield or by using low-value byproducts from the dairy industry, such as buttermilk.

PL-enriched ingredients, such as cream residue powder derived from conversion cream into butter oil, sweet buttermilk powder, or butter serum powder, were found to improve heating-induced structural changes in concentrated milk emulsions (Kasinos et al., 2014). It seemed that the added PLs replaced proteins on the surface of FGs, leading to less protein interaction and causing gelling between serum proteins and FG surface proteins (Kasinos et al., 2014). The presence of phospholipids was found to impair the stability of milk foams, particularly at a temperature at which the milk fat is partially crystalline (e.g., 20 °C), and this is desirable in the reconstitution of milk powders or infant formula (Huppertz, 2013). PLs can compete with milk proteins for the foam bubble interface (Huppertz, 2013). Actually, polar lipids being surfactants on their own can stabilize foams; the combination of proteins and phospholipids on the surface of a foam bubble results in the mutually incompatible means of foam stabilization by proteins and polar lipids, which leads to an unstable foam (Wilde et al., 2004).

Conclusions and Perspectives

Nowadays, dairy polar lipids (PLs) receive much attention due to their health-promoting properties and advantageous technological functionalities. MFGM material can be isolated from dairy byproducts such as buttermilk, butter serum, and whey buttermilk using membrane filtration, and it can be further fractionated to obtain a PL concentrate ingredient using solvent fractionation, solid phase extraction (SPE), supercritical fluid extraction (SFE), or near-critical dimethyl ether (DME) antisolvent fractionation. Several MFGM materials and PL concentrates are commercially available.

Prominent nutraceutical properties of PLs include the contribution to development of brain and intestinal immune systems of infants, protection against gastrointestinal infection, reduction of exercise-induced stress, improvement of memory in the elderly, and lowering cholesterol absorption. Technological functionalities of PL-enriched materials are related to their emulsifying properties and water-binding capacity, making them potential ingredients for development of emulsion-based foods or emulsions and liposomes for protection and delivery of sensitive bioactive nutrients. Because of its health-beneficial properties, PL-enriched material can be fortified to create nutraceuticals and functional foods for infants, children, the elderly, and athletes. An optimized application is where both health-beneficial properties and advantageous technological functionalities are utilized in the same system.

It is quite evident that MFGM components in general, or in PLs in particular, protect against infection and contribute to the development of the brain and gastrointestinal immune systems in the early life of infant mammals. This makes it necessary to consider PLs in formulation of infant foods from dairy ingredients.

References

Anand, B. S.; Romero, J. J.; Sanduja, S. K.; Lichtenberger, L. M. Phospholipid Association Reduces the Gastric Mucosal Toxicity of Aspirin in Human Subjects. *Am. J. Gastroenterol.* **1999,** *94,* 1818–1822.

Anderson, M.; Brooker, B. E.; Andrews, A. T.; Alichanidis, E. Membrane Material Isolated from Milk of Mastitic and Normal Cows. *J. Dairy Sci.* **1974,** *57,* 1448–1458.

Anderson, M.; Brooker, B. E.; Andrews, A. T.; Alichanidis, E. Membrane Material in Bovine Skim-milk from Udder Quarters Infused with Endotoxin and Pathogenic Organisms. *J. Dairy Res.* **1975,** *42,* 401–417.

Angulo, J.; Olivera, M.; Mahecha, L.; Nuernberg, G.; Dannenberger, D.; Nuernberg, K. Distribution of Conjugated Linoleic Acid (CLA) Isomers and Other Fatty Acids in Polar and Neutral Fractions of Milk From Cows Fed Different Lipid Supplements. *Revista Colombiana De Ciencias Pecuarias* **2013,** *26,* 79–89.

Anonymous. Phosphatidylserine. Monograph. *Altern. Med. Rev.* **2008,** *13,* 245–247.

Astaire, J. C.; Ward, R.; German, J. B.; Jiménez-Flores, R. Concentration of Polar MFGM Lipids from Buttermilk by Microfiltration and Supercritical Fluid Extraction. *J. Dairy Sci.* **2003,** *86,* 2297–2307.

Attebery, J. M. Removing Lipid Material from Whey. U.S. Patent 3,560,219, 1968.

Atwal, A. S.; Hidiroglou, M.; Kramer, J. K. G. Effects of Feeding Protec and α-Tocopherol on Fatty Acid Composition and Oxidative Stability of Cow's Milk. *J. Dairy Sci.* **1991,** *74,* 140–145.

Avalli, A.; Contarini, G. Determination of Phospholipids in Dairy Products by SPE/HPLC/ELSD. *J. Chrom. A.* **2005,** *1071,* 185–190.

Bandyopadhyay, A. K.; Ganguli, N. C. Effect of Heating and Chilling Buffalo Milk on the Properties of Fat Globule Membrane Proteins. *J. Food Sci. Tech. India.* **1975,** *12,* 312–315.

Barello, C.; Garoffo, L. P.; Montorfano, G.; Zava, S.; Berra, B.; Conti, A.; Giuffrida, M. G. Analysis of Major Proteins and Fat Fractions Associated with Mare's Milk Fat Globules. *Mol. Nutr. Food. Res.* **2008,** *52,* 1448–1456.

Bartlett, G. R. Phosphorus Assay in Column Chromatography. *J. Biol. Chem.* **1959,** *234,* 466–468.

Baumruck, C. R.; Keenan, T. W. Membranes of Mammary-Gland. 7. Stability of Milk-Fat Globule Membrane in Secreted Milk. *J. Dairy Sci.* **1973,** *56,* 1092–1094.

Benoit, B.; Fauquant, C.; Daira, P.; Peretti, N.; Guichardant, M.; Michalski, M. C. Phospholipid Species and Minor Sterols in French Human Milks. *Food Chem.* **2010,** *120,* 684–691.

Bernhard, W.; Postle, A. D.; Linck, M.; Sewing, K. F. Composition of Phospholipid Classes and Phosphatidylcholine Molecular-Species of Gastric-Mucosa and Mucus. *BBA-Lipid. Lipid Met.* **1995,** *1255,* 99–104.

Bernhard, W.; Postle, A. D.; Linck, M.; Sewing, K. F. Rat Gastric Hydrophobic Barrier: Modulation of Phosphatidylcholine Molecular Species by Dietary Lipids. *Lipids* **1996,** *31,* 507–511.

Berra, B.; Colombo, I.; Sottocornola, E.; Giacosa, A. Dietary Sphingolipids in Colorectal Cancer Prevention. *Eur. J. Cancer Prev.* **2002,** *11,* 193–197.

Bezelgues, J. B.; Morgan, F.; Palomo, G.; Crosset-Perrotin, L.; Ducret, P. Short Communication: Milk Fat Globule Membrane as a Potential Delivery System for Liposoluble Nutrients. *J. Dairy Sci.* **2009,** *92,* 2524–2528.

Bitman, J.; Wood, D. L. Changes in Milk Fat Phospholipids during Lactation. *J. Dairy Sci.* **1990,** *73,* 1208–1216.

Bligh, E. G.; Dyer, W. J. A Rapid Method of Total Lipid Extraction and Purification. *Can. J. Biochem. Physiol.* **1959,** *37,* 911–917.

Blusztajn, J. K. Developmental Neuroscience—Choline, a Vital Amine. *Science* **1998,** *281,* 794–795.

Bo, T.; Pawliszyn, J. Characterization of Phospholipid-protein Interactions by Capillary Isoelectric Focusing with Whole-column Imaging Detection. *Anal. Biochem.* **2006,** *350,* 91–98.

Boersma, E. R.; Offringa, P. J.; Muskiet, F. A. J.; Chase, W. M.; Simmons, I. J. Vitamin-E, Lipid Fractions, and Fatty-Acid Composition of Colostrum, Transitional Milk, and Mature Milk—An International Comparative-Study. *Am. J. Clin. Nutr.* **1991,** *53,* 1197–1204.

Boyd, L. C.; Drye, N. C.; Hansen, A. P. Isolation and Characterization of Whey Phospholipids. *J. Dairy Sci.* **1999,** *82,* 2550–2557.

Britten, M.; Lamothe, S.; Robitaille, G. Effect of Cream Treatment on Phospholipids and Protein Recovery in Butter-making Process. *Int. J. Food Sci. Tech.* **2008,** *43,* 651–657.

Brown, E. M. Interactions of Beta-lactoglobulin and Alpha-lactalbumin with Lipids—A Review. *J. Dairy Sci.* **1984,** *67,* 713–722.

Burling, H.; Graverholt, G. Milk—A New Source for Bioactive Phospholipids for Use in Food Formulations. *Lipid Technol.* **2008,** *20,* 229–231.

Cagalj, M.; Brezovecki, A.; Mikulec, N.; Antunac, N. Composition and Properties of Mare's Milk of Croatian Coldblood Horse Breed. *Mljekarstvo* **2014,** *64,* 3–11.

Cano-Ruiz, M. E.; Richter, R. L. Effect of Homogenization Pressure on the Milk Fat Globule Membrane Proteins. *J. Dairy Sci.* **1997,** *80,* 2732–2739.

Carlson, S. E. Early Determinants of Development: A Lipid Perspective. *Am. J. Clin. Nutr.* **2009,** *89,* S1523–S1529.

Carr, D. J.; Guarcello, V.; Blalock, J. E. Phosphatidylserine Suppresses Antigen-specific IgM Production by Mice Orally Administered Sheep Red Blood Cells. *Proc. Soc. Exp. Biol. Med.* **1992,** *200,* 548–554.

Catchpole, O. J.; Tallon, S. J.; Grey, J. B.; Fenton, K.; Fletcher, K.; Fletcher, A. J. Extraction of Lipids from Aqueous Protein-rich Streams Using Near-critical Dimethylether. *Chem. Eng. Technol.* **2007,** *30,* 501–510.

Catchpole, O. J.; Tallon, S. J.; Grey, J. B.; Fletcher, K.; Fletcher, A. J. Extraction of Lipids from a Specialist Dairy Stream. *J. Supercrit. Fluid.* **2008,** *45,* 314–321.

Cenacchi, T.; Bertoldin, T.; Farina, C.; Fiori, M. G.; Crepaldi, G.; Azzini, C. F.; Girardello, R.; Bagozzi, B.; Garuti, R.; Vivaldi, P., et al. Cognitive Decline in the Elderly—A Double-blind, Placebo-controlled Multicenter Study on Efficacy of Phosphatidylserine Administration. *Aging-Clin. Exp. Res.* **1993,** *5,* 123–133.

Chatteriee, S.; Pandey, A. The Yin and Yang of Lactosylceramide Metabolism: Implications in Cell Function. *Biochimica et Biophysica Acta-General Subjects* **2008,** *1780,* 370–382.

Chilliard, Y.; Ferlay, A.; Doreau, M. Effect of Different Types of Forages, Animal Fat or Marine Oils in Cow's Diet on Milk Fat Secretion and Composition, Especially Conjugated Linoleic Acid (CLA) and Polyunsaturated Fatty Acids. *Livestock Production Science* **2001,** *70,* 31–48.

Christie, W. W.; Noble, R. C.; Davies, G. Phospholipids in Milk and Dairy-Products. *J. Soc. Dairy Tech.* **1987,** *40,* 10–12.

Clandinin, M. T.; Park, E. J.; Suh, M.; Thomson, B.; Thomson, A. B. R.; Ramanujam, K. S. Dietary Ganglioside Decreases Cholesterol Content, Caveolin Expression and Inflammatory Mediators in Rat Intestinal Microdomains. *Glycobiology* **2005,** *15,* 935–942.

Contarini, G.; Povolo, M. Phospholipids in Milk Fat: Composition, Biological and Technological Significance, and Analytical Strategies. *Int. J. Mol. Sci.* **2013,** *14,* 2808–2831.

Corredig, M.; Dalgleish, D. G. Effect of Different Heat Treatments on the Strong Binding Interactions between Whey Proteins and Milk Fat Globules in Whole Milk. *J. Dairy Res.* **1996,** *63,* 441–449.

Corredig, M.; Dalgleish, D. G. Isolates from Industrial Buttermilk: Emulsifying Properties of Materials Derived from the Milk Fat Globule Membrane. *J. Agric. Food Chem.* **1997,** *45,* 4595–4600.

Corredig, M.; Dalgleish, D. G. Buttermilk Properties in Emulsions with Soybean Oil as Affected by Fat Globule Membrane-Derived Proteins. *J. Food Sci.* **1998a,** *63,* 476–480.

Corredig, M.; Dalgleish, D. G. Effect of Heating of Cream on the Properties of Milk Fat Globule Membrane Isolates. *J. Agric. Food Chem.* **1998b,** *46,* 2533–2540.

Corredig, M.; Roesch, R. R.; Dalgleish, D. G. Production of a Novel Ingredient from Buttermilk. *J. Dairy Sci.* **2003,** *86,* 2744–2750.

Costa, M. R.; Elias-Argote, X. E.; Jiménez-Flores, R.; Gigante, M. L. Use of Ultrafiltration and Supercritical Fluid Extraction to Obtain a Whey Buttermilk Powder Enriched in Milk Fat Globule Membrane Phospholipids. *Int. Dairy J.* **2010,** *20,* 598–602.

Crook, T. H.; Tinklenberg, J.; Yesavage, J.; Petrie, W.; Nunzi, M. G.; Massari, D. C. Effects of Phosphatidylserine in Age-associated Memory Impairment. *Neurology* **1991,** *41,* 644–649.

Dalgleish, D. G.; Banks, J. M. The Formation of Complexes between Serum Proteins and Fat Globules during Heating of Whole Milk. *Milchwissenschaft* **1991,** *46,* 75–78.

Damodaran, S. Zinc-induced Precipitation of Milk Fat Globule Membranes: A Simple Method for the Preparation of Fat-free Whey Protein Isolate. *J. Agric. Food Chem.* **2010,** *58,* 11052–11057.

Denise, M. N. Potential for Enhancing the Nutritional Properties of Milk Fat. *J. Dairy Sci.* **1991,** *74,* 4002–4012.

Dewettinck, K.; Le, T. T. Membrane Separations in Food Processing. In *Alternatives to Conventional Food Processing*; Proctor, A., Ed.; RSC Publishing: Cambridge, 2011; pp 184–253.

Dewettinck, K.; Rombaut, R.; Thienpont, N.; Le, T. T.; Messens, K.; Camp, J. V. Nutritional and Technological Aspects of Milk Fat Globule Membrane Material. *Int. Dairy J.* **2008,** *18,* 436–457.

Dial, E. J.; Rooijakkers, S. H. M.; Darling, R. L.; Romero, J. J.; Lichtenberger, L. M. Role of Phosphatidylcholine Saturation in Preventing Bile Salt Toxicity to Gastrointestinal Epithelia and Membranes. *J. Gastroen. Hepatol.* **2008,** *23,* 430–436.

Dillehay, D. L.; Webb, S. K.; Schmelz, E. M.; Merrill, A. H. Dietary Sphingomyelin Inhibits 1,2-Dimethylhydrazine-induced Colon-cancer in CF1 Mice. *J. Nutr.* **1994,** *124,* 615–620.

Duan, J. J.; Sugawara, T.; Sakai, S.; Aida, K.; Hirata, T. Oral Glucosylceramide Reduces 2,4-Dinitrofluorobenzene Induced Inflammatory Response in Mice by Reducing TNF-alpha Levels and Leukocyte Infiltration. *Lipids* **2011,** *46,* 505–512.

Duan, R. D. Physiological Functions and Clinical Implications of Sphingolipids in the Gut. *J. Digest. Dis.* **2011,** *12,* 60–70.

Duivenvoorden, I.; Voshol, P. J.; Rensen, P. C. N.; van Duyvenvoorde, W.; Romijn, J. A.; Emeis, J. J.; Havekes, L. M.; Nieuwenhuizen, W. F. Dietary Sphingolipids Lower Plasma Cholesterol and Triacylglycerol and Prevent Liver Steatosis in APOE*3Leiden Mice. *Am. J. Clin. Nutr.* **2006,** *84,* 312–321.

Dunjic, B. S.; Axelson, J. K.; Bengmark, S. S. Is Resistance to Phospholipase Important for the Gastric-mucosal Protective Capacity of Exogenous Phosphatidylcholine. *Eur. J. Gastroen. Hepat.* **1994,** *6,* 593–598.

Eckhardt, E. R. M.; Wang, D. Q. H.; Donovan, J. M.; Carey, M. C. Dietary Sphingomyelin Suppresses Intestinal Cholesterol Absorption by Decreasing Thermodynamic Activity of Cholesterol Monomers. *Gastroenterology* **2002,** *122,* 948–956.

Ehehalt, R.; Braun, A.; Karner, M.; Fullekrug, J.; Stremmel, W. Phosphatidylcholine as a Constituent in the Colonic Mucosal Barrier—Physiological and Clinical Relevance. *BBA-Mol. Cell Biol. L.* **2010,** *1801,* 983–993.

Engel, R. R.; Satzger, W.; Günther, W.; Kathmann, N.; Bove, D.; Gerke, S.; Münch, U.; Hippius, H. Double-blind Cross-over Study of Phosphatidylserine vs. Placebo in Patients with Early Dementia of the Alzheimer Type. *Eur. Neuropsychopharm.* **1992,** *2,* 149–155.

Erwin, R. E.; Randolph, H. E. Influence of Mastitis on Properties of Milk. XI. Fat Globule Membrane. *J. Dairy Sci.* **1975,** *58,* 9–12.

Evers, J. M. The Milkfat Globule Membrane-compositional and Structural Changes Post Secretion by the Mammary Secretory Cell. *Int. Dairy J.* **2004,** *14,* 661–674.

Fagan, P.; Wijesundera, C. Liquid Chromatographic Analysis of Milk Phospholipids with Online Pre-concentration. *J. Chrom. A.* **2004,** *1054,* 241–249.

Farhang, B.; Kakuda, Y.; Corredig, M. Encapsulation of Ascorbic Acid in Liposomes Prepared with Milk Fat Globule Membrane-derived Phospholipids. *Dairy Sci. Technol.* **2012,** *92,* 353–366.

Fauquant, C.; Briard, V.; Leconte, N.; Michalski, M. C. Differently Sized Native Milk Fat Globules Separated by Microfiltration: Fatty Acid Composition of the Milk Fat Globule Membrane and Triglyceride Core. *Eur. J. Lipid Sci. Technol.* **2005,** *107,* 80–86.

Fauquant, C.; Briard-Bion, V.; Leconte, N.; Guichardant, M.; Michalski, M.-C. Membrane Phospholipids and Sterols in Microfiltered Milk Fat Globules. *Eur. J. Lipid Sci. Technol.* **2007,** *109,* 1167–1173.

Fauquant, J.; Vieco, E.; Brule, G.; Maubois, J. L. Clarification of Sweet Cheese Whey by Thermocalcic Aggregation of Residual Fat. *Lait.* **1985,** *65,* 1–20.

Fittkau, M.; Gerlach, R.; Schmoll, H. J. Phosphatidylcholine Does Not Protect Rats against 5-Fluorouracil/Folinic Acid-induced Damage of the Intestinal Luminal Mucosa. *J. Cancer Res. and Clin.* **2002,** *128,* 80–84.

Folch, J.; Lees, M.; Sloane Stanley, G. H. A Simple Method for the Isolation and Purification of Total Lipids from Animal Tissues. *J. Biol. Chem.* **1957,** *226,* 497–509.

Fong, B. Y.; Norris, C. S.; MacGibbon, A. K. H. Protein and Lipid Composition of Bovine Milk-fat-globule Membrane. *Int. Dairy J.* **2007,** *17,* 275–288.

Fox, P. F.; Guinee, T. P.; Cogan, T. M.; McSweeney, P. L. H. *Fundamentals of Cheese Science.* Aspen Publishers, Inc.: Gaithersburg, MD, 2000.

Fredrick, E.; Walstra, P.; Dewettinck, K. Factors Governing Partial Coalescence in Oil-in-water Emulsions. *Adv. Colloid Interface Sci.* **2010,** *153,* 30–42.

Fuller, K. L.; Kuhlenschmidt, T. B.; Kuhlenschmidt, M. S.; Jimenez-Flores, R.; Donovan, S. M. Milk Fat Globule Membrane Isolated from Buttermilk or Whey Cream and Their Lipid Components Inhibit Infectivity of Rotavirus in Vitro. *J. Dairy Sci.* **2013,** *96,* 3488–3497.

Gagnon, M.; Saragovi, H. U. Gangliosides: Therapeutic Agents or Therapeutic Targets? *Expert Opin. Ther. Pat.* **2002,** *12,* 1215–1223.

Gaitonde, P.; Peng, A.; Straubinger, R. M.; Bankert, R. B.; Balu-Iyer, S. V. Phosphatidylserine Reduces Immune Response against Human Recombinant Factor VIII in Hemophilia A Mice by Regulation of Dendritic Cell Function. *Clin. Immunol.* **2011,** *138,* 135–145.

Gallier, S.; Gragson, D.; Cabral, C.; Jimenez-Flores, R.; Everett, D. W. Composition and Fatty Acid Distribution of Bovine Milk Phospholipids from Processed Milk Products. *J. Agric. Food Chem.* **2010a,** *58,* 10503–10511.

Gallier, S.; Gragson, D.; Jimenez-Flores, R.; Everett, D. Using Confocal Laser Scanning Microscopy to Probe the Milk Fat Globule Membrane and Associated Proteins. *J. Agric. Food Chem.* **2010b,** *58,* 4250–4257.

Garcia, C.; Lutz, N. W.; Confort-Gouny, S.; Cozzone, P. J.; Armand, M.; Bernard, M. Phospholipid Fingerprints of Milk from Different Mammalians Determined by P-31 NMR: Towards Specific Interest in Human Health. *Food Chem.* **2013,** *135,* 1777–1783.

Garcia, C.; Antona, C.; Robert, B.; Lopez, C.; Armand, M. The Size and Interfacial Composition of Milk Fat Globules Are Key Factors Controlling Triglycerides Bioavailability in Simulated Human Gastro-duodenal Digestion. *Food Hydrocolloids* **2014,** *35,* 494–504.

Gassi, J. Y.; Famelart, M. H.; Lopez, C. Heat Treatment of Cream Affects the Physicochemical Properties of Sweet Buttermilk. *Dairy Sci. Technol.* **2008,** *88,* 369–385.

Ghyczy, M.; Torday, C.; Kaszaki, J.; Szabo, A.; Czobel, M.; Boros, M. Oral Phosphatidylcholine Pretreatment Decreases Ischemia-reperfusion-induced Methane Generation and the Inflammatory Response in the Small Intestine. *Shock* **2008,** *30,* 596–602.

Gil, A.; Rueda, R. Interaction of Early Diet and the Development of the Immune System. *Nutr. Res. Rev.* **2002,** *15,* 263–292.

Giuffrida, F.; Cruz-Hernandez, C.; Fluck, B.; Tavazzi, I.; Thakkar, S. K.; Destaillats, F.; Braun, M. Quantification of Phospholipids Classes in Human Milk. *Lipids* **2013,** *48,* 1051–1058.

Govindasamy-Lucey, S.; Lin, T.; Jaeggi, J. J.; Johnson, M. E.; Lucey, J. A. Influence of Condensed Sweet Cream Buttermilk on the Manufacture, Yield, and Functionality of Pizza Cheese. *J. Dairy Sci.* **2006,** *89,* 454–467.

Graves, E. L. F.; Beaulieu, A. D.; Drackley, J. K. Factors Affecting the Concentration of Sphingomyelin in Bovine Milk. *J. Dairy Sci.* **2007,** *90,* 706–715.

Greenbank, G.; Pallansch, M. J. Migration of Phosphatides in Processing Dairy Products. *J. Dairy Sci.* **1961,** *44,* 1597–1602.

Guelseren, I.; Corredig, M. Storage Stability and Physical Characteristics of Tea-polyphenol-bearing Nanoliposomes Prepared with Milk Fat Globule Membrane Phospholipids. *J. Agric. Food Chem.* **2013,** *61,* 3242–3251.

Guinee, T. P.; Mullins, C. G.; Reville, W. J.; Cotter, M. P. Physical-properties of Stirred-curd Unsweetened Yogurts Stabilized with Different Dairy Ingredients. *Milchwissenschaft* **1995,** *50,* 196–200.

Hara, A.; Radin, N. S. Lipid Extraction of Tissues with a Low-toxicity Solvent. *Anal. Biochem.* **1978,** *90,* 420–426.

Hartmann, P.; Szabo, A.; Eros, G.; Gurabi, D.; Horvath, G.; Nemeth, I.; Ghyczy, M.; Boros, M. Anti-inflammatory Effects of Phosphatidylcholine in Neutrophil Leukocyte-dependent Acute Arthritis in Rats. *Eur. J. Pharmacol.* **2009,** *622,* 58–64.

Hata, Y.; Murakami, M.; Okabe, S. Glycoconjugates with NeuAc-NeuAc-GaL-Glc Are More Effective at Preventing Adhesion of *Helicobacter pylori* to Gastric Epithelial Cells Than Glycoconjugates with NeuAc-Gal-Glc. *J. Physiol. Pharmacol.* **2004,** *55,* 607–625.

Heid, H. W.; Keenan, T. W. Intracellular Origin and Secretion of Milk Fat Globules. *Eur. J. Cell Biol.* **2005,** *84,* 245–258.

Hellhammer, J.; Waladkhani, A.-R.; Hero, T.; Buss, C. Effects of Milk Phospholipid on Memory and Psychological Stress Response. *Brit. Food J.* **2010,** *112,* 1124–1137.

Hirayama, S.; Masuda, Y.; Rabeler, R. Effect of Phosphatidylserine Administration on Symptoms of Attention-deficit/Hyperactivity Disorder in Children. *Agro. Food Ind. Hi Tech.* **2006,** *17,* 16–20.

Horn, A. F.; Nielsen, N. S.; Andersen, U.; Sogaard, L. H.; Horsewell, A.; Jacobsen, C. Oxidative Stability of 70% Fish Oil-in-water Emulsions: Impact of Emulsifiers and pH. *Eur. J. Lipid Sci. Technol.* **2011,** *113,* 1243–1257.

Houlihan, A. V.; Goddard, P. A.; Nottingham, S. M.; Kitchen, B. J.; Masters, C. J. Interactions between the Bovine Milk Fat Globule Membrane and Skim Milk Components on Heating Whole Milk. *J. Dairy Res.* **1992,** *59,* 187–195.

Hu, S.; Zhang, L.; Dovichi, N. J. Characterization of the Interaction between Phospholipid and Protein by Capillary Electrophoresis with Laser-induced Fluorescence Detection. *J. Chrom. A.* **2001,** *924,* 369–375.

Huppertz, T. Foaming Properties of Milk: A Review of the Influence of Composition and Processing. *Int. J. Dairy Technol.* **2013,** *63,* 477–488.

Huppertz, T.; Fox, P. F.; Kruif, K. G. D.; Kelly, A. L. High Pressure-induced Changes in Bovine Milk Proteins: A Review. *Biochim. Biophys. Acta* **2006,** *1764,* 593–598.

Hwang, D. C.; Damodaran, S. Selective Precipitation of Fat Globule Membranes of Cheese Whey by Saponin and Bile-salt. *J. Agric. Food Chem.* **1994,** *42,* 1872–1878.

Iametti, S.; Versuraro, L.; Tragna, S.; Giangiacomo, R.; Bonomi, F. Surface Properties of the Fat Globule in Treated Creams. *Int. Dairy J.* **1997,** *7,* 375–380.

Idota, T.; Kawakami, H. Inhibitory Effects of Milk Ganglioside on the Adhesion of *Escherichia Coli* to Human Intestinal Carcinoma Cells. *Biosci. Biotech. Bioch.* **1995,** *59,* 69–72.

Ihara, K.; Ochi, H.; Saito, H.; Iwatsuki, K. Effects of Buttermilk Powders on Emulsification Properties and Acid Tolerance of Cream. *J. Food Sci.* **2011,** *76,* C265–C271.

Iwamori, M.; Takamizawa, K.; Momoeda, M.; Iwamori, Y.; Taketani, Y. Gangliosides in Human, Cow and Goat Milk, and Their Abilities as to Neutralization of Cholera Toxin and Botulinum Type A Neurotoxin. *Glycoconjugate J.* **2008,** *25,* 675–683.

Jager, R.; Purpura, M.; Geiss, K. R.; Weiss, M.; Baumeister, J.; Amatulli, F.; Schroder, L.; Herwegen, H. The Effect of Phosphatidylserine on Golf Performance. *J. Int. Soc. Sports Nutr.* **2007a,** *4.* 1–5.

Jager, R.; Purpura, M.; Kingsley, M. Phospholipids and Sports Performance. *J. Int. Soc. Sports Nutr.* **2007b,** *4.* 1–8.

Jensen, R. G. The Composition of Bovine Milk Lipids: January 1995 to December 2000. *J. Dairy Sci.* **2002,** *85,* 295–350.

Jimenez-Flores, R.; Brisson, G. The Milk Fat Globule Membrane as an Ingredient: Why, How, When? *Dairy Sci. Technol.* **2008,** *88,* 5–18.

Jorissen, B. L.; Brouns, F.; Van Boxtel, M. P.; Riedel, W. J. Safety of Soy-derived Phosphatidylserine in Elderly People. *Nutr. Neurosci.* **2002,** *5,* 337–343.

Kanno, C. Emulsifying Properties of Bovine Milk Fat Globule Membrane in Milk Fat Emulsion: Conditions for the Reconstitution of Milk Fat Globules. *J. Food Sci.* **1989,** *54,* 1534–1539.

Kanno, C.; Shimomura, Y.; Takano, E. Physicochemical Properties of Milk Fat Emulsions Stabilized with Bovine Milk Fat Globule Membrane. *J. Food Sci.* **1991,** *56,* 1219–1223.

Kasinos, M.; Le, T. T.; Van der Meeren, P. Improved Heat Stability of Recombined Evaporated Milk Emulsions upon Addition of Phospholipid Enriched Dairy By-products. *Food Hydrocolloids* **2014,** *34,* 112–118.

Keenan, T. W.; Mather, I. H. Intracellular Origin of Milk Fat Globules and the Nature of the Milk Fat Globule Membrane. In *Advanced Dairy Chemistry*; Fox, P. F., McSweeney, P. L. H., Eds.; Springer: New York, 2006; Vol. 2, pp 137–171.

Keenan, T. W.; Moon, T. W.; Dylewski, D. P. Lipid Globules Retain Globule-Membrane Material after Homogenization. *J. Dairy Sci.* **1983,** *66,* 196–203.

Kidd, P. M. Phosphatidylcholine (Monograph). *Altern. Med. Rev.* **2002a,** *7,* 150–154.

Kidd, P. M. Phospholipids: Versatile Nutraceutical Ingredients for Functional Foods. *Funct. Foods Neutraceut.* **2002b,** *12,* 30–40.

Kim, H. H. Y.; Jimenez-Flores, R. Heat-induced Interactions between the Proteins of Milk-fat Globule-membrane and Skim Milk. *J. Dairy Sci.* **1995,** *78,* 24–35.

Kingsley, M. Effects of Phosphatidylserine Supplementation on Exercising Humans. *Sports Med.* **2006,** *36,* 657–669.

Kitchen, B. J. Fractionation and Characterization of Membranes from Bovine Milk-fat Globules. *J. Dairy Res.* **1977,** *44,* 469–482.

Koops, J.; Tarassuk, N. P. The Effect of Various Processing Treatments on the Partition of Phosphatides between the Fat Phase and the Milk Plasma. *Nederlands Melk-En Zuiveltijdschrift* **1959,** *13,* 180–189.

Kosikowski, F. V. Low Lactose Yogurts and Milk Beverages by Ultrafiltration. *J. Dairy Sci.* **1979,** *62,* 41–46.

Kuchta, A. M.; Kelly, P. M.; Stanton, C.; Devery, R. A. Milk Fat Globule Membrane—A Source of Polar Lipids for Colon Health? A Review. *Int. J. Dairy Technol.* **2012,** *65,* 315–333.

Kullenberg, D.; Taylor, L. A.; Schneider, M.; Massing, U. Health Effects of Dietary Phospholipids. *Lipids in Health and Disease* **2012,** *11,* 1–16.

Kumura, H.; Sawada, T.; Oda, Y.; Konno, M.; Kobayashi, K. Potential of Polar Lipids from Bovine Milk to Regulate the Rodent Dorsal Hair Cycle. *J. Dairy Sci.* **2012,** *95,* 3629–3633.

Lacomba, R.; Salcedo, J.; Alegria, A.; Jesus Lagarda, M.; Barbera, R.; Matencio, E. Determination of Sialic Acid and Gangliosides in Biological Samples and Dairy Products: A Review. *J. Pharm. Biomed. Anal.* **2010,** *51,* 346–357.

Le, T. T. Purification, Analysis and Applications of Bioactive Milk Fat Globule Membrane Material. Ph.D. Thesis, Ghent University, 2012.

Le, T. T.; Van Camp, J.; Rombaut, R.; van Leeckwyck, F.; Dewettinck, K. Effect of Washing Conditions on the Recovery of Milk Fat Globule Membrane Proteins during the Isolation of Milk Fat Globule Membrane from Milk. *J. Dairy Sci.* **2009,** *92,* 3592–3603.

Le, T. T.; Miocinovic, J.; Van Camp, J.; Devreese, B.; Struijs, K.; Van de Wiele, T.; Dewettinck, K. Isolation and Applications of Milk Fat Globule Membrane Material: Isolation from Buttermilk and Butter Serum. *Communications in Agricultural and Applied Biological Sciences, Ghent University* **2010,** *76,* 111–114.

Le, T. T.; Miocinovic, J.; Nguyen, T. M.; Rombaut, R.; Van Camp, J.; Dewettinck, K. Improved Solvent Extraction Procedure and High-performance Liquid Chromatography–Evaporative Light-Scattering Detector Method for Analysis of Polar Lipids from Dairy Materials. *J. Agric. Food Chem.* **2011a,** *59,* 10407–10413.

Le, T. T.; van Camp, J.; Pascual, P. A. L.; Meesen, G.; Thienpont, N.; Messens, K.; Dewettinck, K. Physical Properties and Microstructure of Yoghurt Enriched with Milk Fat Globule Membrane Material. *Int. Dairy J.* **2011b,** *21,* 798–805.

Le, T. T.; Cabaltica, A. D.; Bui, V. M. Membrane Separations in Dairy Processing. *J. Food. Res. Tech.* **2014a,** *2,* 1–14.

Le, T. T.; Van Camp, J.; Dewettinck, K. Milk Fat Globule Membrane Material: Isolation Techniques, Health-beneficial Properties and Potential Applications. In *Studies in Natural Products Chemistry*; Atta-ur-Rahman, F., Ed.; Elsevier: Amsterdam, 2014b; Vol. 41, pp 347–382.

Lee, S. J. E.; Sherbon, J. W. Chemical Changes in Bovine Milk Fat Globule Membrane Caused by Heat Treatment and Homogenization of Whole Milk. *J. Dairy Res.* **2002,** *69,* 555–567.

Lichtenberger, L. M.; Barron, M.; Marathi, U. Association of Phosphatidylcholine and NSAIDs as a Novel Strategy to Reduce Gastrointestinal Toxicity. *Drugs Today* (Barc). **2009,** *45,* 877–890.

Lilbæk, H. M.; Broe, M. L.; Høier, E.; Fatum, T. M.; Ipsen, R.; Sørensen, N. K. Improving the Yield of Mozzarella Cheese by Phospholipase Treatment of Milk. *J. Dairy Sci.* **2006,** *89,* 4114–4125.

Liu, H.-X.; Adachi, I.; Horikoshi, I.; Ueno, M. Mechanism of Promotion of Lymphatic Drug Absorption by Milk Fat Globule Membrane. *Int. J. Pharm.* **1995,** *118,* 55–64.

Lopez, C. Milk Fat Globules Enveloped by Their Biological Membrane: Unique Colloidal Assemblies with a Specific Composition and Structure. *Curr. Opin. Colloid In.* **2011,** *16,* 391–404.

Lopez, C.; Briard-Bion, V.; Menard, O.; Rousseau, F.; Pradel, P.; Besle, J.-M. Phospholipid, Sphingolipid, and Fatty Acid Compositions of the Milk Fat Globule Membrane Are Modified by Diet. *J. Agric. Food Chem.* **2008,** *56,* 5226–5236.

Lopez, C.; Menard, O. Human Milk Fat Globules: Polar Lipid Composition and in Situ Structural Investigations Revealing the Heterogeneous Distribution of Proteins and the Lateral Segregation of Sphingomyelin in the Biological Membrane. *Colloid. Surface. B.* **2011,** *83,* 29–41.

Lopez, C.; Briard-Bion, V.; Menard, O. Polar Lipids, Sphingomyelin and Long-chain Unsaturated Fatty Acids from the Milk Fat Globule Membrane Are Increased in Milks Produced by Cows Fed Fresh Pasture Based Diet during Spring. *Food Res. Int.* **2014,** *58,* 59–68.

Lu, F. S. H.; Nielsen, N. S.; Timm-Heinrich, M.; Jacobsen, C. Oxidative Stability of Marine Phospholipids in the Liposomal Form and Their Applications. *Lipids* **2011,** *46,* 3–23.

Martin, M. J.; Martin-Sosa, S.; Garcia-Pardo, L. A.; Hueso, P. Distribution of Bovine Milk Sialoglycoconjugates during Lactation. *J. Dairy Sci.* **2001,** *84,* 995–1000.

Martin-Sosa, S.; Martin, M. J.; Garcia-Pardo, L. A.; Hueso, P. Sialyloligosaccharides in Human and Bovine Milk and in Infant Formulas: Variations with the Progression of Lactation. *J. Dairy Sci.* **2003,** *86,* 52–59.

Mather, I. H. A Review and Proposed Nomenclature for Major Proteins of the Milk-fat Globule Membrane. *J. Dairy Sci.* **2000,** *83,* 203–247.

Mather, I. H.; Keenan, T. W. Origin and Secretion of Milk Lipids. *Journal of Mammary Gland Biology and Neoplasia* **1998,** *3,* 259–273.

Maubois, J. L.; Pierre, A.; Fauquant, J.; Piot, M. Industrial Fractionation of Main Whey Proteins. *Bull. Int. Dairy Fed.* **1987,** *212,* 154–159.

McJarrow, P.; Schnell, N.; Jumpsen, J.; Clandinin, T. Influence of Dietary Gangliosides on Neonatal Brain Development. *Nutr. Rev.* **2009,** *67,* 451–463.

McPherson, A. V.; Kitchen, B. J. Reviews of the Progress of Dairy Science: The Bovine Milk Fat Globule Membrane—Its Formation, Composition, Structure and Behavior in Milk and Dairy Products. *J. Dairy Res.* **1983,** *50,* 107–133.

Menard, O.; Ahmad, S.; Rousseau, F.; Briard-Bion, V.; Gaucheron, F.; Lopez, C. Buffalo vs. Cow Milk Fat Globules: Size Distribution, Zeta-potential, Compositions in Total Fatty Acids and in Polar Lipids from the Milk Fat Globule Membrane. *Food Chem.* **2010,** *120,* 544–551.

Mesilati-Stahy, R.; Argov-Argaman, N. The Relationship between Size and Lipid Composition of the Bovine Milk Fat Globule Is Modulated by Lactation Stage. *Food Chem.* **2014,** *145,* 562–570.

Michalski, M. C.; Gassi, J. Y.; Famelart, M. H.; Leconte, N.; Camier, B. The Size of Native Milk Fat Globules Affects Physico-chemical and Sensory Properties of Camembert Cheese. *Lait.* **2003,** *83,* 131–143.

Michalski, M. C.; Ollivon, M.; Briard, V.; Leconte, N.; Lopez, C. Native Fat Globules of Different Sizes Selected from Raw Milk: Thermal and Structural Behavior. *Chem. Phys. Lipids.* **2004,** *132,* 247–261.

Michalski, M. C.; Briard, V.; Michel, F.; Tasson, F.; Poulain, P. Size Distribution of Fat Globules in Human Colostrum, Breast Milk, and Infant Formula. *J. Dairy Sci.* **2005,** *88,* 1927–1940.

Miocinovic, J.; Le, T. T.; Fredrick, E.; Van Der Meeren, P.; Pudja, P.; Dewettinck, K. A Comparison of Composition and Emulsifying Properties of MFGM Materials Prepared from Different Dairy Sources by Microfiltration. *Food Sci. Tech. Int.* **2014,** *20,* 441–451.

Mistry, V. V.; Metzger, L. E.; Maubois, J. L. Use of Ultrafiltered Sweet Buttermilk in the Manufacture of Reduced Fat Cheddar Cheese. *J. Dairy Sci.* **1996,** *79,* 1137–1145.

Miura, S.; Tanaka, M.; Suzuki, A.; Sato, K. Application of Phospholipids Extracted from Bovine Milk to the Reconstitution of Cream Using Butter Oil. *J. Am. Oil Chem. Soc.* **2004,** *81,* 97–100.

Miura, S.; Mutoh, T.; Shiinoki, Y.; Yoshioka, T. Emulsifying Properties of Phospholipids in the Reconstitution of Cream Using Butter Oil. *Eur. J. Lipid Sci. Technol.* **2006,** *108,* 898–903.

Morgan, B. L. G.; Winick, M. Effects of Administration of N-Acetylneuraminic Acid (Nana) on Brain Nana Content and Behavior. *J. Nutr.* **1980,** *110,* 416–424.

Morin, P.; Jiménez-Flores, R.; Pouliot, Y. Effect of Temperature and Pore Size on the Fractionation of Fresh and Reconstituted Buttermilk by Microfiltration. *J. Dairy Sci.* **2004,** *87,* 267–273.

Morin, P.; Pouliot, Y.; Jiménez-Flores, R. A Comparative Study of the Fractionation of Regular Buttermilk and Whey Buttermilk by Microfiltration. *J. Food Eng.* **2006,** *77,* 521–528.

Morin, P.; Britten, M.; Jiménez-Flores, R.; Pouliot, Y. Microfiltration of Buttermilk and Washed Cream Buttermilk for Concentration of Milk Fat Globule Membrane Components. *J. Dairy Sci.* **2007,** *90,* 2132–2140.

Morin, P.; Poulio, Y.; Britten, M. Effect of Buttermilk Made from Creams with Different Heat Treatment Histories on Properties of Rennet Gels and Model Cheeses. *J. Dairy Sci.* **2008,** *91,* 871–882.

Mozafari, M. R.; Khosravi-Darani, K.; Borazan, G. G.; Cui, J.; Pardakhty, A.; Yurdugul, S. Encapsulation of Food Ingredients Using Nanoliposome Technology. *Int. J. Food Prop.* **2008,** *11,* 833–844.

Nakamura, T.; Kawase, H.; Kimura, K.; Watanabe, Y.; Ohtani, M.; Arai, I.; Urashima, T. Concentrations of Sialyloligosaccharides in Bovine Colostrum and Milk during the Prepartum and Early Lactation. *J. Dairy Sci.* **2003,** *86,* 1315–1320.

Niederau, C.; Strohmeyer, G.; Heintges, T.; Peter, K.; Göpfert, E. Polyunsaturated Phosphatidyl-choline and Interferon Alpha for Treatment of Chronic Hepatitis B and C: A Multicenter, Randomized, Double-blind, Placebo-controlled Trial. *Hepatogastroenterol.* **1998,** *45,* 797–804.

Noh, S. K.; Koo, S. I. Milk Sphingomyelin Is More Effective than Egg Sphingomyelin in Inhibiting Intestinal Absorption of Cholesterol and Fat in Rats. *Nutr. Metab.* **2004,** *134,* 2611–2616.

Nyberg, L.; Duan, R. D.; Nilsson, A. A Mutual Inhibitory Effect on Absorption of Sphingomyelin and Cholesterol. *J. Nutr. Biochem.* **2000,** *11,* 244–249.

O'Connell, J. E.; Kelly, A. L.; Auty, M. A. E.; Fox, P. F.; de Kruif, K. G. Ethanol-dependent Heat-induced Dissociation of Casein Micelles. *J. Agric. Food Chem.* **2001,** *49,* 4420–4423.

Ohlsson, L.; Burling, H.; Nilsson, A. Long Term Effects on Human Plasma Lipoproteins of a Formulation Enriched in Butter Milk Polar Lipid. *Lipids in Health and Disease* **2009,** *8,* 1–12.

Oshida, K.; Shimizu, T.; Takase, M.; Tamura, Y.; Yamashiro, Y. Effects of Dietary Sphingomyelin on Central Nervous System Myelination in Developing Rats. *Pediatr. Res.* **2003,** *53,* 589–593.

Otnaess, A. B.; Laegreid, A.; Ertresvåg, K. Inhibition of Enterotoxin from *Escherichia coli* and *Vibrio cholerae* by Gangliosides from Human Milk. *Infect. Immun.* **1983,** *40,* 563–569.

Palmquist, D. L.; Schanbacher, F. L. Dietary-fat Composition Influences Fatty-acid Composition of Milk-fat Globule-membrane in Lactating Cows. *Lipids* **1991,** *26,* 718–722.

Pan, X. L.; Izumi, T. Variation of the Ganglioside Compositions of Human Milk, Cow's Milk and Infant Formulas. *Early Human Dev.* **2000,** *57,* 25–31.

Park, C. W.; Drake, M. A. The Distribution of Fat in Dried Dairy Particles Determines Flavor Release and Flavor Stability. *J. Food Sci.* **2014,** *79,* R452–R459.

Park, E. J.; Suh, M.; Thomson, A. B. R.; Ramanujam, K. S.; Clandinin, M. T. Dietary Gangliosides Increase the Content and Molecular Percentage of Ether Phospholipids Containing 20:4n-6 and 22:6n-3 in Weanling Rat Intestine. *J. Nutr. Biochem.* **2006,** *17,* 337–344.

Patton, S.; Keenan, T. W. The Milk Fat Globule Membrane. *Biochim. Biophys. Acta.* **1975,** *415,* 273–309.

Patton, S.; Long, C.; Sokka, T. Effect of Storing Milk on Cholesterol and Phospholipid of Skim Milk. *J. Dairy Sci.* **1980,** *63,* 697–700.

Pepeu, G.; Pepeu, I. M.; Amaducci, L. A Review of Phosphatidylserine Pharmacological and Clinical Effects. Is Phosphatidylserine a Drug for the Ageing Brain? *Pharmacol. Res.* **1996,** *33,* 73–80.

Phan, T. T. Q.; Asaduzzaman, M.; Le, T. T.; Fredrick, E.; Van der Meeren, P.; Dewettinck, K. Composition and Emulsifying Properties of a Milk Fat Globule Membrane Enriched Material. *Int. Dairy J.* **2013,** *29,* 99–106.

Phan, T. T. Q.; Le, T. T.; Van der Meeren, P.; Dewettinck, K. Comparison of Emulsifying Properties of Milk Fat Globule Membrane Materials Isolated from Different Dairy By-products. *J. Dairy Sci.* **2014a,** *97,* 4799–4810.

Phan, T. T. Q.; Moens, K.; Le, T. T.; Van der Meeren, P.; Dewettinck, K. Potential of Milk Fat Globule Membrane Enriched Materials to Improve the Whipping Properties of Recombined Cream. *Intnl. Dairy J.* **2014b,** *39,* 16–23.

Plantz, P. E.; Patton, S. Plasma-membrane Fragments in Bovine and Caprine Skim Milks. *Biochim. Biophys. Acta.* **1973,** *291,* 51–60.

Plantz, P. E.; Keenan, T. W.; Patton, S. Further Evidence of Plasma-membrane Material in Skim Milk. *J. Dairy Sci.* **1973,** *56,* 978–983.

Poduval, V. S.; Mistry, V. V. Manufacture of Reduced Fat Mozzarella Cheese Using Ultrafiltered Sweet Buttermilk and Homogenized Cream. *J. Dairy Sci.* **1999,** *82,* 1–9.

Ponzin, D.; Mancini, C.; Toffano, G.; Bruni, A.; Doria, G. Phosphatidylserine-induced Modulation of the Immune-response in Mice—Effect of Intravenous Administration. *Immunopharmacology* **1989,** *18,* 167–176.

Pouliot, Y. Membrane Processes in Dairy Technology—From a Simple Idea to Worldwide Panacea. *Int. Dairy J.* **2008,** *18,* 735–740.

Puente, R.; GarciaPardo, L. A.; Rueda, R.; Gil, A.; Hueso, P. Seasonal Variations in the Concentration of Gangliosides and Sialic Acids in Milk from Different Mammalian Species. *Int. Dairy J.* **1996,** *6,* 315–322.

Rafalowski, R.; Zegarska, Z.; Kuncewicz, A.; Borejszo, Z. Oxidative Stability of Milk Fat in Respect to its Chemical Composition. *Int. Dairy J.* **2014,** *36,* 82–87.

Ric, R. G. Effect of Feed on the Composition of Milk Fat. *J. Dairy Sci.* **1991,** *74,* 3244–3257.

Rodriguez-Alcala, L. M.; Fontecha, J. Major Lipid Classes Separation of Buttermilk, and Cows, Goats and Ewes Milk by High Performance Liquid Chromatography with an Evaporative Light Scattering Detector Focused on the Phospholipid Fraction. *Journal of Chromatography A* **2010,** *1217,* 3063–3066.

Roesch, R. R.; Rincon, A.; Corredig, M. Emulsifying Properties of Fractions Prepared from Commercial Buttermilk by Microfiltration. *J. Dairy Sci.* **2004,** *87,* 4080–4087.

Rombaut, R. Enrichment of Nutritionally Advantageous Milk Fat Globule Membrane Fragments Present in Dairy Effluents. Ph.D. Thesis, Ghent University, 2006.

Rombaut, R.; Dewettinck, K. Properties, Analysis and Purification of Milk Polar Lipids. *Int. Dairy J.* **2006,** *16,* 1362–1373.

Rombaut, R.; Dewettinck, K. Thermocalcic Aggregation of Milk Fat Globule Membrane Fragments from Acid Buttermilk Cheese Whey. *J. Dairy Sci.* **2007,** *90,* 2665–2674.

Rombaut, R.; Dewettinck, K. Dairy Polar Lipids. In *Handbook of Dairy Foods Analysis*; Nollet, L. M. L., Toldrá, F., Eds.; CRC Press: Boca Raton, FL, 2010; pp 189–209.

Rombaut, R.; Camp, J. V.; Dewettinck, K. Analysis of Phospho- and Sphingolipids in Dairy Products by a New HPLC Method. *J. Dairy Sci.* **2005,** *88,* 482–488.

Rombaut, R.; Dejonckheere, V.; Dewettinck, K. Microfiltration of Butter Serum upon Casein Micelle Destabilization. *J. Dairy Sci.* **2006a,** *89,* 1915–1925.

Rombaut, R.; Van Camp, J.; Dewettinck, K. Phospho- and Sphingolipid Distribution during Processing of Milk, Butter and Whey. *Int. J. Food Sci. Tech.* **2006b,** *41,* 435–443.

Rombaut, R.; Dejonckheere, V.; Dewettinck, K. Filtration of Milk Fat Globule Membrane Fragments from Acid Buttermilk Cheese Whey. *J. Dairy Sci.* **2007a,** *90,* 1662–1673.

Rombaut, R.; Dewettinck, K.; Camp, J. V. Phospho- and Sphingolipid Content of Selected Dairy Products as Determined by HPLC Coupled to an Evaporative Light Scattering Detector (HPLC–ELSD). *Journal of Food Composition and Analysis* **2007b,** *20,* 308–312.

Rueda, R. The Role of Dietary Gangliosides on Immunity and the Prevention of Infection. *Br. J. Nutr.* **2007,** *98,* S68–S73.

Rueda, R.; Puente, R.; Hueso, P.; Maldonado, J.; Gil, A. New Data on Content and Distribution of Gangliosides in Human Milk. *Bio. Chem. Hoppe-Seyler* **1995,** *376,* 723–727.

Rueda, R.; Sabatel, J. L.; Maldonado, J.; Molina-Font, J. A.; Gil, A. Addition of Gangliosides to an Adapted Milk Formula Modifies Levels of Fecal *Escherichia coli* in Preterm Newborn Infants. *J. Pediatr.* **1998,** *133,* 90–94.

Ruiz-Gutierrez, V.; Perez-Camino, M. C. Update on Solid-phase Extraction for the Analysis of Lipid Classes and Related Compounds. *J. Chrom. A.* **2000,** *885,* 321–341.

Sachdeva, S.; Buchheim, W. Recovery of Phospholipids from Buttermilk Using Membrane Processing. *Kieler Milchw. Forsch.* **1997,** *49,* 47–68.

Sanchez-Juanes, F.; Alonso, J. M.; Zancada, L.; Hueso, P. Glycosphingolipids from Bovine Milk and Milk Fat Globule Membranes: A Comparative Study. Adhesion to Enterotoxigenic Escherichia coli Strains. *Biol. Chem.* **2009,** *390,* 31–40.

Schmelz, E. M.; Dillehay, D. L.; Webb, S. K.; Reiter, A.; Adams, J.; Merrill, A. H. Sphingomyelin Consumption Suppresses Aberrant Colonic Crypt Foci and Increases the Proportion of Adenomas versus Adenocarcinomas in CF1 Mice Treated with 1,2-Dimethylhydrazine: Implications for Dietary Sphingolipids and Colon Carcinogenesis. *Cancer Res.* **1996,** *56,* 4936–4941.

Schnaar, R. L.; Gerardy-Schahn, R.; Hildebrandt, H. Sialic Acids in the Brain: Gangliosides and Polysialic Acid in Nervous System Development, Stability, Disease, and Regeneration. *Phys. Rev.* **2014,** *94,* 461–518.

Shimizu, M.; Miyaji, H.; Yamauchi, K. Inhibition of Lipolysis by Milk Fat Globule Membrane Materials in Model Milk Fat Emulsion. *Agric. Biol. Chem.* **1982,** *46,* 795–799.

Singh, H. The Milk Fat Globule Membrane—A Biophysical System for Food Applications. *Curr. Opin. Colloid In.* **2006,** *11,* 154–163.

Singh, H. Aspects of Milk-protein-stabilised Emulsions. *Food Hydrocolloids* **2011,** *25,* 1938–1944.

Smolenski, G.; Haines, S.; Kwan, F. Y. S.; Bond, J.; Farr, V.; Davis, S. R.; Stelwagen, K.; Wheeler, T. T. Characterisation of Host Defence Proteins in Milk Using a Proteomic Approach. *J. Proteome Res.* **2007,** *6,* 207–215.

Snow, D. R.; Jimenez-Flores, R.; Ward, R. E.; Cambell, J.; Young, M. J.; Nemere, I.; Hintze, K. J. Dietary Milk Fat Globule Membrane Reduces the Incidence of Aberrant Crypt Foci in Fischer-344 Rats. *J. Agric. Food Chem.* **2010,** *58,* 2157–2163.

Snow, D. R.; Ward, R. E.; Olsen, A.; Jimenez-Flores, R.; Hintze, K. J. Membrane-rich Milk Fat Diet Provides Protection against Gastrointestinal Leakiness in Mice Treated with Lipopolysaccharide. *J. Dairy Sci.* **2011,** *94,* 2201–2212.

Sodini, I.; Morin, P.; Olabi, A.; Jiménez-Flores, R. Compositional and Functional Properties of Buttermilk: A Comparison Between Sweet, Sour, and Whey Buttermilk. *J. Dairy Sci.* **2006,** *89,* 525–536.

Spector, A. A.; Fletcher, J. E. Binding of Long Chain Fatty Acids to Beta-lactoglobulin. *Lipids.* **1970,** *5,* 403–411.

Spence, A. J.; Jimenez-Flores, R.; Qian, M.; Goddik, L. Phospholipid Enrichment in Sweet and Whey Cream Buttermilk Powders Using Supercritical Fluid Extraction. *J. Dairy Sci.* **2009,** *92,* 2373–2381.

Sprong, R. C.; Hulstein, M. F. E.; Van der Meer, R. Bactericidal Activities of Milk Lipids. *Antimicrob. Agents Chemother.* **2001,** *45,* 1298–1301.

Sprong, R. C.; Hulstein, M. F. E.; van der Meer, R. Bovine Milk Fat Components Inhibit Food-borne Pathogens. *Int. Dairy J.* **2002,** *12,* 209–215.

Starks, M. A.; Starks, S. L.; Kingsley, M.; Purpura, M.; Jager, R. The Effects of Phosphatidylserine on Endocrine Response to Moderate Intensity Exercise. *J. Int. Soc. Sports Nutr.* **2008,** *5,* 1–6.

Stremmel, W.; Braun, A.; Hanemann, A.; Ehehalt, R.; Autschbach, F.; Karner, M. Delayed Release Phosphatidylcholine in Chronic-active Ulcerative Colitis a Randomized, Double-blinded, Dose Finding Study. *J. Clin. Gastroenterol.* **2010a,** *44,* E101–E107.

Stremmel, W.; Hanemann, A.; Ehehalt, R.; Karner, M.; Braun, A. Phosphatidylcholine (Lecithin) and the Mucus Layer: Evidence of Therapeutic Efficacy in Ulcerative Colitis? *Digest. Dis.* **2010b,** *28,* 490–496.

Suh, M.; Belosevic, M.; Clandinin, M. T. Dietary Lipids Containing Gangliosides Reduce Giardia Muris Infection in Vivo and Survival of Giardia Lamblia Trophozoites in Vitro. *Parasitology.* **2004,** *128,* 595–602.

Surel, O.; Famelart, M. H. Ability of Ceramic Membranes to Reject Lipids of Dairy-products. *Aust. J. Dairy Technol.* **1995,** *50,* 36–40.

Thompson, A. K.; Singh, H. Preparation of Liposomes from Milk Fat Globule Membrane Phospholipids Using a Microfluidizer. *J. Dairy Sci.* **2006,** *89,* 410–419.

Thompson, A. K.; Haisman, D.; Singh, H. Physical Stability of Liposomes Prepared from Milk Fat Globule Membrane and Soya Phospholipids. *J. Agric. Food Chem.* **2006,** *54,* 6390–6397.

Thompson, A. K.; Mozafari, M. R.; Singh, H. The Properties of Liposomes Produced from Milk Fat Globule Membrane Material Using Different Techniques. *Lait.* **2007,** *87,* 349–360.

Tokes, T.; Eros, G.; Varszegi, S.; Hartmann, P.; Bebes, A.; Kaszaki, J.; Gulya, K.; Boros, M. Protective Effects of Phosphatidylcholine Pretreatment in Endotoxin-induced Systemic Inflammation in the Rat Hippocampus. *Acta Physiologica Hungarica* **2010,** *97,* 483–483.

Trachoo, N.; Mistry, V. V. Application of Ultrafiltered Sweet Buttermilk and Sweet Buttermilk Powder in the Manufacture of Nonfat and Low Fat Yogurts. *J. Dairy Sci.* **1998,** *81,* 3163–3171.

Treede, I.; Braun, A.; Sparla, R.; Kuhnel, M.; Giese, T.; Turner, J. R.; Anes, E.; Kulaksiz, H.; Fullekrug, J.; Stremmel, W., et al. Anti-inflammatory Effects of Phosphatidylcholine. *J. Biol. Chem.* **2007,** *282,* 27155–27164.

Trenerry, V. C.; Akbaridoust, G.; Plozza, T.; Rochfort, S.; Wales, W. J.; Auldist, M.; Ajlouni, S. Ultra-high-performance Liquid Chromatography-ion Trap Mass Spectrometry Characterisation of Milk Polar Lipids from Dairy Cows Fed Different Diets. *Food Chem.* **2013,** *141,* 1451–1460.

Tsioulpas, A.; Grandison, A. S.; Lewis, M. J. Changes in Physical Properties of Bovine Milk from the Colostrum Period to Early Lactation. *J. Dairy Sci.* **2007,** *90,* 5012–5017.

Udabage, P.; McKinnon, I. R.; Augustin, M. A. Mineral and Casein Equilibria in Milk: Effects of Added Salts and Calcium-chelating Agents. *J. Dairy Res.* **2000,** *67,* 361–370.

Vaghela, M. N.; Kilara, A. A Rapid Method for Extraction of Total Lipids from Whey-protein Concentrates and Separation of Lipid Classes with Solid-phase Extraction. *J. Am. Oil Chem. Soc.* **1995,** *72,* 1117–1121.

Vaisman, N.; Kaysar, N.; Zaruk-Adasha, Y.; Pelled, D.; Brichon, G.; Zwingelstein, G.; Bodennec, J. Correlation between Changes in Blood Fatty Acid Composition and Visual Sustained Attention Performance in Children with Inattention: Effect of Dietary n-3 Fatty Acids Containing Phospholipids. *Am. J. Clin. Nutr.* **2008,** *87,* 1170–1180.

van Boekel, M. A. J. S.; Walstra, P. Physical Changes in the Fat Globules in Unhomogenized & Homogenized Milk. *Bulletin of the IDF* **1989,** *238,* 13–16.

Vandana, V.; Karuna, M. S. L.; Vijayalakshmi, P.; Prasad, R. B. N. A Simple Method to Enrich Phospholipid Content in Commercial Soybean Lecithin. *J. Am. Oil Chem.* **2001,** *78,* 555–556.

Vanderghem, C.; Bodson, P.; Danthine, S.; Paquot, M.; Deroanne, C.; Blecker, C. Milk Fat Globule Membrane and Buttermilks: From Composition to Valorization. *Biotechnologie Agronomie Societe et Environnement* **2010,** *14,* 485–500.

Vazquez, E.; Gil, A.; Rueda, R. Dietary Gangliosides Positively Modulate the Percentages of Th1 and Th2 Lymphocyte Subsets in Small Intestine of Mice at Weaning. *Biofactors* **2001,** *15,* 1–9.

Veereman-Wauters, G.; Staelens, S.; Rombaut, R.; Dewettinck, K.; Deboutte, D.; Brummer, R. J.; Boone, M.; Le Ruyet, P. Milk Fat Globule Membrane (INPULSE) Enriched Formula Milk Decreases Febrile Episodes and May Improve Behavioral Regulation in Young Children. *Nutrition* **2012,** *28,* 749–752.

Vila, A. S.; Castellote-Bargallo, A. I.; Rodriguez-Palmero-Seuma, M.; Lopez-Sabater, M. C. High-performance Liquid Chromatography with Evaporative Light-scattering Detection for the Determination of Phospholipid Classes in Human Milk, Infant Formulas and Phospholipid Sources of Long-chain Polyunsaturated Fatty Acids. *J. Chrom. A.* **2003,** *1008,* 73–80.

Wada, A.; Hasegawa, M.; Wong, P. F.; Shirai, E.; Shirai, N.; Tan, L. J.; Llanes, R.; Hojo, H.; Yamasaki, E.; Ichinose, A., et al. Direct Binding of Gangliosides to *Helicobacter pylori* Vacuolating Cytotoxin (VacA) Neutralizes its Toxin Activity. *Glycobiology* **2010,** *20,* 668–678.

Walstra, P. Some Comments on the Isolation of Fat Globule Membrane Material. *J. Dairy Res.* **1985,** *52,* 309–312.

Walstra, P.; Wouters, J. T. M.; Geurts, T. J. *Dairy Science Technology,* 2nd ed; CRC Press: Boca Raton, FL, 2006.

Wang, B. Sialic Acid Is an Essential Nutrient for Brain Development and Cognition. *Annu. Rev. Nutr.* **2009,** *29,* 177–222.

Wang, B.; Miller, J. B.; McNeil, Y.; McVeagh, P. Sialic Acid Concentration of Brain Gangliosides: Variation among Eight Mammalian Species. *Comp. Biochem. Phys. A.* **1998,** *119,* 435–439.

Wang, B.; Brand-Miller, J.; McVeagh, P.; Petocz, P. Concentration and Distribution of Sialic Acid in Human Milk and Infant Formulas. *Am. J. Clin. Nutr.* **2001a,** *74,* 510–515.

Wang, X.; Hirmo, S.; Willen, R.; Wadstrom, T. Inhibition of *Helicobacter pylori* Infection by Bovine Milk Glycoconjugates in a BALB/cA Mouse Model. *J. Med. Microbiol.* **2001b,** *50,* 430–435.

Wang, B.; Yu, B.; Karim, M.; Hu, H.; Sun, Y.; McGreevy, P.; Petocz, P.; Held, S.; Brand-Miller, J. Dietary Sialic Acid Supplementation Improves Learning and Memory in Piglets. *Am. J. Clin. Nutr.* **2007,** *85,* 561–569.

Ward, R. E.; German, J. B.; Corredig, M. Composition, Applications, Fractionation, Technological and Nutritional Significance of Milk Fat Globule Membrane Material. In *Advanced Dairy Chemistry*; Fox, P. F., McSweeney, P. L. H., Eds.; Springer: New York, 2006; Vol. 2, pp 213–138.

Watanabe, S.; Takahashi, T.; Tanaka, L.; Haruta, Y.; Shiota, M.; Hosokawa, M.; Miyashita, K. The Effect of Milk Polar Lipids Separated from Butter Serum on the Lipid Levels in the Liver and the Plasma of Obese-model Mouse (KK-A(y)). *J. Funct. Foods.* **2011,** *3,* 313–320.

Welsch, U.; Buchheim, W.; Schumacher, U.; Schinko, I.; Patton, S. Structural, Histochemical and Biochemical Observations on Horse Milk-fat-globule Membranes and Casein Micelles. *Histochemistry* **1988,** *88,* 357–365.

Wiking, L.; Nielsen, J. H.; Båvius, A.-K.; Edvardsson, A.; Svennersten-Sjaunja, K. Impact of Milking Frequencies on the Level of Free Fatty Acids in Milk, Fat Globule Size, and Fatty Acid Composition. *J. Dairy Sci.* **2006,** *89,* 1004–1009.

Wilde, P.; Mackie, A.; Husband, F.; Gunning, P.; Morris, V. Proteins and Emulsifiers at Liquid Interfaces. *Adv. Colloid Interface Sci.* **2004,** *108,* 63–71.

Wong, P. Y. Y.; Kitts, D. D. A Comparison of the Buttermilk Solids Functional Properties to Nonfat Dried Milk, Soy Protein Isolate, Dried Egg White, and Egg Yolk Powders. *J. Dairy Sci.* **2003,** *86,* 746–754.

Yamazaki, M.; Inoue, A.; Koh, C.-S.; Sakai, T.; Ishihara, Y. Phosphatidylserine Suppresses Theiler's Murine Encephalomyelitis Virus-induced Demyelinating Disease. *J. Neuroimmunol.* **1997,** *75,* 113–122.

Ye, A.; Singh, H.; Taylor, M. W.; Anema, S. Interactions of Whey Proteins with Milk Fat Globule Membrane Proteins during Heat Treatment of Whole Milk. *Lait.* **2004,** *84,* 269–283.

Yuyama, Y.; Yoshimatsu, K.; Ono, E.; Saito, M.; Naiki, M. Postnatal Change of Pig Intestinal Ganglioside Bound by *Escherichia Coli* with K99 Fimbriae. *J. Biochem-Tokyo.* **1993,** *113,* 488–492.

Zadow, J. G. Alcohol-mediated Temperature-induced Reversible Dissociation of the Casein Micelle in Milk. *Aust. J. Dairy Technol.* **1993,** *48,* 78–81.

Zancada, L.; Perez-Diez, F.; Sanchez-Juanes, F.; Alonso, J. M.; Garcia-Pardo, L. A.; Hueso, P. Phospholipid Classes and Fatty Acid Composition of Ewe's and Goat's Milk. *Grasas y Aceites* **2013,** *64,* 304–310.

Zhu, D.; Damodaran, S. Dairy Lecithin from Cheese Whey Fat Globule Membrane: Its Extraction, Composition, Oxidative Stability, and Emulsifying Properties. *J. Am. Chem. Soc.* **2013,** *90,* 217–224.

Zou, X. Q.; Guo, Z.; Huang, J. H.; Jin, Q. Z.; Cheong, L. Z.; Wang, X. G.; Xu, X. B. Human Milk Fat Globules from Different Stages of Lactation: A Lipid Composition Analysis and Microstructure Characterization. *J. Agric. Food Chem.* **2012,** *60,* 7158–7167.

6

Phosphatidylserine: Biology, Technologies, and Applications

Xiaoli Liu ▪ *R&D Center, Nagase Co., Ltd., Hyogo, Japan*

Misa Shiihara, Naruyuki Taniwaki, Naoki Shirasaka, Yuta Atsumi, and Masatoshi Shiojiri ▪ *R&D Section, Biochemical Division, Nagase ChemteX Corporation, Kyoto, Japan*

Introduction

Like other glycerophospholipids, phosphatidylserine (PS) exists as one of the components of the bilayer membrane of a cell. Though not as abundant as phosphatidylcholine (PC) and phosphatidylethanolamine (PE or cepharin), PS is one of the most important phospholipids, both from biological function and dietary nutrition points of view. PS is located asymmetrically in the cell membrane and mostly in the inner leaflet of the cell. Besides maintaining membrane integrity, it is also involved in various cell activities through binding to and activating on several classes of enzymes and supplying a negatively charged surface in order to recruit positively charged molecules such as proteins. The loss of asymmetricity signals events such as cell apoptosis and blood clotting in mammals.

As a member of the glycerophospholipid class, the PS molecule has a glycerol moiety esterified by fatty acids at the first and second positions of the glycerol molecule, and the head group at the third position is the amino acid serine esterified to phosphate (see Figure 6.1). Therefore, it is negatively charged under physiological conditions, unlike PC and PE that are zwitterionic. This physical property is very important for PS as it lays the foundation for the biological functions of PS within the cell. The acyl groups at the first and second position are often different fatty acids—at the first position a saturated fatty acid and the second position an unsaturated one. The fatty acid compositions of PS are origin dependent and tissue specific. For example, PS from soy origin (SOY-PS) is rich in linoleic acid, whereas that from egg yolk contains primarily palmitic acid. PS in marine organisms is enriched with polyunsaturated fatty acids (PUFA) (Boselli et al., 2012; Uddin et al., 2011). The distribution of PS varies among animal tissues, being more abundant in retina and brain tissues than other organs (Hicks et al., 2006), and the fatty acid composition of PS in brain is rich in docosahexaenoic acids (Abbott et al., 2013).

From late 1980s, studies on the therapeutic importance of PS as a functional diet material became popular; these studies continue today. The early studies involved using materials extracted from bovine brain (BC-PS) (Pepeu et al., 1996). Administration of BC-PS on aged rats showed improvement on behavior related to memory (Nunzi et al., 1992); BC-PS intake by elder people also showed positive impact on

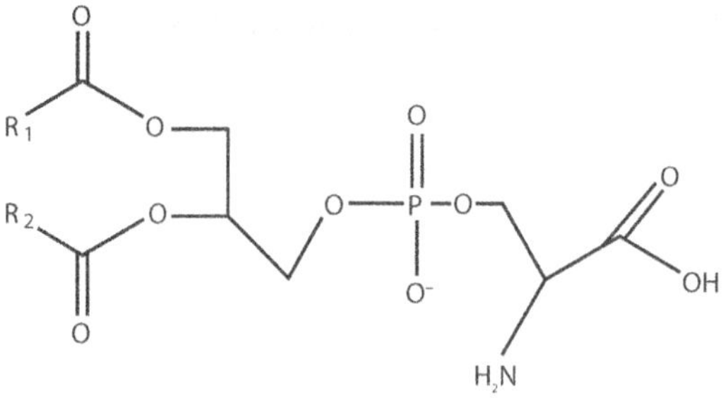

Figure 6.1 Molecular structure of PS.

improvement of age-related brain dysfunction (Cenaachi et al., 1993; Delwaide et al., 1986) and depressive symptoms (Maggioni et al., 1990); and intake by healthy subjects showed improvement on brain performance (Rosadini et al., 1990) and stress alleviation (Monteleone et al., 1992). It was not until late the 1990s that alternatives of BC-PS emerged on the market as a result of solving supply problems caused by bovine spongiform encephalopathy (BSE, commonly mad cow disease). From then on, technologies have been developed to produce PS from readily-available and safer plant lecithins, such as that from soy. Clinical studies on soy-derived PS also showed positive effects on improving age-related memory impairment (Richter et al., 2013).

In recent years, with the help of emerging technologies, elucidation of the biological function of PS has made great progress. For example, locations of PS within a living cell can be monitored through genetically encoded probes (Kay and Grinstein, 2011). PS was detected not only in the plasma membrane, but also in endosomal membranes, leading to further elucidation of its metabolism and functions. In this chapter, rather than doing a thorough scientific research review, the authors would like to concentrate on new findings about the functionality of this important phospholipid, including the authors' works, from a practical point of view.

Occurrence in Nature

PS is found not only in animals, within which it is distributed unevenly among tissues, but also in plants and microbial resources. It is typically extracted together with other phospholipids by solvents (e.g., chloroform:methanol, 2:1) and separated from other solvent-dissolving materials such as triglycerides and fatty acids by precipitation with acetone. PS can be easily differentiated from other glycerophospholipids such as PC, PE, PI, and PA by UV or evaporative light scattering detected silica gel chromatography. Different species of PS can be detected through mass spectrometry. Nowadays methods for extraction and quantification of PS are well established and can be referred to in other chapters or literatures (Myers et al., 2011).

The most commonly available sources are soy and milk. Soy lecithin contains approximately 0.5% PS (Weber, 1981), whereas in milk the content is about 5–10% of the total phospholipids (Contarini and Povolo, 2013). There are trace amounts of PS in vegetable oils such as corn oil, almond oil, and sunflower oil (Boukhchina et al., 2004), as well as in grains (Liu et al., 2013). The existence of PS is more abundant in animal tissues, especially in organs, retinas, and brains. The origin of human consumption for PS relies on diet such as meat, egg, milk, grains, and other materials (see Table 6.A). Fish milt (or seminal fluid), a common food in Japan, contains a very high percentage of PS, with docosahexaenoic acid as the main fatty acid composition (Bell et al., 1997).

Biosynthesis and Metabolism

PS is synthesized by base-exchange of PC and PE in mammalian cells through enzymatic reactions. The enzymes involved are phosphatidylserine (PtdSer) synthetase I and II. In microorganisms, it is synthesized through cytidine diphosphatediacylglycerol from phosphatidic acid. Studies suggest that the substrate specificity of PtdSer synthetase prefers PUFA species in the brain, indicating the existence of an enrichment mechanism of PUFA in the PS molecule (Kim et al., 2004). This might have close relationship with its biofunction. On the other hand, catabolism of PS occurs in two ways. The first is hydrolysis by secretory enzyme phospholipase A1 and/or A2, producing monoacyl phosphatidylserine (lysophosphatidylserines) or glycerol-3-phosphoserine.

Table 6.A Phospholipid Content in Food

	Total Lipid g/100g Food	Total Phospholipid mg/100g Food	PC	PE	PI	PS	SM	LPC	PA
Soybean	20.8	2,038	917	536	287	120[a]	—	—	102
Egg yolk	31.8	10,306	6771	1917	64	0	486	419	—
Milk	3.7	34	12	19	1	2	9	—	—
Beef	4.1	660	407	207	—	—	46	—	—
Wheat	1	251	47	10	—	—	78	—	—
Apple	0.09	40	21	10	0.4	6	—	—	0.6
Potato	0.15	76	38	22	1	12	—	—	0.4

Notes: PC: phosphatidylcholine; PE: phosphatidylethanolamine; PI: phosphatidylinositol; PS: phosphatidylserine; SM: sphingomyerine; LPC: lyso-phosphatidylcholine; PA: phosphatidic acid.

[a]Negishi et al. (1967); Weber (1981).

Source: Weihrauch and Son (1983).

Lysophosphatidylserines are important lipid mediators in resolving inflammation and restoring tissue functions (Frasch and Bratton, 2012). The second is by a decarboxylase conversion to PE. The metabolisms of PE and PS are closely related. The fact that PS can be synthesized from PE through PtdSer synthetase and convert back to PE through decarboxylase indicates that the amount of this molecule is tightly regulated inside the cell (Kuge and Nishijima, 2003), which again implicates a close relationship with its biological functions.

Biological Functions

PS, being the main negatively charged phospholipid and located in the inner leaflet in the plasma membrane, plays important biological roles. The unsaturated fatty acids at the *sn*-2 position give fluidity to the cell membrane (Nojima et al., 1994). When PS exists in substantial amount, the negatively charged molecule provides electrostatic forces that attract cations or positively charged domains (such as the C2 domain) of proteins (Yeung et al., 2008, 2009). Once bound to the positively charged domains of an enzyme (protein), PS can change the molecular structure of the protein and modulate its activities, indirectly participating in cell metabolism. The most well-known enzyme modulated by PS is Na^+/K^+ ATPase that binds to PS and is activated by it (Tsakiris and Deliconstantinos, 1984). Others include Ras and Rho-family GTPase and the tyrosine kinase and other calcium-dependant protein kinases (Yeung et al., 2008).

PS is also reported to have antioxidant activity on iron ion–induced lipid peroxidation and an inhibitory effect on hypochlorous acid (HOCl)–derived membrane modification, which may suggest a self-protection mechanism of cells from oxidative stress and inflammation (Dacaranhe and Terao, 2001; Kawai et al., 2006). Recently, the significance of a high concentration of PS enriched with DHA at the *sn*-2 position in the brain and retina was elucidated (Palacios-Pelaez et al., 2010). The high concentration of polyunsaturated fatty acid (PUFA)-conjugated PS seemed to serve as a reservoir of DHA which acts as a precursor for protectin (NPD1) formation, a process resulting from cell response to oxidative stress. NPD1 is responsible for protecting neural cells and restoring retinal cell integrity.

PS is a key signal in the blood-clotting process. Activated platelet cells expose PS to the cell surface and then trigger a series of factors related to clotting by a recognition process on PS (Zwaal, 1998). Another extracellular modulation activity of PS is being involved in cell apoptosis. Apoptotic cells expose PS to their surface and PS is then recognized by macrophage receptors, which triggers engulfment of the dead cell, hence reducing further damage caused by factors secreted from these cells (Fadok et al., 2001). PS is also found on the surface of cancer cells and stressed tumor cells (Utsugi et al., 1991), and it is related to various pathological conditions including chronic autoimmunity and infections (Zwaal et al., 2005).

In mammalian cells, PS serves as a precursor for PE, which also has numerous functions, such as membrane fusion/fission and cell signaling, in addition to keeping the integrity of the plasma membrane. The physical property of PS is also important in bone formation, in which case PS is potently chelated by calcium ion to form a tightly structured lipid–calcium complex (Merolli and Santin, 2009).

Nutritional Applications

Commercially Available Products

Commercially available PS products are mostly derived from soy. For example, Leci-PS™ 90 is a brand of highly purified PS (~90%) produced from soy PC by enzymatic technology. PS can also be obtained by using lecithin from less common resources such as marine organisms (Pinsolle et al., 2013) and squid skin (Hosokawa et al., 2000), which yields products rich in PUFA. There are other products with less content of PS in the form of complexes and mixtures with other functional glycerophospholipids such as PA (Lipogen PAS, manufactured by Lipogen) or PI (PIPS™ NAGASE, manufactured by Nagase ChemteX Corporation). There are also products formulated with docosahexaenoic acid (DHA), gingko extracts, and other functional ingredients.

Semisynthesized PS with homogeneous fatty acid molecules attached at the first and second positions of the glycerol backbone, such as dioleoylphosphatidylserine (DOPS) and dimyristoylphosphatidylserine (DMPS), are used as liposome agents for drug delivery (Sen et al., 2002).

Supplements for Neurodisorders

Age-Related Cognitive Dysfunction

Clinical studies on PS extracted from bovine cortex (BC-PS) strongly supported the use of this material in brain supplements. However, the availability issue of the material due to the risk of BSE has initiated efforts to find safe and accessible alternatives. PS obtained by converting soy lecithin, a byproduct from soybean oil processing, through an enzyme reaction in the presence of L-serine emerged as the primary choice because of the abundance of soy lecithin. World soy lecithin production is estimated in the range of 100,000 tons, which is only part of the total recoverable amount (Nieuwenhuyzen, 2014). However, because BC-PS and soy-derived PS (SOY-PS) only partly share their molecular structure and are very different in the fatty acid compositions on the *sn*-1 and *sn*-2 positions, questions arose about SOY-PS: Does it have the same functionality as BC-PS?

Several preclinical studies performed on rodents with SOY-PS gave positive results for improvement of age-related cognitive performance (Blokland et al., 1999;

Sakai et al., 1996; Suzuki et al., 2001). However, early clinical studies on human subjects showed controversial results. Jorissen et al. (2001) performed tests on 81 subjects with age-associated memory impairment (AAMI) by administrating of 300–600 mg/day of SOY-PS for a period of 12 weeks. The test was a double-blind, placebo-controlled study that turned out not to be able to demonstrate the efficacy of SOY-PS on improving cognitive performance. On the contrary, studies conducted by Gindin et al. (1998) and Schreiber et al. (2000) have showed positive effects of SOY-PS on improving age-associated cognitive functions.

Recently, two clinical studies on the administration of soy-derived PS on healthy elderly subjects with possible AAMI were conducted and both showed its effectiveness in improving cognitive performance. One is a pilot study conducted by Richter et al. (2013). The other is a double-blind, random-controlled study performed by Kato-Kataoka et al. (2010). In both cases, subjects over 50 years old and with subjective memory complaints were chosen. These studies conclusively suggest that SOY-PS has likely beneficial effects on improving the symptoms of age-related memory decline.

To summarize, the effect of SOY-PS on age-related cognitive dysfunction is probably not as strong as that of BS-PS, but it can be recommended for use as food supplement for non-demented elderly for maintaining a state of well-being. Combining soy-derived phosphatidylserine with other antioxidants or brain tonic agents might give synergetic effects on improving age-related memory impairment (Zanotta et al., 2014).

Brain phospholipids are enriched with PUFAs. It is clear that dietary PUFAs directly influence the membrane integrity of the brain and consequently its functioning (Filburn, 2000). The difference of the fatty acid species between bovine cortex–originated PS and soy-derived PS is that the BC-PS is rich in PUFAs, especially DHA, and the soy-derived PS has none at all (Chen and Li, 2008). Therefore, it was postulated that a PS of the same molecular structure like that present in bovine cortex could be the solution for a BC-PS substitute. Oral administration of PS obtained from marine resources, such as krill, improved the learning ability and memory of both aged and young rats (Park et al., 2012, 2013). Krill oil is an abundant marine resource rich in phospholipids (~30%) according to Winther et al. (2011). Krill PS can be obtained by a transphosphatidylation reaction of phospholipase D with L-serine. The effect of krill PS has also been studied with a brain disease model mouse, senescence-accelerated mouse (SAM) (Wang et al., 2012). Oral treatment of krill PS partly repaired age-related neurodegenerative symptoms by inhibition of ionized calcium binding adaptor molecule 1 (IBA-1) and by up-regulating insulin-like growth factor 1 (IGF-1). Antioxidation function is also observed, such as increased glutathione peroxidase activities corresponding to reduced lipid oxidation. Recently, Vakhapova et al. (2010, 2014) performed a double-blind, placebo-controlled clinical trial with krill-originated PS followed by an open-label extension study. These studies were performed on subjects carefully chosen and belonging to the category of non-demented subjects that, in other words, are healthy elderly with naturally age-related memory complaints. The results are promising, demonstrating significant improvements on cognitive performances such as immediate verbal recall, learning abilities, and complex figure copying. The following open-label extension study confirmed these favorable results. To conclude, collective studies so far suggest that PS, a major phospholipid in the brain, could be used as a food supplement in helping age-related cognitive function. PUFA-enriched PS, such as that from animal origins, especially that from marine organisms, is probably a better choice. The easily obtainable SOY-PS may have lesser effects, but could be compensated by combining it with other brain food supplements such as Gingko extracts and/or DHA. Technologies for converting safe and sustainable plant raw materials into PUFA-enriched animal PS molecules need to be developed in order to meet the growing demands due to an increasing aging population.

Attention-Deficit Hyperactivity Disorder (ADHD)

Attention-deficit hyperactivity disorder (ADHD) is a common mental disorder diagnosed in childhood. The symptoms are typically lack of ability to concentrate or to focus, rather impulsive actions, and hyperactivity. Often the symptoms start before 7 years of age and can continue through adulthood. The cause of this disorder is not completely elucidated, but several reasons are speculated, such as genetic biological deficit in the brain structure, infection, and nutritional deficiencies (Pellow et al., 2011). There are no sufficiently effective medicines to cure ADHD, other than some that usually bring unfavorable side effects. Health care for children with ADHD symptoms is mostly carried out at school together with supports from both physicians and family units. There is a strong need for safer and more effective materials both of pharmacological and supplemental importance. Collective evidences of PS administration on the improvement of brain functions inspired efforts to elucidate the possibility that PS may play such a role. Hirayama et al. (2006, 2013) conducted studies on the effect of oral administration of SOY-PS on children with suspected ADHD. The randomized, double-blind, placebo-controlled clinical trial on 36 children with a dosage of 200 mg/day SOY-PS capsules showed significant improvements in ADHD, Attention Deficiency (AD), and Hyperactivity Disorder (HD) symptoms and further improved short-term verbal memory, inattention, and impulsivity compared to the placebo group. It was then concluded that SOY-PS can be used as a food supplement for children suffering from ADHD. Because soy-derived PS has been on the market for more than 20 years and no claims have been made so far concerning its tolerability, the material can be one of the choices to meet the criteria of market needs.

A double-blind, placebo-controlled clinical trial followed by open-label extension at a much larger scale was conducted on n-3 enriched PS by Manor et al. (2012). This clinical study demonstrated that a daily intake of 150 mg of the material significantly improved the ADHD index, and subsequent extension studies confirmed these positive effects. The trial lasted for 30 weeks, and standard blood biochemical and hematological safety parameters, blood pressure, heart rate, weight, and height were evaluated. No adverse side effects were found, indicating that the material is well tolerated. Scientific evidence so far strongly supports the efficacy of PS on ADHD no matter the differences of the PS molecular species.

Anti-Stress

Maggioni et al. (1990) reported clinical results of PS from bovine cortex on improving depressive disorders on geriatric patients. It was also suggested that parenteral administration of PS could improve mental conditions and relieve anxiety for women with severe menopause symptoms (Rachev et al., 2001). Supplementation of PS at 300 mg/day was shown to relieve mental stress and improve mood in young healthy adults (Benton et al., 2001). Oral administration of a soy-derived product containing PS and phosphatidic acid (PAS) also gave a similar effect (Hellhammer et al., 2004). Recently, a double-blind, placebo-controlled clinical trial on the administration of n-3 enriched PS suggested that it could be effective in improving chronic stress, especially on subjects in a high chronic stress state (Hellhammer et al., 2012). In another case, when healthy subjects were exposed to cognitive tasks as the induced stressor, for those who had taken phosphatidylserine before (for 42 days), the electroencephalography (EEG) analysis of their brains revealed a decreased beta-power in the right hemispheric frontal regions, suggesting a more relaxed mood state than placebo groups (Baumeister et al., 2008). It seems that PS, no matter the origin, is likely to have positive effects on mood control, although the mechanism is not fully understood.

We have carried out an open trial to study the effect of a soy-derived PS complex enriched with PI (PIPS) on improving mood or reducing mental stress. The complex is obtained by converting soy lecithin with L-serine by transphosphatidylation reaction of phospholipase D. Table 6.B and Table 6.C show the phospholipid composition and fatty acid profile of the material. The open trial was conducted with a total of 22 volunteers. The volunteers were given 400 mg of the material for 8 weeks. At the start and after 4 weeks, as well as at the end of the trial, the volunteers were subjected to the POMS test (Yokoyama et al., 1990). The answers were collected and analyzed. Results are shown in Figure 6.2 (p. 154). Oral administration of PIPS seemed to reduce fatigue and release mental tension.

Sports Supplements

Acute exercises induce active oxygen species in the body and damage cells and even cause cell death. PS is a known antioxidant (Kawai et al., 2006); therefore, it can be expected to improve athletes' mental conditions. On the other hand, oral administration of PS may help athletes to perform better by controlling mental stress. Jäger

Table 6.B PIPS™ Composition (Lot NX01013-083-01)[a]

Component	% (w/w)	Component	% (w/w)
PC	1.3	PI	20.9
LPC	2.6	PS	31.2
PE	8.6	Others	18.9
PA	16.2	Total	100
LPE	0.3		

[a]Soy phosphatidylserine as standard.

Table 6.C Fatty Acids Composition of PIPS™

	PIPS (Lot NX01013-083-021)	Ultralec P[a] (Lot 0501022U085)
C16:0	21.5	20.8
C18:0	3.3	3.1
C18:1	11.4	12.0
C18:2	56.5	56.6
C18:3	5.5	5.7
Others	1.8	1.8
Total	100	100

[a]Source of lecithin.

et al. (2007) reported a trial with golf players. Although not statistically significant, a daily intake of 200 mg PS showed a tendency for better stress management of the players compared with the placebo group. As a result, the performance scores were significantly better. Kinsley et al. (2006) suggested the possible ergogenic effects of PS by demonstrating that 750 mg daily intake of PS prolonged athletes' time to exhaustion. Although the mechanism on how PS influences on athletes in a positive way is not utterly clear, there is evidence to support that PS intake reduces cortisol in blood (Starks et al., 2008) and athletes showed improved cognitive functioning.

Alleviation of Metabolic Syndrome

Metabolic syndrome can be diagnosed in patients who are overweight, have fatty liver, or have lipid metabolism dysfunction, all of which are closely related to oxidation stress on the body. There are numerous causes of metabolic syndrome, such as alcohol abuse, overeating, and sometimes chronic inflammatory in body organs. Fatty

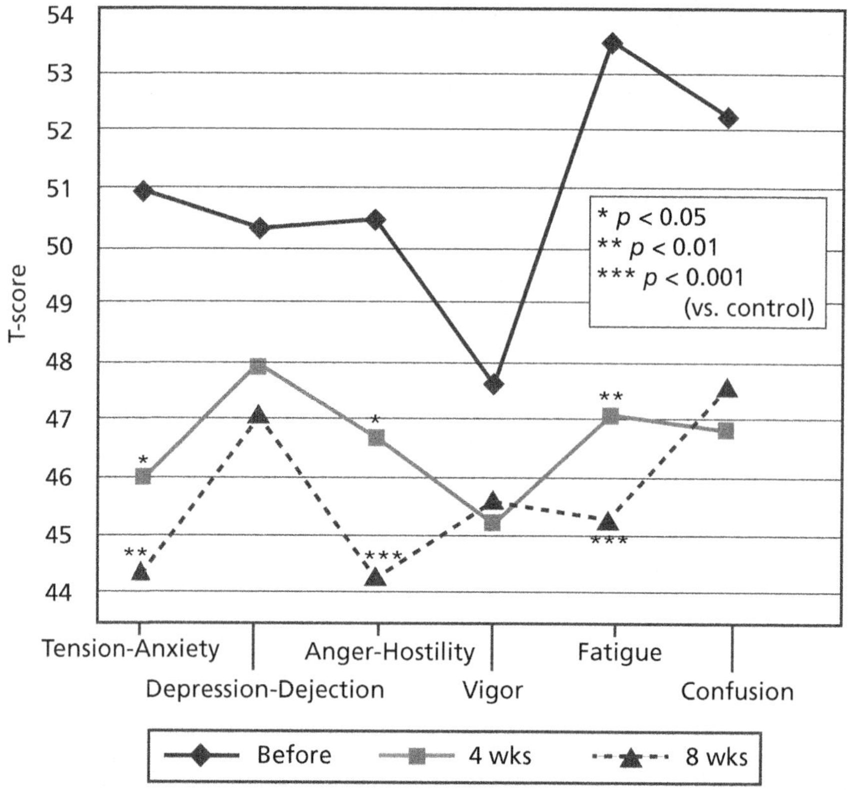

Figure 6.2 POMS test result.

liver, complicated with diabetic disease, is often observed in people with metabolic syndrome. Here we would like to report on a single-blind, placebo-controlled clinical trial of PIPS on subjects with metabolic syndrome.

A preclinical study of the complex performed on obese Zucker rats was reported by Shirouchi et al. (2010), demonstrating its efficacy on lipid metabolism and related liver functions. A follow-up preclinical study performed on obese Zucker rats with PS (>90%) and PI (>75%) separately indicated that PI is the primary factor for improving blood adiponectin, whereas PS is effective on adipocytes. Oral administration of PS significantly reduced the sizes of adipocytes (see Figure 6.3). PI is related to lipid metabolism and has been reported to prevent diet-induced obesity in mice and to lower cholesterol (Shimizu et al., 2010), while PS is an antioxidant. Therefore, it is interesting to find out the combined effect of PS and PI on improving metabolic syndromes.

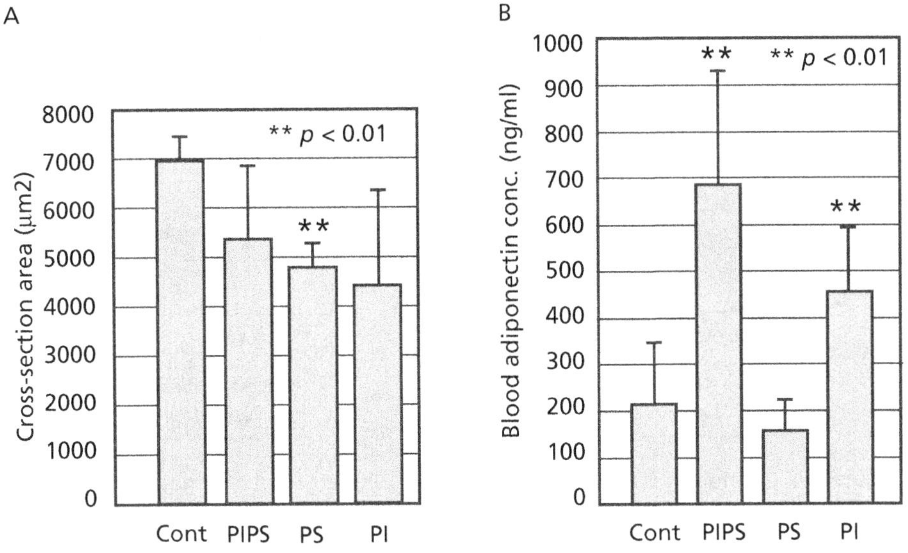

Figure 6.3 (A) Average cross-section is per 100 adipocytes; (B) blood adiponectin concentration.

Design of the Study

Method: Randomized, single-blind, placebo-controlled clinical trial

Subject: Total number of 40 (placebo $n = 20$, test $n = 20$) adults over 20 years of age with blood LDL over 140 mg/dl, chosen by physicians with professional training. The subjects were further grouped into male ($n = 11$) and female subgroups ($n = 29$), as well as metabolic syndrome subgroup according to the classification by Grundy et al. (2005) ($n = 13$) and obesity subgroup (body mass index [BMI] > 25) ($n = 21$). These subjects were evenly allocated to the test group and placebo group.

Dosage: 200mg PIPS in two tablets or placebo tablets, once daily after breakfast.

Trial period: 4 weeks

Monitoring index: Blood parameters before (baseline) and after 4 weeks.

Improvement on Lipid Metabolism

The ratio of serum LDL-cholesterol:HDL-cholesterol has been suggested as an efficient index for assessing the risk of arterial sclerosis rather than total cholesterol (Ingelsson et al., 2007). In our study, administration of PIPS at a level of 200 mg/day

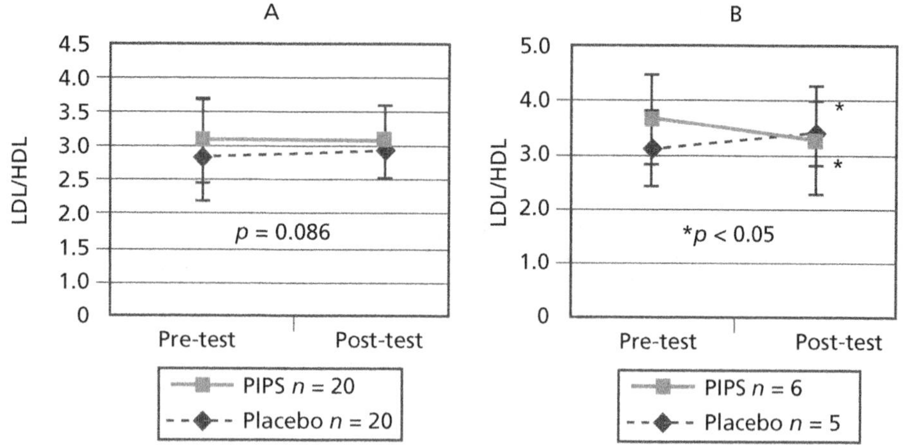

Figure 6.4 Change of LDL/HDL cholesterol (A) whole group; (B) male group.

showed a tendency of reduced LDL:HDL ratio compared to the placebo and, interestingly, it is statistically significant with male subjects (Figure 6.4).

Possibility of Protecting Liver

The index for liver function was also monitored. The test group showed a lowered aspartate aminotransferase (AST) level and a statistically significantly reduced lactate dehydrogenase (LDH) level (Figure 6.5). AST exists in heart and liver, and AST level indicates the functioning of these organs. On the other hand, LDH is an enzyme involved in energy metabolism and thus is ubiquitously expressed in the heart, liver, skeleton muscle, kidney, and other organs and is also an index of tissue damage. Abnormally high AST and LDH indicate possibilities of disorders in internal organs such as the liver and heart. Moreover, blood thymol turbidity tests (TTTs) also showed significant decreased levels in test group and not in the placebo group (Figure 6.6). (TTT is a screening tool for liver malfunction.) Therefore, the PI-enriched PS may be useful for improvement of liver function.

Reduced Risk of Cardiovascular Disease

Chronic inflammation is the cause of all sorts of diseases. Therefore, it is important to monitor the level of inflammation in order to find disease signs at the earliest possible stage. C-reactive protein (CRP) is produced in the liver and is an index for inflammatory conditions of the body. For example, CRP is related to type 2 diabetes and cardiac infarction. In this study, we found that high CRP in the PIPS test group was

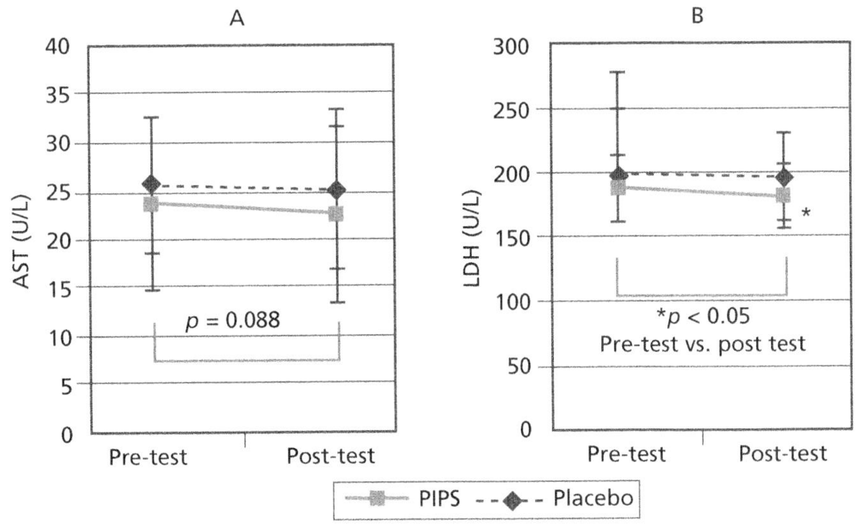

Figure 6.5 Results of PIPS (A) blood AST and (B) blood LDH analysis.

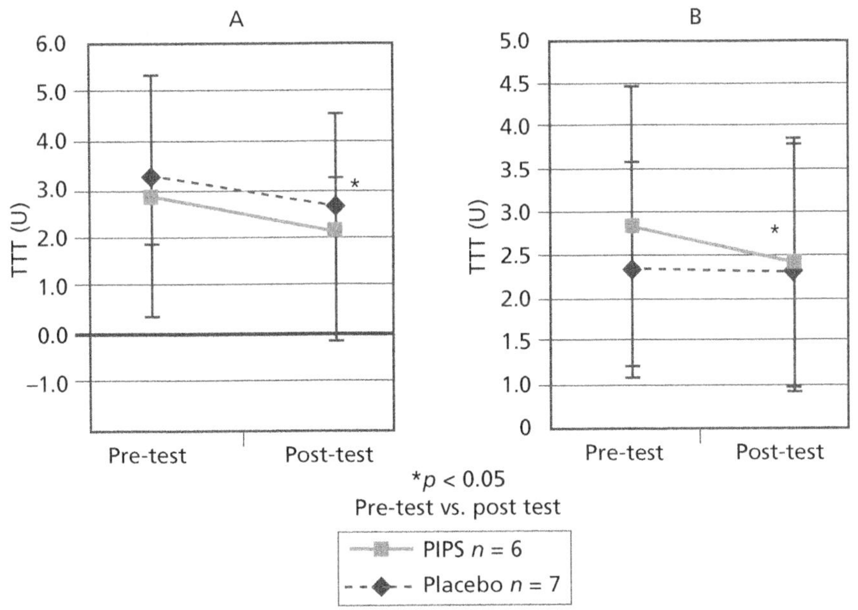

Figure 6.6 Results of blood TTT anaysis in (A) metabolic syndrome group and (B) obesity group (over BMI 25).

Table 6.D Changes on Parameters with High CRP Group

Parameters	Group	Before	After	Change
CRP (mg/ml)	PIPS	3.60 ± 2.73	2.37 ± 1.51[a]	−1.23 ± 1.30[a]
	Placebo	2.11 ± 1.13	2.99 ± 3.90	0.88 ± 2.96
Body fat rate (%)	PIPS	38.87 ± 7.85	38.29 ± 7.82[a]	−0.59 ± 0.50[a]
	Placebo	40.50 ± 8.53	40.27 ± 8.08	−0.23 ± 0.93
TG	PIPS	176.33 ± 74.56	138.00 ± 59.39[a]	−38.33 ± 32.97[a]
	Placebo	139.80 ± 71.89	146.20 ± 61.84	6.40 ± 30.69
TTT	PIPS	3.10 ± 2.00	2.31 ± 1.67[b]	−0.79 ± 0.53[b]
	Placebo	1.91 ± 1.00	1.88 ± 0.77	−0.03 ± 0.42

[a] $p < 0.05$
[b] $p < 0.01$

significantly reduced compared to the placebo group. At the same time, the body fat rate, blood triglyceride, and the TTT value were also reduced (Table 6.D).

Reduced Risk of Diabetes

Glucoalbumin (GA) is carbohydrated albumin caused by high concentration of blood glucose. It is an important index for blood glucose level in the previous weeks, and it is now used extensively for assessing risk of diabetic disease. Although the mechanism is unknown, we found in this study that administration of PIPS can significantly reduce the GA in blood compared to the placebo group, suggesting that using PIPS to prevent diabetes might be a good strategy (Figure 6.7).

To summarize, this clinical study gave encouraging results and promotes future investigations on the functions of this phospholipid complex. It is well known that both PS and PI are located in the inner leaflet of the plasma membrane. The functional role of these phospholipids is not only in maintaining the membrane integrity (physically), but also modulating cell metabolism in elaborated ways (chemically). PS is the activator of numerous intracellular proteins and enzymes, and the metabolic intermediates of phosphatidylinositol, such as PIP, PIP2, and PIP3, are important in cell signaling. Therefore, we presumed that the complex of PS and PI may have synergetic effects. Our study showed some interesting phenomena when this complex was taken as a dietary supplement. To certify it as an important dietary supplement, more integrated clinical trials are necessary, and elucidation of the mechanisms behind the phenomena is needed for future work. Although human consumption of soy products has a long history and there is little doubt about their safety, we conducted safety studies on PIPS nevertheless, including acute toxicity, sub-acute toxicity, chro-

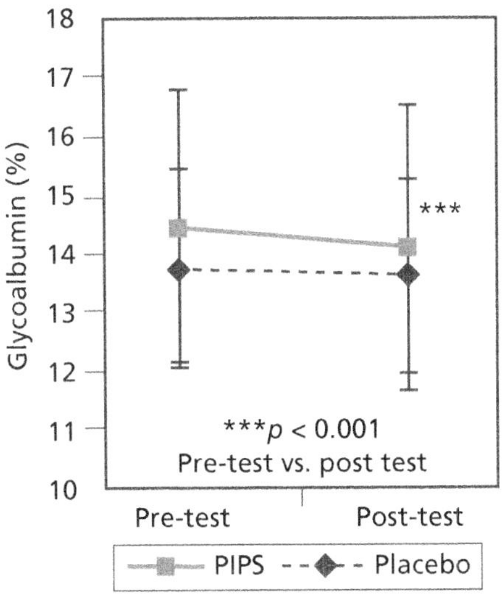

Figure 6.7 Results of glycoalbumin analysis.

mosome abbreviations, as well as the Ames test. These test results showed no adverse effects of the complex, suggesting its safety for consumption. It is noteworthy to mention that 200 mg of PIPS, 40mg PI, and 60mg PS contain the equivalent of 24 g (in the case of PI) and 600 g (in the case of PS) of dry soy. It is not always possible to consume this amount of soy every day; therefore, intake as supplements may help to maintain a healthy diet.

The clinical study confirmed the safety acknowledgement because the results of doctor inspections, physiological analysis of biological samples (blood, urine, etc.), as well as daily physical records, all supported the conclusion that PI-enriched PS is well-tolerated and there are no adverse effects related to its intake.

Skin Care

Skin is the largest organ of the human body and is exposed to the external environment and prone to all kinds of damages from both internal and external causes. PS is located in the inner leaflet of the cell membrane and plays important roles in skin protection. Besides maintaining the integrity of the cell membrane, it modulates enzyme activities that are important for maintaining healthy skin. For example, PS activates protein kinase C, which is involved in collagen synthesis and fibroblast cell proliferation (Zhang et al., 2004). One of the factors of skin aging is ultraviolet-induced reactive oxygen species. According to studies, PS on cell membrane is a preferred substrate of oxidation. The oxidation occurs not only on the *sn*-1 and *sn*-2 fatty acid chains but also on the polar head group, resulting in PS hydroxide and hydroperoxide derivatives and truncated *sn*-2 fatty acyl species, as well as glycerophosphoacetic acid derivatives (GPAAs) (Maciel et al., 2014). Oxidized PS derived from the fatty acid oxidation is then exposed on the cell surface to become the target for receptors of macrophage, initiating the process of apoptosis. This oxidation scavenging activity of PS suggests its physiological role in both cell apoptosis and maintaining cell health. Cho et al. (2008) gave the first report of PS as a possible skin therapeutic agent. Topical administration of PS on human skin significantly increased collagen, an important structural substance in the cell matrix, by inhibiting the activities of matrix metalloproteinase I (MMP-I) in skin fibroblasts. This effect was confirmed in vivo on both aged and young human skins. Moreover, when young skin was exposed to UV irradiation, PS was able to suppress the UV-induced expression of MMP-I activity, suggesting its usefulness in both intrinsic aging and aging caused by external UV irradiation. Here we would like to report our findings of PIPS complex for possible applications as skin care ingredients.

Enzyme Modulation

Inhibition of Hyaluronidase Activity in Vitro

Hyaluronic acid (HA) is an extracellular substance produced in the dermis and secreted in the skin matrix; it is important for moisture maintenance and skin elasticity. Aged skin has a decreased amount of hyaluronic acid (Oh et al., 2011). Although Terazawa et al. (2014) recently published that the decrease of HA in aged skin is attributed to down-regulation of hyaluronic acid syntheses, not hyaluronidase, inhibition of hyaluronidase may help to maintain the necessary HA level in the skin because HA is dynamically metabolized and has high turnover rates in tissues. To investigate the effect of phospholipids on hyaluronidase activity, hyaluronidase from bovine testis (Type-VI-S, SIGMA) was used as the probe enzyme. Various lecithin materials were suspended in 1% dimethylsulfoxide (DMSO) acetate buffer pH 4.0 to desired test concentrations. Activity was measured with or without lecithin using appropriately prepared enzyme solutions and according to the recommended protocols. The results are shown in Figure 6.8 and Table 6.E. PI exhibits high inhibition on hyaluronidase activity, with an IC50 at 25 μg/ml. Lysophosphatidic acid and phosphatidic acid are the second strongest inhibitors of the enzyme, while other lecithins such as PC or PE have no effect.

Activation of Hyaluronic Acid Synthetase in Vivo

Hyaluronic acid syntheses, especially hyaluronan synthase 2 (HAS2), are the key enzymes for the production of hyaluronic acids in skin (Terazawa et al., 2014). To investigate the effect of PIPS complex on the up-regulation of HAS2, human fibroblast cells (NHDF, Kurabo) were inoculated into six well plates at 8×10^5/ml concentra-

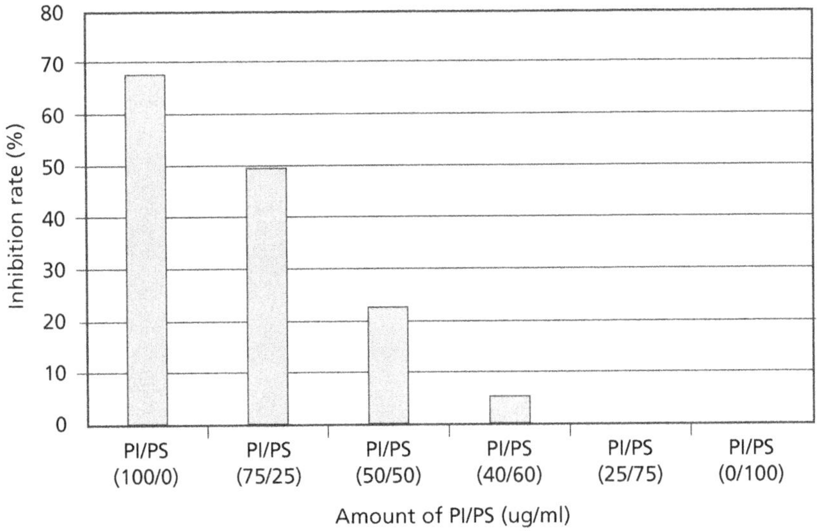

Figure 6.8 Inhibition of hyaluronidase by PI.

tion and pre-cultured at standard conditions for 24 hours. The medium was then discarded and exchanged with a medium containing PIPS (106S, Kurabo) and the cells were cultured for another 24 hours. The culture medium was discarded and RNA was extracted from the cells according to standard procedure (ISOGEN). RT-PCR was carried out for determination of the expressed HAS2 using a properly designed primer. β-Actin was used as internal standard. After the amplification reaction, products were differentiated on electrophoresis gel and the ratio of HAS2 to β-actin was calculated. The result is presented in Figure 6.9. The result shows that PIPS up-regulates hyaluronic acid synthetase activity significantly in vivo.

Boosting of Hyaluronic Acid Production in Vivo
The boosted hyaluronic acid production in vivo was confirmed by using human fibroblast cells. Human fibroblast cells (NHDF, Kurabo) were inoculated into 24-well

Table 6.E Comparison of the Inhibition Activities of Phospholipids on Hyaluronidase

Phospholipids	PC	PE	PI	PS	PA	LPC	LPA	PIPS	Lecithin	Lysolecithin
IC50 (μg/ml)	ND	ND	25	1000	50	800	40	25	75	75

Note: ND: not detectable.

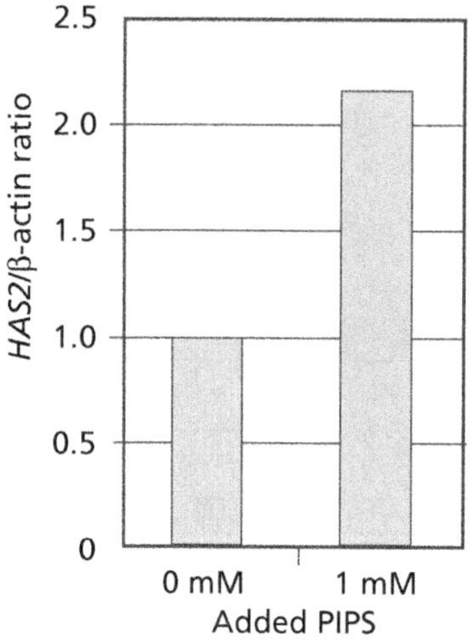

Figure 6.9 Up-regulation of hyaluronic acid synthetase gene (*HAS2*).

plate at 1 × 105/well concentration and pre-cultured at standard conditions for 24 hours. The medium was then discarded and exchanged with PIPS (0–10 μg/ml) or another phospholipid-containing medium (106S, Kurabo) and the cells were cultured for another 72 hours. After boiling and centrifuging to recover the supernatant, a hyaluronan test kit (Seikagaku Corporation) was used for the detection of the amount of hyaluronic acid. Results are presented in Figure 6.10. All phospholipids tested expressed effects on cell hyaluronic acid promotion, and PS had the strongest activity.

A 3-D human skin model was also used to test the consistency of the effect of PIPS on hyaluronic acid promotion. A 3-D human skin model was purchased from Toyobo (Testskin LSE-high). First, 5 mg/ml of solution containing PIPS in citrate buffer (control: citrate buffer only) was added to the cell, and the cell cultured at 37 °C in a CO_2 incubator for 24 hours. Afterward, cells were collected and actinase (KAKEN Pharma) was used to hydrolyze the cells, and after centrifugation, supernatant-containing hyaluronic acid was recovered and the amount of hyaluronic acid was determined by enzyme linked immunosorbent assay (ELISA). The result is shown in Figure 6.11. Compared with the amount of hyaluronic acid in control tissue, which is 620 ng/mg dry tissue, the PIPS-cultured cell tissue gave 910 ng/mg dry tissue of hyaluronic acid (~1.5 times increase).

Open Trial on Humans

Two males and two females aged 30–40 years took part in the trial. A lotion (glycerol 5% w/v, ethanol 5% v/v, methylparaben 0.1% w/v, propylparaben 0.05% w/v) with

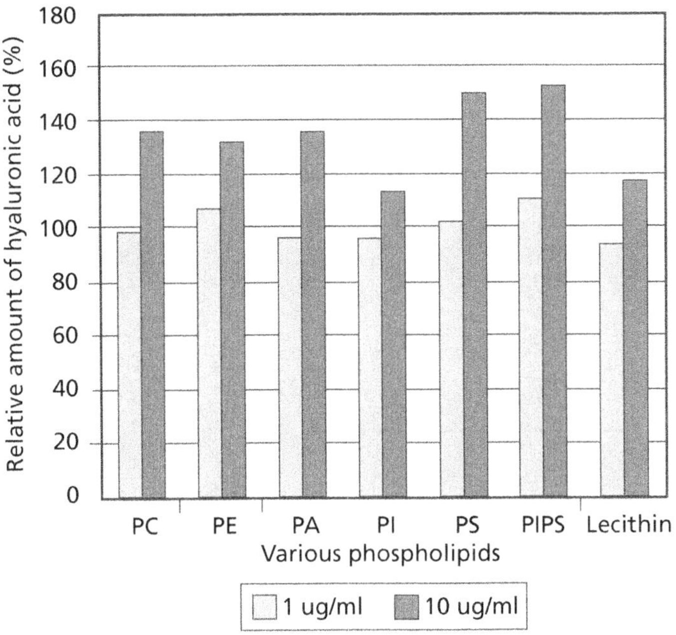

Figure 6.10 Boosting of hyaluronic acid production in fibroblast cells.

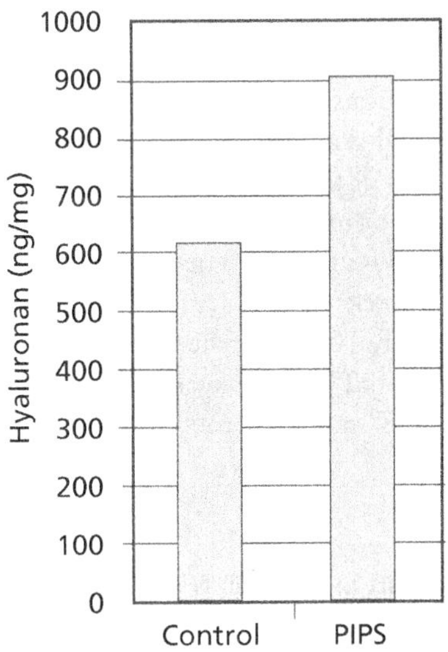

Figure 6.11 Boosting of hyaluronic acid production in 3D human skin model cells.

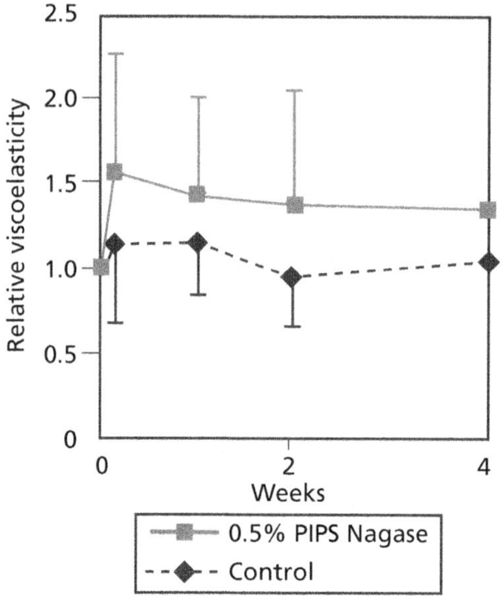

Figure 6.12 Result of open trial on humans.

PIPS (0.5%) was prepared according to a standard procedure. The subjects applied the lotion twice daily (morning and night) on the back of the hands for 4 weeks. The elasticity of the skin was measured before and after using a Cutometer (Cutometer SEM575; Courage-Khazaka). The relative elasticity was calculated and analyzed by paired t-test. Results are shown in Figure 6.12. There is a significant increase in the elasticity of the skin after applying the lotion containing PIPS ($p = 0.04685$). Subjective assessment was also monitored by answering questions, and the results are shown in Figure 6.13. The lotion containing PIPS gives good sense of use, and application on skin for 4 weeks significantly increased the softness of the skin.

As the result of increased skin elasticity, skin smoothness is improved and fewer wrinkles are expected. One such example is shown in Figure 6.14. The sample, a replica taken from around the eyes before and after 2 weeks of application of a cream containing 0.5% PIPS (compositions: 82.10% water, 10% caprylic, 5% glycerin, 1.4% sodium hydroxide, 0.6% acrylates, 0.5% PIPS, 0.2% phenoxyethanol, 0.15% methylparabene, 0.05% tocopherol), clearly indicates improvements on the depth and areas of wrinkles.

Effects on Skin Elasticity by Oral Administration

Three males and two females aged 30–40 years took part in the open trial. A daily oral administration of 400 mg PIPS was given for 4 weeks. The elasticity of the back of the hands at starting and finishing points was measured using a Cutometer (Cutometer

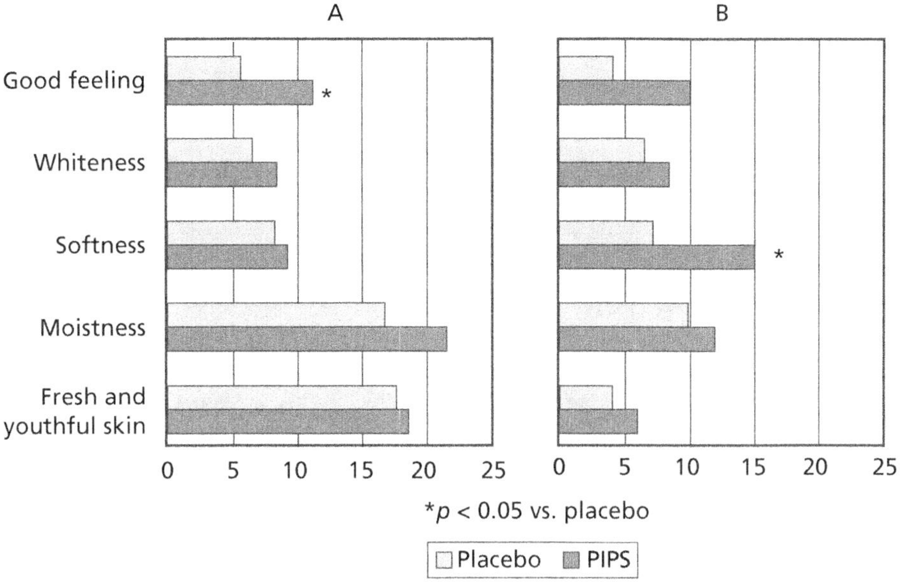

Figure 6.13 Sensory evaluation results (A) immediately after first application and (B) after two weeks of application.

SEM575; Courage+Khazaka). Data was analyzed by paired t-test. The result is shown in Figure 6.15. There is a clear increase in skin elasticity with administration of PIPS.

As demonstrated, various studies indicate the multifunctional possibilities of PS and PI complex. Most of these studies were carried out from a practical approach, emphasizing evidence collection. The results of these tests encourage controlled clinical studies in the future to confirm the positive effects, and further mechanism studies on how PS and PI works *in vivo* are needed to fully understand the complex's beneficial effects. Nevertheless, these are convincing results related to the application of the complex as a skin care ingredient. To promote this new application, a mutagenicity study, chromosome aberration analysis, a skin irritation test, and a standardized human skin patch test were performed on 13 volunteers. There were no abnormalities or adverse effects observed with topical application of the material.

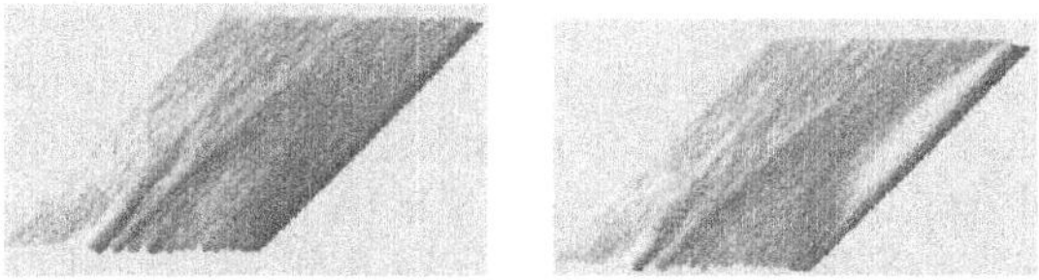

Figure 6.14 Effect of lotion containing PIPS on improving wrinkles (an example).

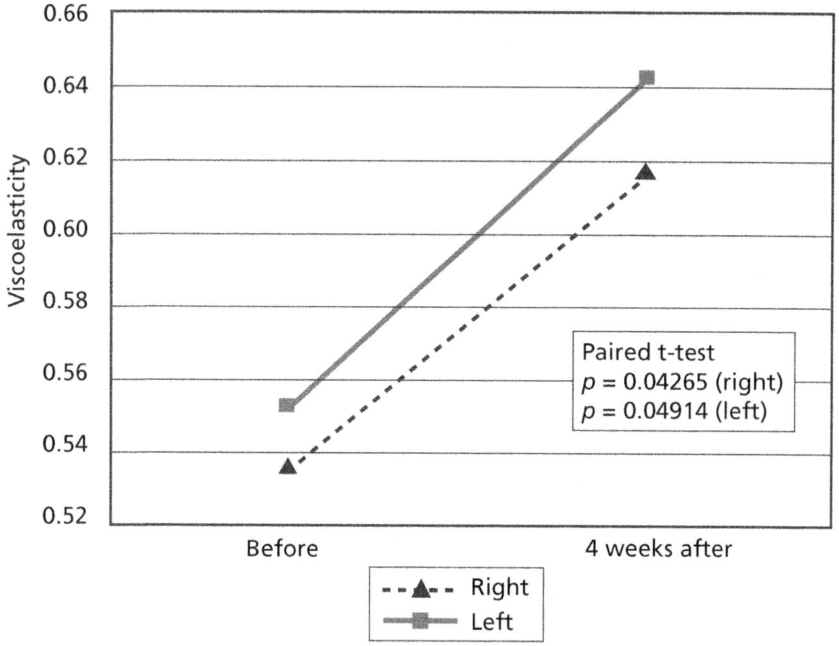

Figure 6.15 Improvement of skin elasticity by oral administration.

Other Applications

Other main applications of PS include liposome agents in drug delivery systems and as novel osteointegrative biomaterials. Several examples for using PS liposome showed improved efficacy of drugs (Sen et al., 2002; Tempone et al., 2004). Merolli and Santin (2009) showed the potential of using PS as a biomaterial in promoting calcification, taking advantage of its calcium-binding property.

Processing Technologies

Materials

Enzymes

Because of the scarcity of PS in natural resources, it is not economical to obtain PS through extraction, which involves using solvents and tedious purification procedures. Chemical synthesis is also not practical considering that phospholipids including PS have complicated structures. Alternatively, it is becoming common to produce PS through an enzymatic reaction catalyzed by phospholipase D (PLD) with lecithins as starting materials, which are abundantly available from plants and animal resources.

PLD hydrolyzes phospholipids into phosphatidic acid in vivo, but in the presence of alcohols, it also catalyzes the transphosphatidylation reaction between phospholipids and the alcohol, resulting in phospholipids with a modified polar head. If the alcohol is serine, lecithin is then converted into phosphatidylserine. PLD can be obtained from cabbage and several microorganisms. The transphosphatidylation rate depends on enzymes and reaction conditions (Juneja et al., 1988), but the cabbage-originated PLD is known to have a higher hydrolytic activity over transphosphatidylation activity, therefore a lesser conversion rate to PS is usually obtained. PLD from microorganisms, especially from *Streptomyces,* has been reported to have high transphosphatidylation activities (Carrea et al., 1995). Our group has successfully cloned a gene of phospholipase D from *Streptomyces cinnamoneum* and expressed the gene in a very closely related species of the same genus, *Streptomyces violaceoruber,* with several thousand times production rate than the original donor strain (Fukuda et al., 2011).

Lipase can be used to hydrolyze the fatty acid in the *sn*-1 position of PS and to further introduce a different fatty acid to the same position with the sum reaction described as transesterification. Totani and Hara (2003) reported a good transesterification rate by using mono- or diacyl glycerin as the fatty acid donor for the production of PUFA-enriched phospholipids, though the position of the introduced fatty acids is not clarified. The fatty acid at the *sn*-2 position is not easily transformed, probably due to structural steric hindrance. No lipase reported so far possesses such a property. The authors had experience in trying to esterify the fatty acid at the *sn*-2 position of PS by using a microbial enzyme of phospholipase A2, also isolated from a *Streptomyces sp*. We first hydrolyzed soy-originated PS in a water system. After the reaction, the free fatty acid released from the *sn*-2 position was removed from the system. Then a second fatty acid, together with newly added phospholipase A2, was incubated to allow the esterification reaction to occur. A very low conversion rate was observed (data not shown). A membrane-bound *o*-acyltransferase from the yeast, *Saccharomyces cerevisiae* (Tamaki et al., 2007), having a broad substrate specificity of lysophospholipids, might be useful for producing modified PS such as PUFA-enriched PS. However, the membrane-bound characteristics would probably make it difficult for application and, to our best knowledge, no commercial enzymes are available yet to efficiently modify PS at the *sn*-2 position, and it remains a challenge to produce 2-docosahexanoyl phosphatidylserine, the speculated species of PS related to cognitive function, from cheap and abundant lecithins such as soy.

Enzymatic modification is preferred over the synthetic method not only for economic reasons but also due to the suitability of the products as food materials from both a safety point of view and meeting global compliance. However, enzymes are usually expensive to produce. New technologies for enzyme production will facilitate the modification of lecithin and further broaden the application of phospholipid products. Our group has been engaged in developing a technology platform using the microorganism *Streptomyces*. *Streptomyces* dominates the soil microbial population and has the biggest genome in prokaryotes. Therefore, it is a rich source of gene-encoding enzymes. Furthermore, *Streptomyces* has developed several secretory systems through which enzymes produced are secreted outside the cells and into the culture medium. Therefore, the process for enzyme recovery is shortened, hence the production costs reduced. We have successfully developed enzymes such as PLD and lyso-phospholipase D, phospholipase C (a PI-specific phospholipase C and a sphingomyelinase), and phospholipase A2. These enzymes are used in developing technologies for phospholipid modification as well as directly in processing phospholipid-containing foods, which often results in improved food qualities and added value products.

Serine

Historically, the amino acid serine is produced by hydrolyzing proteins that are rich in serine content, such as animal hair, silk protein, and others, followed by separation using chromatography. The hydrolysis of these proteins involves using strong acids; therefore, it is neither environmentally friendly nor user friendly for operators. It is now the mainstream to produce serine by fermentation technology (Gu et al., 2014). Biotechnology has been developed to produce amino acids by using microorganisms at very economic rates (Becker and Wittmann, 2012).

Lecithin

Lecithin, the raw material for manufacturing PS, is easier to obtain from plants than from animal tissues. The most common ones are from soy, rapeseed, and sunflower. Other sources, such as rice, peanut, corn, and palm, are also available. Egg lecithin is the most common one from animal origin. In recent years, marine lecithin has received much attention because of its high content of n-3 fatty acids, especially lecithin from krill. Krill oil has a high content of phospholipids, mainly phosphatidylcholine. Table 6.F lists the composition of the major lecithins on the market.

The enzyme for converting lecithin to PS, namely PLD, preferably reacts on PC, and to a lesser extent on PE as substrates; highly-purified PC is often used in producing high content PS. Leci-PS™ is produced by using lecithin that has a content of PC as high as 90%. In soy lecithin powder, besides the major components (PC and PE), minor phospholipids such as PI and PA together exist at substantial amounts. These phospholipids are poor substrates for PLD. Naturally occurring PLD does not react at all on PI. The existence of these phospholipids in commercial soy lecithin powder inhibits the conversion rate of PC and PE to PS; therefore, more enzymes are needed when using such substrates. PS produced by using crude lecithin is a mixture of PS with residual substrates and nonreactive phospholipids, unless applied to an extra

Table 6.F Compositions of Major Commercial Lecithins

			Rapeseed Oil[d]					
Composition (%)[a]	**Soy[b]**	**Sunflower[c]**	**Hybridol**	**Pactol**	**Egg Yolk[e]**	**Milk[f]**	**Marine[g]**	**Squid[h]**
PC	23	65	6	19	83.2	26	68	80.5
LPC	—	—	—	—	0.7	—	—	—
PE	20	10	38	2	14.2	31.5	23	13.2
PA	8	6	1	25	—	—	—	—
PI	14	3	55	24	—	4.9	2	—
PS	—	—	—	—	—	8.8	2	—
SM	—	—	—	—	—	—	1	—
PG	—	—	—	30	—	—	—	—
Others	35	19	—	—	1.9	—	4	6.3

[a]Total phospholipids.
[b]Soy lecithin UntralecP, ADM.
[c]Hollo et al. (1993).
[d]Boukhchina et al. (2004).
[e]Egg yolk lecithin PL-100P, Kewpie Corporation.
[f]Rombaut et al. (2006).
[g]Nacka et al. (2001).
[h]Uddin et al. (2011).

purification process. For example, Lipogen PAS typically contains 20% PS and 15% PA; PIPS typically contains 30% PS and 20% PI.

Reaction Systems

Aqueous System

There are several systems for producing PS by PLD enzyme conversion. One is the aqueous system, which usually does not involve using solvents. Substrates of lecithin and serine, together with a buffer at an appropriate pH, are dispensed in water, and then the enzyme is added to the system and the mixture is allowed to react for a period of time at appropriate temperatures. Because PLD does not only catalyze the reaction of transphosphatidylation but also catalyzes the reaction of hydrolysis and results in the production of phosphatidic acid as a byproduct, a low conversion rate of PS is consequently observed. The aqueous system, in which the hydrolysis reaction tends to act over transphosphatidylation due to the excess amount of water, is not the preferred choice. However, by adding calcium ion, PS and calcium ion form an insoluble complex and this is precipitated out of the system, hence facilitating the reaction toward the formation of PS to achieve a higher yield. Although the solubility of the PS calcium complex is low, there is no difference in hydrolysis rate by a phospholipase A2 from porcine pancreas (data not shown), indicating it possibly has the same bioavailability as its sodium salt form, which has a higher solubility because phospholipase A2 hydrolysis plays critical roles in phospholipid digestion in mammals (Pruzanski et al., 2007). Recovery of the calcium-phosphatidylserine product is easily performed by centrifugation, and further washing can be carried out in order to remove residual raw materials, such as enzymes. On the other hand, the limited solubility of lecithin in water is another obstacle for high conversion. Therefore, a surfactant is often included in this system to increase the efficiency of enzyme-substrate contact. In the case of PS for food consumption, a food-grade emulsifier is the preferred choice (Pinsolle et al., 2013).

The aqueous system has the advantage of not only being environmental conscious but also of better capacity efficiency from an industrial point of view; often, a high yield of the product per batch can be obtained. Here we report a method for producing phosphatidylserine by using a semiaqueous system.

Methods

A paste lecithin containing approximately 50% PC was used as the raw material. Twenty percent lecithin was mixed with serine in acetate buffer (pH 5.5) containing 0.75 M sodium chloride at 3:1 mol ratio; 125U/g lecithin of phospholipase D (PLD Nagase^{TM}, Nagase ChemteX Corp.) and acetone at 0, 10, 18, 25, 30% (v/v) was then added to the system and the mixture were kept at 32 °C for 20 hours with agitation. After the reaction, phospholipids were extracted with chloroform and methanol (2:1) and analyzed on HPLC. The HPLC conditions are: column: UnisilQ NH2 (GL Sciences), column size: 4.6mm i.d. × 250 mm; effluent: acetonitoril/methanol/10mMN $H_4H_2PO_4$ (ratio: 1856 to 874 to 270); rate: 1.3 ml/min.; detection wavelength: UV 205 nm.

Results

The resulting yield of PS has a typical bell-shaped curve to added acetone, with the highest conversion rate at 18% acetone (v/v) and the lowest conversion rate to PA at greater than 25% acetone (v/v) (see Figure 6.16). By controlling the acetone concentration, the highest ratio of PS to PA can be obtained. This effect is probably due to emulsifying characteristics of acetone at low concentration and control of water activity at high concentration. It is worth mentioning that using acetone is quite convenient because acetone is generally used in the subsequent recovering process for PS products.

Water-Solvent System

A biphasic system is usually used in manufacturing PS. This system uses a nonpolar solvent to dissolve the lecithin substrate (commonly called the *oil phase*), enzyme, serine and buffer are prepared with water (commonly called the *water phase*). The two phases are then mixed and the mixture is kept stirring at an appropriate temperature in order for the catalytic reaction of substrate in both phases to occur. Agitation is

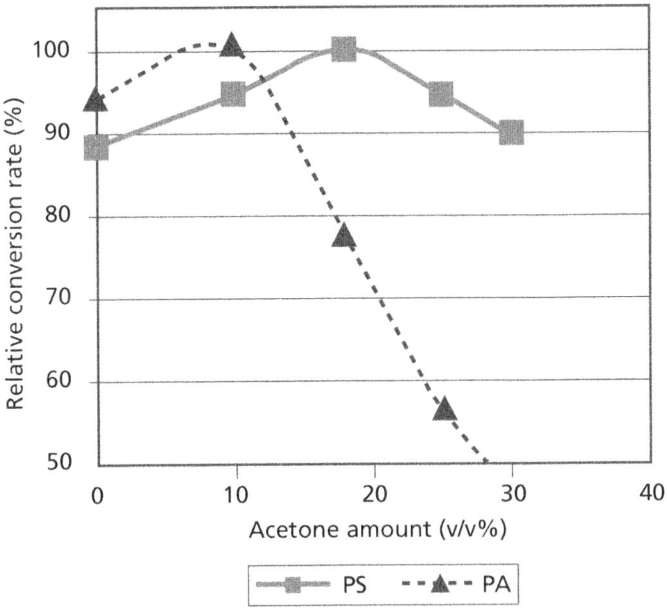

Figure 6.16 Acetone effect on conversion of lecithin to PS.

necessary for high efficient reaction; on the other hand, too much agitation tends to inactivate some enzymes. State-of-the-art methodology is applied in determining the optimum reaction conditions, such as substrate concentration in the oil phase, the ratio of lecithin and serine, buffer pH, reaction temperature, enzyme amount, agitation speed, and reaction time. In order to facilitate the reaction, an emulsifier can be added to the system; however, this may pose problems in the purification process afterward. We found that acetone has a remarkable effect in improving the reaction toward the synthesis of PS, not only in the previously described aqueous system but also in the biphasic system. Here is one example of producing PS using the biphasic system.

Methods

Solvent phase: A powder lecithin containing approximately 20% PC was suspended in heptane or hexane to a concentration of 15%. Acetone was included in the solvent phase at various volume concentrations (5%, 10%, 15%, 20%, and 25%).

Water phase: Serine (three times more in mol amount to lecithin), buffer at various pH (3.0, 3.5, 4.0, 4.5, 5.0, 5.5, 6.0), and enzyme phospholipase D were prepared accordingly. The volume ratio of solvent phase to water phase is controlled at 9:1.

Results

Figure 6.17 shows the results of the influence of acetone. At low concentrations, the reactions favor hydrolysis rather than synthesis. When the amount of acetone increased to approximately 15%, the equal points of the relative conversion rate of hydrolysis and transphosphatidylation meet; the transphosphatidylation reaction rate peaks at the concentration of 15–20% and declines when the acetone amount is further increased. The optimum concentration of acetone is approximately 20%, at which the highest PS to PA is achieved.

Figure 6.18 shows the optimum pH for PS production. The transphosphatidylation rate and the hydrolysis rate have different pH preferences. By shifting the pH toward transphosphatidylation, the hydrolysis to phosphatidic acid is minimized and gives a higher content of PS.

Purification Process

The key for purification of PS after PLD transphosphatidylation lies not only in increasing the content of PS but also in the removal of unfavorable ingredients. For example, in the case of soy-derived PS, the lecithin raw material often contains original proteins that may cause allergic reactions. Removal of such proteins results in quality products. Another important factor is the removal of the enzyme PLD from the product. Residual PLD can lead to decreased product stability during storage, and when formulated in soft capsules, the PS component is often hydrolyzed into PA or converted to phosphatidylglycerol (PG) with glycerol used as formulating agents, thus

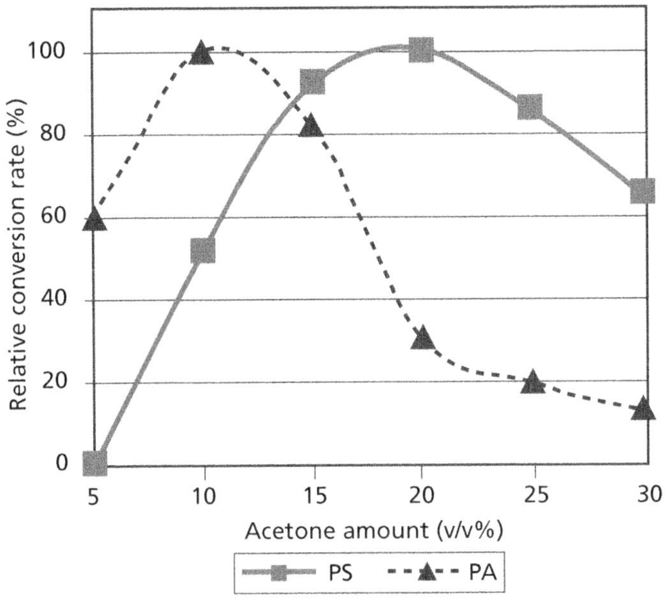

Figure 6.17 Influence of acetone on PS and PA production.

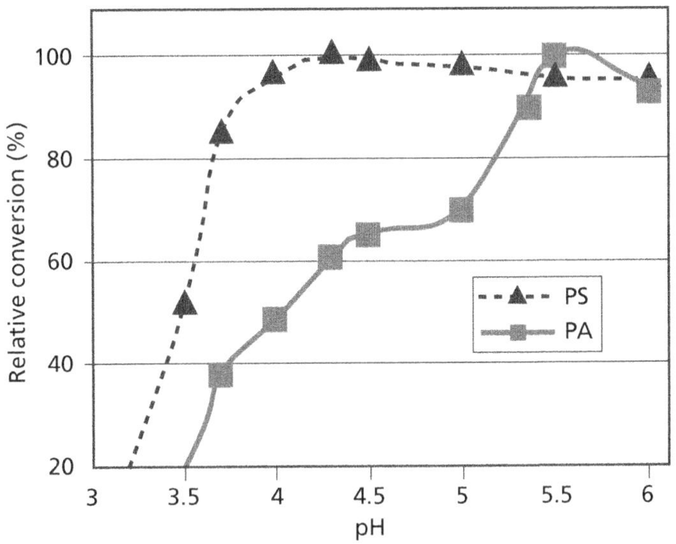

Figure 6.18 Influence of pH on PS and PA production.

its content is reduced during storage, resulting in shortened shelf life of consumer products (Plat et al., 2012). We found that after the reaction, residual phospholipase D is very difficult to remove, possibly due to strong binding of the enzyme to PS molecules. Heating is not a good strategy because, although it may be effective in denaturing some of the enzyme, it affects the quality of the product, often by increasing peroxide value. Here we would like to report an effective way to remove the residual phospholipase D from PS products (Liu and Taniwaki, 2010).

In the biphasic system previously described (heptane or hexane/water, acetone added), after the reaction is completed, the water phase was discarded and a solution containing salt and a polar solvent was added. The volume of the solution added has an optimum at a ratio of three to five of the solvent phase (in volume). After adding the solution, mix well and allow to stand to separate the two phases. The water phase is then discarded and the solvent phase is collected and dried to obtain the product. Residual PLD in the product is then analyzed. Results are summarized in Table 6.G. This is a very efficient way to not only remove the residual PLD, but also to remove extra proteins originated from the raw material after repeated treatment.

Applications

Dietary Supplements

Although its working mechanisms are not completely elucidated, evidence has accumulated in support of PS as a dietary supplement for various health issues, such as cognitive dysfunction, ADHD, learning capability, mental well-being, and athletic performance. Oral supplementation to acquire a sufficient amount is unavoidable due to the rarity of PS in common foods. It is probably easier to enclose the material in a hard capsule, in which case no extra additives are necessary. PS dissolved into oil, such as medium chain triglyceride (MCT) together with vitamin E, is suitable for

Table 6.G Removal of Residual Enzymes and Proteins

Samples	Amount of Residual Protein (μg/g lecithin)	Amount of Residual PLD (U/g lecithin)
Raw material	23	ND
After reaction	100.1	18.75
Treatment 1	45.3	ND
Treatment 2	20	ND
Treatment 3	ND	ND

Note: ND: not detectable.

making soft capsules. It is also possible to make tablets out of powdered PS, provided a proper filling is available.

Food Ingredients

An alternative to taking daily supplements is to include the ingredient in processed food, for example, bakery foods, dairy foods, processed meats, and others. Under these circumstances, survival of PS during processing is often the primary concern. Surprisingly, PS can survive the baking process (data not shown), perhaps because the combination with starch, sugar, oils, and other ingredients stabilizes it at heated conditions. PS is usually not stable in a water solution, especially at an acidic pH. When added into a dairy food such as milk or yogurt, there is not much degradation during storage. Phosphatidylserine can also be included in processed meat such as ham and sausages.

Skin Care Ingredients

Collective evidence of PS on improving skin conditions encourages its application in the cosmetic industry. However, it may be difficult to use the material directly in a lotion due to its stability problem in water. PS in water tends to be degraded, even at mild acidic conditions. Some forms of PS, such as calcium salt, maybe better in stability but they have dispersion problems due to poor solubility. Nevertheless, emulsifying technology can be applied to cope with the problem. PS in an oil-in-water emulsion is relatively stable. Table 6.H shows a cream formulation using a PI-enriched PS product (PIPS). This formulation stands stable at 50 °C for at least 2 weeks.

Neutraceutical Agents for Pets

With the growth of the global population, the pet industry will see growth in most of the world. Cognitive dysfunction syndrome is the main age-related feature of pets such as dogs; therefore, care products in this category are facing a growing market demand. Osella et al. (2008) summarized the perception of using PS to prevent and treat neurodegerative disease symptoms of aged pets (canine and feline), which is based on collective evidences of clinical studies on assorted animal species and on humans. PS has good bioavailability and is distributed ubiquitously in various tissues upon oral administration. The broad tolerability supports the expansion of its application.

Future Trends

The early studies of the pharmacological effects of dietary PS on improving brain function were conducted with bovine cortex originated material, rich in PUFA at the

Table 6.H Cream Formulations[a]

Phase	Component	Composition (%) (w/w)
Water	Water	60.60
	Citric acid	0.33
	Glycerin	0.25
	Betain	2.00
	1,3-Butylene glycol	10.00
	Xanthan gum	0.20
Oil	Glyceryl (caprylate/Caprate)	20.00
	PIPS	0.50
	Stearic acid	2.00
	Biphenyl alcohol	2.00
	Beeswax	2.00
	dl-α-tocopherol	0.10
	Total	100

[a]Mixed well by homogenizer (pH 5.5–5.9).

sn-2 position of the molecule. Recent study results on improving brain functions with marine PS, also rich in PUFA at the *sn*-2 position, showed good alignment. However, results of clinical studies on the dietary effects of PS that is originated from plants (e.g., soy), which shows a very different fatty acid profile (Chen and Li, 2008), are not always consistent. It seems that the PUFAs on the *sn*-2 position play important roles (Filburn, 2000). On the other hand, evidence did show there are no differences between animal origin and plant origin (Lee et al., 2010).

Further investigations on the effect of different species of PS are needed from academic points of view. In addition, the dietary habits of the experimental subjects should be put into consideration because dietary habits could affect results, especially concerning PUFAs. There is a clear bias on habitual intake of PUFA-rich foods, such as marine animals, in different regions of the world.

There is a potentially huge market in the segment of functional drinks, such as energy drinks and beauty drinks, that deserves better attention. Although it is challenging to formulate PS into drinks due to its degradation problem, studies on stabilizing it in aqueous solutions have been attempted. For example, by processing phospholipids into nanoparticles, the phospholipid solution is stabilized in both physical and chemical ways (Edris, 2012). We have attempted to develop an effective method of using PI-enriched PS in drink formulations. The method involved two steps, first making an emulsion out of PIPS, then use the emulsion in making a drink.

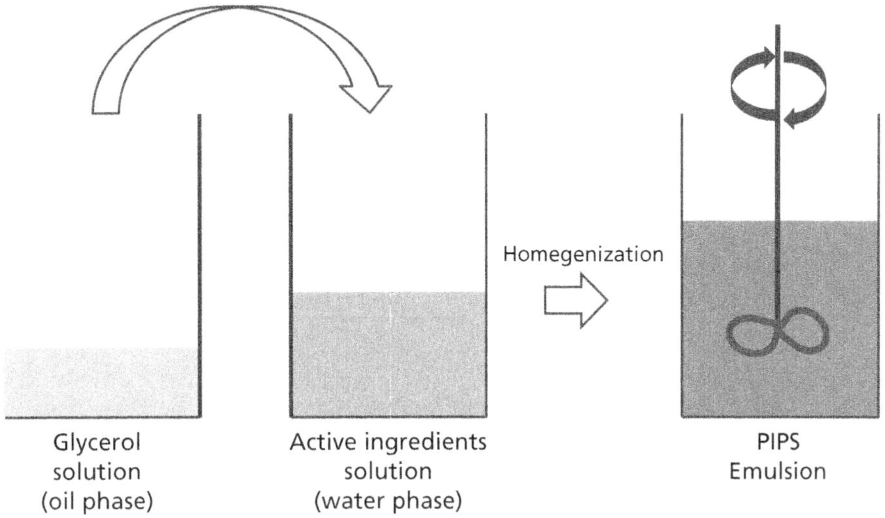

Figure 6.19 Diagram for making PIPS emulsion for preparing drink formulations.

The protocol is described below and Figure 6.19 shows a diagram of the process. Tables 6.I and 6.J show the compositions of the emulsion and the drink. The drink prepared this way has stability for 6 months at ambient temperatures, which is a sufficient stability requirement for most regional distribution.

Table 6.I Composition of PIPS Emulsion

Phase	Component	Composition (%) (w/w)
Water	Water	44.50
	Glycerin fatty acid ester (DP-95RF)	5
	Sodium caseinate	2
	Calcium oxide	0.30
	Trehalose	5.00
	Dextrin or α-cyclodextrin	2.00
	Carrageenan	0.20
Oil	dl-α-Tocopherol	1.00
	PIPS	5.00
	Glycerol	35.00
	Total	100

Table 6.J Composition of PIPS Drink

Component	Composition (%) (w/v)
Granulated sugar	8
Citric acid	0.02
Sodium citrate	0.55
Vitamin C	0.02
PIPS emulsion	2
Water	Up to 100 ml

Protocol for Making a PS-Containing Emulsion

1. Mix PIPS in glycerol at 40 °C, then add vitamin E and mix thoroughly to make the oil phase.
2. Mix trehalose, carrageenan, sodium caseinate, and α-cyclodextrin in water at 60 °C, then add the solution to glycerin fatty acid ester; further add calcium oxide to make the water phase.
3. Add oil phase to water phase and homogenize for 2 minutes.

There is no doubt that industrial efforts for PS production will pursue efficient low-cost and eco-friendly processes to ensure sustainable growth. This includes using renewable solvents or no solvents, achieving better yield per production batch, and reducing the cost of raw materials such as serine and others. New technologies are awaiting development, especially to modify PS efficiently in the nonpolar tail in order to create new value-added PS products with functional fatty acids. Together with the accumulation of more evidence and development of methods for applications, more people will benefit from this important phospholipid.

References

Abbott, S.; Jenner, A.; Mitchell, T.; Brown, S.; Halliday, G.; Garner, B. An Improved High-Throughput Lipid Extraction Method for the Analysis of Human Brain Lipids. *Lipids* **2013,** *48,* 307–318.

Baumeister, J.; Barthel, T.; Geiss, K. R.; Weiss, M. Influence of Phosphatidylserine on Cognitive Performance and Cortical Activity after Induced Stress. *Nutr. Neurosci.* **2008,** *11* (3), 103–110.

Becker, J.; Wittmann, C. Systems and Synthetic Metabolic Engineering for Amino Acid Production—The Heartbeat of Industrial Strain Development. *Current Opin. Biotech.* **2012,** *23* (5), 718–726.

Bell, M. V.; Dick, J. R.; Buda, C. S. Molecular Speciation of Fish Sperm Phospholipids: Large Amounts of Dipolyunsaturated Phosphatidylserine. *Lipids* **2001,** *32* (10), 1085–1091.

Benton, D.; Donohoe, R. T.; Sillance, B.; Nabb, S. The Influence of Phosphatidylserine Supplementation on Mood and Heart Rate When Faced with an Acute Stressor. *Nutr. Neurosci.* **2001,** *4* (3), 169–178.

Blokland, A.; Honig, W.; Brouns, F.; Jolles, J. Cognition-Enhancing Properties of Subchronic Phosphatidylserine (PS) Treatment in Middle-Aged Rats: Comparison of Bovine Cortex PS with Egg PS and Soybean PS. *Nutr.* **1999,** *15* (10), 778–783.

Boselli, E.; Pacetti, D.; Lucci, P.; Grega, N. Characterization of Phospholipid Molecular Species in the Edible Parts of Bony Fish and Shellfish. *J. Agric. Food Chem.* **2012,** *60* (12), 3234–3245.

Boukhchina, S.; Sebai, K.; Cherif, A.; Kallel, H.; Mayer, P. M. Identification of Glycerolphospholipids in Rapeseed, Olive, Almond, and Sunflower Oils by LC-MS and LC-MS-MS. *Can. J. Chem.* **2004,** *82,* 1210–1215.

Carrea, G.; D'Arrigo, P.; Piergianni, V.; Roncaglio, S.; Secudo, F.; Servi, S. Purification and Properties of Two Phospholipase D from *Streptomyces sp. Biochim. Biophys. Acta* **1995,** *1255* (3), 273–279.

Cenacchi, T.; Bertoldin, T.; Farina, C.; Fiori, M. G.; Crepaldi, G. Cognitive Decline in the Elderly: A Double-Blind, Placebo-Controlled Multicenter Study on Efficacy of Phosphatidylserine Administration. *Aging* (Milano) **1993,** *5* (2), 123–133.

Chen, S.; Li, K. W. Comparison of Molecular Species of Various Transphosphatidylated Phosphatidylserine (PS) with Bovine Cortex PS by Mass Spectrometry. *Chem. Phys. Lipids.* **2008,** *152,* 46–56.

Cho, S.; Kim, H. H.; Lee, M. J.; Lee, S.; Park, C.-S.; Nam, S.-J.; Han, J.-J.; Kim, J.-W.; Chung, J. H. Phosphatidylserine Prevents UV-Induced Decrease of Type I Procollagen and Increase of MMP-1 In Dermal Fibroblasts and Human Skin in Vivo. *J. Lipid Res.* **2008,** *49,* 1235–1245.

Contarini, G.; Povolo, M. Phospholipids in Milk Fat: Composition, Biological and Technological Significance, and Analytical Strategies. *Int. J. Mol. Sci.* **2013,** *14,* 2808–2831.

Dacaranhe, C. D.; Terao, J. A Unique Antioxidant Activity of Phosphatidylserine on Iron-induced Lipid Peroxidation of Phospholipid Bilayers. *Lipids* **2001,** *32* (10), 1105–1109.

Delwaide, P. J.; Gyselynck-Mambourg, A. M.; Hurlet, A.; Ylieff, M. Double-blind Randomized Controlled Study of Phosphatidylserine in Senile Demented Patients. *Acta Neurol. Scand.* **1986,** *73* (2), 136–140.

Edris, A. E. Formulation and Shelf Life Stability of Water-borne Lecithin Nanoparticles for Potential Application in Dietary Supplements Field. *J. Diet. Supp.* **2012,** *9* (2), 211–222.

Fadok, V. A.; de Cathelineau, A.; Daleke, D. L.; Henson, P. M.; Bratton, D. L. Loss of Phospholipid Asymmetry and Surface Exposure of Phosphatidylserine is Required for Phagocytosis of Apoptotic Cells by Macrophages and Fibroblasts. *J. Biol. Chem.* **2001,** *276,* 1071–1077.

Filburn, C. R. Dietary Supplementation with Phospholipids and Docosahexaenoic Acid for Age-Related Cognitive Impairment. *JANA* **2000,** *3* (3), 45–55.

Frasch, S. C.; Bratton, D. Emerging Roles for Lysophosphatidylserine in Resolution of Inflammation. *Prog. Lipid Res.* **2012,** *51* (3), 199–207.

Fukuda, H.; Kondo, A.; Kuroda, S.; Tanizawa, K.; Tokuyama, S. Promoter, Vector and Recombinant Microorganism Having the Same and Process for Producing Protein. Japanese Patent 4834900, October 7, 2011.

Gindin, J.; Novikov, M.; Dedar, A.; Walter-Ginzburg, S.; Naor, S. *The Effect of Plant Phosphatidylserine on Age-associated Memory Impairment and Mood in the Functioning Elderly.* The Geriatric Institute for Education and Research, and Department of Geriatrics, Kaplan Hospital: Rehovot, Israel, 1998.

Grundy, S. M.; Cleeman, J. I.; Daniels, S. R.; Donato, K. A.; Eckel, R. H.; Franklin, B. A.; Gordon, D. J.; Krauss, R. M.; Savage, P. J.; Smith, S. C., Jr.; et al. Diagnosis and Management of the Metabolic Syndrome: An American Heart Association/National Heart, Lung, and Blood Institute Scientific Statement. *Circulation* **2005,** *112,* 2735–2752.

Gu, P.; Yang, F.; Su, T.; Li, F.; Li, Y.; Qi, Q. Construction of an l-Serine Producing Escherichia Coli Via Metabolic Engineering. *J. Ind. Microbio. Biotech.* **2014,** *41,* 1–8.

Hellhammer, J.; Fries, E.; Buss, C.; Engert, V.; Tuch, A.; Rutenberg, D.; Hellhammer, D. Effects of Soy Lecithin Phosphatidic Acid and Phosphatidylserine Complex (PAS) on The Endocrine and Psychological Responses to Mental Stress. *Stress: Int. J. Biol. Stress* **2004,** *7* (2), 119–126.

Hellhammer, J.; Hero, T.; Franz, N.; Contreras, C.; Schubert, M. Omega-3 Fatty Acids Administered in Phosphatidylserine Improved Certain Aspects of High Chronic Stress in Men. *Nutr. Res.* **2012,** *32* (4), 241–250.

Hicks, A. M.; DeLong, C. J.; Thomas, M. J.; Samuel, M; Cui, Z. Unique Molecular Signatures of Glycerophospholipids Species in Different Rat Tissues Analyzed by Tandem Mass Spectrometry. *Biochim. Biphys. Acta* **2006,** *1761,* 1022–1029.

Hirayama, S.; Masuda, Y.; Rabeller, R. Effect of Phosphatidylserine Administration on Symptoms of Attention-deficit/Hyperactivity Disorder in Children. *AgroFOOD Industry Hi-Tech* **2006,** *17* (5), 16–20.

Hirayama, S.; Terasawa, K.; Rabeler, R.; Hirayama, T.; Inoue, T.; Tatsumi, Y.; Purpura, M. The Effect of Phosphatidylserine Administration on Memory and Symptoms of Attention-deficit Hyperactivity Disorder: A Randomized, Double-blind, Placebo-controlled Clinical Trial. *J. Hum. Nutr. Dietetics* **2013,** 1–8.

Hollo, J.; Peredi, J.; Ruzics, A.; Jeranek, M.; Erdelyi, A. Sunflower Lecithin and Possibilities for Utilization. *J. Am. Oil Chem. Soc.* **1993,** *70* (10), 997–1001.

Hosokawa, M.; Shimatani, T.; Kanada, T.; Inoue Y.; Takahashi, K. Conversion to Docosahexaenoic Acid-Containing Phosphatidylserine from Squid Skin Lecithin by Phospholipase D-mediated Transphosphatidylation. *J. Agric. Food Chem.* **2000,** *48* (10), 4550–4554.

Ingelsson, E.; Schaefer, E. J.; Contois, J. H.; McNamara, J. R; Sullivan, L.; Keyes, M. J.; Pencina, M. J.; Schoonmaker, C.; Wilson, P. W.; D'Agostino, R. B.; et al. Clinical Utility of Different Lipid Measures for Prediction of Coronary Heart Disease in Men and Women. *JAMA* **2007,** *298,* 776–785.

Jäger, R.; Purpura, M.; Geiss, K.-R.; Weiβ, M.; Baumeister, J.; Amatulli, F.; Schröder, L.; Herwegen, H. The Effect of Phosphatidylserine on Golf Performance. *J. Int. Soc. Sports Nutr.* **2007,** *4,* 23.

Jorissen, B. L.; Brouns, F.; Van Boxtel, M. P. J.; Ponds, R.W. H. M.; Verhey, F. R. J.; Jolles, J.; Riedel, W. J. The Influence of Soy-Derived Phosphatidylserine on Cognition in Age-associated Memory Impairment. *Nutr. Neurosci.* **2001,** *4* (2), 121–134.

Juneja, L. R.; Kazuoka, T; Yamane, T.; Shimizu, S. Kinetic Evaluation of Conversion of Phosphatidylcholine to Phosphatidylethanolamine by Phospholipase D from Different Sources. *Biochim. Biophys. Acta* **1988,** *960* (3), 334–341.

Kato-Kataoka, A.; Sakai, M.; Ebina, R.; Nonaka C.; Asano, T.; Miyamori, T. Soybean-derived Phosphatidylserine Improves Memory Function of the Elderly Japanese Subjects with Memory Complaints. *J. Clin. Biochem. Nutr.* **2010,** *47* (3), 246–255.

Kawai, Y.; Kiyokawa, H.; Kimura, Y.; Kato, Y.; Tsuchiya, K.; Terao, J. Hypochlorous Acid-derived Modification of Phospholipids: Characterization of Aminophospholipids as Regulatory Molecules for Lipid Peroxidation. *Biochemistry* **2006,** *45* (47), 14201–14211.

Kay, J. G.; Grinstein S. Sensing Phosphatidylserine in Cellular Membranes. *Sensors* **2011,** *11,* 1744–1755.

Kim, H. Y.; Bigelow, J.; Kevala, J. H. Substrate Preference in Phosphatidylserine Biosynthesis for Docosahexaenoic Acid Containing Species. *Biochem.* **2004,** *43* (4), 1030–1036.

Kingsley, M. I.; Miller, M.; Kilduff, L. P., McEneny, J.; Benton, J. Effects of Phosphatidylserine on Exercise Capacity during Cycling in Active Males. *Med. Sci. Sports Exerc.* **2006,** *38* (1), 64–71.

Kuge, O.; Nishijima, M. Biosynthetic Regulation and Intracellular Transport of Phosphatidylserine in Mammalian Cells. *J. Biochem.* **2003,** *133,* 397–403.

Lee, B.; Sur, B.-J.; Han, J.-J.; Shim, I.; Her, S.; Lee, H.-J.; Hahm, D.-H. Krill Phosphatidylserine Improves Learning and Memory in Morris Water Maze in Aged Rats. *Prog. Neuro-Psychopharm. Biol. Psych.* **2010,** *34,* 1085–1093.

Liu, L.; Waters, D.; Rose, T.; Bao, J.; King, G. Phospholipids in Rice: Significance in Grain Quality and Health Benefits: A Review. *Food Chem.* **2013,** *139,* 1133–1145.

Liu, X. L.; Taniwaki, N. Method of Removing Enzyme and Method of Base Exchange or Hydrolysis of Phospholipid Using the Same. Japanese Patent 4650746, December 24, 2010.

Maciel, E.; Neves, B. M.; Santinha, D.; Reis, A.; Domingues, P.; Cruz, M. T.; Pitt, A. R.; Spickett, C. M.; Domingues, M. R. M. Detection of Phosphatidylserine with a Modified Polar Head Group in Human Keratinocytes Exposed to the Radical Generator AAPH. *Arch. Biochem. Biophys.* **2014,** *548,* 38–45.

Maggioni, M.; Picotti, G. B.; Bondiolotti, G. P.; Panerai, A.; Cenacchi, T.; Norbile, P.; Brambilla, F. Effect of Phosphatidylserine Therapy in Geriatric Patients with Depressive Disorders. *Acta Psychiatr. Scand.* **1990,** *81,* 265–270.

Manor, I.; Magen, A.; Keidar, D.; Rosen, S.; Tasker, H.; Cohen, T.; Richter, Y.; Zaaroor-Regev, D.; Manor, Y.; Weizman, A. The Effect of Phosphatidylserine Containing Omega3 Fatty-acids on Attention-deficit Hyperactivity Disorder Symptoms in Children: A Double-blind Placebo-controlled Trial, Followed by an Open-label Extension. *Eur. Psych.* **2012,** *27* (5), 335–342.

Merolli, A.; Santin, M. Role of Phosphatidyl-serine in Bone Repair and Its Technological Exploitation. *Molecules* **2009,** *14,* 5367–5381.

Monteleone, P.; Maj, M.; Beinat, L.; Natale, M.; Kemali, D. Blunting by Chronic Phosphatidylserine Administration of the Stress-induced Activation of the Hypothalamo-pituitary-adrenal Axis in Healthy Men. *Eur. J. Clin. Pharmacol.* **1992,** *42* (4), 385–388.

Myers, D.; Ivanova, P.; Milne, S.; Brown, A. Quantitative Analysis of Glycerophospholipids by LC-MS: Acquision, Data Handling, and Interpretation. *Biochim Biophys Acta* **2011,** *1811* (11), 748–757.

Nacka, F.; Cansell, M.; Gouygou J. P.; Gerbeaud, C.; Méléard, P.; Entressangles, B. Physical and Chemical Stability of Marine Lipid-based Liposomes under Acid Conditions. *Colloids and Surfaces B: Biointerfaces* **2001,** *20(3)*, 257–266.

Negishi, T.; Hayashi, H.; Ito, S.; Fujino, Y. Chemical Composition of the Phospholipids Prepared from Commercial "Soy Bean Lecithin." *Res. Bull. Obihiro Univ.* **1967,** *5,* 97–101.

Nieuwenhuyzen, W. The Changing World of Lecithins. *Inform* **2014,** *25* (4), 254–259.

Nojima, M.; Hosokawa, M.; Takahashi, K.; Hatano, M. Effect of EPA and DHA Containing Glycerophospholipids Molecular Species on the Fluidity of Erythrocyte Cell Membranes. *Fisheries Science* **1994,** *60* (6), 729–734.

Nunzi, M. G.; Guidolin, D.; Petrelli, L.; Polato, P.; Zanotti, A. Behavioral and Morphofunctional Correlates of Brain Aging: A Preclinical Study with Phosphatidylserine. *Adv. Exp. Med. Biol.* **1992,** *318,* 393–398.

Oh, J. H.; Kim, Y. K.; Jung, J. Y.; Shin, J. E.; Chung, J. H. Changes in Glycosaminoglycans and Related Proteoglycans in Intrinsically Aged Human Skin in Vivo. *Exp. Dermatol.* **2011,** *20* (5), 454–456.

Osella, M.; Giovanni, Re.; Badino, P.; Bergamasco, L.; Miolo, A. Phosphatidylserine (PS) as a Potential Nutraceutical for Canine Brain Aging: A Review. *J. Vet. Behav.* **2008,** *3,* 41–51.

Palacios-Pelaez, R.; Lukiw, W. J.; Bazan, N. G. Omega-3 Essential Fatty Acids Modulate Initiation and Progression of Neurodegenerative Disease. *Mol. Neurobiol.* **2010,** *41,* 367–374.

Park, H.-J.; Lee, S. Y.; Shim, H. S.; Kim, J. S.; Kim, K. S.; Shim, I. Chronic Treatment with Squid Phosphatidylserine Activates Glucose Uptake and Ameliorates TMT-induced Cognitive Deficit in Rats via Activation of Cholinergic Systems. *Evidence-Based Complementary and Altenative Medicine* **2012,** *2012,* 1–8.

Park, H.-J.; Shim, H. S.; Kim, K. S.; Han, J.-J.; Kim, J. S.; Yu, A. R.; Shim, I. Enhanced Learning and Memory of Normal Young Rats by Repeated Oral Administration of Krill Phosphatidylserine. *Nutritional Neroscience* **2013,** *16* (2), 47–53.

Pellow, J.; Solomon, E. M.; Barnard, C. N. Complementary and Alternative Medical Therapies for Children with Attention-deficit/Hyperactivity Disorder (ADHD). *Altern. Med. Rev.* **2011,** *16* (4), 323–337.

Pepeu, G.; Pepeu, I. M.; Amaducci, L. A Review of Phosphatidylserine Pharmacological and Clinical Effects. Is Phosphatidylserine a Drug for the Ageing Brain? *Pharmacol Res.* **1996,** *33* (2), 73–80.

Pinsolle, A.; Roy, P.; Bure, C.; Thienpont, A.; Cansell, M. Enzymatic Synthesis of Phosphatidylserine Using Bile Salt Mixed Micelles. *Colloids and Surfaces B: Biointerfaces* **2013,** *106,* 191–197.

Plat, D.; Shulman, A.; Dror, G. B.; Scheinman, N.; Twito, Y.; Zuabi, R. Stabilized Formulations of Phosphatidylserine. U.S. Patent 8324187, December 4, 2012.

Pruzanski, W.; Lambeau, G.; Lazdunski, M.; Cho, W.; Kopilov, J.; Kuksis, A. Hydrolysis of Minor Glycerophospholipids of Plasma Lipoproteins by Human Group IIA, V and X Secretory Phospholipase A2. *Biochim Biophys Acta* **2007,** *1771* (1), 5–19.

Rachev, E.; Nalbansky, B.; Kolarov, G.; Agrosi, M. Efficacy and Safety of Phospholipid Liposomes in the Treatment of Neuro-psychological Disorders Associated with the Menopause: A Double-blind, Randomized, Placebo-controlled Study. *Curr. Med. Res. Opin.* **2001,** *17* (2), 105–110.

Richter, Y.; Herzog, Y.; Lifshitz, Y.; Hayun, R.; Zchut S. The Effect of Soy-bean-derived Phosphatidylserine on Cognitive Performance in Elderly with Subjective Memory Complaints: A Pilot Study. *Clin. Interven. Aging* **2013,** *8,* 557–563.

Rosadini, G.; Sannita, W. G.; Nobili, F.; Cenacchi, T. Phosphatidylserine: Quantitative EEG Effects in Healthy Volunteers. *Neuropsych.* **1990,** *24* (1), 42–48.

Sakai, M.; Yamatoya, H.; Kudo, S. Pharmacological Effects of Phosphatidylserine Enzymatically Synthesized from Soybean Lecithin on Brain Functions in Rodents. *J. Nutr. Sci. Vitaminol.* **1996,** *42,* 47–54.

Schreiber, S.; Kampf-Sherf, O.; Gorfine, M.; Kelly, D.; Oppenheim, Y.; Lerer, B. An Open Trial of Plant-source Derived Phosphatidylserine for Treatment of Age-related Cognitive Decline. *Isr. J. Psychiatry Relat. Sci.* **2000,** *37* (4), 302–307.

Sen, A.; Zhao, Y.-L.; Hui, S. W. Saturated Anionic Phospholipids Enhance Transdermal Transport by Electroporation. *Biophys. J.* **2002,** *83,* 2064–2073.

Shimizu, K.; Ida, T.; Tsutsui, H.; Asai, T.; Otsubo, K.; Oku, N. Anti-Obesity Effect of Phosphatidylinositol on Diet-induced Obesity in Mice. *J. Agric. Food Chem.* **2010,** *58,* 11218–11225.

Shirouchi, B.; Nagao, K.; Furuya, K.; Shiojiri, M.; Liu, X.; Yanagita, T. Physiological Effects of Dietary PIPS Soybean-derived Phospholipid in Obese Zucker (fa/fa) Rats. *Biosci. Biotechnol. Biochem.* **2010,** *74* (11), 2333–2335.

Starks, M.; Starks, S.; Kingsley, M.; Purpura, M.; Jäger, R. The Effects of Phosphatidylserine on Endocrine Response to Moderate Intensity Exercise. *J. Int. Sports. Nutr.* **2008,** *5,* 11.

Suzuki, S.; Yamatoya, H.; Sakai, M.; Kataoka, A.; Furushiro, M.; Kuda, S. Oral Administration of Soybean Lecithin Transphosphatidylated Phosphatidylserine Improves Memory Impairment in Aged Rats. *J. Nut.* **2001,** *131,* 2951–2956.

Tamaki, H.; Shimada, A.; Ito, Y.; Ohya, M.; Takase, J.; Miyashita, M.; Miyagawa, H.; Nozaki, H.; Nakayama, R.; Kumagai, H. *LPT1* Encodes a Membrane-bound *O*-acyltransferase Involved in the Acylation of Lysophospholipids in the Yeast *Saccharomyces cerevisiae. J. Biol. Chem.* **2007,** *282* (47), 34288–34298.

Terazawa, S.; Nakajima, H.; Tobita, K.; Imokawa, G. The Decreased Secretion of Hyaluronan by Older Human Fibroblasts under Physiological Conditions Is Mainly Associated with Down-regulated Expression of Hyaluronan Synthases but Not with the Expression Levels of Hyaluronidases. *Cytotech.* **2014,** Mar 4.

Tempone, A. G.; Perez, D.; Rath, S.; Vilarinho, A. L.; Mortara, R. A.; de Andrade, H. F., Jr. Targeting Leishmania (L.) Chagasi Amastigotes through Macrophage Scavenger Receptors: The Use of Drugs Entrapped in Liposomes Containing Phosphatidylserine. *J. Antimicro. Chemo.* **2004,** *54* (1), 60–68.

Totani, Y.; Hara, S. Preparation of Functional Phospholipids with Lipases [in Japanese]. *Sci. Ind.* **2003,** *77* (5), 225–232.

Tsakiris, S.; Deliconstantinos, G. Influence of Phosphatidylserine on (Na^+ + K^+)-stimulated ATPase and Acetylcholinesterase Activities of Dog Brain Synaptosomal Plasma Membranes. *Biochem. J.* **1984,** *220,* 301–307.

Uddin, M. S.; Kishimura, H.; Chun, B. S. Isolation and Characterization of Lecithin from Squid (*Todarodes pacificus*) Viscera Deoiled by Supercritical Carbon Dioxide Extraction. *J. Food Sci.* **2011,** *76* (2), 350–354.

Utsugi, T.; Schroit, A. J.; Connor, J.; Bucana, C. D.; Fidler, I. J. Elevated Expression of Phosphatidylserine in the Outer Membrane Leaflet of Human Tumor Cells and Recognition by Activated Human Blood Monocytes. *Cancer Res.* **1991,** *51,* 3062–3066.

Vakhapova, V.; Cohen, T.; Rithter, Y.; Herzog, Y.; Korczyn, A. Phosphatidylserine Containing ω-3 fatty Acids May Improve Memory Abilities in Non-Demented Elderly with Memory Complaints: A Double-blind Placebo-controlled Trial. *Dementia Geriatr. Cogn. Disord.* **2010,** *29,* 467–474.

Vakhapova, V.; Cohen, T.; Rithter, Y.; Herzog, Y.; Kam, Y.; Korczyn, A. Phosphatidylserine Containing Omega-3 Fatty Acids May Improve Memory Abilities in Nondemented Elderly Individuals with Memory Complaints: Results from an Open-label Extension Study. *Dementia Geriatr. Cogn. Disord.* **2014,** *38,* 39–45.

Wang, J.; Lei, H.; Li, P.; Han, L.; Hou, J.; Yan, Y.; Zhao, H.; Tsuji, T. The Anti-brain Ageing Effects of Krill Phosphatidylserine in SAMP10 Mice. *J. Agri. Sci.* **2012,** *4* (9), 196–208.

Weber, E. J. Compositions of Commercial Corn and Soy Lecithins. *J. Am. Oil Chem.* **1981,** *58* (10), 898–901.

Weihrauch, J. L., Son, Y.-S. The Phospholipid Content of Foods, *J. Am. Oil Chem. Soc.* **1983,** *60,* 1971–1978.

Winther, B.; Hoem, N.; Berge, K.; Reubsaet, L. Elucidation of Phosphatidylcholine Composition in Krill Oil Extracted from *Euphausia superb. Lipids* **2011,** *46,* 25–36.

Yeung, T.; Gilbert, G.; Shi, J.; Silvius, J.; Kapus, A.; Grinstein, S. Membrane Phosphatidylserine Regulates Surface Charge and Protein Localization. *Science* **2008,** *319,* 210–213.

Yeung, T.; Heit, B.; Dubuisson, J.-F.; Fairn, G. D.; Chiu, B.; Inman, R.; Kapus, A.; Swanson, M.; Grinstein, S. Contribution of Phosphatidylserine to Membrane Surface Charge and Protein Targeting During Phagosome Maturation. *J. Cell Biol.* **2009,** *185,* 917–928.

Yokoyama, K.; Araki, S.; Kawakami, N.; Takeshita, T. Production of the Japanese Edition of Profile of Mood States (POMS): Assessment of Reliability and Validity. *Nihon Koshu Eisei Zasshi (Japanese Journal of Public Health)* **1990**, *37 (11),* 913–918.

Zanotta, D.; Puricelli, S.; Bonoldi, G. Cognitive Effects of A Dietary Supplement Made from Extract of *Bacopa Monnieri*, Astaxanthin, Phosphatidylserine, and Vitamin E in Subjects with Mild Cognitive Impairment: A Noncomparative, Exploratory Clinical Study. *Neuropsych. Dis. Treat.* **2014,** *10,* 225–230.

Zhang, X. F.; Guo, S. Z.; Lu, K. H.; Li, H. Y.; Li, X. D.; Zhang, L. X.; Yang, L. Different Roles of PKC and PKA in Effect of Interferon-Gamma on Proliferation and Collagen Synthesis of Fibroblasts. *Acta. Pharmacol Sin.* **2004,** *25* (10), 1320–1326.

Zwaal, R. F.; Comfurius, P.; Bevers, E. M. Lipid-protein Interactions in Blood Coagulation. *Biochim. Biophys. Acta* **1998,** *1376,* 433–453.

Zwaal, R. F.; Comfurius, P.; Bevers, E. M. Surface Exposure of Phosphatidylserine in Pathological Cells. *Cell. Mol. Life Sci.* **2005,** *62,* 971–988.

7

Phenolipids as New Antioxidants: Production, Activity, and Potential Applications

Derya Kahveci ▪ *Department of Food Engineering, Faculty of Engineering, Yeditepe University, Istanbul, Turkey*

Mickaël Laguerre and Pierre Villeneuve ▪ *UMR IATE, Montpellier, France*

Introduction

Phenolics are secondary metabolites widely found in plants; they have several biological roles and they contribute to the defense system of the host. Among them, the phenolic acid family is one of the most important classes and is mainly composed of the cinnamic and benzoic acid derivatives (Figure 7.1). The interest in these compounds from a nutritional point of view has risen due to their potential properties, including antioxidant, anti-inflammatory, antiallergic, antimicrobial, antiviral, and anticarcinogenic (Fernandez-Panchon et al., 2008; Haminiuk et al., 2012; Sun-Waterhouse, 2011).

Because phenolic acids have a rather low solubility in oils, improvement of hydrophopicity of these compounds by chemical or enzymatic lipophilization has been applied extensively in order to render these functionalized compounds, so-called phenolipids, active in the oil–water interphase. This chapter focuses on the synthesis of phenolipids derived from phenolic acids, their physicochemical and biological activity, and their potential applications.

Production Technology of Phenolipids

One of the most reliable methods to improve antioxidant activity of phenolics is to incorporate properly positioned lipophilic groups to obtain phenolipids. Because it is difficult to manipulate the polar phenolic head(s), the traditional experience is that the phenolic has the polar group in the correct position and that antioxidant activity is improved by correctly positioned lipophilic groups (Lipinsky et al., 2001). Accordingly, lipophilization appears more and more as a crucial step in the design of new antioxidant additives and drugs. It basically consists of grafting lipophilic moieties to a given molecule to make it more active. The molecules so prepared, with fine-tuned lipophilicity, have improved bioavailability in vivo over their parent antioxidants. Phenolipids also show greater miscibility and incorporation into lipid phases and lipocarriers, offering an advantage for their use in drug delivery systems, pharmaceuticals, nutraceuticals, foods, and cosmetic formulations. Lipophilization of phenolic acids can be performed chemically, enzymatically, or chemo-enzymatically, most

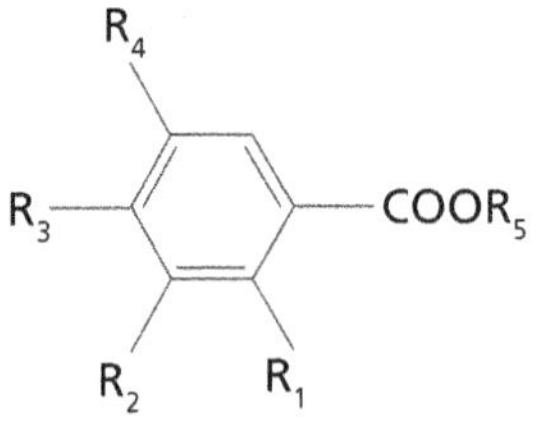

Hydroxybenzoic Acid Derivatives

Phenolic Acid	**R_1**	**R_2**	**R_3**	**R_4**	**R_5**
Salicylic	OH	H	H	H	H
p-Hydroxybenzoic	H	H	OH	H	H
Protocatechuic	H	OH	OH	H	H
Gentisic	OH	H	H	OH	H
Vanillic	H	OMe	OH	H	H
Veratric	H	OMe	OMe	H	H
Gallic	H	OH	OH	OH	H
Syringic	H	OMe	OH	OMe	H

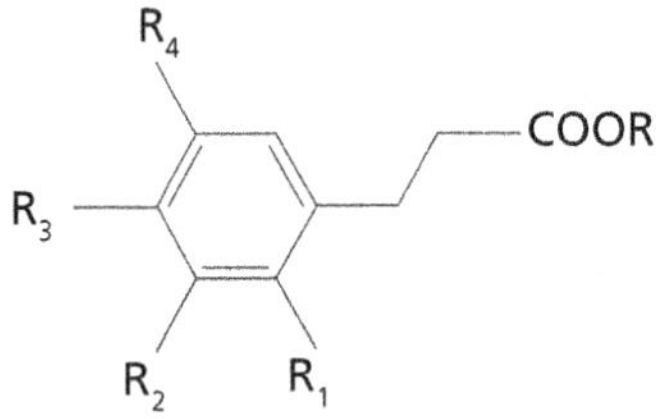

Hydroxyphenylpropionic Acid Derivatives

Phenolic Acid	**R_1**	**R_2**	**R_3**	**R_4**	**R_5**
Dihydrocaffeic	H	OH	OH	H	H
Dihydroferulic	H	OMe	OH	H	H

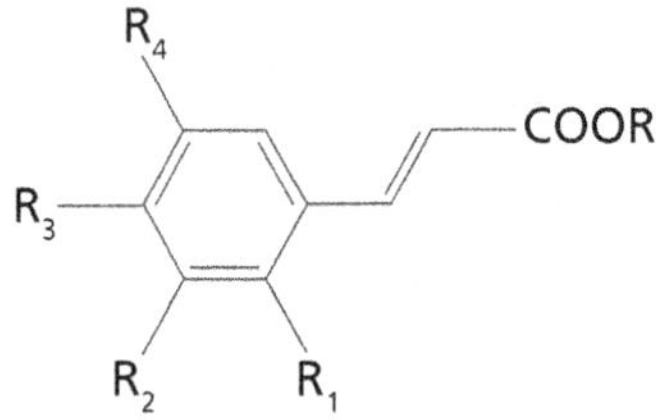

Hydroxycinnamic Acid Derivatives

Phenolic Acid	**R_1**	**R_2**	**R_3**	**R_4**	**R_5**
o-Coumaric	OH	H	H	H	H
m-Coumaric	H	OH	H	H	H
p-Coumaric	H	H	OH	H	H
p-Coumaroyl Quinic	H	H	OH	H	*
Caffeic	H	OH	OH	H	H
4-*O*/5-*O* Caffeoyl Quinic	H	OH	OH	H	*
Ferulic	H	OMe	OH	H	H
Sinapic	H	OMe	OH	OMe	H
Rosmarinic	H	OH	OH	H	**

Figure 7.1 Structure of some of the main naturally occurring phenolic acids. *Quinic acid; **3-(3,4-dihydroxy phenyl) lactic acid.

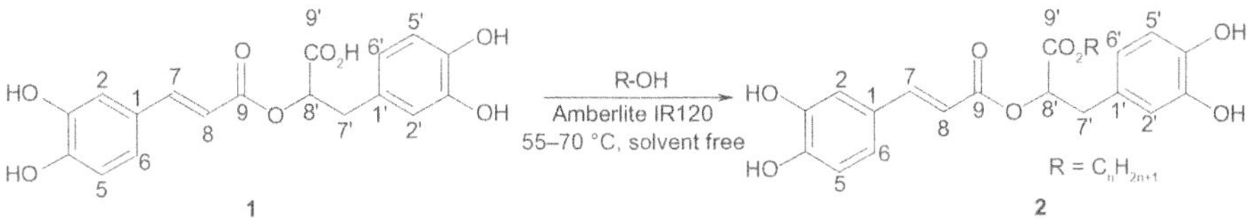

Figure 7.2 Heterogeneous acid-catalyzed esterification of rosmarinic acid with aliphatic alcohols of various chain lengths (1, 4, 8, 12, 16, 18, and 20 carbon atoms) (from Lecomte et al., 2010).

often through esterification of (1) the –COOH group with fatty alcohols or (2) the phenolic–OH with fatty acids.

Chemical esterification is usually performed under drastic conditions of temperature and pH. Phenolic acids being unstable in alkali (phenolates are much more prone to oxidation than phenols), the reaction is carried out with homogeneous or heterogeneous strongly acidic catalysts. In the first case, hydrochloric, sulfuric, or *para*-toluene sulfonic acids (*p*-TSA) are currently used; sulfonic resins are good catalysts in the second case.

When phenolics inhibit lipases, chemical lipophilization is the only way to graft lipophilic moiety to them. Such is the case with rosmarinic acid, for which esterification must be performed using sulfonic resins (heterogeneous catalysis). In the Lecomte et al. (2010) procedure, the reaction mixtures containing rosmarinic acid and fatty alcohol were stirred at 250 rpm at 55–70 °C prior to the addition of the strongly acidic sulfonic resin Amberlite® IR-120H (5% w/w, total weight of both substrates) previously dried at 110 °C for 48 hours. The water generated during the reaction was removed by absorption on molecular sieves. After complete (4–21 days) conversion of rosmarinic acid into the corresponding ester, the latter was purified in a two-step procedure (Figure 7.2). First, a liquid–liquid extraction using hexane and acetonitrile was achieved to remove the excess of alcohol. Then, the remaining traces of the alcohol and rosmarinic acid were eliminated by flash chromatography. Separation was achieved on a silica column using an elution gradient of hexane and ether. The reaction was achieved at a low temperature (55–70 °C) to avoid rosmarinic degradation in an excess of the alcohol that played the role of both substrate and solvent when melted. Despite slow reaction rates (from 4 to 21 days of reaction), the pure esters were obtained in high yields ranging from 82% to 99.5%.

Although quick and simple, chemical processes are generally not selective, they result in unwanted side reactions, and they involve many purification steps to remove byproducts and catalyst residues, generating extra wastes. Enzymatic synthesis, however, shows a better selectivity, implies milder reaction conditions, requires fewer purification steps, and has lower waste production. However, besides the higher cost of biocatalysts compared to conventional chemical catalysts, enzymatic synthesis may require longer reaction times and a good knowledge of the selected enzyme's behavior toward the chosen operating conditions. Still, biocatalysis is now regarded as a real alternative to conventional chemical processes in various sectors such as the pharmaceutical, cosmetic, food, and detergents industries. In that context, not surprisingly, enzymatic processes have been evaluated for the synthesis of phenolipids with mainly the use of lipases (triacylglycerol hydrolases, E.C 3.1.1.3) (Figueroa-Espinoza and Villeneuve, 2005; Villeneuve, 2007). In most of these processes, the lipases are used as immobilized biocatalysts in order to improve their activity and operational stability (Villeneuve et al., 2000). Moreover, the use of immobilized enzymes allows an easy removal and recovery of the biocatalyst once the reaction is over.

Actually, the main difficulty in such a reaction is to overcome the great difference of polarity of the two substrates, namely the hydrophilic phenolic compound and the highly hydrophobic aliphatic chain. Therefore, various parameters have to be considered in order to guarantee an optimized contact between those two substrates as well as optimized reaction kinetics and yields. We will discuss herein these crucial parameters, with a special focus on the lipase-catalyzed synthesis of phenolipids corresponding to the grafting of an aliphatic chain to a phenolic acid. Some other examples of lipophilization with more complex phenolic structures (e.g., lipophilization of flavonoids with fatty acids) have been described and reviewed (Chebil et al., 2006), but these are outside of the scope of this chapter, which is dedicated to phenolipids derived from phenolic acids.

Influence of the Selected Organic Medium

Solvent-Free Systems

Among the different factors governing lipase activity and efficiency, the nature of the medium is obviously crucial. Ideally, the most attractive strategy would be to carry out such lipophilization reactions in a solvent-free system. Typically, in such systems, the fatty alcohol substrate is used in large excess to play the role of both substrate and solvating medium. Such systems have the advantage of avoiding organic solvents and can be considered environmentally friendly. However, they also have some disadvantages, such as difficulties in obtaining good solubilization of both substrates, slow reaction kinetics, and a large excess of one substrate that can be problematic for further purification of final product. Such solvent-free systems have already been described in the literature. For example, Guyot et al. (1997) studied the lipase-catalyzed lipophilization of various phenolic acids with various fatty alcohols using *Candida antarctica* lipase B and without added solvent. Depending on the nature of the phenolic moieties (cinnamic, caffeic, ferulic, and dimethoxycinnamic), different reaction yields were achieved. Results showed that the lipase-catalyzed esterification by *n*-butanol was possible in the cinnamic series, provided the aromatic cycle was not *para*-hydroxylated. Indeed, while cinnamic acid and 3,4-dimethoxycinnamic acid were esterified with satisfactory yields (97% and 60%, respectively), no ester formation was observed in the case of caffeic acid (3,4-dihydroxycinnamic acid) and ferulic acid (4-hydroxy-3-ethoxycinnamic acid). Moreover, it was also observed that in the presence of a saturated side chain on the phenolic acid (dihydroxycaffeic acid), *para*-hydroxylation had no effect. The authors stated that the simultaneous presence of (1) a double bond on the side chain conjugated with the cycle, and (2) a parahydroxyl, totally inhibits the *C. antarctica* lipase. Concerning the nature of fatty alcohol, for cinnamic acid, highest yields were reached with butanol and octanol. For hydrocaffeic acid, yield was virtually always higher than 80%, irrespective of the alcohol's nature, and the best one was obtained with oleyl alcohol. Curiously, ferulic acid esterification was only possible using alcohols with a chain length equal to or higher than eight carbons, though yields remained low (10–14%). The performance of 3,4-dimethoxycinnamic acid was somewhat different; the maximum yield occurred with *n*-butanol only. However, for all the previously mentioned reactions, kinetics were very slow in such a solvent-free system. Therefore, the same group extended all this knowledge for the enzymatic synthesis of fatty esters of chlorogenic acid (Guyot et al., 2000). The esterification was studied with various alcohols (C8, C12, C16, and C18:1) using the same lipase from *C. antarctica*, and the authors compared the solvent-free system to the reaction using 2-methyl 2-butanol as the solvent. In this case, esterification by fatty alcohols was possible, with conversion depending on the length of the carbon chain (around 60% for C8 alcohol and 40% for C12 to C18). Here again, the lipophilization in a solvent-free system was very slow, taking more than 30 days of reaction to reach the equilibrium. In the presence of 2-methyl 2-butanol, reactions were faster and conversions were systematically higher than those obtained in a solvent-free medium, ranging from 55% to 75% depending on the alcohol chain length. In all cases, the esterification kinetics increased rapidly after 4 days. Stamatis et al. (1999) also made a comparison of a solvent-free system versus reaction in an organic medium; in the esterification of cinnamic acid derivatives with 1-octanol using *C. antarctica* and *Rhizomucor miehei* lipases, they observed that for both lipases, higher yields were actually obtained in the solvent-free system than in organic solvents (acetone, 2-methyl-2-propanol, and 2-methyl-2-butanol). Lately, solvent-free systems have been studied for lipophilization of phenolic compounds where the acyl donor is present as a glyceridic form. For example, Xin et al. (2009) evaluated the lipase-catalyzed transesterification of ethyl ferulate with triolein; the same strategy was applied with castor bean oil (Sun et al., 2014). Similarly, Sorour et al. (2012) studied the transesterification of flaxseed oil with 3,4-dihydrophenyl acetic acid. For all these examples, yields and kinetics were quite satisfactory.

Reactions in Organic Medium

Because reactions in solvent-free systems often result in slower reaction kinetics and a large excess of the lipophilic substrate, a large amount of work has been performed in order to identify an appropriate medium that would allow better kinetics and yield for the production of phenolipids. Typically the selected medium, preferably harmless, has to guarantee a good activity of the enzyme over time and no deactivation effect. This is generally the case with solvents having log $P > 3$. The negative influence of solvent polarity on lipase activity and stability seems to be due to competition of the solvent and protein structure of the enzyme for water. Such a competition governs the protein hydration state, which is crucial for enzyme activity. Solvents with log $P > 3$ are widely used in reactions such as the lipase-catalyzed restructuration of oils and fats; however, they are not appropriate for reactions in which the two substrates greatly differ in terms of polarity.

As an example, Buisman et al. (1998) studied the effect of the nature of the selected organic solvent for the esterification of simple phenolic acids by various fatty alcohols in the presence of *C. antarctica* lipase B. The authors showed that in apolar solvents, such as pentane or cyclohexane, conversions up to 85% were observed within 5 days. Conversely, reaction rates were very slow with more polar solvents such as diethyl ether, *t*-butylmethyl ether, and 1-butanol. Such results were also confirmed by Priya et al. (2002), who showed that solvents with higher log *P* values, such as cyclohexane, heptane, or hexane, gave the best results for *Pseudomonas cepacia* lipase-catalyzed transesterification of cinnamate derivatives with various alcohols. In all these cases, a conversion of 90% was observed in 48 hours. Di-isopropylether and toluene were not found to be appropriate, however. Stamatis et al. (1999) investigated the ability of *C. antarctica* and *R. miehei* lipases to catalyze the synthesis of octyl cinnamate in various solvents such as such as acetone, 2-methyl 2-propanol, and 2-methyl 2-butanol. It was shown that reaction yield was dependent on the solvent and lipase used. With *C. antarctica* lipase, octyl cinnamate was formed in good yields (up to 80%), whereas *R. miehei* showed moderate to low efficiency depending on the solvent used. Many different examples exist on lipophilization of phenolic compounds in solvent systems. Examples include the use of used *tert*-butanol for the production of octyl dihydrocaffeate (Feddern et al., 2011), isooctane for the synthesis of cinnamic acid derivatives (Lee et al., 2006) and octyl caffeate (Chen et al., 2010), and cyclooctane to obtain 2-ethylhexyl-p-methoxycinnamate using *R. oryzae* lipase (Kumar et al., 2014).

Some authors studied the efficiency of binary solvent systems. Sabally et al. (2006) performed the lipase-catalyzed transesterification of dihydrocaffeic acid with flaxseed oil in hexane/2-butanone media and showed that the yield of phenolic diacylglycerols increased from 25.1% to 55.8%, when the ratio of the hexane/2-butanone was changed from 65:35 to 85:25 (v/v). Similarly, Yang et al. (2012a) investigated the potentiality of the same binary solvent system for the lipophilization of various phenolic acids. It was found that the conversion of phenolic acids strongly depended on the proportions of hexane and butanone with the optimal mixture ratio of hexane to butanone being 65:35 (v/v). Aissa et al. (2012) evaluated a binary mixture made of 2-methyl-2-propanol and hexane for the synthesis lipophilic tyrosol esters. Some groups evaluated the possibilities to perform the reaction in organogel systems. For instance, Zoumpanioti et al. (2010) immobilized lipases from *R. miehei* and *C. antarctica* in hydroxypropylmethyl cellulose organogels based on surfactant-free microemulsions consisting of n-hexane, 1-propanol, and water and used such systems for the esterification of various phenolic acids with 1-octanol. Both lipases kept their catalytic activity, catalyzing the esterification reactions in high yields (up to 94%).

Reactions in Ionic Liquids or Deep Eutectic Solvents

The potentialities of ionic liquids (ILs) as new media for biocatalytic processes have been largely investigated. Indeed, enzymes generally show increased selectivity and stability in such media in comparison with more classical organic solvents (De Diego et al., 2005; Kaar et al., 2003; Kim et al., 2003). Moreover, ILs have the advantage of being nonvolatile, thermally stable, and having high solvation properties. However, ILs can have some drawbacks. Recently, their toxicity has been questioned, their production cost can be prohibitive for using them on an industrial scale, and their high viscosity can make it difficult to recover reaction products and immobilized enzymes. Some examples of lipase-catalyzed lipophilization of phenolic compounds in IL have been documented. Katsoura et al. (2009) carried out the lipophilization of ferulic acid in several BF_4- and PF_6- imidazolium ILs using immobilized lipases and showed that the reaction efficiency was strongly dependent on the ion composition of ILs. Conversions and initial reaction rates were significantly higher in PF_6 as compared with BF_4 ILs. Yang et al. (2012b), using trioctylmethylammonium trifluoroacetate, observed very good yields in the synthesis of octyl dihydrocaffeic acid using *C. antarctica* lipase B. More recently, Wang et al. (2013) evaluated the feasibility of producing caffeic acid phenethyl ester from caffeic acid and phenylethanol using *C. antarctica* lipase and compared 16 different ILs. The results indicated that ILs containing weakly coordinating anions and cations with adequate alkyl chain lengths improved the synthesis, with [Emim][Tf_2N] giving the optimum results. The highest caffeic acid conversion in [Emim][Tf_2N] reached 98.76% with a final product yield of 63.75%.

Recently, deep eutectic solvents (DESs) have appeared as a promising alternative to ILs as efficient "green" media in lipase-catalyzed reactions (Gorke et al., 2010; Lindberg et al., 2010; Zhao et al., 2011, 2013). These solvents, which share many characteristics with ILs (nonvolatile, thermally stable up to nearly 200 °C, nonflammable, etc.), have many other advantages: They are relatively inexpensive, are environmentally benign, and have a very low toxicity. Moreover, unlike ILs, these solvents do not require a preliminary purification step. Indeed, they result from the association

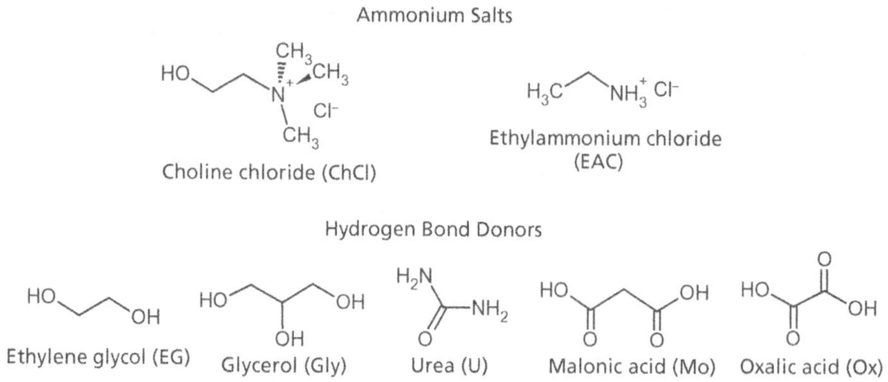

Figure 7.3 Different ammonium salts and hydrogen bond donors (HBDs) used for deep eutectic solvents production (adapted from Durand et al., 2013).

of a cationic salt (ammonium or phosphonium) with a hydrogen bond donor (HBD) (Figure 7.3). This results in a DES with a room temperature melting point. The strong interaction between the HBD and the anion leads to a considerable reduction in the melting point of the mixture. In addition, the strong association between the components drastically decreases their reactivity, making them inert in most cases. The term *eutectic point* is used to characterize the lower melting point of the mixture, which is often much lower than that of the pure constituents. Thus, the mixture can be used at a temperature that allows biocatalytic reactions.

Recently, our group has made significant progress in the application of these solvents in the lipase modification of polar substrates (Durand et al., 2013a). Indeed, we demonstrated that DES based on choline chloride in a binary mixture with water could be effectively used for the alcoholysis of phenolic esters with 1-octanol, using immobilized *C. antarctica* lipase B (Durand et al., 2013b). It was also shown that the lipase-catalyzed reactions of dissolved substrates in DES are extremely difficult to perform without the addition of water. The best results were obtained in DES based on choline chloride associated with urea as a hydrogen-bond donor after water addition (Durand et al., 2014).

Reactions in Supercritical Fluids

Supercritical fluids, and especially supercritical carbon dioxide ($ScCO_2$), have been described as promising media for enzymatically catalyzed reactions by Hammond et al. (1985). The main advantage of supercritical fluids for biocatalysis is their flexibility in their solvent properties. $ScCO_2$ has a dipole moment close to the one of pentane or hexane and therefore allows the solubilization of many low molecular weight apolar compounds. In addition, supercritical fluids are considered environmental friendly and their handling and elimination at the completion of the reactions are rather easy. Therefore, much research has been carried out in which lipases are used in such fluids (Knez, 2009; Matsuda, 2013). Despite the advantages of supercritical fluids, their use as media for lipase-catalyzed lipophilization of phenolic derivatives is still rare. Compton and King (2001) studied the transesterification of ethyl ferulate with triolein in a batch $ScCO_2$ reactor between temperatures of 45 °C and 80 °C at a pressure range of 103–345 bar and showed that a maximum yield of 74% was reached at 80 °C and 241 bar after 48 hours for *C. antarctica* lipase B. Ciftci and Saldana (2012) evaluated the enzymatic synthesis of phenolipids from flaxseed oil and ferulic acid in $ScCO_2$ with the same lipase. Reactions at moderate pressures resulted in higher yields compared to lower and higher pressures. The highest yield (57.6%) was obtained at 80 °C, 215 bar, 6:1 substrate molar ratio, and 27.5 hours of reaction.

Effect of Water Activity

Lipase activity is influenced by various parameters such as temperature or pH, but the most important parameter to control in enzymatic synthesis is the aqueous microenvironment of the enzyme, which results in thermodynamic water activity (a_w) (Halling, 1994). Indeed, a_w determines the mass action effects of water in hydrolytic equilibrium as well as the distribution of water between the various phases that can compete in binding water. The water content and activity of the catalytic environment greatly influence the reaction rates and the balance between the hydrolysis or synthesis reaction (Villeneuve, 2007). Although moisture, or water content, measures the total amount of water present, a_w evaluates the vapor pressure of water generated by the product. By definition, the a_w of a product is in fact the ratio of the vapor pressure of water on the surface of the product to the vapor pressure of pure water. Thus, this parameter reflects the content of unbound or free water available as a substrate. Therefore, enzymatic reaction rates and yields are better described and anticipated by a_w than water content of the medium (Valivety et al., 1992). Consequently, the development of lipase-catalyzed reactions requires the study of the influence of the enzyme a_w on the catalyzed reaction efficiency. Indeed, the study of a_w and water content must be done to define the optimal reaction conditions that allow the enzyme to catalyze acyl transfers while excluding hydrolysis (Caro et al., 2002; Chowdary and Pradulla, 2002; Schmid et al., 1998; Svensson et al., 1994; Xia et al., 2009).

The study of the sorption isotherms of the enzyme is also used to determine the water content corresponding to the optimum a_w of the biocatalyst (Cambon et al., 2006; Caro et al., 2002; Perignon et al., 2013). In addition, it is of paramount importance to control a_w during the reaction course. Different methods have been employed for this purpose. Saturated salts solutions are generally used for adjusting initial water activity (Kaur et al., 1997; Svensson et al., 1994; Wehtje et al., 1993) with individual pre-equilibration of enzymes and substrates or during reaction (Dudal and Lortie, 1995). Some other authors have proposed the equilibration of reaction medium by equilibrated air with a saturated solution (Ujang and Vaidya, 1998).

The influence of the a_w parameter on the efficiency of phenolic compounds lipophilization has been studied by Sorour et al. (2012). Transesterification of 3,4-dihydroxyphenyl acetic acid with flaxseed oil was investigated in solvent-free medium using *C. antarctica* lipase B. Salt hydrate pairs were used to control a_w in the reaction medium. The authors showed that increasing the a_w of the reaction mixture from 0.18 to 0.38 resulted in a significant increase in the bioconversion yield (from 62% to 77%). The same group evaluated the influence of water activity on the synthesis of oleoyl cinnamate (Luen et al., 2005). The results showed that lower initial a_w values (0.05) resulted in a higher enzymatic activity and bioconversion yield. Xin et al. (2009) carried out the transesterification of ethyl ferulate with triolein in a solvent-free medium. Both enzyme and substrate were pre-equilibrated at given water activities by various saturated salts before initiating the reaction. The authors observed that a_w had an obvious influence on transesterification efficiency. Reaction under the lowest a_w condition (<0.01) was slower than at a higher a_w (0.75). When a_w was changed from <0.01 to 0.75, the combined yield of ferulyl diolein and ferulyl monoolein increased from 7.6% to 15.5%. Similarly, Lopez Giraldo et al. (2007) studied the influence of water activity pre-equilibration of *C. antarctica* lipase on the synthesis of chlorogenate fatty esters. In both direct esterification (of free chlorogenic acid) or transesterification (of methyl chlorogenate) with 1-dodecanol, the enzymatic activities were favored at the lower a_w values, indicating that *C. antarctica* lipase B requires minimal amounts of water to maintain its active conformation. For high a_w values, the enzymatic activities decreased. Yu et al. (2010) studied the effect of a_w in the synthesis of feruloylated lipids by transesterification of alkyl ferulates with triolein. Results showed that that a minimum amount of water was required to activate the enzyme.

Effect of the Nature of Substrates

Preliminary Modification of Polar Substrate

Because the main difficulty in lipase-catalyzed lipophilization reactions resides in the very different polarities of the two substrates involved, some strategies have been based on the preliminary modification of the phenolic substrate in order to increase its solubility in the chosen medium and improve its contact with the other substrate, apolar acyl donor.

For example, López Giraldo et al. (2007) proposed a chemo-enzymatic strategy to obtain fatty esters of chlorogenic acid (Figure 7.4). In the first step, chlorogenic

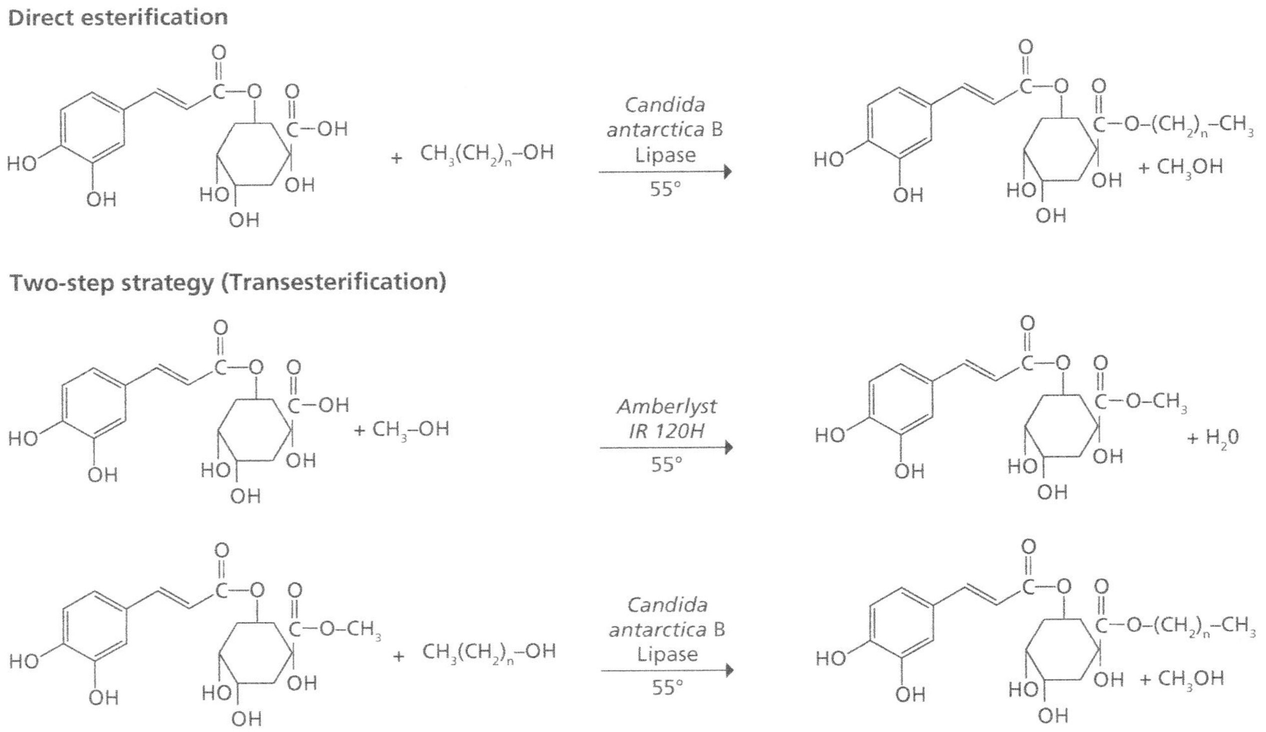

Figure 7.4 Different strategies for production of fatty esters of chlorogenic acid (adapted from Lopez-Giraldo et al., 2007).

acid was chemically transformed to its methyl chlorogenate by methanol esterification using a sulfonic resin as catalyst. This chlorogenate methyl ester showed a better solubility in fatty alcohols than the unmodified molecule, favoring consequently its alcoholysis reaction by various alcohols (from C4 to C16) in the presence of *C. antarctica* lipase B. The two-step reaction overall yield was between 61% and 93%, depending on the alcohol chain length, whereas it was 40% to 60% for the direct esterification of unmodified free chlorogenic acid with the same alcohols.

In a comparable strategy, Vosmann et al. (2008) carried out the synthesis of long-chain alkyl hydroxybenzoates either by transesterification of methyl hydroxybenzoates or direct esterification of the free phenolic acid. All the tested lipases, namely *C. antarctica* lipase B, *R. miehei,* and *Thermomyces lanuginosus*, showed at least a 10-fold higher activity in the transesterification strategy, with a maximum observed for *C. antarctica* lipase B showing a 25-fold higher activity for the preparation of oleyl 4-hydroxybenzoate. Similarly, Priya et al. (2002) used ethyl cinnamates instead of free cinnamic acid to perform its lipophilization, and Compton et al. (2001) compared direct esterification of ferulic acid or alcoolysis of ethyl ferulate by octanol and showed the clear advantage of using the ethyl derivative in terms of reaction kinetics and yields. Also, Xin et al. (2009) performed lipase-catalyzed transesterification of ethyl ferulate with triolein to form ferulyl oleins in solvent-free systems. Others carried out the same kind of reaction, in which ethyl ferulate was also used for transesterification with castor bean oil to compare the efficiency of *R. miehei*, *T. lanuginosus,* and *C. antarctica* lipase B, showing that the latter was the most effective in performing the reaction (Sun et al., 2014). Besides the fact that the so-modified polar substrate shows a better solubility in the chosen medium, such a transesterification strategy has also the great advantage of producing a short-chain alcohol (methanol or ethanol) as a coproduct that can be easily removed from the reaction medium in order to shift the reaction equilibrium toward the synthesis.

Effect of the Nature of the Acyl Donor

The efficiency of a lipase-catalyzed lipophilization reaction can also be modulated depending on the nature of the acyl donor. This is typically the case when such lipophilization corresponds to the grafting of the aliphatic chain to a nonaromatic hydroxyl function of the phenolic substrate (e.g., tyrosol and hydroxytyrosol). In this reaction, the acyl donor corresponding to a fatty acid moiety can be used as free fatty acid such as methyl, ethyl, or even vinyl esters. When methyl or ethyl esters are used, a transesterification occurs, and methanol and ethanol are respectively produced, which can be eliminated by working at reduced pressure. The use of such esters is generally advantageous compared to free fatty acids because the removal of the alcoholic byproducts is easier than the elimination of water. Vinyl esters are also attractive due to the fact that they are irreversible acyl donors. Indeed, the vinyl alcohol byproduct that is obtained during the transesterification is transformed by tautomerization into acetaldehyde, which is easily eliminated from the reaction medium due to its very low boiling point. Consequently, the equilibrium is rapidly displaced toward the synthesis. However, depending on the used lipase, acetaldehyde formation can be problematic due to a deactivation effect on the enzyme through Schiff bases formation with some amino groups of the protein residues (Weber et al., 1995).

Some authors have also used glyceridic derivatives as acyl donors (Figure 7.5). For example, Laszlo et al. (2013) carried out the synthesis of medium-chain alkyl esters of tyrosol and hydroxytyrosol by immobilized *C. antarctica* lipase B using cuphea oil as the acylating agent and 2-methyl 2-butanol as the cosolvent. Various tyrosol:oil ratios were tested, and at all mole ratios the reaction reached equilibrium within 2–4 hours. At low tyrosol:oil ratios (1:3 and 1:2), the conversion of tyrosol to product was greater than 98%. The extent of tyrosol conversion to product progressively lowered as the tyrosol:oil ratio increased. Conversely, the residual cuphea oil triglyceride contents of the reaction were 56%, 42%, and 19%, respectively, for 1:3, 1:2, and 1:1 tyrosol:oil substrate ratios. At higher tyrosol:oil ratios (2:1 and 3:1), the triglyceride content was negligible and only monoacylglycerols and diacylglycerols remained. The same tendencies were observed when using hydroxytyrosol, and the authors concluded that the catalytic performance of the enzyme was not impaired by the ortho dihydroxy groups of hydroxytyrosol.

Effects of Other Parameters

Many others parameters can also influence the efficiency of a lipase-catalyzed reaction. Indeed, the effects of experimental conditions such as temperature, pressure, enzyme load, and type of the bioreactor (batch, packed bed, etc.) are also important. The influences of such parameters have been previously reviewed (Adachi and Kobayashi, 2005; Plou et al., 2002). Because lipase-catalyzed reactions are reversible, it is advantageous to shift the equilibrium toward the synthesis, and various strategies can be employed to accomplish this. First, the molar ratio between the two initial compounds can be adjusted in order to have one of them in excess. In such conditions, the esterification is favored over the reverse hydrolysis reaction. The nature of the reaction itself also influences its efficiency. It is admitted that performing reactions that generate a short-chain alcohol (transesterification) instead of water (direct esterification) are preferable. Indeed, methanol or ethanol can be continuously and easily removed from the reaction medium under pressure, whereas addition of molecular sieves or water trapping systems are generally necessary for direct esterification that forms water as a coproduct.

Concerning lipophilization of phenolic compounds, many examples are available for which such parameters have been adjusted for optimized yields and kinetics.

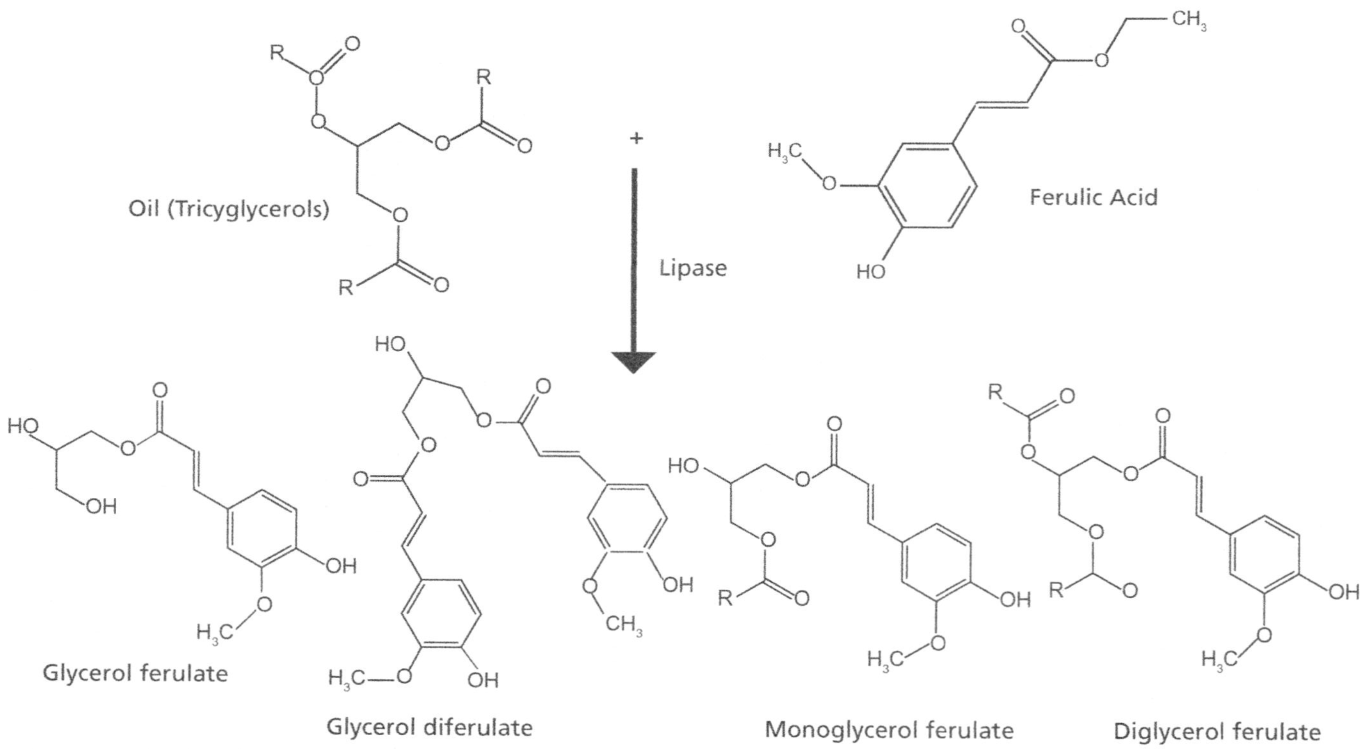

Figure 7.5 Example of glyceridic phenolipids that can be obtained in transesterification reaction of phenolic acid (e.g., ferulic acid) with triacylglcycerols.

Widjaja et al. (2008) used a very large molar excess of phenylethyl alcohol (1:92) for the esterification of caffeic acid with *C. antarctica* lipase B. The authors also studied the recyclability of the lipase, which is a crucial parameter for subsequent scaling up of the process, and they observed that this lipase could maintain more than 90% of its original activity up to the third batch. Sun et al. (2013) used response surface methodology (RSM) to study the effects of reaction variables (reaction temperatures, enzyme load, and reaction time) on the transesterification of ethyl ferulate and monostearin. The optimum conditions were a reaction temperature of 74 °C, a reaction time of 23 hours, and an enzyme load of 20%. A similar experimental design (RSM) was performed by Chen et al. (2010) for the synthesis of octyl caffeate with the study of parameters such as reaction temperature (40–80 °C), reaction time (24–72 h), and substrate molar ratio of caffeic acid to octanol (1:20–1:100). The conversion results showed that reaction temperature and time had significant effects, with optimum conditions being a reaction time of 55 hours, a reaction temperature of 75 °C, and a substrate molar ratio of 1:78. The molar conversions of predicted and actual experimental values were 94% and 90%, respectively. Finally, one can also cite the work by Lee et al. (2010), who studied the ultrasound-accelerated enzymatic synthesis of octyl hydroxyphenylpropionate with *C. antarctica* lipase B. A three-level, three-factor Box-Behnken experiment design and RSM were used to evaluate the effects of temperature, reaction time, and enzyme activity. The results indicated that temperature and enzyme activity significantly affected yield, whereas reaction time did not. Based on a ridge max analysis, the optimum conditions for octyl hydroxyphenylpropionate synthesis were predicted to use a reaction temperature of 58.8 °C, a reaction time of 14.6 hours, and an enzyme activity of 410.5 propyllaurate units (PLU) with a yield of 98.5%.

The lipase-catalyzed synthesis of phenolipids is more and more documented and can appear advantageous owing to the mild operating conditions of lipases and their selectivities. However, many parameters have to be considered and optimized in order to achieve optimum reaction kinetics and yields. Among these parameters, appropriate choice of reaction medium, control of water activity and water content of the systems, nature of acyl donor, and substrate ratio are crucial. Lately, a lot of progress has been made in order to propose new enzymatic routes to phenolipids that can be considered as real alternatives to chemical catalysis.

Physicochemical and Biological Activity of Phenolipids

Phenolipids are molecules exhibiting various activities in foods, cosmetics, nutraceuticals, and pharmaceuticals, as well as in biological samples whether they are living or not. One can cite the activity of phenolipids to counteract oxidation in physicochemical systems such as emulsions, micelles, and liposomes. They are also known to display a large spectrum of biological activities, especially against oxidative degradation of lipoproteins or DNA (as biological antioxidants) and against microorganisms (as antimicrobials). Here we present a brief panorama of the diverse performances of such multifunctional compounds.

Physicochemical Antioxidant Activity

The antioxidant activity of natural or artificial phenolipids is conveyed by the same mechanistic pathways as common phenolics. The main pathway is the reducing ability, consisting of donating electron(s) to pro-oxidant species that lack them, either as a single electron (fast reduction) or as a hydrogen atom (slow reduction). This reduction, conveyed by phenolic hydroxyls, often leads to the transformation of a species of high oxidizing potential into a new one of lower potential. Such a phenolic indeed acts as an antioxidant. But the term *often* does not mean *always*. In some situations, reduction can lead to the transformation of pro-oxidant species into "super pro-oxidant" ones. This is the case with the reduction of Fe^{3+} into Fe^{2+}.

In terms of the structure–antioxidant activity relationship, the first determinant trait is the number and the position of the phenolic hydroxyls. A monophenol is less of a reducer than a diphenol, which, in turn, is less of a reducer than a triphenol. Position is also important because the *ortho* position between phenolic hydroxyls allows the establishment of an intramolecular H-bond between them that stabilizes the phenoxy free radical, thus enhancing reducing properties. In the *meta* and *para* positions, phenolic hydroxyls are too far for such an intramolecular hydrogen bond to be formed between them. Consequently, catechol (*o*-diphenol) and, above all, pyrrogallol (*o*-triphenol) are the most efficient phenolic structures regarding reducing performances. The presence of some phenolic ring substituents, such as methoxy ether, can also stabilize the phenoxy radical and can consequently enhance reducing properties. The electronic delocalization area (conjugated double bonds system) is another crucial factor: The more extensive this area, the better the reducer. A large electronic delocalization area will allow the unpaired electron—resulting from the antioxidative action of a phenolic—to be stabilized, which enhances the reducing properties.

If phenolipids *react* as common phenolics, they do not *behave* as common phenolics. In other words, if their chemistry is identical, their physics are totally different, because the lipid moiety confers the additional functionalities of lipids to phenolics. The main functionality of phenolipids is their ability to interact with polar compounds through their polar part—the phenolic ring(s)—all the while interacting with nonpolar components through hydrophobic contacts with their lipid moiety. Phenolipids will not distribute, locate, organize, and diffuse as common phenolics. They are more lipophilic. As such, they tend to move into a more lipophilic environment to minimize the entropic penalty resulting from their contact with water. However, due to their phenolic part, they cannot enter environments that are too hydrophobic. Interfacial space between oil and water, when present in a system, represents a good trade-off for phenolipids having the appropriate lipid chain. These multifunctional molecules can also self-organize into micelles, co-micelles, or reverse micelles, depending on their environment. It is obvious that such supramolecular assemblies modulate their interaction with the reaction centers of the lipid oxidation and antioxidation. In terms of diffusion, the chain length can hinder the phenolipid mobility and make the long phenolipid molecules very bulky, evolving slowly from one point of the system to another one. All these parameters are physical ones; that is why we said that the physics are different between phenolics and phenolipids, but the chemistry is not.

The Polar Paradox Hypothesis

Having delineated the main traits of phenolipid antioxidants, we now have to answer the question of how the phenolipid chain impacts their antioxidant activity in quantitative terms. To do so, we will focus on natural media such as lipid-based systems. Phenolipid antioxidants are often thought to be more active in these systems than their nonlipophilized counterparts. This common belief, which is partially false, probably comes from the fact that phenolipids bear a lipophilic moiety that is supposed to orient the antioxidant in a location and position that is beneficial to neutralize pro-oxidant species in lipid-based systems. In 1989, however, American researchers from Porter's group questioned the relevancy of considering lipid-based systems as a sort of generic system behaving in a universal manner (Porter et al., 1989). They dichotomized nonbiological lipid-based systems in two groups: those having a large lipid–water interface, such as emulsified, micellar, and membrane systems, and those supposedly having a low one, such as bulk oils. Using these two different behavioral systems, they stated that in oil-in-water emulsions, micelles, and membranes (further referred to as *lipid dispersions*), nonpolar antioxidants tend to be more active than polar ones, whereas the reverse is generally observed in bulk oils. This hypothesis is known as the *polar paradox,* and it remained an unexplained observation until Frankel et al. (1994) introduced the concept of interfacial antioxidation. Indeed, they postulated that the better activity of nonpolar antioxidants in lipid dispersions was due to a better location because it was also the case for polar antioxidants in bulk oils. For lipid dispersions, the less polar they are, the closer they become to the interfacial region where lipid oxidation primarily occurs. On the contrary, in bulk oils, in which nonpolar antioxidants are solubilized in the bulky phase of oil, they are distributed far away from the air–oil interface and are consequently less active than nonpolar antioxidants located near the interface where oxidation of bulk oils is supposed to primarily occur (Figure 7.6).

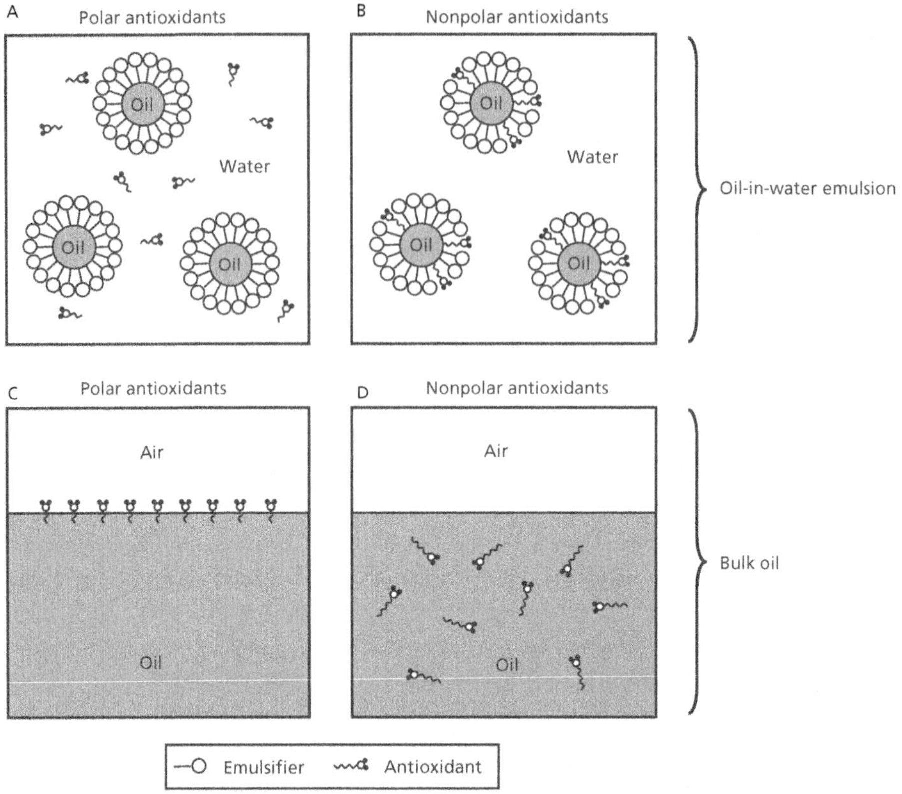

Figure 7.6 Interfacial phenomena as a possible mechanism of action of the polar paradox in oil-in-water emulsion (A and B) and in bulk oil (C and D) (adapted from Frankel et al., 1994).

Regarding lipid dispersions, however, when researchers started to assess complete homologous series of phenolipid antioxidants, inconsistencies contradicting the polar paradox accumulated. Porter et al. found a rather nonlinear effect of the chain length—thus the hydrophobicity—on the antioxidant activity of gallate alkyl esters (0, 1, 2, 3, 8, and 12 carbon atoms length in the aliphatic chain) in lipid dispersions. The same homologous series has been used by Stöckmann et al. (2000) to test the polar paradox in a stripped corn oil-in-water emulsion stabilized by soy lecithin. Their results confirm those obtained by Porter et al. (1989) in lecithin-stabilized emulsions. The order of effectiveness was approximately butyl gallate (not tested by Porter et al.) ≅≅ propyl gallate > octyl gallate ≅≅ ethyl gallate ≅≅ methyl gallate >> gallic acid. Both studies found gallic acid to be the worst antioxidant and propyl (and butyl) gallate to be the best one, with octyl gallate falling in between. The authors stated that "there was a nonlinear relationship between decreasing polarity of the gallates and antioxidant activity in SDS, Brij 58, and PHLC [lecithin] emulsions" (Stöckmann et al., 2000). Although such studies showed results challenging the polar paradox in lipid dispersions, to the best of our knowledge, this rule was never openly contradicted (for the emulsion part) until new evidence of a more global phenomenon described as the *cut-off effect* was brought (Laguerre et al., 2009).

The Cut-Off Hypothesis

A lot of reviews and original articles have been recently released commenting on the cut-off effect hypothesis (Costa et al., 2014; Laguerre et al., 2013, 2015); therefore, we will not spend too much time on it. Briefly, the cut-off effect states that activity of phenolic antioxidants will increase in lipid dispersions when their hydrophobicity (chain length) increases until a critical chain length is attained for which the activity is maximal. Beyond this threshold, antioxidant activity collapses as the lipid chain is lengthened (Figure 7.7). The cut-off effect on antioxidant activity on lipid dispersion was first mentioned by Laguerre et al. in 2009. (We coined this nonlinear trend as the "cut-off effect," a term already used by biologists since the 1930s.) Using chlorogenate alkyl esters in oil-in-water microemulsions, they found a parabolic-like fashion regarding the influence of the chain length on antioxidant activity. In emulsions, this trend has now been confirmed in numerous cases, mostly on synthesized phenolipids, such as chlorogenate alkyl esters (Laguerre et al., 2009), rosmarinate alkyl esters (Laguerre et

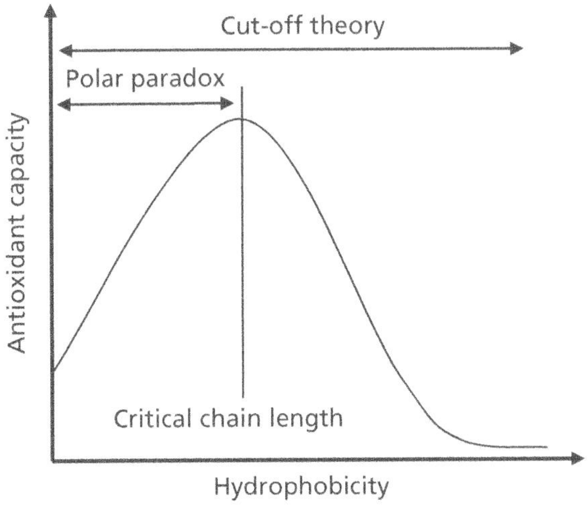

Figure 7.7 The polar paradox is a linear section of a broader nonlinear effect (cut-off effect).

al., 2010; Lee et al., 2013; Panya et al., 2012), hydroxytyrosol alkyl esters (Medina et al., 2009), ferulate alkyl esters (Sørensen 2014a, 2014b), gallate alkyl esters (Losada et al., 2013), rutin alkyl esters (Sørensen, 2012), caffeate alkyl esters (Aleman et al., 2015; Costa et al., 2014; Sørensen et al., 2014a), coumarate alkyl esters (Sørensen et al., 2014a), and coniferyl alcohol alkyl esters, called capsiconiate analogues (Reddy and Kanjilal, 2014). A cut-off effect for phenolipids or ascorbate antioxidants has also been reported in a multitude of studies performed in liposomal and cellular systems. Even more promising are the results recently published by Munoz-Marin et al. (2013) showing that a cut-off effect occurs in the antioxidant activity of hydroxytyrosol alkyl ethers orally administered to rats. At 20 mg/kg/day, the maximal effect was obtained for the hexyl derivative. If confirmed with molar comparison of phenolipids, and not just mass comparison, it would open an avenue for studying the fascinating in vivo implications of the antioxidant cut-off phenomenon.

Finally, this discussion dealing with prediction models of the antioxidant activity of phenolipids in lipid dispersions lead to two conclusions: a practical one and a theoretical one. First, lipophilization appeared to be a double-edged sword technique: If the grafted chain is too long or too short, the corresponding phenolipid will not be optimal in terms of antioxidant activity in lipid dispersions. When looking for the best phenolipid antioxidant to protect a lipid dispersion from being oxidized, one should find the phenolipid with the critical chain length (the optimal chain), which is often a medium one (6 to 12 carbon atoms) for saturated chains of monoesters or monoethers.

The second conclusion is more general. Since Plato, we have been asking whether models—like the polar paradox or the cut-off effect—exist independently of our minds. Do the polar paradox or the cut-off effect really exist independently of the way we are thinking of them? Of course not; both of them are mind constructions that are intended to fit with the real data from the real world as much as possible. We do not have to stick with them when more and more experimental contradictions accumulate. If available, we should consider new paradigms with better prediction capacities. However, we sometimes behave as if models have a real existence, as if data are subordinate to models and not the reverse. This probably explains why the polar paradox was not been questioned and replaced by another model before 2009.

In Vitro Biological Activity

Esterification of phenolic acids leads to products with improved hydrophobicity, which has been the driving interest toward production and use of these compounds as antioxidants with increased solubility in lipid-based systems. Recently, a few research groups have been exploiting this increase in lipophilicity in order to apply lipophilized phenolic acids as antimicrobial and/or cytotoxic agents, as well as to prevent human low-density lipoprotein (LDL) oxidation and cell damage in vitro. Literature accumulated so far is briefly reviewed next.

Ferulic acid–based structured lipids were synthesized via a chemo-enzymatic route (Kanjilal et al., 2008). Briefly, the olefinic side chain of ferulic acid was modified to generate 1,2-diol, which was then esterified with fatty acids. Final products involved either butyric acid or palmitic acid together with ferulic acid within the same structure. Free radical scavenging activity of both compounds was found to be approximately half of that of butylated hydroxytoluene (BHT). When tested for antibacterial (against two representative Gram [+] organisms: *Bacillus subtilis* and *Staphaylococcus aureus*; and four Gram [–] organisms: *Escherichia coli*, *Pseudomonas aeroginosa*, *P. oleovorians*, and *Klebsiella*) and antifungal (against *Candida albicans*, *Saccharomyces cerevisiae*, *Rhizopus oryzae*, and *Aspergillus niger*) activity, the phenolipid with palmitic acid did not show any inhibitory activity toward the tested microorganisms. The authors linked this result to the bulky structure of the molecule, which would make it difficult to penetrate into the cell. Phenolipid with butyric acid, on the other hand, showed moderate antibacterial activity, but showed no antifungal activity. The inhibitory effect against both Gram (+) and Gram (–) organisms were from one-sixth to one-third of that shown by the control (streptomycin).

Lipoconjugates of methyl ricinoleate and methyl 12-hydroxystearate, prepared from castor oil with ferulic and vanillic acids, were synthesized by the Mitsunobu protocol (Reddy et al., 2012). The synthesized compounds have different structures compared to the other reported phenolipids due to their connectivity through the secondary hydroxyl moiety positioned at C-12. None of the compounds showed any antibacterial activity (against three representative Gram [+] organisms: *B. subtilis*, *S. aureus*, and *Staphylococcus epidermidis*; and three Gram [–] organisms: *E. coli*, *P. aeruginosa*, and *Klebsiella pneumoniae*). On the other hand, their antifungal activities were moderate (against *C. albicans*, *C. rugosa*, *S. cerevisiae*, *R. oryzae*, and *A. niger*), although still lower than the control (amphotericin B). For example, the most effective compound against *C. albicans*, ferulic acid linked to methyl ricinoleate, was only as effective as 30% of the control. Compounds with ferulic acid showed better antifungal activity against all strains compared to those with vanillic acid, which was suggested to be linked to the side-chain double bond present in the former group of compounds.

A recent study aimed to combine the positive effects of phenolic compounds with those of sapienic acid (C16:1 *n*-10) by lipase-catalyzed synthesis (Kaki et al., 2013). One-third of skin's endogenous lubricant (sepum) consists of sapienic acid. The deficiency of this fatty acid is reported to render the skin more susceptible to colonization by *S. aureus*. Sapienic acid alone did not show any cytotoxicity on the cell lines tested (derived from human alveolar adenocarcinoma epithelial cells, human cervical cancer cells, human prostate adenocarcinoma cells, and human breast adenocarcinoma cells). On the other hand, sapienic acid esters comprising hydroxybenzyl alcohol, tyrosol, and coniferyl alcohol exhibited some cytotoxicity toward all the tested cell lines (in the range of 1/25 to 1/52 of that shown by the control, 5-fluoro uracil). Applying similar tests using skin-based cancer cell lines is projected to be helpful in order to evaluate the potential cosmetic applications of these compounds.

The effect of phenolipids on LDL oxidation has been of interest recently. Katsoura et al. (2009) synthesized various phenolipids by lipase-catalyzed acidolysis of cinnamic acid derivatives with aliphatic alcohols in ionic liquids. Methyl, ethyl, and octyl ferulate had significantly higher antioxidant capacity toward LDL, high-density lipoprotein (HDL), and total serum oxidation. Moreover, this study showed that the lipophilic ferulates improved the antioxidant efficiency of HDL_{3c}, a HDL subfraction with antioxidant potential, toward the protection of LDL in vitro. In both cases, the protection effectiveness increased with the increasing length of the alkyl chain from methyl to ethyl then to octyl. Shahidi's group extended their experience on the lipase-catalyzed synthesis of phenolipids further so as to test these compounds' potential inhibitory effect on LDL oxidation as well as radical-induced DNA cleavage (Wang and Shahidi, 2014). Acidolysis of *p*-coumaric acid with seal blubber oil and menhaden oil was carried out to synthesize phenolipids in the form of mono- and diacylglycerols. The inhibitory activity of the products against copper-induced LDL oxidation was compared with that of *p*-coumaric acid alone. Both products showed higher inhibitory capacity than the control because their improved lipophilicity led to greater affinity to LDL; however, the difference was not significant. In a similar fashion, the synthesized phenolipids had a higher protective effect against radical-induced DNA cleavage compared to *p*-coumaric acid. In both tests, *p*-coumaric acid linked to menhaden oil resulted in higher protection against oxidation.

Phenolipids undoubtedly present an interesting potential for extending their use from lipophilic antioxidants toward functional, bioactive ingredients. Nevertheless, the field lacks in-depth research so far and awaits further interest.

Potential Applications of Phenolipids

Owing to their many biological and physicochemical properties, mentioned previously, phenolipids are a class of multifunctional molecules that is of particular interest in various sectors such as the food, cosmetics, and pharmaceutical industries. This multifunctionality is attractive for formulators because the use of a single molecule with several functions reduces the risk of incompatibility between ingredients along with the number of chemical additives that are rather distrusted by consumers, especially in foods. Using multifunctional molecules also saves time and facilitates production processes.

In terms of current industrial applications, the most-known phenolipids are gallate alkyl esters and parabens (alkyl esters of *p*-hydroxybenzoic acid), which are authorized as food additives by the Codex Alimentarius. Regarding the former, propyl (E-310), octyl (E-311), and dodecyl (E-312) gallates are classified as antioxidants, but they also possess antifungal and antibacterial activities, essentially on Gram (+) bacteria (Kubo et al., 2004). They are, however, not recommended for consumption by children.

Alkyl *p*-hydroxybenzoates, mainly methyl (E-218), ethyl (E-214), and propyl (E-216) parabens and their corresponding sodium salts (E-219, E-215, and E-217, respectively) are widely used as antimicrobial preservatives in foods, beverages, pharmaceuticals, cosmetics, toothpastes, and toiletries. Nevertheless, even though these molecules have low toxicity and no carcinogenicity at currently used concentrations, they are increasingly controversial due to their estrogenic potency (from 100,000 to 1 million times lower than that of 17 β-estradiol) that rises with increasing length (and branching) of the alkyl chain. Proposals have been made to ban or restrict parabens for some applications in European countries such as in France (2011), even if it is not yet followed at the practical level. Meantime, products labeled "paraben-free" are increasingly released by the skin care industry. However, the replacement of parabens by other antimicrobials is problematic, especially with the methylisothiazolinone. A huge replacement market for parabens is thus virtually open for other phenolipids that are not estro-mimetics. Parabens are found in more than 80% of the commercially available skin care products, such as shampoos, moisturizing creams, shaving creams, cleaning gels, and others. More than 400 pharmaceuticals contain parabens, such as creams (Biafine), cough syrups (Clarix, Codotussyl, Drill, Hexapneumine, Humex, Pectosan, Rhinathiol), throat pain syrups (Ergix), stomach-coating medication (Maalox, Gaviscon), antiemetics (Prinperan, Vogalene). Methylparaben is the most used local preservative in local anesthesia. Therefore, paraben replacement, if any, may be the most important potential application of phenolipids (in volume) in the near future.

Apart from parabens, other phenolipids that are used in industry have been shown to mimic estrogens. This is the case with alkylresorcinols (natural phenolipids found in the bran of wheat, rye, or barley, for instance) such as 4-hexylresorcinol, a food additive (E-586) used as an antioxidant and color-stabilizer agent. In the EU, it is exclusively found in fresh or frozen crustaceans, except in organic foods, for which it is not authorized (both in the EU and the U.S.). Hexylresorcinol is also a mild anesthetic introduced in the formulation of some pills (Strepsils, for example) designed to relieve pain caused by sore throats. In addition, this medicine helps lubricate and soothe the throat and kills the germs responsible for sore throats. Overall, hexylresorcinol can serve as antimicrobial in cosmetics, an antiseptic agent in pharmaceuticals to treat minor external infections, and a medicine for counteracting intestinal worms. Another application of alkylresorcinols is found in the dye industry. One can cite 2-methylresorcinol, which is a fine chemical for hair dyes.

In addition to the previous examples, a few phenolic esters are available on the market and used in mass consumer goods, mainly in cosmetics: some ferulates (ethyl,

ethyl hexyl, isononyl, isostearyl) are used as UV filters, skin conditioners, or antioxidants; ethyl caffeate is a skin conditioner; and vanillates (methyl, ethyl) and salicylates (methyl to octyl) are widely applied in the perfumery and flavoring sectors (Figueroa-Espinoza et al., 2013).

For the years to come, future applications of phenolipids appear to be promising, first because their use has been very limited so far, especially for medium- and long-chain phenolipids (as we noted, most of the industrial applications involve short-chain phenolipids); and second, because they offer to formulators an attractive alternative to phenolics that are rather hydrophilic (for nonpolymeric ones) and thus poorly soluble in lipid products. This is of particular importance when considering that as industries use more and more lipid-based systems, the need for phenolipids consequently grows. These last decades have seen a boom in the use of lipid-based products, especially in nanomedicine and functional food sectors, and to a lesser extent in conventional pharmaceutical, food, and cosmetics industries. One can cite the potential applications in oil-in-water microemulsions, nanoemulsions, solid lipid nanoparticles, colloidosomes, liposomes, multiple-layered emulsions, and so forth. In all of these systems, the lipids forming the product architecture cannot be totally saturated or the products will be too stiff. The formulator has to introduce unsaturated lipids, thus introducing a risk of an oxidative process. To counteract lipid oxidation, phenolipids, which are oil-soluble antioxidants, are beneficial compared to the current strategy to protect unsaturated lipids. In addition, these lipid-based products are not only sensitive to oxidation, they are also prone to hydrolytic or microbial attacks, whether it is from single lipases or esterases or from bacteria and yeasts. Viruses can also contaminate these oily products. In such situations, phenolipids may provide an efficient way to fight any deleterious effects by killing bacteria and viruses. Finally, the third reason why using phenolipids appears more and more as a promising alternative to other vitamins (esters of vitamin C, vitamin E, tocotrienols), additives, or drugs is the huge number of findings that have been released over the last years from research laboratories. These have boosted industrial and academic interest in these molecules for a diverse range of activities and situations. The systematic scrutiny of alkyl chain-length effects has highlighted some very promising bioactive molecules, such as dodecyl chlorogenate, which displays spectacular antimicrobial activity on the Gram (+) bacteria *Staphylococcus carnosus* (Suriyarak et al., 2014). It is worth noting that, owing to their cost and availability, phenolipids (either natural or chemio-enzymatically synthesized) should find applications in niche sectors in complement to their current use in the food industry. In pharmaceuticals, cosmetics, or functional foods, phenolipids can be used as a drug that can have a targeted action in humans, not just as a simple preservative for the product. Alone or associated with a drug delivery system, phenolipids have a good membrane crossing ability and a high bioavailability that make them potent "drug" candidates for these kinds of high-added-value products. Recently, it has been shown that octyl rosmarinate is a powerful mitochondria-targeting antioxidant, paving the way for new therapeutic strategies against oxidative stress-related disorders (Bayrasy et al., 2013).

References

Adachi, S.; Kobayashi, T. Synthesis of Esters by Immobilized-lipase Catalyzed Condensation Reaction of Sugars and Fatty Acids in Water-miscible Organic Solvent. *J. Biosci. Bioeng.* **2005,** *99,* 87–94.

Aissa, I.; Sghair, R. B.; Bouaziz, M.; Laouini, D.; Sayadi, S.; Gargouri, Y. Synthesis of Lipophilic Tyrosyl Esters Derivatives and Assessment of Their Antimicrobial and Antileishmania Activities. *Lipids Health Dis.* **2012,** *11,* 13.

Aleman, M.; Bou, R.; Guardiola, F.; Durand, E.; Villeneuve, P.; Jacobsen, C.; Sørensen, A. D. M. Antioxidative Effect of Lipophilized Caffeic Acid in Fish Oil Enriched Mayonnaise and Milk. *Food Chem.* **2015**, *167,* 236–244.

Bayrasy, C.; Chabi, B.; Laguerre, M.; Lecomte, J.; Jublanc, E.; Villeneuve, P.; Wrutniak-Cabello, C.; Cabello, G. Boosting Antioxidants by Lipophilization: A Strategy to Increase Cell Uptake and Target Mitochondria. *Pharm Res.* **2013,** *30,* 1979–1989.

Buisman, G. J. H.; van Helteren, C. T. W.; Kramer, G. F. H.; Veldsink, J. W.; Derksen, J. T. P.; Cuperus, F. P. Enzymatic Esterifications of Functionalized Phenols for the Synthesis of Lipophilic Antioxidants. *Biotechnol Lett.* **1998,** *20,* 131–136.

Cambon, E.; Gouzou, F.; Pina, M.; Baréa, B.; Barouh, N.; Lago, R.; Ruales, J.; Tsai, S. W.; Villeneuve, P. Comparison of the Lipase Activity in Hydrolysis and Acyltranfer Reactions of Two Latex Plant Extracts from Babaco (Vasconcellea × Heilbornii cv.) and Plumeria Rubra: Effect of the Aqueous Microenvironment. *J. Agric. Food Chem.* **2006,** *54,* 2726–2731.

Caro, Y.; Pina, M.; Turon, F.; Guilbert, S.; Mougeot, E.; Fetsch, D. V.; Attwool, P.; Graille, J. Plant Lipases: Biocatalyst Aqueous Environment in Relation to Optimal Catalytic Activity in Lipase-catalyzed Synthesis Reactions. *Biotechnol. Bioeng.* **2002,** *77,* 693–703.

Chebil, L.; Humeau, C.; Falcimaigne, A.; Engasser, J. M.; Ghoul, M. Enzymatic Acylation of Flavonoids. *Process Biochem.* **2006,** *41,* 2237–2251.

Chen, H. C.; Twu, Y. K.; Chang, C. M. J.; Liu, Y. C.; Shieh, C. J. Optimized Synthesis of Lipase-catalyzed Octyl Caffeate by Novozym® 435. *Ind. Crop. Prod.* **2010,** *32,* 522–526.

Chowdary, G. V.; Prapulla, S. G. The Influence of Water Activity on the Lipase Catalyzed Synthesis of Butyl Butyrate by Transesterification. *Process. Biochem.* **2002,** *38,* 393–397.

Ciftci, D.; Saldana, M. D. A. Enzymatic Synthesis of Phenolic Lipids Using Flaxseed Oil and Ferulic Acid in Supercritical Carbon Dioxide Media. *J. Supercrit. Fluid.* **2012,** *72,* 255–262.

Compton, D. L.; King, J. W. Lipase-catalyzed Synthesis of Triolein-based Sunscreens in Supercritical CO_2. *J. Am. Oil Chem. Soc.* **2001,** *78,* 43–47.

Costa, M.; Losada-Barreiro, S.; Paiva-Martins, F.; Bravo-Díaz, C.; Romsted, L. S. A Direct Correlation between the Efficiencies of Caffeic Acid Alkyl Esters and Their Mole Fractions in the Interfacial Region of Olive Oil Emulsions. The Pseudophase Model Interpretation of the "Cut-off" Effect. *Food Chem.* **2014,** in press.

de Diego, A.; Lozano, P.; Gmouh, S.; Vaultier, M.; Ibbora, J. L. Understanding Structure–Stability Relationships of *Candida antarctica* Lipase B in Ionic Liquids. *Biomacromolecules* **2005,** *6,* 1457–1464.

Dudal, Y.; Lortie, R. Influence of Water Activity on the Synthesis of Triolein Catalyzed by Immobilized *Mucor miehei* Lipase. *Biotechnol. Bioeng.* **1995,** *45,* 129–134.

Durand, E.; Lecomte, J.; Villeneuve, P. Deep Eutectic Solvents: Synthesis, Application and Focus on Lipase-catalyzed Reactions. *Eur. J. Lipids Sci. Tech.* **2013a,** *115,* 379–385.

Durand, E.; Lecomte, J.; Baréa, B.; Dubreucq, E.; Lortie, R.; Villeneuve, P. Evaluation of Deep Eutectic Solvent-Water Binary Mixtures for Lipase-catalyzed Lipophilization of Phenolic Acids. *Green Chem.* **2013b,** *15,* 2275–2282.

Durand, E.; Lecomte, J.; Baréa, B.; Villeneuve, P. Towards a Better Understanding of How to Improve Lipase-catalyzed Reactions Using Deep Eutectic Solvents Based on Choline Chloride. *Eur. J. Lipids Sci. Tech.* **2014,** *116,* 16–23.

Feddern, V.; Yang, Z.; Xu, X.; Badiale-Furlong, E.; Almeida de Souza-Soares, L. Synthesis of Octyl Dihydrocaffeate and Its Transesterification with Tricaprylin Catalyzed by *Candida antarctica* Lipase. *Ind. Eng. Chem. Res.* **2011,** *50,* 7183–7190.

Fernandez-Panchon, M. S.; Villano, D.; Troncoso, A. M.; Garcia-Parrilla, M. C. Antioxidant Activity of Phenolic Compounds: From in Vitro Results to in Vivo Evidence. *Crit. Rev. Food Sci. Nutr.* **2008,** *48,* 649–671.

Figueroa-Espinoza, M. C.; Villeneuve, P. Phenolic Acids Enzymatic Lipophilization. *J. Agric. Food Chem.* **2005,** *53,* 2779–2787.

Figueroa-Espinoza, M. C.; Laguerre, M.; Villeneuve, P.; Lecomte, J. From Phenolics to Phenolipids: Optimizing Antioxidants in Lipid Dispersions. *Lipid Technology* **2013,** *25,* 131–134.

Frankel, E. N.; Huang, S.-W.; Kanner, J.; German, J. B. Interfacial Phenomena in the Evaluation of Antioxidants: Bulk Oils vs Emulsion. *J. Agric. Food Chem.* **1994,** *42,* 1054–1059.

Gorke, J. T.; Srienc, F.; Kazlauskas, R. J. Deep Eutectic Solvents for *Candida antarctica* Lipase B-catalyzed Reactions. In *Ionic Liquid Applications: Pharmaceuticals, Therapeutics, and Biotechnology*; Malhotra, S. V., Ed.; American Chemical Society: Washington, DC, 2010; pp 169–180.

Guyot, B.; Bosquette, B.; Pina, M.; Graille, J. Esterification of Phenolic Acids from Green Coffee with an Immobilized Lipase from *Candida antarctica* in Solvent-free Medium. *Biotechnol. Lett.* **1997,** *19,* 529–532.

Guyot, B.; Gueule, D.; Pina, M.; Graille, J.; Farines, V.; Farines, M. Enzymatic Synthesis of Fatty Esters in 5-Caffeoyl Quinic Acid. *Eur. J. Lipid Sci. Technol.* **2000,** *102,* 93–96.

Halling, P. J. Thermodynamic Predictions for Biocatalysis in Non-conventional Media: Theory, Tests and Recommendations for Experimental Design and Analysis. *Enzyme Microb. Technol.* **1994,** *16,* 178–206.

Haminiuk, C. W. I.; Maciel, G. M.; Plata-Oviedo, M. S. V.; Peralta, R. M. P. Phenolic Compounds in Fruits—An Overview. *Int. J. Food Sci. Technol.* **2012,** *47,* 2023–2044.

Hammond, D. A.; Karel, M.; Klibanov, A. M.; Krukonis, V. J. Enzymatic Reactions in Supercritical Gases. *Appl. Biochem. Biotech.* **1985,** *11,* 393–400.

Kaar, J. L.; Jesionowski, A. M.; Berberich, J. A.; Moulton, R.; Russel, A. J. Impact of Ionic Liquid Physical Properties on Lipase Activity and Stability. *J. Am. Chem. Soc.* **2003,** *125,* 4125–4131.

Kaki, S. S.; Gopal, S. C.; Rao, B. V. S. K.; Poornachandra, Y.; Kumar, C. G.; Prasad, R. B. N. Chemo-enzymatic Synthesis of Sapienic Acid Esters of Functional Phenolics and Evaluation of Their Antioxidant and Cytotoxicity Activities. *Eur. J. Lipid Sci. Technol.* **2013,** *115,* 1123–1129.

Kanjilal, S.; Shanker, K. S.; Rao, K. S.; Reddy, K. K.; Rao, B. V. S. K.; Kumar, K. B. S.; Kantam, M. L.; Prasad, R. B. N. Chemo-enzymatic Synthesis of Lipophilic Ferulates and Their Evaluation for Antioxidant and Antimicrobial Activities. *Eur. J. Lipid Sci. Technol.* **2008,** *110,* 1175–1182.

Katsoura, M. H.; Polydera, A. C.; Tsironis, L. D.; Petraki, M. P.; Rajacic, S. K.; Tselepis, A. D.; Stamatis, H. Efficient Enzymatic Preparation of Hydroxycinnamates in Ionic Liquids Enhances Their Antioxidant Effect on Lipoproteins Oxidative Modification. *New Biotech.* **2009,** *26,* 83–91.

Kaur, J.; Wehtje, E.; Adlercreutz, P.; Chand, S.; Mattiasson, B. Water Transfer Kinetics in a Water Activity Control System Designed for Biocatalysis in Organic Media. *Enzyme Microb. Technol.* **1997,** *21,* 496–501.

Kim, M. J.; Choi, M. Y.; Lee, J. K.; Ahn, Y. Enzymatic Selective Acylation of Glycosides in Ionic Liquids: Significantly Enhanced Reactivity and Regioselectivity. *J. Mol. Catal. B. Enz.* **2003,** *26,* 115–118.

Knez, Z. Enzymatic Reactions in Dense Gases. *J. Supercrit. Fluid.* **2009,** *47,* 357–372.

Kubo, I.; Fujita, K.; Nihei, K.; Nihei A. Antibacterial Activity of Akyl Gallates against *Bacillus subtilis. J. Agric. Food Chem.* **2004,** *52,* 1072–1076.

Kumar, V.; Jahan, F.; Kameswaran, K.; Mahajan, R. V.; Kumar Saxena, R. Eco-friendly Methodology for Efficient Synthesis and Scale-up of 2-Ethylhexyl-*p*-methoxycinnamate Using *Rhizopus oryzae* Lipase and its Biological-evaluation. *J. Ind. Microbiol. Biotechnol.* **2014,** *41,* 907–912.

Laguerre, M.; López Giraldo, L. J.; Lecomte, J.; Figueroa-Espinoza, M.-C.; Baréa, B.; Weiss, J.; Decker, E. A.; Villeneuve, P. Chain Length Affects Antioxidant Properties of Chlorogenate Esters in Emulsion: The Cut-off Theory behind the Polar Paradox. *J. Agric. Food Chem.* **2009,** *57,* 11335–11342.

Laguerre, M.; López Giraldo, L. J.; Lecomte, J.; Figueroa-Espinoza, M.-C.; Baréa, B.; Weiss, J.; Decker, E. A.; Villeneuve, P. Relationship between Hydrophobicity and Antioxidant Ability of Phenolipids in Emulsion: A Parabolic Effect of the Chain Length of Rosmarinate Esters. *J. Agric. Food Chem.* **2010,** *58,* 2869–2876.

Laguerre, M.; Bayrasy, C.; Lecomte, J.; Chabi, B.; Decker, E. A.; Wrutniak-Cabello, C.; Cabello, G.; Villeneuve, P. How to Boost Antioxidants by Lipophilization? *Biochimie* **2013,** *95,* 20–26.

Laguerre, M.; Bayrasy, C.; Panya, A.; Weiss, J.; McClements, D. J.; Lecomte, J.; Decker, E.A.; Villeneuve, P. What Makes Good Antioxidants in Lipid-based Systems? The Next Theories beyond the Polar Paradox. *Crit. Rev. Food Sci. Nutr.* **2015,** *55,* 183–201.

Laszlo, J. A.; Cermak, S. C.; Evans, K. O.; Compton, D. L.; Evangelista, R.; Berhow, M. A. Medium-chain Alkyl Esters of Tyrosol and Hydroxytyrosol Antioxidants by Cuphea Oil Transesterification. *Eur. J. Lipid Sci. Technol.* **2013,** *115,* 363–371.

Lecomte, J.; López Giraldo, L. J.; Laguerre, M.; Baréa, B.; Villeneuve, P. Synthesis, Characterization, and Free Radical Scavenging Properties of Rosmarinic Acid Fatty Esters. *J. Am. Oil Chem. Soc.* **2010,** *87,* 615–620.

Lee, C. C.; Chen, H. C.; Ju, H. Y.; Liu, H. C.; Chen, J. H.; Shie, S. Y.; Chang, C.; Shieh, C. J. Optimization of Ultrasound-accelerated Synthesis of Enzymatic Octyl Hydroxyphenylpropionate by Response Surface Methodology. *Biotechnol. Prog.* **2010,** *26,* 1629–1634.

Lee, G. S.; Widjaja, A.; Ju, Y. H. Enzymatic Synthesis of Cinnamic Acid Derivatives. *Biotech. Lett.* **2006,** *28,* 581–585.

Lee, J. H.; Panya, A.; Laguerre, M.; Bayrasy, C.; Lecomte, J.; Villeneuve, P.; Decker, E. A. Comparison of Antioxidant Capacities of Rosmarinate Alkyl Esters in Riboflavin Photosensitized Oil-in-water Emulsions. *J. Am. Oil Chem Soc.* **2013,** *90,* 225–232.

Lindberg, D.; Revenga, M. D.; Widersten, M. Deep Eutectic Solvents (DESs) Are Viable Cosolvents for Enzyme-catalyzed Epoxide Hydrolysis. *J. Biotechnol.* **2010,** *147,* 169–171.

Lipinski, C. A.; Lombardo, F.; Dominy, B. W.; Feeney, P. J. Experimental and Computational Approaches to Estimate Solubility and Permeability in Drug Discovery and Development Settings. *Adv. Drug Delivery Rev.* **2001,** *46,* 3–26.

Lopez Giraldo, L. J.; Laguerre, M.; Lecomte, J.; Figueroa-Espinoza, M. C.; Barouh, N.; Baréa, B.; Villeneuve, P. Lipase Catalyzed Synthesis of Chlorogenate Fatty Esters in Solvent-free Medium. *Enz. Microb. Tech.* **2007,** *41,* 721–726.

Losada-Barreiro, S.; Bravo-Diaz, C.; Paiva-Martins, F.; Romsted, L. S. Maxima in Antioxidant Distributions and Efficiencies with Increasing Hydrophobicity of Gallic Acid and Its Alkyl Esters. The Pseudophase Model Interpretation of the "Cut-off Effect." *J. Agric. Food Chem.* **2013,** *61,* 6533–6543.

Matsuda, T. Recent Progress in Biocatalysis Using Supercritical Carbon Dioxide. *J. Biosc. Bioeng.* **2013,** *115,* 233–241.

Medina, I.; Lois, S.; Alcántara, D.; Lucas, R.; Morales, J. C. Effect of Lipophilization of Hydroxytyrosol on Its Antioxidant Activity in Fish Oils and Fish Oil-in-water Emulsions. *J. Agric. Food Chem.* **2009,** *57,* 9773–9779.

Munoz-Marin, J.; De La Cruz, J. P.; Reyes, J. J.; Lopez-Villodres, J. A.; Guerrero, A.; Lopez-Leiva, I.; Espartero, J. L.; Labajos, M. T.; Gonzalez-Correa, J. A. Hydroxytyrosyl Alkyl Ether Derivatives Inhibit Platelet Activation after Oral Administration to Rats. *Food Chem. Toxicol.* **2013,** *58,* 295–300.

Panya, A.; Laguerre, M.; Bayrasy, C.; Lecomte, J.; Villeneuve, P.; McClements, D. J.; Decker, E. A. An Investigation of the Versatile Antioxidant Mechanism of Action of Rosmarinate Alkyl Esters in Oil-in-water Emulsions. *J. Agric. Food Chem.* **2012,** *60,* 2692–2700.

Pérignon, M.; Lecomte, J.; Pina, M.; Renault, A.; Simonneau-Deve, C.; Villeneuve, P. Activity of Immobilized *Thermomyces lanuginosus* and *Candida antarctica* B Lipases in Interesterification Reactions: Effect of the Aqueous Microenvironment. *J. Am. Oil Chem. Soc.* **2013,** *90,* 1151–1156.

Plou, F. J.; Cruces, M. A.; Ferrer, M.; Fuentes, G.; Pastor, E.; Bernabé, M. Enzymatic Acylation of Di- and Trisaccharides with Fatty Acids: Choosing the Appropriate Enzyme, Support and Solvent. *J. Biotechnol.* **2002,** *96,* 55–66.

Porter, W. L.; Black, E. D.; Drolet, A. M. Use of Polyamide Oxidative Fluorescence Test on Lipid Emulsions: Contrast in Relative Effectiveness of Antioxidants in Bulk versus Dispersed Systems. *J. Agric. Food Chem.* **1989,** *37,* 615–624.

Priya, K.; Venugopal, T.; Chadha, A. *Pseudomonas cepacia* Lipase Mediated Transesterification Reactions of Hydrocinnamates. *Indian J. Biochem. Biophys.* **2002,** *39,* 259–263.

Reddy, K. K.; Kanjilal, S. Evaluation of the Antioxidant Activity of Capsiconiate Analogues: Effect of Lipophilization. *J. Agric. Food Chem.*, submitted for publication, 2014.

Reddy, K. K.; Ravinder, T.; Kanjilal, S. Synthesis and Evaluation of Antioxidant and Antifungal Activities of Novel Ricinoleate-Based Lipoconjugates of Phenolic Acids. *Food Chem.* **2012,** *134,* 2201–2207.

Sabally, K.; Karboune, S.; Saint-Louis, R.; Kermasha, S. Lipase-Catalyzed Transesterification of Dihydrocaffeic Acid with Flaxseed Oil for the Synthesis of Phenolic Lipids. *J. Biotechnol.* **2006,** *127,* 167–176.

Schmid, U.; Bornscheuer, U. T.; Soumanou, M. M.; McNeill, G. P.; Schmid, R. D. Optimization of the Reaction Conditions in the Lipase-Catalyzed Synthesis of Structured Triglycerides. *J. Am. Oil Chem. Soc.* **1998,** *75,* 1527–1531.

Sorour, N.; Karboune, S.; Saint-Louis, R.; Kermasha, S. Enzymatic Synthesis of Phenolic Lipids in Solvent-free Medium Using Flaxseed Oil and 3,4-Dihydroxyphenyl Acetic Acid. *Process Biochem.* **2012,** *47,* 1813–1819.

Sørensen, A. D. M.; Petersen, L. K.; de Diego, S.; Nielsen, N. S.; Lue, B. M.; Yang, Z.; Xu, X.; Jacobsen, C. The Antioxidative Effect of Lipophilized Rutin and Dihydrocaffeic Acid in Fish Oil Enriched Milk. *Eur. J. Lipid Sci. Technol.* **2012,** *114,* 434–445.

Sørensen, A. D. M., Durand, E.; Laguerre, M.; Bayrasy, C.; Lecomte, J.; Villeneuve, P.; Jacobsen, C. Antioxidant Properties and Efficacies of Synthesized Caffeates, Ferulates, and Coumarates. *J. Agric. Food Chem.* **2014a,** *62,* 12553–12562.

Sørensen, A. D. M.; Lyneborg, K. S.; Villeneuve, P.; Jacobsen, C. Alkyl Chain Length Impacts the Antioxidative Effect of Lipophilized Ferulic Acid in Fish Oil Enriched Milk. *J. Func. Foods* **2014b,** in press.

Stamatis, H.; Sereti, V.; Kolisis, F. N. Studies on the Enzymatic Synthesis of Lipophilic Derivatives of Natural Antioxidants. *J. Am. Oil Chem. Soc.* **1999,** *76,* 1505–1510.

Stöckmann, H.; Schwarz, K.; Huynh-Ba, T. The Influence of Various Emulsifiers on the Partitioning and Antioxidant Activity of Hydroxybenzoic Acids and Their Derivatives in Oil-in-water Emulsions. *J. Am. Oil Chem. Soc.* **2000,** *77,* 535–542.

Sun, S.; Song, F.; Bi, Y.; Yang, G.; Liu, W. Solvent-free Enzymatic Transesterification of Ethyl Ferulate and Monostearin: Optimized by Response Surface Methodology. *J. Biotech.* **2013,** *164,* 340–345.

Sun, S.; Zhu, S.; Bi, Y. Solvent-free Enzymatic Syntheis of Feruloylated Structured Lipids by the Transesterification of Ethyl Ferulate with Castor Oil. *Food Chem.* **2014,** *158,* 292–295.

Sun-Waterhouse, D. The Development of Fruit-Based Functional Foods Targeting the Health and Wellness Market: A Review. *Int. J. Food Sci. Technol.* **2011,** *46,* 899–920.

Suriyarak, S.; Gibis, M.; Schmidt, H.; Villeneuve, P.; Weiss, J. Antimicrobial Mechanism and Activity of Dodecyl Rosmarinate against *Staphylococcus carnosus* LTH1502 as Influenced by Addition of Salt and Change in pH. *J. Food Prot.* **2014,** *77,* 444–452.

Svensson, I.; Wehtje, E.; Adlercreutz, P.; Mattiasson, B. Effects of Water Activity on Reaction Rates and Equilibrium Positions in Enzymatic Esterifications. *Biotechnol. Bioeng.* **1994,** *44,* 549–556.

Ujang, Z.; Vaidya, A.M. Stepped Water Activity Control for Efficient Enzymatic Interesterification. *Appl. Microbiol. Biotechnol.* **1998,** *50,* 318–322.

Valivety, R. H.; Halling, P. J.; Macrae, A. R. Reaction Rate with Suspended Lipase Catalysts Shows Similar Dependence on Water Activity in Different Organic Solvents. *Biochim. Biophys. Acta* **1992,** *1198,* 218–222.

Villeneuve, P. Lipases in Lipophilization Reactions. *Biotech. Adv.* **2007,** *25,* 515–536.

Villeneuve, P.; Muderhwa, J.; Haas, M.J.; Graille, J. Customizing Lipases for Biocatalysis: A Survey of Chemical, Physical and Molecular Biological Approaches. *J. Molec. Cat: B Enz.* **2000,** *9,* 113–148.

Vosmann, K.; Wiege, B.; Weitkamp, P.; Weber, N. Preparation of Lipophilic Alkyl (Hydroxy) Benzoates by Solvent-free Lipase-catalyzed Esterification and Transesterification. *Appl. Microbiol. Biotechnol.* **2008,** *80,* 29–36.

Wang, J.; Shahidi, F. Acidolysis of *p*-Coumaric Acid with Omega-3 Oils and Antioxidant Activity of Phenolipid Products in in Vitro and Biological Model Systems. *J. Agric. Food Chem.* **2014,** *62,* 454-461.

Wang, J.; Li, J.; Zhang, L.; Gu, S.; Wu, F. Lipase-catalyzed Synthesis of Caffeic Acid Phenethyl Ester in Ionic Liquids: Effect of Specific Ions and Reaction Parameters. *Chinese J. Chem. Eng.* **2013,** *21,* 1376–1385.

Weber, H. K.; Stecher, H.; Faber, K. Sensitivity of Microbial Lipases to Acetaldehyde Formed by Acyl-transfer Reactions from Vinyl Esters. *Biotechnol. Lett.* **1995,** *17,* 803–808.

Wehtje, E.; Svensson, I.; Adlercreutz, P.; Mattiasson, B. Continuous Control of Water Activity during Biocatalysis in Organic Media. *Biotechnol. Lett.* **1993,** *7,* 873–878.

Widjaja, A.; Yeh, T. H.; Ju, Y. H. Enzymatic Synthesis of Caffeic Acid Phenethyl Ester. *J. Chin. Inst. Chem. Eng.* **2008,** *39,* 413–418.

Xia, X.; Wang, C.; Yang, B.; Wang, Y. H.; Wang, X. Water Activity Dependence of Lipases in Non-aqueous Biocatalysis. *Appl. Biochem. Biotechnol.* **2009,** *159,* 759–767.

Xin, J. Y.; Zhang, L.; Chen, L. L.; Zheng, Y.; Wu, X. M.; Xi, C. H. Lipase-catalyzed Synthesis of Ferulyl Oleins in Solvent-free Medium. *Food Chem.* **2009,** *112,* 640–645.

Yang, Z.; Guo, Z.; Xu, X. Enzymatic Lipophilisation of Phenolic Acids through Esterification with Fatty Alcohols in Organic Solvents. *Food Chem.* **2012a,** *132,* 1311–1315.

Yang, Z.; Guo, Z.; Xu, X. Ionic Liquid-assisted Solubilization for Improved Enzymatic Esterification of Phenolic Acids. *J. Am. Oil Chem. Soc.* **2012b,** *89,* 1049–1055.

Yu, Y.; Zheng, Y.; Quan, J.; Wu, C. Y.; Wang, Y. J.; Branford-White, C.; Zhu, L. M. Enzymatic Synthesis of Feruloylated Lipids: Comparison of the Efficiency of Vinyl Ferulate and Ethyl Ferulate as Substrates. *J. Am. Oil Chem. Soc.* **2010,** *87,* 1443–1449.

Zhao, H.; Baker, G. A.; Holmes, S. Protease Activation in Glycerol-based Deep Eutectic Solvents. *J. Mol. Catal. B Enzym.* **2011,** *72,* 163–167.

Zhao, H.; Zhang, C.; Crittle, T. D. Choline-based Deep Eutectic Solvents for Enzymatic Preparation of Biodiesel from Soybean Oil. *J. Mol. Catal. B Enzym.* **2013,** *85–86,* 243–247.

Zoumpanioti, M.; Merianou, E.; Karandreas, T.; Stamatis, H.; Xenakis, A. Esterification of Phenolic Acids Catalyzed by Lipases Immobilized in Organogels. *Biotechnol. Lett.* **2010,** *32,* 1457–1462.

8

Sugar Fatty Acid Esters

Yan Zheng, Minying Zheng, Zonghui Ma, Benrong Xin, Ruihua Guo, and Xuebing Xu ▪ *Wilmar (Shanghai) Biotechnology Research and Development Center Co., Ltd., Shanghai, China*

Introduction

Sugar fatty acid esters, usually called sugar esters (SEs), are non-ionic and biodegradable surfactants. Because of their good stabilizing and conditioning properties, they have broad applications in the food, pharmaceutical, detergent, agricultural, fine chemical, and personal care industries (H-Kittikun et al., 2012; Magg, 1984; Nakamura et al., 1997). SEs can be synthesized by an esterification reaction between sugar/sugar alcohols (e.g., sucrose, fructose, glucose, sorbitol, xylitol) and nonpolar fatty acids. The higher substitution esters (HSEs), hexa, hepta, and octa, find use as fat replacers. The lower s ubstitution esters (LSEs), mono-, di-, and tri-esters, find use as oil-in-water (O/W) as well as water-in-oil (W/O) emulsifiers.

In one typical example, sucrose esters are non-ionic compounds synthesized by esterification of fatty acids (or natural glycerides) with sucrose. Sucrose (α-D-glucopyranosyl β-D-fructofuranoside) is a polyhydric alcohol with eight hydroxyl groups (Figure 8.1): three primary hydroxyls (C6, C1', C6') and five secondary hydroxyls. The three primary hydroxyls on the sucrose molecule are the most reactive and are the easiest to substitute with fatty acids, forming mono-, di-, and tri-esters. Hence, compounds ranging from mono- to octa- esters are theoretically possible, and different ester substitutions can determine the properties of the resultant sucrose esters.

LSEs used as emulsifiers, which depends on their hydrophilic-lipophilic balance (HLB), has been widely researched. For HSEs, the most extensively studied and publicized of the fat substitutes is olestra (Olean®, Procter & Gamble). It is prepared

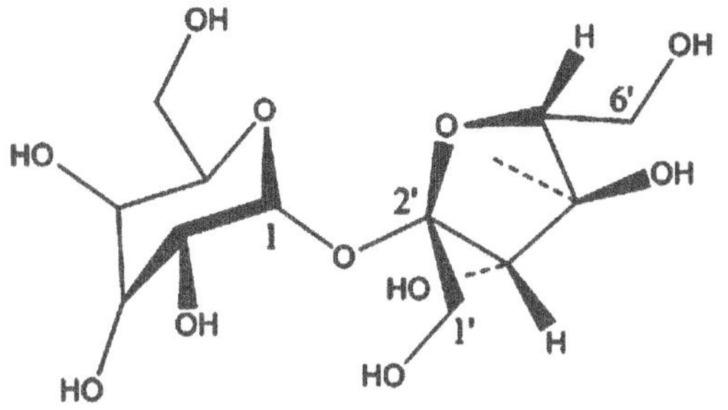

Figure 8.1 Structure of sucrose.

from the reaction of sucrose with long-chain fatty acid methyl esters. Digestive enzymes do not release the fatty acids, so olestra is noncaloric (Mattson and Volpenhein, 1972).

This chapter will focus on various methods of synthesis of SEs, physiochemical properties, and the uses as emulsifiers and fat replacers.

Properties of Sugar Fatty Acid Esters

Physicochemical Properties

It is well known that the usage of fatty acid esters in various areas is based on their special physicochemical properties. SEs are non-ionic surface-active agents consisting of sugar as a hydrophilic group and fatty acids as lipophilic groups. The carbon chain length and nature of the sugar head group, together with the many possibilities for linkage between the hydrophilic sugar head group and the hydrophobic alkyl chain, contribute to the unique physicochemical properties of sugar fatty acid esters (El-Laithy et al., 2011). Depending on the composition, SEs exist as solids, waxy materials, or liquids (Szüts and Szabó-Révész, 2012). Depending on the degree of esterification, SEs decrease the surface tension of water. Therefore, they can exhibit different surface-active properties including HLB values, critical micelle concentration (CMC), emulsifying stability, and foaming ability. In addition, the sugar esters have special thermal properties.

Hydrophilic–Lipophilic Balance

Hydrophilic–lipophilic balance (HLB) is the balance of the size and strength of the hydrophilic and lipophilic moieties of a surfactant molecule. The HLB scale ranges from 0 to 20. In the range of 3.5 to 6.0, surfactants are more suitable for use in W/O emulsions. Surfactants with HLB values in the 8 to 18 range are most commonly used in O/W emulsions (Griffin, 1949). The longer the fatty acid chains in the SEs and the higher the degree of esterification, the lower the HLB value (Figure 8.2).

Micelle Formation and Critical Micelle Concentration

Due to their amphipathic structure, sugar esters in aqueous solutions tend to form thermodynamically stable molecular aggregates called *micelles*. Micelles begin to form at a specific concentration called *critical micelle concentration* (CMC), which is dependent on the surfactant structure and experimental conditions. Below the CMC, the surfactants are solubilized as monomers in the solution. Once the CMC is reached, all additional surfactants that have been added are employed either in the formation of new micelles or for promoting the growth of the aggregates (Könnecker et al., 2011).

The glucose molecule has a highly hydrophilic head group, and when the degree of glucosidation increases, the surface-active properties will decrease. Research (Joshi et al., 2007) shows that, in the micelle, a longer hydrocarbon chain results in a larger

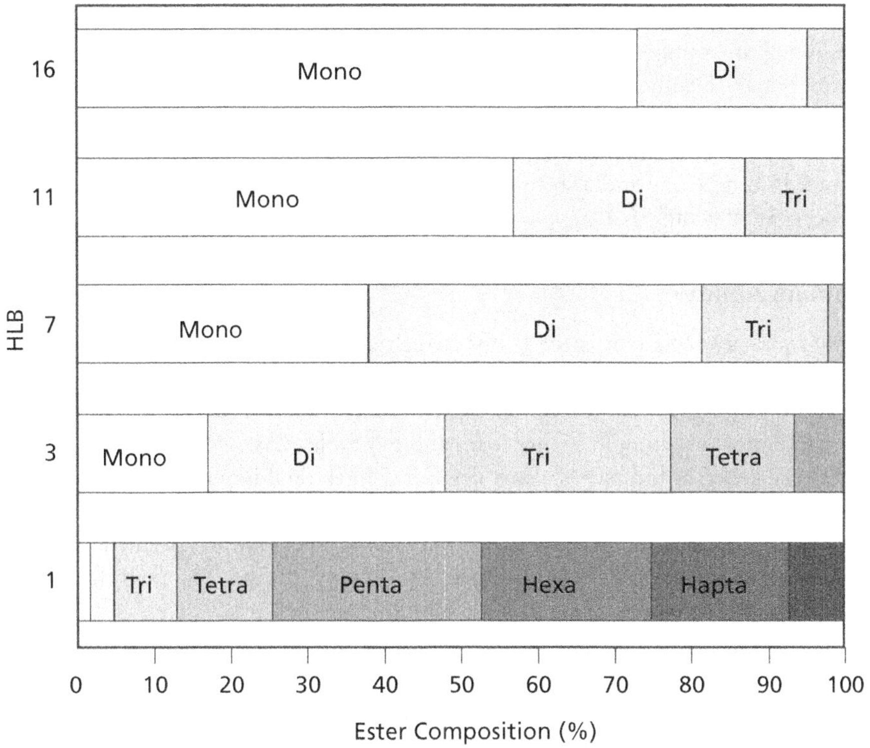

Figure 8.2 Ester compositions and HLB values of SEs. HLB: hydrophilic-lipophilic balance; SE: sugar ester.

hydrocarbon region. Furthermore, the CMC decreases as the alkyl chain length increases, so that at a given total surfactant concentration, a longer chain surfactant generally has a larger concentration of surfactant in the micellar state.

The temperature also affects the micellar formation and surface activity, and the SEs are very sensitive to temperature change. The increase of the temperature leads to larger micelle sizes and lower CMC (Molina-Bolívar and Ruiz, 2008).

Emulsification Properties

The surface-active properties of sugar fatty acid esters are derived from the original hydrophilic group of sugar/sugar alcohol and the original lipophilic group of fatty acids. By varying the degree of substitution or the fatty acid chain lengths, wide ranges of functionality can be obtained. Emulsification is the most important function of sugar esters.

Sugar fatty acid esters with three or fewer fatty acids reduce the surface tension of water. For example, sophorolipid esters can effectively reduce the surface tension of water to values below 38.7 mN/m^{-1} (Zhang et al., 2004). Moreover, Neta et al. (2012) found that fructose esters, synthesized from oleic acid, fructose, and ethanol by CALB (*Candida antarctica* lipase B), are able to reduce the surface tension to 35.8 mN/m^{-1} and also to stabilize an emulsion (emulsification indexes [EI] between 54.4% and 58.4%).

Foaming Ability

Many food products consist of foams, which are thermodynamically unstable systems. Foam stability can be improved by the addition of SEs, which decreases the surface tension. A lower surface tension facilitates the enlargement of the air–water interfacial area, resulting in higher foam ability (Table 8.A).

Coalescence of bubbles is also a key factor in the stability of foams. The coalescence of air bubbles is greatly influenced by the type and concentration of SEs present in the solution. In pure water, air bubbles coalesce almost instantly. However, coalescence is not instantaneous in aqueous solutions of SEs. The main factors that determine the coalescence time in these solutions are the surface excess concentration of the surfactant and the repulsive surface forces (steric forces). So, the SEs stabilize the films mainly by steric force (Samanta and Ghosh, 2011).

On the other hand, some of the SEs, especially sucrose laurate, do not typically form interfaces with high viscoelasticity. Instead, they stabilize the interface by the

Table 8.A Effects of SEs with Different HLB Values on Foam Characteristics

Emulsifier	Mono-Ester %	HLB Value	Surface Tension[a] (mN/m)	Foam Height[b] (ml)	
				0 min	5 min
Sucrose laurate	70	15	28.5	127	1245
Sucrose palmitate	75	16	34.0	29	36
Sucrose stearate	70	15	34.5	31	29
Sucrose stearate	50	11	36.7	12	9
Sucrose stearate	30	6	46.8	4	2
Distilled water	—	—	72.8	—	—

Note: SE: sugar fatty acid ester; HLB: hydrophilic–lipophilic balance. All tested in 0.1% solutions.

[a]Du Nouy method.

[b]Ross and Miles method.

Gibbs-Marangoni mechanism. This mechanism relies on rapid surface diffusion of sucrose laurate that will reduce surface concentration gradients that may develop as a result of deformations of the interface (Kempen et al., 2013). During diffusion, sucrose laurate drags along some of the continuous phase. This slows down drainage of liquid from the films, which will increase foam stability.

Thermal Properties

The thermal properties of SEs can be used to evaluate their applicability in hot-melt technology. The melting points of most sugars are high, but SEs, depending upon their degree of esterification, have melting points of 40–79 °C and are quite stable to heat; hence, they can be employed to prepare solid dispersions by the melt technology. It is very important to know the thermal behavior of these materials so that the changes in the base materials can be predicted during storage and technological processes such as the preparation of solid dispersions by melting (Szüts et al., 2007). Evaluation of the thermal properties of sugar fatty acid esters can also help to observe and analyze the time-dependent solid-state changes.

SEs with high or moderate HLB values have a glass transition temperature (Tg) instead of a melting point. They soften during heating, whereas SEs with low HLB values melt and then quickly recrystallize from their melts. However, the original structure does not return for SEs with high, moderate, or low HLB values; after melting and solidification, their melts continuously change. In addition, the SEs with various HLB values can also be used to influence (increase or decrease) the rate of dissolution of other materials, and hence to change the development of the effect. Due to their low melting points, they are promising carriers for the melting method.

Biological Properties

The most prominent biological property of SEs is the antimicrobial property. Conley and Kabara (1973) conducted several antimicrobial experiments using SEs. They determined the minimal inhibitory concentration of a series of sucrose fatty acid esters against gram-negative and gram-positive organisms. Gram-positive organisms were affected. Sucrose esters had greater inhibitory effect than the free fatty acids except lauric acid. The antimicrobial activity of sucrose esters comes from the interaction of the esters with cell membranes of bacteria, causing autolysis. The lytic action is assumed to be due to stimulation of autolytic enzymes rather than to actual solubilization of cell membranes of bacteria (Wang, 2004). However, although the effectiveness of fatty acids increased when esterified, the spectrum of antimicrobial action of the esters is narrower when compared with the free acids. In addition, very similar organisms do not have the same susceptibility to comparable sucrose esters (Marshall and Bullerman, 1996). Marshall and Bullerman (1986) examined the antimicrobial properties of six sucrose esters substituted to different degrees with a mixture of fatty acids (palmitic and stearic). Antimycotic activity was detected against several mold species from *Aspergillus*, *Penicillium*, *Cladosporium*, and *Alternaria*. The least-substituted sucrose ester was the most active in reducing mold growth. Investigations in vitro and using refrigerated beef showed that sucrose fatty acid esters were also active against *Escherichia coli* O157, *L. monocytogenes*, *Staph. aureus*, and psychrotrophic spoilage microorganisms (Hathcox and Beuchat, 1996; Kabara, 1993; Monk and Beuchat, 1995; Monk et al., 1996).

Fructose esters could be used as antimicrobial agents that suppressed the cell growth of *Streptococcus mutans*, a causative organism of dental caries. Among the different carbohydrate esters, fructose laurate showed the highest growth inhibitory effect (Watanabe et al., 2000).

In food hygiene, biofilms were also a very important potential hazard and a source of microbial contamination of foods. Furukawa et al. (2010) investigated the effects of food additives on biofilm formation by food-borne pathogenic bacteria. They found that sucrose monomyristate and sucrose monopalmitate significantly inhibited biofilm formation by *Staphylococcus aureus* and *Escherichia coli* at a low concentration (0.001% w/w). The addition of sucrose monopalmitate at the early growth stage of *S. aureus* exhibited a strong inhibitory effect, suggesting that the ester inhibited the initial attachment of the bacterial cells to the abiotic surface.

Production of Sugar Fatty Acid Esters

Chemical Synthesis

There have been a significant number of publications on sugar fatty acid esters from the mid-1950s, in particular the LSEs that have been commercially manufactured and used as emulsifiers since the early 1960s.

The initial synthesis of LSEs by transesterification used dimethyl formamide (DMF) as the mutual solvent for solubilizing sugar and free fatty acid. This process was first commercialized by Dai-Nippon Sugar Manufacturing Co., Ltd., in Japan in the late 1960s to produce sucrose fatty acid esters as food additives (Ryoto, 1987; Yamada et al., 1980). Sucrose fatty acid esters manufactured by this process were not approved for use in the United States because of the odor and toxic materials present in the product. Osipow et al. (1956) described the first commercial process for the preparation of sucrose fatty acid ester from methyl ester in DMF. A relatively safer process, known as the Nebraska-Snell process, was developed that involved reacting a microemulsion of sucrose in propylene glycol as solvent with the fatty acid methyl ester in the presence of potassium carbonate as the catalyst (Osipow and Rosenblatt, 1967).

A lot of workers have modified the process to avoid using a toxic solvent. A solvent-free interesterificaiton process was developed by Feuge et al. (1970) at the Southern Regional Research Center (SRRC), and this was licensed to Ryoto Co. in Japan (now Mitsubishi-Kasei Food Corporation). This process involved the reaction between molten sucrose (mp 185 °C) and fatty acid methyl esters in the presence of lithium, potassium, and sodium soaps as solubilizers and catalysts at temperatures between 170 °C and 187 °C. A combination of lithium oleate with sodium or potassium oleate at 25% of total soap, based on the weight of sugar, gave the best sucrose fatty acid ester product. The drawback of this process is that molten sucrose is rapidly degraded to a black tarry mass at 170–187 °C. In both solvent and solvent-free processes, distillation is often required to remove the unreacted methyl esters, fatty acids, and alcoholic byproducts (Akoh, 1994).

Other studies on the syntheses of sucrose fatty acid esters have also been reported. Lemieux and McInnes (1962) reported a preparative method for the synthesis of sucrose mono-esters in anhydrous DMF, of which 80% of the total product was monoester and 20% di-ester based on the amount of methyl ester that reacted. Jones (1981) described a process for preparing mono- and di-esters of sucrose from sucrose and alkenyl ester of fatty acid in a polar aprotic solvent.

Another hot topic is that of chemical synthesis of HSEs. They are mixtures of sucrose esters formed by chemical interesterification of sucrose with six to eight fatty acids. The HSE commonly known as olestra (Olean®) is manufactured from saturated and unsaturated fatty acids of chain length C12 and higher, obtained from conventional edible fats and vegetable oils (Shieh et al., 1996). The first step of the process is methylation of fatty acids. The second step is transesterification of sucrose and fatty acid methyl ester using catalysts, such as alkali metals or their soaps, under anhydrous conditions and high vacuum. The resulting crude olestra product is purified by washing, bleaching, and deodorizing to remove free fatty acids and odors, followed by distillation to remove unreacted fatty acid methyl esters and sucrose esters with low degrees of fatty acid substitution.

Rizzi and Taylor (1978) described a solvent-free two-stage reaction sequence for synthesizing HSEs. In the first stage, sucrose and fatty acid methyl ester (FAME) with a mole ratio of 1:3 were reacted in the presence of potassium soaps to form a homogenous melt in 2–3.5 hours. The product of this stage contained mainly esters of sucrose with a low degree of substitution (DS = 1 – 3). In the second stage, excess FAME and more NaH were added and reacted for an additional 6 hours to produce HSEs in yields up to 90%. This process was modified by Hamm (1984) by adding methyl oleate at the beginning of the reaction and sucrose and NaH in increments, and HSEs with a DS of 4–8 were obtained in 42% theoretical yield.

Akoh and Swanson (1990) reported an optimized synthesis of HSEs that gave yields between 99.6% and 99.8% of the purified HSEs based on the initial weight of sucrose octa-acetate (SOAC). This was a one-stage, solvent-free process involving admixing of sucrose octa-acetate, 1–2% sodium metal as catalyst, and FAME of vegetable oils in a three-neck reaction flask prior to application of heat. Formation of a one-phase melt was achieved 20–30 minutes after heat was applied. High yields of HSEs were obtained at temperatures as low as 105 °C and synthesis times as short as 2 hours by pulling a vacuum of 0–5 mm Hg pressure.

Enzymatic Synthesis

In recent years, utilization of enzymes as biocatalysts for preparing SEs has attracted great attention because of growing consumer demand for green processes and products (Table 8.B on p. 224). The enzymatic process will bring the advantage of the high specificity and regioselectivity of the reaction at a lower temperature and an easier downstream process, which will generate various products with controlled structure and functionality. On the other hand, due to the steric hindrance, only LSEs can be achieved through the enzymatic method. In the following section, solvents', enzymes', and substrates' properties and how they affect each other on the productivity of enzymatic synthesis of LSEs are briefly discussed.

Solvents are essential for LSE production due to the different solubilities of sugar and fatty acids. Looking for a suitable organic medium for enzymatic synthesis of LSEs requires full consideration of solvent toxicity to the biocatalyst and the solubility of substrates. Proteases such as substilisin can maintain catalytic activity in strongly polar solvents like dimethylformamide (DMF) (Pedersen et al., 2003). Ferrer et al. (2002) synthesized 6-O-lauroyl sucrose from sucrose and vinyl laurate using a lipase in mixed solvents of tert-amyl alcohol and Dimethylsulfoxide (DMSO) to improve substrates solubility, and the yield to 6-O-lauroyl sucrose was up to 70%. Degn and Zimmermann (2001) reported the synthesis of myristate ester of different carbohydrates using Novozyme 435; the production rate of myristyl glucose was improved from 222 to 1212 mmol g^{-1} h^{-1} by changing the solvent from pure tert-butanol to a mixture of tert-butanol and pyridine. Besides toxicity and solubility, parameters such as the solvent dielectric constant (Affleck et al., 1992), polarity and partition coefficient (Lu et al., 2008), as well as electron acceptance index (Valivety et al., 1994) have been used to study the effect different solvents on the rate of lipase-mediated synthesis of SEs.

Besides organic solvents, a new kind of solvents called *ionic liquids* (ILs) are being increasingly used in lipase-mediated synthesis of SEs for their nonvolatile character and thermal stability. They exhibit excellent physicochemical characteristics, including the ability to dissolve polar, nonpolar, organic, inorganic, and polymeric compounds. It was also observed that ILs enhanced the reactivity, selectivity, and stability of enzymes. ILs containing dicyanamide anion were found to be able to dissolve considerable amount of glucose and sucrose (Forsyth and Macfarlane, 2003). Park and Kazlauskas (2001) synthesized 10 ILs and investigated the acylation of sugar with vinyl acetate catalyzed by *Candida antarctica* lipase B (CALB). Their result showed that enzymes' catalytic efficiency and selectivity could be changed by using different ILs. Lee et al. (2008) observed that enzyme activities in ILs were highly enhanced by using a supersaturated solution under ultrasound irradiation in lipase-catalyzed esterifications of sugar with vinyl laurate or lauric acid, but the stability of the enzyme in ILs was not influenced by ultrasound irradiation. Although these studies about ILs mainly deal with the synthesis of monosaccharide esters, these samples provide useful insights considered to be appropriate for an application of enzymatic synthesis of HSEs in the future.

Lipase (triacylglycerol acylhydrolase, EC 3.1.1.3) has been widely used in enzymatic synthesis of LSEs for its properties in nonaqueous media. To avoid the side effect of water and enhance the activity and regioselectivity of a given lipase, immobilization and modification technologies are widely studied. Cao et al. (1999) reported the conversion in the synthesis of 6-O-glucose palmitate increased with decreasing hydrophilicity of the support.

Tsuzuki et al. (1999) found that use of a detergent-modified lipase powder afforded greater yields of SEs compared to the use of untreated powder of the same lipase. Ferrer et al. (2002) evaluated the *Thermomyces lanuginosus* lipase (TLL) immobilized by three methods: adsorption on polypropylene (Accurel EP100), covalent attachment to Eupergit C, and silica-granulation. The lipase adsorbed on Accurel showed an extraordinary initial activity and the highest selectivity to 6-O-lauroyl sucrose was found with granulated lipase. Other methods like preparing surfactant-lipase complex (Maruyama et al., 2002), using different ionic liquids as coating materials (Mutschler et al., 2009), as well as rational design of lipases (Magnusson et al., 2005) to improve lipase's property have been fully discussed.

Fatty acids of different chain length and lipase from different sources used to synthesize SEs have a great influence on the efficiency of a catalyzing process. Some lipases exhibit a high selectivity for long- and medium-chain fatty acids, whereas others are selective toward short and branched fatty acids. Research showed that the conversion increased with increasing chain length of the fatty acid when the reaction was catalyzed by Hog pancreas lipase (HPL). When the synthesis reaction was catalyzed by TLL, the conversion decreased with increasing chain length of the acid. The conversion is nearly independent of the chain length of the acid for the reaction catalyzed by Novozym 435 (Kumar et al., 2005). In the case of synthesis of sucrose 6-mono-esters, TLL is more effective than lipase from CALB, which is more interesting for the synthesis of the corresponding 6,6'-di-esters (Ferrer et al., 2005). The type of acyl donor also can affect the reaction. Among the methyl, ethyl, and vinyl ester substrates, vinyl ester is extensively used because its byproduct, vinyl alcohol, can be converted to acetaldehyde irreversibly, thereby increasing the yield of the desired ester (Yadav and Trivedi, 2003).

Table 8.B Enzymatic Synthesis of SEs by Lipases in Various Organic Solvents

Substrates							
Acyl acceptor	**Acyl donor**	**Enzymes**	**Solvents**	**Time (h)**	**Products**	**Yield**	**References**
D-fructose	Stearic acid	Lipozyme SP 382	tert-Butanol	46	2 or 6-O-stearate fructose	8.6%	Chang and Shaw (2009)
D-Fructose	PFAD	Novozym 435	Acetone	72	6-O-palmitoyl-α-D-fructopyranose	38.8%	Chaiyaso et al. (2006)
D-Galactose					6-O-palmitoyl-α-D-glucopyranose	6.8%	
D-Glucose					6-O-palmitoyl-α-D-glucopyranose	76.0%	
a-Methyl-glucose					6-O-palmitoyl-α-D-glucopyranose	78.0%	
Glucose	Lauric acid	TLL	tert-amyl alcohol:DMSO	20	6-O-lauroylglucose	98.0%	Ferrer et al. (2005)
D-Glucose	L-Alanine	Lipozyme IM20 RML	CH2Cl2:DMF (90:10, v/v; 40°C)	72	Mixture of mono-esters	2.0%	Somashekar and Divakar (2007)
D-Fructose					Mixture of di-esters	17.0%	
					Mixture of mono-esters	6.0%	
Lactose					Mixture of di-esters	6.0%	
					Mixture of mono-esters	14.0%	
Sucrose					Mixture of mono-esters	8.0%	

Fructose	Oleic acid	Novozym 435	Ethanol	37.8	Mixture of esters	88.4%	Neta et al. (2011)
Sucrose	Dodecanoic	CALB	ILs		Mixture of mono-esters	25.0%	Liu et al. (2005)
D-Glucose	Myristic acid	Novozym 435	tert-butanol: pyridine (55:45, v/v)	24	Myristyl glucose	34 mg/mL	Degn and Zimmermann (2001)
D-Fructose					Myristyl fructose	22.3 μmol/min.g	
Maltose					Myristyl maltose	1.9 μmol/min.g	
Maltotriose					Myristyl maltotriose	ND	
Cellobiose					Myristyl cellobiose	ND	
Sucrose					Myristyl sucrose	ND	
Lactose					Myristyl lactose	ND	
α- and β-D-Glucopyranoside					Myristyla-D-glucopyranoside	26.9 μmol/min.g	
Sucrose	Vinyl laurate	Metalloprotease Thermolysin (*Bacillus thermoproteolyticus*)	Dimethylsulfoxide (DMSO)	12	2-O-Lauroyl-sucrose	53.0 nmol/min.g	Pedersen et al. (2002)

Note: DMSO: dimethylsulfoxide; PFAD: palm fatty acid distillates; ND: not detected; CALB: Candida antarctica lipase B; TLL: Thermomyces lanuginosus lipase; RML: Rhizomucor miehei lipase; DMF: dimethylformamide; SE: sugar ester.

The solubility of the substrates is affected by the concentration of the other substrate because dissolution of a substrate affects the polarity of the reaction medium; thus, the molar ratio of the two substrates has a great impact on the esterification process. For example, for glucose stearic acid ester synthesis using CALB as the catalyst, an excess of the fatty acid in the reaction medium led to a significant increase of sugar ester synthesis. On the other hand, the sugar ester yield was reduced by decreasing the chain length of the fatty acid in the reaction involving glucose (Yan et al., 2001). The esterification of fructose in tert-amyl alcohol was favored by an excess of short-chain fatty acids; however, an excess of long-chain fatty acids decreased the conversion rate (Soultani et al., 2001). In general, the conversion rate of lipase-catalyzed transesterification involving fatty acids with less than 10 carbons can be improved by supplying an excess of fatty acid. On the other hand, a lower molar ratio of fatty acid to sugar is preferable if the reaction involves longer fatty acids, because a high concentration of a nonpolar fatty acid decreases the solubility of sugar in the reaction medium (Gumel et al., 2011).

Applications of Sugar Fatty Acid Esters

As we know, there are many kinds of sugar fatty acid esters such as sucrose esters, maltose esters, fructose esters, raffinose esters, and so forth. The most widely used sugar esters, sucrose esters, are synthesized by esterification of fatty acids (or natural glycerides) with sucrose. In this part, we mainly focus on the application of various sucrose esters that have been approved for use in the food, pharmaceutical, and cosmetic industries.

Applications in the Food Industry

Without any safety or laws and regulations issues, only sucrose esters have broad applications in the food industry (Nakamura, 1997; Watanabe et al., 2000). During food processing, sucrose esters could be utilized as food emulsifiers, wetting and dispersing agents, and antibiotics. Moreover, they could retard staling, solubilize flavor oils, improve organoleptic properties in bakery and ice cream formulations, and could be used as fat stabilizers during the cooking of fats (Banat et al., 2010).

Emulsifiers

Sucrose esters have been manufactured commercially as food emulsifiers since the early 1960s. Their wide range of HLB, depending on degree of esterification of fatty acids and sucrose, provides ultimate application of sucrose esters to each product type.

Megahed (1999) prepared sucrose fatty acid esters as food emulsifiers and evaluated their surface active and emulsification properties. They compared the surface tension, interfacial tension, and HLB of sucrose esters with different fatty acids. Results indicated the presence of unsaturated fatty acid moieties (oleic), and the hydrophilic carbohydrate backbone in the carbohydrate polyesters contributed to a much lower surface tension. Based on their emulsion stability experiments, they found that sucrose laurate and palmitate have high O/W emulsion stability.

The synthesis of long-chain fatty acid sucrose mono-esters was one of the first major achievements of the Sugar Research Foundation (Polat and Linhardt, 2001). Ariyaprakai et al. (2013) investigated the interfacial and emulsifier properties of sucrose monostearate in comparison with Tween 60 in coconut milk emulsions. They found that sucrose monostearate was slightly better in lowering the interfacial tension between the coconut oil and water interface. The complex between coconut protein and sucrose monostearate seemed to be an emulsifying membrane that could better protect coconut milk emulsions from heat and freeze damage.

Surfactant in Colloidal Delivery Systems

Another important application of sucrose esters surfactant is the manufacture of food-grade colloidal delivery systems, namely microemulsions and nanoemulsions. Microemulsions and nanoemulsions are composed of oil, surfactant, and water and can be easily fabricated from food ingredients via relatively simple operation procedures, such as agitation, mixing, shearing, and homogenization. However, the widespread application of microemulsions and nanoemulsions in many food products is limited by several technical and practical problems, including the limited number of food-grade surfactants available for preparing and stabilizing these systems (Kralova and Sjoblom, 2009). Consequently, sucrose ester surfactants seem to be a good alternative for the manufacture of such colloidal delivery systems. Edris and Malone (2011) used sucrose laurate (containing 78–81% of sucrose monolaurate) to deliver flavoring to food or beverages without using organic solvents. The sucrose laurate surfactant exerted a good emulsification property in this food nano-delivery system. Sucrose monopalmitate was also investigated in its capability of forming colloidal dispersions, in which lemon oil was the oil phase (Rao and McClements, 2011). This study provided important information for optimizing the application of sucrose mono-esters to form colloidal dispersions in food and beverage products. Fanun (2009) reported the properties of microemulsions manufactured from sucrose monolaurate, sucrose dilaurate, and peppermint oil (edible oil suitable for food, pharmaceutical, and cosmetics applications).

Fat Substitutes

Fat substitutes are compounds that physically and chemically resemble triglycerides and are stable to cooking and frying temperatures. Sugar or sugar alcohol fatty acid esters such as sucrose polyester (olestra), sorbitol polyester, and raffinose polyester are among the most studied fat substitutes. Olestra is industrialized by Procter & Gamble. Because olestra is not hydrolyzed by fat-splitting enzymes in the small intestine, it is not absorbed from the small intestine into blood and tissues, and therefore provides no energy that can be utilized by the body (Jandacek, 2012).

The physical and hedonic properties of olestra are very similar to those of triacylglycerol fats with similar fatty acid composition. The combination of nonabsorbability as well as fat-like perception in the mouth contributes to its application as a noncaloric substitute for dietary fat. Olestra has been used in the commercial production of savory snacks such as potato and corn chips and crackers for nearly two decades (Slough et al., 2001). It could replace up to 100% of the conventional fats used in the preparation of those savory snacks. Although olestra is not toxic, carcinogenic, genotoxic, or teratogenic and is neither absorbed nor metabolized by the body, it is associated with gastrointestinal tract symptoms such as cramping or loose stools. In addition, olestra will affect the absorption of fat-soluble vitamins, thus marketed olestra products should include added levels of vitamins A, D, E, and K to counteract potential vitamin deficiency (Prince and Welschenbach, 1998).

Pharmaceutical Applications

As we previously mentioned, sucrose esters are widely used as additives in the food industry, and they have also been noted as good emulsifying and stabilizing agents in the pharmaceutical field. What's more, they can modify the drug bioavailability and release process in the pharmaceutical industry.

Emulsification and Stabilization

Table 8.C summarizes the application of sucrose esters as emulsifiers and stabilizers in conventional and advanced drug delivery systems. For example, Yokoi et al. (2005) demonstrated that sucrose monopalmitate and sucrose monostearate inhibited the crystal growth and nucleation of amorphous cefditoren pivoxil, which improved the physicochemical stability of the suspension.

Various nanosystems in pharmaceutical technology have recently gained increasing attention. Sucrose esters act as important emulsifiers and stabilizing agents in nanosystems such as nanoemulsions, nanosuspension, and nanoparticles. Klang et al. (2011) constructed different nanoemulsion delivery systems using sucrose esters instead of lecithin. They prepared progesterone nanoemulsion by sucrose stearate and different cyclodextrins as stabilizing agents. In their study, a thorough comparison between the novel sucrose stearate-based nanoemulsions and the corresponding lecithin-based nanoemulsions revealed that the sucrose stearate was superior in terms of emulsifying efficiency, droplet formation, and physical and chemical stability.

Bioavailability Modification

Currently, investigations regarding the pharmaceutical application of sucrose esters mainly focus on their modification of bioavailability. As a consequence of their wide HLB spectrum, they can influence drug dissolution, drug absorption/penetration, and drug release in different ways.

Table 8.C Applications of Sucrose Esters as Emulsifiers and Stabilizers in Various Formulation Systems

Formulation	Sucrose Esters	Function	References
Emulsions, suspensions	Sucrose stearate (HLB 16); sucrose palmitate (HLB 15)	Emulsifiers, stabilizers	Akoh (1992); Yokoi et al. (2005)
Microemulsions	Sucrose laurates (HLB 5, 16); sucrose myristate (HLB: 16); ucrose stearates (HLB: 7-15); sucrose oleate (HLB: 15), sucrose palmitate (HLB: 15)	Emulsifiers	Thevenin et al. (1996); Fanun (2008, 2009)
Vesicles	Sucrose stearate (HLB: 5,7,15)	Emulsifiers, stabilizers	Valdés et al. (2014)
Microspheres, microparticles,	Sucrose palmitate (HLB: 15)	Emulsifiers	Diab et al. (2012)
Nanoparticles, lipid nanoparticles	Sucrose stearates (HLB: 15)	Emulsifiers, stabilizers	Huang and Moriyoshi (2008)
Nanoemulsions, nanosuspensions, nanodispersions	Sucrose palmitate (HLB: 15), sucrose erucate (HLB: 2), sucrose stearate (HLB: 9)	Emulsifiers, stabilizers	Tagekami et al. (2008); Tahara et al. (2008); Klang et al. (2011)

Note: HLB: hydrophilic–lipophilic balance.

Solubilization is the prerequisite condition for improving the bioavailability of drugs, especially for those poorly water-soluble medicines. Many research articles have investigated sucrose esters as drug solubilization–modifying agents. Ntawukulilyayo et al. (1993) evaluated the dissolution rate–enhancing properties of sucrose esters. Hülsmann et al. (2000) utilized hydrophilic sucrose monopalmitate (HLB 15) as an additive in melt extrusion to enhance the dissolution of 17-β-oestradiol hemihydrates, a poorly water-soluble drug.

Drug release is a crucial and rate-limiting step for oral bioavailability, particularly for drugs with low gastrointestinal solubility and high permeability (Vasconcelos et al., 2007). Many researchers have concluded that an improvement in the release profiles of these drugs by suitable formulation can enhance their bioavailability and reduce side effects (Leuner and Dressman, 2000; Tanaka et al., 2006). There has recently been great interest in the use of sucrose esters as controlled-release agents in various drug delivery systems. Table 8.D lists different drugs and dosage forms where sucrose esters used as an important composition.

Besides the modification of drug dissolution, other properties of sucrose esters result in interactions with biological barriers, and the effects on oral, nasal, and ocular absorption are therefore widely investigated. In peroral formulations of cyclosporin A, sucrose laurate enhanced the absorption of the drug in an ex vivo experiment involving the use of normal gut epithelial tissue and Peyer's patch tissue of guinea pigs (Lerk and Sucker, 1993). Regarding the high degree of absorption, sucrose laurate may be considered as a new promising excipient for peroral formulations. Ganem-Quintanar et al. (1998) studied the drug absorption–enhancing effect of a series of sucrose fatty acid esters (sucrose stearate, HLB: 16; sucrose oleate, HLB: 15; sucrose palmitate, HLB: 15; sucrose laurate, HLB: 16) in two different regions of the oral mucosa: pig palate (keratinized) and cheek (nonkeratinized).

In the evaluation of the potential of the use of sucrose esters as oral absorption enhancers, Kis et al. (2010) tested the toxicity and paracellular permeability of three water soluble sugar esters, sucrose laurate, sucrose myristate, and sucrose palmitate, with the same HLB value (16) on human Caco-2 epithelial monolayers. Similarly, Szüts et al. (2011) found that sucrose laurate (HLB: 16) significantly reduced the transepithelial electrical resistance and increased the paracellular transport of the marker molecule fluorescein in Caco-2 cell layers at a concentration of 200 μg/mL. This indicated the possible use of sucrose laurate as an absorption enhancer in oral formulations.

Cosmetic Applications

Sucrose esters are also an important surfactant in the cosmetic industry. They elicit little or no irritation, suggesting applications in cosmetics including skin preparations, hair treatments, eyelash products, cosmetic oil gels, and deodorants.

Sucrose cocoate, a mixture of sucrose esters of coconut fatty acids in aqueous ethanol solution, is a common emulsifier widely employed in emollient, skin-moisturizing cosmetic formulations (Ahsan et al., 2003). Kilpatrick-Liverman and Miller (1997) disclosed a clear, soap-gelled cosmetic stick composition including the sucrose cocoate as a clarifying agent. The compositions had superior clarity as originally formed and over extended periods of time after being manufactured. Carroll and Cataneo (2002) disclosed a skin-refining composition using the sucrose cocoate as one of the five surfactants. The composition was used to clean, exfoliate, and condition the skin. Park et al. (2006) produced a skin external preparation having a hexagonal gel structure, in which sucrose cocoate was an emulsifying agent. This preparation played a very important role in skin protection and moisturizing functions. Lim et al. (2011) provided a method for manufacturing a cosmetic emulsion

Table 8.D Applicability of Sucrose Esters in Various Drug Delivery Formulations

Drug	Formulation	Sucrose Esters	Reference
Ibuprofen	Tablet	Sucrose stearates (HLB: 1, 7, 15), sucrose palmitate (HLB: 15)	Ntawukulilyayo et al. (1995)
Glybuzole	Ground product, physical mixture, melted product	Sucrose stearates (HLB: 16)	Otsuka et al. (1998)
17-β-Oestradiol hemihydrate	Tablet from melt extrudate	Sucrose palmitate (HLB: 15)	Hülsmann et al. (2000)
Cromolyn sodium	Proniosome-derived noisome	Sucrose stearates (HLB: 11, 16)	Abd-Elbary et al. (2008)
Furosemide	Directly-compressed fast-disintegrating tablets	Sucrose stearates (HLB: 11, 15, 16)	Koseki et al. (2009)
Paracetamol	Physical mixture	Sucrose palmitate (HLB: 16), sucrose stearates (HLB: 9)	Szüts et al. (2010a, 2010b)
Metoprolol tartrate	Directly compacted matrix tablets	Sucrose stearates (HLB: 0, 1, 3, 5, 9, 11, 15, 16)	Chansanroj and Betz (2010)
Vitamins	Granules prepared by wet granulation, melt granulation, or compression and grinding	Sucrose stearates (HLB: 3, 11)	Seidenberger et al. (2011)

Note: HLB: hydrophilic–lipophilic balance.

containing a high content of silicone derivative to enhance formulation stability and skin stability and to ensure excellent skin texture. This cosmetic emulsion composition contained sucrose cocoate as the emulsifier.

Sucrose esters with high HLB values provided a low viscosity emulsion suitable for formulation as milks or thin lotions, or they can be thickened. Naomi (2000) disclosed a transparent or translucent cosmetic, excellent in power solubilizing poorly water-soluble components. Sucrose ester of a fatty acid was one of the surfactants that suppressed the formation of sediments and clouding with time. Jacobus and Jozef (2007) provided an aerated cosmetic mousse containing at least one sucrose ester as the surfactant. The mousse was stable at room temperature for at least 30 days, and heat stable at 40 °C for more than two weeks. Marlene et al. (2012) disclosed a hair care agent containing at least one sucrose ester. They used this agent to condition keratin fibers, in particular hair. In the same year, Mirko (2012) disclosed an easily removable nail varnish composition made by at least 5% sucrose ester.

Regulatory Status and Toxicological Status of Sugar Fatty Acid Esters

Regulatory Status

SEs are widely produced and used as food additives and fat substitutes. Among these SEs, sucrose esters were first approved as food additives and are widely used in the food industry. In 1992, the Scientific Committee for Food established an acceptable daily intake (ADI) of 0–20 mg/kg body weight per day for sucrose esters of fatty acids. In 2004, the European Food Safety Authority (EFSA) reexamined the safety of these food additives and established a group ADI of 40 mg/kg body weight per day for sucrose esters of fatty acids.

Olestra has physical and organoleptic properties similar to those of conventional dietary fats. It was synthesized by Procter & Gamble in 1968. The Food and Drug Administration (FDA) originally approved olestra for use as a replacement for fats and oils in prepackaged ready-to-eat snacks in 1996. The environmental assessment for olestra (Procter & Gamble Company, 1995) was reviewed by both the FDA and the U.S. Environmental Protection Agency (EPA), both of which concluded that manufacture of olestra, and production and consumption of savory snacks containing olestra, will not result in adverse environmental effects (U.S. Food and Drug Administration, 1996).

Toxicity

The increased use of sucrose fatty acid esters in the food field has led to a number of detailed investigations into toxicology by the World Health Organization (WHO). In particular, SEs were reviewed in terms of their pharmacokinetics, pharmacodynamics, acute toxicity, and short- and long-term tolerability (Holmberg, 2003).

Toxicity in Animal Models

The acute toxicity for a number of different compounds in this general class has been studied in a range of animal models. The effects of oral ingestion of different sucrose esters over short periods of time have also been studied in rat and dog models, which indicated that short-term ingestion has no consequence in these animal models

(Tokita, 1958). Studies of the chronic toxicity of sucrose esters (Table 8.E) showed no adverse effects on the animals, and no evidence of increased level of tumors or other abnormalities. Recently, a chronic combined toxicity and carcinogenicity study of S-170, a sucrose fatty acid ester, was performed in male and female F344 rats (Yoshida et al., 2004). Major components of S-170 were tetra-, penta-, hexa-, and hepta-sucrose fatty acid esters (75.49% of total esters). As for other components, S-170 contained mono-, di-, and tri-esters at 13.79%, and octo-esters at 10.73%. They were given ad libitum in the diet at levels of 0%, 1.25%, 2.5%, or 5% to 10 rats/sex/group for 12 months to determine chronic toxicity and at levels of 0%, 2.5%, or 5% to 50 rats/sex/group for 2 years in the carcinogenicity study. Treatment with S-170 exerted no effect on survival in either sex. In the 12-month chronic toxicity study, no treatment-related effects on body weights or hematological, blood biochemical, urinary, and pathological parameters were demonstrated in any of the treated groups. In the carcinogenicity study, S-170 did not cause any dose-related significant increase in the incidences of tumors in any organs or tissues. Taken together, the results clearly demonstrate that S-170 has neither toxic nor carcinogenic activity in F344 rats under the conditions of the study. As a consequence of these results, the human ADI for sucrose esters as a general class of compound was set by the WHO at 0–16 mg/kg, which was based on a no observable effect level in animals of 50 g/kg (JECFA, 1996).

Table 8.E Studies of the Chronic Toxicity of Sucrose Esters Administered Orally in Diet for Animal Models

Sucrose Ester	Test Animal	Effects of Exposure (Percentage by Weight in Diet of Maximum Dose Given)
Sucrose monopalmitate	Rat	<2% no adverse effects for 60 days[a]
Sucrose monostearate	Rat	<2% no adverse effects for 60 days[a]
Mixed palmitic and stearic acid	Rat	<2% no adverse effects for 60 days[a]
Esters of sucrose	Rat	<10% no adverse effects for 91 days[b]
The blends of olestra and vegetable oil	Dog	<10% no adverse effects for 20 months[c]
Olestra	Mice	<10% no toxic and carcinogenic effects for 2 years[d]

[a]Hara (1959).

[b]Miller and Long (1990).

[c]Miller et al. (1991).

[d]Lafranconi et al. (1994).

A variety of approaches have been used to investigate and quantify the absorption and metabolism of olestra. In vitro experiments demonstrated that sucrose with 6–8 fatty acid is not hydrolyzed to free fatty acids by pancreatic lipases in the presence of bile acids, regardless of fatty acid chain length or unsaturation (Mattson and Nolen, 1972). By administering olestra intravenously to rats and monkeys, it was demonstrated that if olestra were absorbed it would be very rapidly taken up by the liver and slowly excreted via the bile into the intestinal lumen (Fallat et al., 1976; Mattson and Nolen, 1972). Results of toxicity testing demonstrate that olestra is not toxic, carcinogenic, or mutagenic and is not a reproductive toxin (Bergholz, 1992). These studies include two lifetime toxicity studies in both rats and mice and a chronic toxicity study in dogs with olestra fed at levels up to 10% (w/w) of the diet (Lafranconi et al., 1994; Miller et al., 1991; Wood et al., 1991). Because of its large molecular size, olestra is not absorbed across the gastrointestinal tract following consumption (Miller et al., 1991, 1995). In the 20-month dietary study, the only observed digestive change in the dogs was isolated incidences of soft stools. In a 6-month dietary study, no effects were observed on growth of pigs that ate olestra unsupplemented with vitamins at levels up to 5% (w/w) of the diet (Cooper et al., 1994). Olestra is neither genotoxic (Skare and Skare, 1990) nor teratogenic (Nolen et al., 1987).

Toxicity in Humans

Some studies on the toxicology of sucrose esters in humans are reported. A pharmacokinetic study of human volunteers fed 1 g of sucrose tallowate showed no adverse effects, and plasma and urine analyses indicated rapid hydrolysis of the ester in the gastrointestinal tract. This indicated that the sucrose ester had little accumulation in body tissues.

Clinical trials have been studied with humans (Mitsubishi Institute, 1994) on the effects of ingestion of up to 7.5 g/day of S-1170 (sucrose ester mixture). Analysis of blood and feces showed that around 80% of the esters were hydrolyzed, and the hydrolysis degree was variable for mono-, di-, and tri-esters and alkyl chain length. Plasma levels were extremely low and close to the limit of assay sensitivity (~0.1 μg/mL) or not detectable. However, the group receiving 7.5 g/day of S-1170 reported adverse gastrointestinal effects such as diarrhea, soft stools, and flatulence, which were not reported in animal models at similar or greater doses (Holmberg, 2003).

As a result of these adverse effects and perceived deficiencies in the prior studies, a further clinical trial was conducted at a dose level of approximately 30 mg/kg/day (Mitsubishi-Kagaku Foods Corporation, 1996). There were no apparent differences between the treatment or control groups, and as such the ADI for sucrose esters was adjusted by the WHO to 0–30 mg/kg in 1988. By way of comparison, the WHO has set an ADI for Tween 80, another nonionic surfactant commonly used in food, at 25 mg/kg. The ADI level indicates the general safety of this class of compound.

Future Perspectives

This chapter has presented a short review of various types of sugar fatty acid esters including their production methods, properties, and primary applications in food, pharmaceutical, and cosmetics industries. Even though sugar fatty acid esters have been known for many years and a lot of work has been done on the application properties, many questions concerning the nature of these compounds are still open for future scientific research and applied studies.

The use of LSEs as emulsifiers will remain central for SEs in practical uses. There have been a variety of applications, as discussed herein. Compared to other surfactants or emulsifiers, SEs have unreplaceable functions in many applications such as in food systems, liposome systems, and pharmaceutical and cosmetic systems. Chemical synthesis usually produces a mixture of different sugar esters. The separation of the mixture into individual sugar esters with different esterification degrees can be a challenge in practice or with a high process cost. In this case, enzymatic synthesis demonstrates a potential value. Relatively pure mono sugar esters can be produced using lipases as the catalyst. The system has been demonstrated in pilot production systems. This could be a useful strategy when more pure mono sugar esters are needed for high value applications.

The use of HSEs as fat substitutes gives a great initiative in tailor-making fats. We long have a dream to design a fat to meet our needs, not only functionally but also nutritionally. The esters of sugars with full esterification or close to full esterification of all –OH groups will give the property similar to triacylglycerols in terms of general oils and fats properties. In particular, the required properties can be designed with different fatty acid profiles. Furthermore, the esters are not digested in the human body, while the triacylglycerols are digested. Thus, low calorie or zero calorie fats can be achieved through such a design. The concept has been initiated for the last two decades. The acceptance of the market is not strong, even though it has been pushed by industry. The biological concerns may also need more work for better understanding. The idea of designing oils and fats is still exciting for further exploration. Esters such as SEs give a great inspiration for such exploration.

References

Abd-Elbary, A.; El-Laithy, H. M.; Tadros, M. I. Sucrose Stearate-based Proniosome-derived Niosomes for the Nebulisable Delivery of Cromolyn Sodium. *Int. J. Pharm.* **2008,** *357,* 189–198.

Affleck, R.; Haynes, C. A.; Clark, D. S. Solvent Dielectric Effects on Protein Dynamics. *Proc. Natl. Acad. Sci. U.S.* **1992,** *89,* 167–170.

Ahsan, F.; Arnold, J. J.; Meezan, E.; Pillion, D. J. Sucrose Cocoate, a Component of Cosmetic Preparations, Enhances Nasal and Ocular Peptide Absorption. *Int. J. Pharm.* **2003,** *251,* 195–203.

Akoh, C. C. Emulsification Properties of Polyesters and Sucrose Ester Blends I: Carbohydrate Fatty Acid Polyesters. *J. Am. Oil Chem. Soc.* **1992,** *69,* 9–13.

Akoh, C. C. Synthesis of Carbohydrate Fatty Acid Polyesters. In *Carbohydrate Polyesters as Fat Substitutes*; Akoh, C. C., Swanson, B. G., Eds.; Marcel Dekker: New York, 1994; pp 12–25.

Akoh, C. C.; Swanson, B. G. Optimization of Sucrose Polyester Synthesis: Comparison of Properties of Sucrose Polyesters, Raffinose Polyesters and Salad Oil. *J. Food Sci.* **1990,** *55,* 236–243.

Ariyaprakai, S.; Limpachoti, T.; Pradipasena, P. Interfacial and Emulsifying Properties of Sucrose Ester in Coconut Milk Emulsions in Comparison with Tween. *Food Hydrocolloid.* **2013,** *30,* 358–367.

Banat, I. M.; Franzetti, A.; Gandolfi, I.; Bestetti, G.; Martinotti, M. G.; Fracchia, L.; Smyth, T. J.; Marchant, R. Microbial Biosurfactants Production, Applications and Future Potential. *Appl. Microbiol. Biotechnol.* **2010,** *87,* 427–444.

Bergholz, C. M. Safety Evaluation of Olestra, a Nonabsorbed, Fatlike Fat Replacement. *Crit. Rev. Food Sci. Nutr.* **1992,** *32,* 141–146.

Cao, L.; Bornscheuer, U. T.; Schmid, R. D. Lipase-catalyzed Solid-phase Synthesis of Sugar Esters. Influence of Immobilization on Productivity and Stability of the Enzyme. *J. Mol. Catal. B: Enzymatic* **1999,** *6,* 279–285.

Carroll, T. E.; Cataneo, R. J. Skin Refining Composition and Applicator. U.S. Patent 6,565,838, October 24, 2002.

Chaiyaso, T.; H-Kittikun, A.; Zimmermann, W. Biocatalytic Acylation of Carbohydrates with Fatty Acids from Palm Fatty Acid Distillates. *J. Ind. Microbiol. Biotechnol.* **2006,** *33,* 338–342.

Chang, S. W.; Shaw, J. F. Biocatalysis for the Production of Carbohydrate Esters. *New Biotechnol.* **2009,** *26,* 109–116.

Chansanroj, K.; Betz, G. Sucrose Esters with Various Hydrophilic-lipophilic Properties: Novel Controlled Release Agents for Oral Drug Delivery Matrix Tablets Prepared by Direct Compaction. *Acta Biomater.* **2010,** *6,* 3101–3109.

Conley, A. J.; Kabara, J. J. Antimicrobial Action of Esters of Polyhydric Alcohols. *Antimicrob. Agents Ch.* **1973,** *4,* 501–506.

Cooper, D. A.; Berry, D.; Jones, M.; Spendel, V.; Peters, J.; King, D.; Aldridge, D.; Kiorpes, A. An Assessment of the Nutritional Effects of Olestra in the Domestic Pig. In *Abstracts,* Annual Meeting of the Federation of American Societies of Experimental Biology, Anaheim, CA, April 26, 1994, 8, A191.

Degn, P.; Zimmermann, W. Optimization of Carbohydrate Fatty Acid Ester Synthesis in Organic Media by a Lipase from *Candida antarctica. Biotechnol. Bioeng.* **2001,** *74,* 483–491.

Diab, R.; Brillault, J.; Bardy, A.; Gontijo, A. V. L.; Olivier, J. C. Formulation and in Vitro Characterization of Inhalable Polyvinyl Alcohol-free Rifampicin-loaded PLGA Microspheres Prepared with Sucrose Palmitate as Stabilizer: Efficiency for ex Vivo Alveolar Macrophage Targeting. *Int. J. Pharm.* **2012,** *436,* 833–839.

Edris, A. E.; Malone, C. R. Formulation of Banana Aroma Impact Ester in Water-based Microemulsion Nano-delivery System for Flavoring Applications Using Sucrose Laurate Surfactant. *Procedia-Food Sci.* **2011,** *1,* 1821–1827.

El-Laithy, H. M.; Shoukry, O.; Mahran, L. G. Novel Sugar Esters Proniosomes for Transdermal Delivery of Vinpocetine: Preclinical and Clinical Studies. *Eur. J. Pharm. Biopharm.* **2011,** *77,* 43–55.

Fallat, R. W.; Glueck, C. J.; Lutmer, R.; Mattson, F. H. Short-term Study of Sucrose Polyester, a Nonabsorbable Fat-like Material as a Dietary Agent for Lowering Plasma Cholesterol. *Am. J. Clin. Nutr.* **1976,** *29,* 1204–1215.

Fanun, M. Surfactant Chain Length Effect on the Structural Parameters of Nonionic Microemulsions. *J. Dispers. Sci. Technol.* **2008,** *29,* 289–296.

Fanun, M. Properties of Microemulsions with Sugar Surfactants and Peppermint Oil. *Col. Pol. Sci.* **2009,** *287,* 899–910.

Ferrer, M.; Plou, F. J.; Fuentes, G.; Cruces, M. A.; Andersen, L.; Kirk, O. Effect of the Immobilization Method of Lipase from *Thermomyces lanuginosus* on Sucrose Acylation. *Biocatal. Biotransform.* **2002,** *20,* 63–71.

Ferrer, M.; Soliverib, J.; Plou, F. J.; López-Cortés, N.; Reyes-Duarte, D.; Christensenc, M.; Copa-Patiño, J. L.; Ballesteros, A. Synthesis of Sugar Esters in Solvent Mixtures by Lipases from *Thermomyces lanuginosus* and *Candida antarctica* B, and Their Antimicrobial Properties. *Enz. Microb. Technol.* **2005,** *36,* 391–398.

Feuge, R. O.; Zeringue, H. J.; Weiss, T. J.; Brown, M. Preparation of Sucrose Esters by Interesterification. *J. Am. Oil Chem. Soc.* **1970,** *47,* 56–60.

Forsyth, S. A.; Macfarlane, D. R. 1-Alkyl-3-methylbenzotriazolium Salts: Ionic Solvents and Electrolytes. *J. Mater. Chem.* **2003,** *13,* 2451–2456.

Furukawa, S.; Akiyoshi, Y.; O'Toole, G. A., Ogihara, H.,; Morinaga, Y. Sugar Fatty Acid Esters Inhibit Biofilm Formation by Food-borne Pathogenic Bacteria. *Int. J. Food Microbiol.* **2010,** *138,* 176–180.

Ganem-Quintanar, A.; Quintanar-Guerrero, D.; Falson-Rieg, F.; Buri, P. Ex Vivo Oral Mucosal Permeation of Lidocaine Hydrochloride with Sucrose Fatty Acid Esters as Absorption Enhancers. *Int. J. Pharm.* **1998,** *173,* 203–210.

Griffin, W. C. Classification of Surface-Active Agents by HLB. *J. Soc. Cosmet. Chem.* **1949,** 311–326.

Gumel, A. M.; Annuar, M. S. M.; Heidelberg, T.; Chisti, Y. Lipase Mediated Synthesis of Sugar Fatty Acid Esters. *Process Biochem.* **2011,** *46,* 2079–2090.

Hamm, D. J. Preparation and Evaluation of Trialkoxytricarballylate, Trialkoxycitrate, Trialkoxyglycerylether, Jojoba, Oil and Sucrose Polyesters as Low Calorie Replacements of Edible Fats and Oils. *J. Food Sci.* **1984,** *49,* 419–428.

Hara, S. Sub-acute Toxicity Test (Sucrose Palmitic Acid Ester; Sucrose Stearic Acid Ester). Unpublished report from the Tokyo Medical College submitted to the World Health Organisation by Seiyaku Co. Ltd., Shimokyo-ku, Kyoto, Japan, 1959.

Hathcox, A. K.; Beuchat, L. R. Inhibitory Effects of Sucrose Fatty Acid Esters, Alone and in Combination with Ethylenediaminetetracetic Acid and Other Organic Acids, on Viability of *Escherichia coli* O157: H7. *Food Microbiol.* **1996,** *13,* 213–225.

H-Kittikun, A.; Prasertsan, P.; Zimmermann, W.; Seesuriyachan, P.; Chaiyaso, T. Sugar Ester Synthesis by Thermostable Lipase from *Streptomyces thermocarboxydus* ME168. *Appl. Biochem. Biotech.* **2012,** *166,* 1969–1982.

Holmberg, K. Novel Surfactants Preparation, Applications, and Biodegradability. In *Sugar Fatty Acid Esters*; Holmberg, K., Ed.; Marcel Dekker: New York, 2003; p 111.

Huang, J.; Moriyoshi, T. Preparation of Stabilized Lidocaine Particles by a Combination of Supercritical CO_2 Technique and Particle Surface Control. *J. Mater. Sci.* **2008,** *43,* 2323–2327.

Hülsmann, S.; Backensfeld, T.; Keitel, S.; Bodmeier, R. Melt Extrusion—An Alternative Method for Enhancing the Dissolution Rate of 17 β-Estradiol Hemihydrate. *Eur. J. Pharm. Biopharm.* **2000,** *49,* 237–242.

Jacobus, B. J. P.; Jozef, W. C. L. Cosmetic Mousse. E.P. Patent 2,042,154, September 28, 2007.

Jandacek, R. J. Review of the Effects of Dilution of Dietary Energy with Olestra on Energy Intake. *Physiol. Behav.* **2012,** *105,* 1124–1131.

Jones, H. F. Process for the Preparation of Sucrose Monoesters. U.S. Patent 4,306,062, December 15, 1981.

Joshi, J. V. Aswalb, V. K. Goyal, P. S. Effect of Sodium Salicylate on the Structure of Micelles of Different Hydrocarbon Chain Lengths. *Physica B.* **2007,** *391,* 65–71.

Kabara, J. J. Medium Chain Fatty Acids and Esters. In *Antimicrobials in Food;* Davidson, P. M., Barnes, A. L., Eds.; Marcel Dekker: New York, 1993; pp 307–342.

Kempen, S. E. H. J.; Maas, K.; Schols, H. A.; Linden, E.; Sagis, L. M. C. Interfacial Properties of Air/Water Interfaces Stabilized by Oligofructose Palmitic Acid Esters in the Presence of Whey Protein Isolate. *Food Hydrocolloid.* **2013,** *32,* 162–171.

Kilpatrick-Liverman, L.; Miller, L. A. Clear Cosmetic Stick Composition Containing Sucrose Esters and Method of Use. U.S. Patent 5,776,475, February 13, 1997.

Kis, L.; Szüts, A.; Otomo, N.; Szabó-Révész, P.; Deli, M. A. The Pontential of Sucrose Esters to Be Used as Oral Absorption Enhancers. In *8th Central European Symposium on Pharmaceutical Technology,* Graz, Austria, 2010.

Klang, V.; Matsko, N.; Raupach, K.; El-Hagin, N.; Valenta, C. Development of Sucrose Stearate-Based Nanoemulsions and Optimisation through Gammacyclodextrin. *Eur. J. Pharm. Biopharm.* **2011,** *79,* 58–67.

Könnecker, G.; Regelmann, J.; Belanger, S.; Gamon, K.; Sedlak, R. Environmental Properties and Aquatic Hazard Assessment of Anionic Surfactants: Physico-chemical, Environmental Fate and Ecotoxicity Properties. *Ecotoxicol. Environ. Saf.* **2011,** *74,* 1445–1460.

Koseki, T.; Onishi, H.; Takahashi, Y.; Uchida, M.; Machida, Y. Preparation and Evaluation of Novel Directly-compressed Fast-disintegrating Furosemide Tablets with Sucrose Stearic Acid Ester. *Biol. Pharm. Bull.* **2009,** *32,* 1126–1130.

Kralova, I.; Sjoblom, J. Surfactants Used in Food Industry: A Review. *J. Disp. Sci. Technol.* **2009,** *30,* 1361–1383.

Kumar, R.; Modak, J.; Madras, G. Effect of the Chain Length of the Acid on the Enzymatic Synthesis of Flavors in Supercritical Carbon Dioxide. *Biochem. Eng. J.* **2005,** *23,* 199–202.

Lafranconi, W. M.; Long, P. H.; Atkinson, J. E.; Knezevich, A. L.; Wooding, W. L. Chronic Toxicity and Carcinogenicity of Olestra in Swiss CD-1 Mice. *Food Chem. Toxicol.* **1994,** *32,* 789–798.

Lee, S. H.; Nguyen, H. M.; Koo, Y. M.; Ha, S. H. Ultrasound-enhanced Lipase Activity in the Synthesis of Sugar Ester Using Ionic Liquids. *Process Biochem.* **2008,** *43,* 1009–1012.

Lemieux, R. U.; McInnes, A. G. The Preparation of Sucrose Monoesters. *Can. J. Chem.* **1962,** *40,* 2376–2393.

Lerk, P. C.; Sucker, H. Application of Sucrose Laurate, a New Pharmaceutical Excipient, in Peroral Formulations of Cyclosporine A. *Int. J. Pharm.* **1993,** *92,* 197–202.

Leuner, C.; Dressman, J. Improving Drug Solubility for Oral Delivery Using Solid Dispersions. *Eur. J. Pharm. Biopharm.* **2000,** *50,* 47–60.

Lim, T. H.; Choi, K. B.; Kim, T. S.; Yeo, I. H.; Nam, S. W. Manufacturing Method for Cosmetic Emulsion Using High Content Silicone Derivatives and the Cosmetic Composition Containing this Method. K. R. Patent 20120085361, January 24, 2011.

Liu, Q.; Janssen, M. H. A.; Rantwijk, F.; Sheldon, R. A. Room-temperature Ionic Liquids That Dissolve Carbohydrates in High Concentrations. *Green Chem.* **2005,** 7, 39–42.

Lu, J.; Nie, K.; Wang, F.; Tan, T. Immobilized Lipase *Candida* sp. 99–125 Catalyzed Methanolysis of Glycerol Trioleate: Solvent Effect. *Bioresources Technol.* **2008,** *99,* 6070–6074.

Magg, H. Fatty Acid Derivatives: Important Surfactants for Household, Cosmetics and Industrial Purposes. *J. Am. Oil Chem. Soc.* **1984,** *61,* 259–267.

Magnusson, A. O.; Rotticci-Mulder, J. C.; Santagostino, A.; Hult, K. Creating Space for Large Secondary Alcohols by Rational Redesign of *Candida antarctica* Lipase B. *Eur. J. Chem. Biol.* **2005,** *6,* 1051–1056.

Marlene, B.; Edith, V. A.; Thomas, H. Hair Care Agent. W.O. Patent 2013083349, November 9, 2012.

Marshall, D. L.; Bullerman, L. B. Antimicrobial Activity of Sucrose Fatty Acid Ester Emulsifiers. *J. Food Sci.* **1986,** *51* (2), 468–470.

Marshall, D. L.; Bullerman, L. B. Antimicrobial Properties of Sucrose Fatty Acid Esters. In *Carbohydrate Polyesters as Fat Substitutes*; Akoh, C. C., Swanson, B. G., Eds.; Marcel Dekker: New York, 1996; pp 149–168.

Maruyama, T.; Nagasawa, S.; Goto, M. Enzymatic Synthesis of Sugar Amino Acid Esters in Organic Solvents. *J. Biosci. Bioeng.* **2002,** *94,* 357–361.

Mattson, F. H.; Nolen, G. A. Absorbability by Rats of Compounds Containing from One to Eight Ester Groups. *J. Nutr.* **1972,** *102,* 1171–1175.

Mattson, F. H.; Volpenhein, R. A. Hydrolysis of Fully Esterified Alcohols Containing from One to Eight Hydroxyl Groups by the Lipolytic Enzymes of Rat Pancreatic Juice. *J. Lipid Res.* **1972,** *13,* 325–328.

Megahed, M. G. Preparation of Sucrose Fatty Acid Esters as Food Emulsifiers and Evaluation of Their Surface Active and Emulsification Properties. *Grasas y Aceites* **1999,** *50,* 280–282.

Miller, K. W.; Long, P. H. A 91-day Feeding Study in Rats with Heated Olestra/Vegetable Oil Blends. *Food Chem. Toxicol.* **1990,** *28,* 307–315.

Miller, K. W.; Wood, F. E.; Stuart, S. B.; Alden, C. L. A 20-month Olestra Feeding Study in Dogs. *Food Chem. Toxicol.* **1991,** *29,* 427–435.

Miller, K. W.; Lawson, K. D.; Tallmadge, D. H.; Madison, B. L.; Okenfuss, J. R.; Hudson, P.; Wilson, S.; Thorstenson, J.; Vanderploeg, P. Disposition of Ingested Olestra in the Fisher 344 Rat. *Fundam. Appl. Toxicol.* **1995,** *24,* 229–237.

Mirko, K. Easily Removable Nail Varnish Composition. W.O. Patent 2012163851, May 25, 2012.

Mitsubishi Institute. *Clinical and Pharmacokinetic Studies of Sucrose Esters of Fatty Acids in Human;* Report No. 3B159. Yokohama, Japan, 1994.

Mitsubishi-Kagaku Foods Corporation. Study of Sucrose Esters of Fatty Acids: Laxative Study of S-1170. Unpublished report submitted to JECFA by Mitsubishi Chemical Corporation, Tokyo, Japan, 1996.

Molina-Bolívar, J. A.; Ruiz, C. C. Self-assembly and Micellar Structures of Sugar-based Surfactants: Effect of Temperature and Salt Addition. In *Sugar-based Surfactants Fundamentals and Applications*; Ruiz, C. C, Ed.; CRC: Boca Raton, FL, 2008; Vol. 143, pp 61–104.

Monk, J. D.; Beuchat, L. R. Viability of *Listeria monocytogenes*, *Staphylococcus aureus* and Psychrotrophic Spoilage Micro-organisms in Refrigerated Ground Beef Supplemented with Sucrose Esters of Fatty Acids. *Food Microbiol.* **1995,** *12,* 397–404.

Monk, J. D.; Beuchat, L. R.; Hathcox, A. K. Inhibitory Effects of Sucrose Monolaurate, Alone and in Combination with Organic Acids, on *Listeria rnonocytogenes* and *Staphylococcus aureus*. *J. Appl. Bacteriol.* **1996,** *81,* 7–18.

Mutschler, J.; Rausis, T.; Bourgeois, J.; Bastian, C.; Zufferey, D.; Mohrenz, I.V. Ionic Liquid-coated Immobilized Lipase for the Synthesis of Methyl Glucose Fatty Acid Esters. *Green Chem.* **2009,** *11,* 1793–1800.

Nakamura, S. Using Sucrose Esters as Food Emulsifiers. *Oleochemicals* **1997,** *8,* 866–874.

Naomi, T. Transparent or Translucent Cosmetic. J.P. Patent 2002087933, September 8, 2000.

Neta, N. S.; Peres, A. M.; Teixeira, J. A.; Rodrigues, L. R. Maximization of Fructose Esters Synthesis by Response Surface Methodology. *New Biotechnol.* **2011,** *28,* 349–355.

Neta, N. S.; Santos, J. C.; Sancho, S. O.; Rodrigues, S.; Gonçalves, L. R. B; Rodrigues, L. R.; Teixeira, J. A. Enzymatic Synthesis of Sugar Esters and Their Potential as Surface-active Stabilizers of Coconut Milk Emulsions. *Food Hydrocolloid.* **2012,** *27* (2), 324–331.

Nolen, G. A.; Wood, F. E.; Dierckman, T. A. A Two Generation Reproductive and Developmental Toxicity Study of Sucrose Polyester. *Food Chem. Toxicol.* **1987,** *25,* 1–8.

Ntawukulilyayo, J. D.; Bouckaert, S.; Remon, J. P. Enhancement of Dissolution Rate of Nifedipine Using Sucrose Ester Coprecipitates. *Int. J. Pharm.* **1993,** *93,* 209–214.

Ntawukulilyayo, J. D.; Demuynck, C.; Remon, J. P. Microcrystalline Cellulose-sucrose Esters as Tablet Matrix Forming Agents. *Int. J. Pharm.* **1995,** *121,* 205–210.

Osipow, L.; Snell, F. D.; York, W. C.; Finchler, A. Methods of Preparation of Fatty Acid Esters of Sucrose. *Ind. Eng. Chem.* **1956,** *48,* 1459–1462.

Osipow, L. I.; Rosenblatt, W. Micro-emulsion Process for the Preparation of Sucrose Esters. *J. Am. Oil. Chem. Soc.* **1967,** *44,* 307–309.

Otsuka, M.; Ofusa, T.; Matsuda, Y. Dissolution Improvement of Water-insoluble Glybuzole by Co-grinding and Co-melting with Surfactants and Their Physicochemical Properties. *Colloids Surface. B.* **1998,** *10,* 217–226.

Park, B. D.; Youm, J. K.; Lee, S. H. Skin External Preparation Having Hexagonal Gel Structure. U.S. Patent 20070286835, July 28, 2006.

Park, S.; Kazlauskas, R. J. Improved Preparation and Use of Room Temperature Ionic Liquids in Lipase Catalyzed Enantio- and Regioselective Acylations. *J. Org. Chem.* **2001,** *66,* 8395–8401.

Pedersen, N. R.; Halling, P. J.; Pedersen, L. H.; Wimmer, R.; Matthiesen, R.; Veltman, O. R. Efficient Transesterification of Sucrose Catalysed by the Metalloprotease Thermolysin in Dimethylsulfoxide. *FEBS Lett.* **2002,** *519,* 181–184.

Pedersen, N. R.; Wimmer, R.; Matthiseen, R.; Pedersen, L. H.; Gessese, A. Synthesis of Sucrose Laurate Using a New Alkaline Protease. *Tetrahedron: Asymmetry* **2003,** *14,* 667–673.

Polat, T.; Linhardt, R. J. Synthesis and Applications of Sucrose-based Esters. *J. Surf. Deterg.* **2001,** 4, 415–421.

Prince, D. M.; Welschenbach, M. A. Olestra: A New Food Additive. *J. Am. Diet Assoc.* **1998,** *98,* 565–569.

Procter & Gamble Company. *Food Additive Petition Environmental Assessment for Olestra;* Food Additive Petition 7A 3997. U.S. Food and Drug Administration: Washington, DC, 1995.

Rao, J.; McClements, J. Food-grade Microemulsions, Nanoemulsions and Emulsions: Fabrication from Sucrose Monopalmitate and Lemon Oil. *Food Hydrocolloid.* **2011,** *25,* 1413–1423.

Rizzi, G. P.; Taylor, H. M. A Solvent-free Synthesis of Sucrose Polyesters. *J. Am. Oil Chem. Soc.* **1978,** *55,* 398–401.

Ryoto. *Ryoto Sugar Ester Technical Information; Nonionic Surfactant/Sucrose Fatty Acid Ester/ Food Additive.* Mitsubishi-Kasei Food Corporation: Tokyo, Japan, 1987; p 8.

Samanta, S.; Ghosh, P. Coalescence of Bubbles and Stability of Foams in Aqueous Solutions of Tween Surfactants. *Chem. Eng. Res. Des.* **2011,** *89,* 2344–2355.

Seidenberger, T.; Siepmann, J.; Bley, H.; Maeder, K.; Siepmann, F. Simultaneous Controlled Vitamin Release from Multiparticulates: Theory and Experiment. *Int. J. Pharm.* **2011,** *412,* 68–76.

Shieh, C. J.; Koehler, P. E.; Akoh, C. C. Optimization of Sucrose Polyester Synthesis Using Response Surface Methodology. *J. Food Sci.* **1996,** *61,* 97–100.

Skare, K. L.; Skare, J. A.; Thompson, E. D. Evaluation of Olestra in Short-term Genotoxicity Assays. *Food Chem. Toxicol.* **1990,** *28,* 69–73.

Slough, C. L.; Miday, R. K.; Zorich, N. L.; Jones, J. K. Postmarketing Surveillance of New Food Ingredients: Design and Implementation of the Program for the Fat Replacer Olestra. *Regul. Toxicol. Pharm.* **2001,** *33,* 218–223.

Somashekar, B.R.; Divakar, S. Lipase Catalyzed Synthesis of l-Alanyl Esters of Carbohydrates. *Enzyme Microb. Technol.* **2007,** *40,* 299–309.

Soultani, S.; Engasser, J. M.; Ghoul, M. Effect of Acyl Donor Chain Length and Sugar/Acyl Donor Molar Ratio on Enzymatic Synthesis of Fatty Acid Fructose Esters. *J. Mol. Catal. B: Enzym.* **2001,** *11,* 725–731.

Szüts, A.; Szabó-Révész, P. Sucrose Esters as Natural Surfactants in Drug Delivery Systems—A Mini-review. *Int. J. Pharm.* **2012,** *433,* 1–9.

Szüts, A.; Pallagi, E.; Regdon, G., Jr.; Aigner, Z.; Szabó-Révész, P. Study of Thermal Behaviour of Sugar Esters. *Int. J. Pharm.* **2007,** *336,* 199–207.

Szüts, A.; Budai-Szücs, M.; Erös, I.; Otomo, N.; Szabó-Révész, P. Study of Gel-forming Properties of Sucrose Esters for Thermosensitive Drug Delivery Systems. *Int. J. Pharm.* **2010a,** *383,* 132–137.

Szüts, A.; Budai-Szücs, M.; Erös, I.; Ambrus, R.; Otomo, N.; Szabó-Révész, P. Study of Thermosensitive Gel-forming Properties of Sucrose Stearates. *J. Excipients Food Chem.* **2010b,** *1,* 13–20.

Szüts, A.; Láng, P.; Ambrus, R.; Kiss, L.; Deli, M. A.; Szabó-Révész, P. Applicability of Sucrose Laurate as Surfactant in Solid Dispersions Prepared by Melt Technology. *Int. J. Pharm.* **2011,** *410,* 107–110.

Tagekami, S.; Kitamura, K.; Kawada, H.; Matsumoto, Y.; Kitade, T.; Ishida, H.; Nagata, C. Preparation and Characterization of a New Lipid Nano-emulsion Containing Two Cosurfactants, Sodium Palmitate for Droplet Size Reduction and Sucrose Palmitate for Stability Enhancement. *Chem. Pharm. Bull.* **2008,** *56,* 1097–1102.

Tahara, Y.; Honda, S.; Kamiya, N.; Piao, H.; Hirata, A.; Hayakawa, E.; Fujii, T.; Goto, M. A Solid-in-oil Nanodispersion for Transcutaneous Protein Delivery. *J. Control. Release,* **2008,** *131,* 14–18.

Tanaka, N.; Imai, K.; Ueda, S.; Kimura, T. Development of Novel Sustained-release System, Disintegration Controlled Matrix Tablet (DCMT) with Solid Dispersion Granules of Nilcadipine (II): In Vivo Evaluation. *J. Controlled Release* **2006,** *122,* 51–56.

The Forty-Fourth Meeting of the Joint FAO/WHO Expert Committee on Food Additives (JECFA). Sucrose Esters of Fatty Acids and Sucroglycerides. In *WHO Food Additives Series 35: Toxicological Evaluation of Certain Food Additives and Contaminants in Food;* World Health Organization: Geneva, 1996; pp 129–138.

Thevenin, M. A.; Grossiord, J. L.; Poelman, M. C. Sucrose Esters/Cosurfactant Microemulsion Systems for Transdermal Delivery: Assessment of Bicontinuous Structures. *Int. J. Pharm.* **1996,** *137,* 177–186.

Tokita, K. Acute Toxicity Test (Sucrose Stearic Acid Ester; Sucrose Palmitic Acid Ester and Mixture of Sucrose Stearic Acid Ester and Sucrose Palmitic Acid Ester). Unpublished report from Dept of Pharmacology, Toho University, submitted to the World Health Organisation by Dai-Nippon Sugar Manufacturing Company Ltd., Japan, 1958.

Tsuzuki, W.; Kitamura, Y.; Suzuki, T.; Kobayashi, S. Synthesis of Sugar Fatty Acid Esters by Modified Lipase. *Biotechnol. Bioeng.* **1999,** *64,* 267–271.

U.S. Food and Drug Administration. Department of Health and Human Services, Fed. Register, Vol. 61 (20), 21 CFR Part 172 (Docket 87F-0179), January 30, pp 3118–3173. Washington, DC, 1996.

Valdés, K.; Morilla, M. J.; Romero, E.; Chávez, J. Physicochemical Characterization and Cytotoxic Studies of Nonionicsurfactant Vesicles Using Sucrose Esters as Oral Delivery Systems. *Colloids Surface B.* **2014,** *117,* 1–6.

Valivety, R. H.; Halling, P. J.; Peilow, A. D.; Macrae, A. R. Relationship between Water Activity and Catalytic Activity of Lipases in Organic Media. Effects of Supports, Loading and Enzyme Preparation. *Eur. J. Biochem.* **1994,** *222,* 461–466.

Vasconcelos, T.; Sarmento, B.; Costa, P. Solid Dispersions as Strategy to Improve Oral Bioavailability of Poorly Water Soluble Drugs. *Drug Discov. Today* **2007,** *12,* 1068–1075.

Wang, Y. J. Saccharides: Modifications and Applications. In *Chemical and Functional Properties of Food Saccharides*; Tomasik, P., Tomasik, T., Eds.; CRC Press: New York, 2004; pp 35–46.

Watanabe, T.; Katayama, S.; Matsubara, M.; Honda, Y.; Kuwahara, M. Antibacterial Carbohydrate Monoesters Suppressing Cell Growth of Streptococcus Mutans in the Presence of Sucrose. *Curr. Microbiol.* **2000,** *41,* 210–213.

Wood, F. E.; Tierney, W. J.; Knezevich, A. L.; Bolte, H. F.; Maurer. J. K.; Bruce, R. D. Chronic Toxicity and Carcinogenicity Studies of Olestra in Fischer 344 Rats. *Food Chem. Toxicol.* **1991,** *29,* 223–230.

Yadav, G. D.; Trivedi, A. H. Kinetic Modeling of Immobilized Lipase Catalyzed Transesterification of N-octanol with Vinyl Acetate in Non-aqueous Media. *Enzyme Microb. Technol.* **2003,** *32,* 783–789.

Yamada, T.; Kawase, N.; Ogimoto, K. Sucrose Esters of Long-chain Fatty Acids. *J. Jpn. Oil Chem. Soc.* **1980,** *29,* 543–547.

Yan, Y.; Bornscheuer, U. T.; Stadler, G.; Lutz-Wahl, S.; Reuss, M.; Schmid, R. D. Production of Sugar Fatty Acid Esters by Enzymatic Esterification in a Stirred-tank Membrane Reactor: Optimization of Parameters by Response Surface Methodology. *J. Am. Oil Chem. Soc.* **2001,** *78,* 147–153.

Yokoi, Y.; Yonemochi, E.; Terada, K. Effects of Sugar Ester and Hydroxypropyl Methylcellulose on the Physicochemical Stability of Amorphous Cefditoren Pivoxil in Aqueous Suspension. *Int. J. Pharm.* **2005,** *290,* 91–99.

Yoshida, M.; Katsuda, S.; Nakae, D.; Maekawa, A. Lack of Toxicity or Carcinogenicity of S-170, a Sucrose Fatty Acid Ester, in F344 Rats. *Food Chem. Toxicol.* **2004,** *42,* 667–676.

Zhang, L.; Somasundaran, P.; Singh, S. K.; Felse, A. P.; Gross, R. Synthesis and Interfacial Properties of Sophorolipid Derivatives. *Colloid. Surface A.* **2004,** *240,* 75–82.

9

Production and Utilization of Natural Phospholipids

Willem van Nieuwenhuyzen ■ *Lecipro Consulting, 1906 AH Limmen, The Netherlands*

Introduction

Phospholipids (PLs) are the major components of biological membranes and are important biochemical intermediates in the growth and functioning of cells, both in plants and in animals. PLs mainly from vegetable origin (vegetable lecithins) are derived commercially from oil-bearing seeds such as soybeans, sunflower kernels, and canola (rapeseed) and are widely used for their emulsifying and structural improvement properties in food matrices. In biochemistry and medicine, the name *lecithin* is exclusively given to the *sn*-3 phosphatidylcholine. Here we define lecithin as a mixture of PLs with adherent glycolipids and oil. Vegetable lecithins contain predominantly phosphatidylcholine (PC), phosphatidylethanolamine (PE), phosphatidylinositol (PI), phosphatidic acid (PA), small amounts of lysophosphatidylcholine (LPC), and other glycerol PLs of complex fatty acid composition. Animal lecithins are sourced from egg, dairy, and marine species, including krill.

Due to their surface-active properties, PLs are used as emulsifiers in foods as well as in pharmaceutical, cosmetic, feed, and technical applications. PLs also have nutritional functions because they contain organic bound choline, inositol, and fatty acids. Both vegetable and animal lecithins are gaining interest due to their nutritional and technological qualities.

In this chapter, the current state of industrial knowledge toward the production and utilization of natural phospholipids (which is widely scattered in the literature) is summarized. Special focus is given to the production of canola lecithin and egg lecithin because other sources are described in separate chapters.

Occurrence

Oilseed volumes grow steadily because increasing agricultural areas are sown with seed varieties with higher yields per acre/hectare. Oilseeds are important as a vegetable oil source for the growing world population and as a protein source for animal and fish feed, resulting in supplying people with meat and farmed fish. Only a small fraction of the oilseed protein is directly incorporated in meat analogues in a versatile human diet. Vegetable oil from the seeds is refined, making it suitable and pleasant for human consumption. The removal of the phospholipids as lecithin gum from the extracted crude oil is beneficial for economic oil refining. Vegetable lecithin gums are available in abundance worldwide, and some of these are produced into dried commercial lecithin with dedicated high-quality technological properties.

Soybeans are the largest oilseed source in the world, provide an abundance of soy lecition gums, and are the primary source of soy lecithin, which serve as food emulsifiers Large crops of soybeans grow in North America, Latin America, China, and India. Crushing plants are spread all over the world. Quite a number of crushing and refinery plants have production units for drying gums into dried lecithin. Basic lecithin with a 0.5% yield is a coproduct of the soy crushing industry.

However, European retailers and consumer requirements for non-GMO food induced the search for identity preserved (IP) non-GMO lecithins beginning in 1996. This changed lecithin sourcing worldwide. Traditional non-GMO soybeans are scarcer, which presents market opportunities for high-quality IP soy, canola, and sunflower lecithins.

Canola lecithin represents the lecithin derived from rapeseed. In the 1970s, new rapeseed varieties with low erucic fatty acid (C22:0) composition were bred; these were marketed in Canada as Canadian Low Erucic Acid Rapeseed (CLEAR). Later, varieties with low glucosinolate content followed as "double zero" rapeseed with positive nutritive performance in food (oil) and feed (meal). The products are traded as "canola" in North America, but elsewhere the name "rapeseed" is often used.

Interestingly, lecithin from the traditional high erucic varieties was already processed in Germany and Eastern European countries from the 1940s for use in technical industries and increasingly in feed and food. The low dosage of below 1% in feed and food recipes made the product an attractive lower-priced alternative to soy lecithin. Today, canola seed crops grow in Canada, Europe, China, and India in the areas with good soil and favorable weather conditions. Canola lecithin is produced in those regions depending on market demand. Sunflower lecithin is processed from the black oil-bearing sunflower seed varieties. Whereas high wax content in the lecithin gums caused viscous handling problems in the past, improved seed treatment and degumming technology allow the processing of good qualities of sunflower lecithin with low wax content. Sunflower oil crushing is often located in the harvesting regions. Sunflower lecithin is produced in increasing quantities in Argentina, Central Europe (Hungary), and Eastern Europe (Ukraine, Russia).

Corn lecithin can be processed from the gums of corn (maize) germ oil. The potential availability is small because corn lecithin never reached a breakthrough on the food emulsifier market.

Cottonseed lecithin can be processed from cottonseed oil. The potential availability is small because residual gossypol content does not make this lecithin attractive for commercial application.

Palm oil lecithin is not available commercially due to the fact that water-extracted palm oil contains only a low amount (<0.1%) of phospholipids. During the oil refining process, the residual nonhydratable phospholipids are often separated with the phosphoric acid pretreatment step.

Rice bran oil lecithin is investigated in older research projects. The lecithin may have interesting emulsifying properties, but for commercial exploitation, large volumes of high quality nonoxidized rice bran oil, a coproduct from the bran separation of rice, should be preserved and processed adequately. To date constant production is not secured.

Algae lecithin may become a future potential source if large quantities of algae oil can be harvested and processed in commercial plants. Varying quantities of phospholipids may be extracted from the algae oil, depending on the type of algae and the oil recovery process.

Table 9.A lists vegetable lecithin sources in alphabetical order. Lecithin can also be derived from animal sources. In the last century, the egg yolk was the focus for phospholipid research and the search for a cost-effective emulsifier. Other animal and marine sources are investigated primarily for nutritional performance (Schneider, 2008).

Egg Lecithin

Hen eggs and egg yolks are full of PL, which are active emulsifying components. The phospholipids are the basic elements in the cell membrane, regulating cell metabolism. Eggs and egg yolks have been used for centuries as nutritional foods and as emulsifying ingredients in baked products. The polar lipids also made egg a suitable emulsifier for paint dispersions. Liquid chicken egg yolk contains 30% lipids with 9% phospholipids; on a basis of dry matter, egg yolk contains 60% lipids, including 18% phospholipids (Palacios and Wang, 2005).

Milk Phospholipids

The milk fat globule membranes contain phospholipids. In liquid cow's milk the PL content is only 0.1%, whereas the content in dry ultra-filtered butter serum is about 20%. Milk phospholipids contain equal amounts of PC, PE, and sphingomyelin (SPM), with small amounts of phosphatidylserine (PS).

Fish Phospholipids

Fish roe contains the fish eggs that are full of phospholipids. With the exception of caviar from sturgeons, fish roe is often handled as a byproduct of the fish catch and fish farming. Increasingly, attempts are made to extract the fish phospholipids, predominantly PC, as an essential source of polar omega fatty acids DHA and EPA. Herring, squid, and cod roe are potential sources.

Table 9.A Lecithins from Vegetable Sources

Lecithin Source	Market Availability	Potential	Limitation
Algae	None	Depends on many conditions for oil recovery	Needs intensive research and development and sources up-scaling
Canola/rapeseed	Limited commercial quantities	Sufficient raw material; process optimization possible	Greenish color; critical sensory/flavor
Corn	Very small quantities; corn gums	Small	—
Cottonseed	Very small quantities; cottonseed gums	Small	Oft GMO varieties; no food "image"
Palm oil	No lecithin gums available	—	—
Rice bran oil	Small quantities RBO lecithin gums	Perfect RB-Oil source needed	Logistics
Soybeans	Large during decades	Soy (GM quality) is largest oil seed crop; gums available in surplus	IP non-GMO lecithin quality is scarce for food use in some regions
Sunflower seeds	Growing availability	High; some state-of-the-art plants	High food use interest

Table 9.B Lecithins from Animal Sources

Lecithin Source	Market Availability	Potential	Limitation
Egg	Balanced supply	Effective emulsifier; DHA, ARA, choline supplier for infant food, parenteral nutrition	High price limits food uses
Milk	Small	Supplier of PC, PE, PS, sphingomyelin; health food segment	Health claim (HC) approval pending
Fish	Small	Omega-PC supplier for health foods segment	Fish co-product valorization needs commercialization; HC requirement
Krill	Growing	Omega-PC supplier for health foods segment; intensive clinical research ongoing	HC requirement

Krill Phospholipids

Krill are small crustaceans that are fished in the Antarctic under agreed conditions and quota. Krill is specifically caught as a source of essential polar omega fatty acids DHA and EPA. Krill oil is extracted and is a natural mixture of omega triglycerides and omega phospholipids.

Table 9.B lists the animal sources of lecithins with information on composition, market potential, and current limitations.

Functional Properties

Phospholipid Structures

The molecular structure in Figure 9.1 shows the amphiphilic character of the main PLs, such as PC, PE, PI, and PA. The PLs contain both hydrophobic (water repelling) and hydrophilic (water loving) groups, and are therefore called *amphipathic* (both hydrophilic and hydrophobic). The hydrophilic side forms the head and the hydrophobic side forms the tail. These result in diverse phase behavior in aqueous solutions. At a neutral pH, the head group may be electrically neutral with zwitterionic character or may be negatively charged. However, at a lower pH, the net negative charge facilitates stronger swelling.

At the interface of oil and water, the PC forms a lamellar layer, which is different from the reversed hexagonal phase of PE or the hexagonal phase of lysophospholipids,

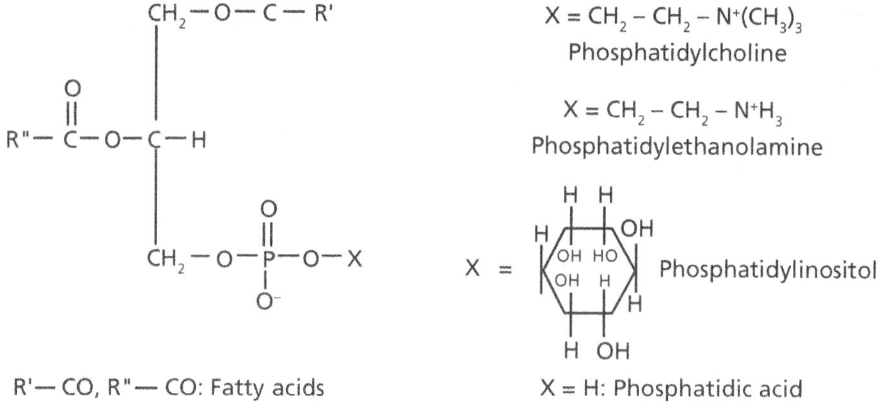

Figure 9.1 Molecular structure of main phospholipids.

as shown in Figure 9.2 (Bergenståhl, 1990). This knowledge is used for production of lecithins with adapted, modified, or fractionated PL composition. The different phase structures at the interface influence the emulsion formation and its stability. The enzymatic hydrolysis, acetylation, and alcohol fractionation processes all have in common that the zwitterion group of the PE is modified or removed.

Technological Function as Emulsifier

Surface Activity

Emulsifiers, including lecithins, concentrate at the interface between oil and water and subsequently reduce the interfacial tension. The formation of interfaces and emulsions usually requires energy input by mixers and homogenizers. Lecithins and

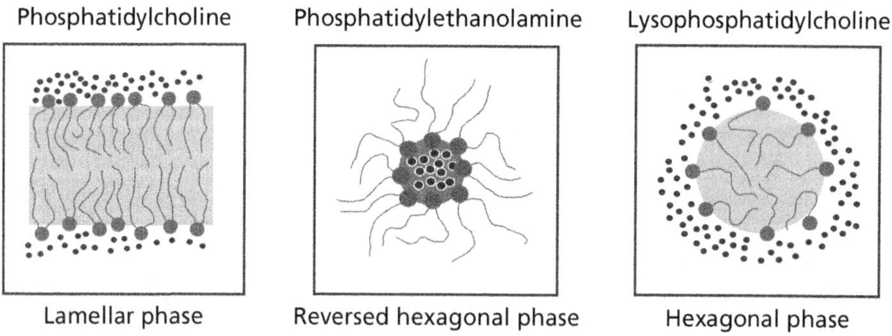

Figure 9.2 Phospholipid structures at the interface.

other emulsifiers at a dosage of 0.1–0.5% facilitate the formation and stabilization of the emulsion. In food emulsion recipes, the emulsifiers support emulsion stability because of the interaction with the three main food components: fat, protein, and carbohydrates. The stable emulsions should not cream, coalesce in larger droplets, sediment, or separate during product shelf life.

Evaluation of Lecithin Emulsifying Properties

Numerous emulsifying tests are available nowadays, varying from simple laboratory tests to sophisticated instruments requiring specific skills. In dispersion tests, the hydration speed of lecithin in distilled or tap water is observed. Modified hydrophilic lecithins show a short dispersion time, an indication of relatively strong oil-in-water (O/W) surfactant performance. In an emulsion capacity test, a lecithin/oil mixture (ratio 1:9) is emulsified in water, and the oil and water separation is measured over a period of time. Hydrolyzed lecithins give more stable emulsions with low fat creaming compared to standard liquid lecithin. Proteins or other ingredients are incorporated in the lecithin/oil/water system, to measure the stability of the emulsion (fat creaming, protein sedimentation) over a period of time. Modified hydrophilic lecithins usually give specific stability performance; in particular, lysophospholipid–protein interactions may enhance emulsion stability of food products. The preparation procedure may require specific processing conditions. Suboptimal processing may lead to undesired, unstable products.

These tests can be carried out in a simple manner for internal quality control and initial product development; therefore, the conditions (mixing time, temperature, stirring equipment, ingredients, time of measurement) should be carefully controlled in order to obtain reproducible results. Hydrophilic lecithins facilitate the formation of smaller droplets in O/W emulsions. However, more detailed scientific information will require sophisticated equipment, reliable and reproducible test procedures, and skilled scientists (van Nieuwenhuyzen and Tomás, 2008). Some techniques for measuring stability are:

- *Particle size distribution (PSD):* Information about the size of droplets in diluted emulsions.
- *Turbidity:* Measurement of emulsions by the optical Turbiscan or Beckman QuickSCANR Vertical Scan Analyzer follow the physical evolution of liquid dispersions during resting time (Pan et al., 2004; Scuriatti et al., 2003).
- *Microscopy:* Visualizing the particle size, shape, and structure of emulsion droplets through various microscope techniques such as phase contrast light microscopy, confocal scanning laser microscopy (CSLM), Raman Microscopy, scanning electron microscopy (SEM), transmission electron microscopy (TEM), X-ray micro-tomography (XRT), atomic force microscopy (AFM), and other imaging techniques.

Table 9.C Hydrophilic–Lipophilic Balance Values of Modified Soy Lecithins

Lecithin Type	HLB Value
PE-enriched (PC-depleted) fraction	2–3
Standard liquid lecithin	4
PC-30 fraction	5
Acetylated lecithin	6–7
Enzymatically hydrolyzed lecithin	7
Hydroxylated lecithin	9
Acetylated + hydroxylated lecithin	11
Hydrolyzed + hydroxylated lecithin	11

All of these procedures can be used to classify the lecithins in the hydrophilic–lipophilic balance (HLB) system by comparing the hydrophilic properties of modified lecithins with known non-ionic emulsifiers. Originally this system was developed as an indicator for selecting appropriate non-ionic emulsifiers in a range of HLB values between 0 and 20 (Griffin, 1949). The HLB ranking for lecithins is scientifically problematic because it provides little information about the emulsion stability properties and synergistic effects. Nevertheless, the HLB ranking in Table 9.C is still a valuable support for selecting emulsifiers for development of emulsions (Bonekamp, 2008).

Antioxidant Performance

Interestingly, soy lecithin has antioxidant functions, although it does not have the chemical reaction properties for antioxidative performance of refined oils and fats. In the oil refining industry it is an experienced fact that crude oil containing residual phospholipids can be stored for months without severe oxidative damage, whereas the refined oil has a shorter shelf life. Lecithin contains tocopherols by nature, and it may scavenge metal and alkali ions and absorb moisture. Lecithin shows synergistic effects with γ- and Δ-tocopherols, but not with α-tocopherols when the tested oil was rich in linoleic acid (Judde et al., 2003). All these attributes facilitate use of lecithin in improving refined oil shelf life. In addition, lecithin emulsifies natural and synthetic antioxidant mixtures efficiently in refined oils.

Excipient

Soy lecithin fractions with PC greater than 80%, and preferably greater than 90%, are used as excipients in pharmaceutical drug preparation and cosmetic products. The PC isolates are produced under good manufacturing practice (GMP) conditions and meet pharmaceutical purity criteria. In addition to PC fractions with the natural fatty acid composition, PC products with a fully hydrogenated fatty acid profile are approved and used. These phospholipids facilitate the emulsification of drug recipes and enhance the efficacy of drug supplementation, targeted drug transport, and release. PC forms liposomal bylayer structures in unilamellar and multilamellar vesicles. The liposome stability increases with the PC purity content and the degree of stretching of hydrogenated saturated fatty acids (Fricker et al., 2010).

Nutritional Properties

Soy and egg phospholipids are used as a source for choline and essential fatty acids. Oil-free soy lecithin powders and PC fractions are used in various regions as health food supplements. These uses require further substantiation of health effects in clinical trials.

Choline Supply

The Food and Nutrition Board of the Institute of Medicine (United States) published an authoritative report with recommendations for the intake of choline. The recommended daily intake (RDI) for choline is 550 mg for men and 425 mg for women, with a tolerable upper intake level (UL) of 3500 mg. PC is an excellent biological source for choline (Cho and Zeisel, 2006). In 2001 the U.S. Food and Drug Administration (FDA) accepted the choline claim, which can be used by food processors in labeling the required amount of choline per serving. In Europe, the choline claim is not yet approved.

Omega Fatty Acid Supply

Animal and vegetable oils with a high content of essential omega fatty acids are often also potential sources for phospholipids, particularly PC, which has a similar beneficial omega fatty acid profile.

Egg phospholipids contain choline and DHA and arachidonic acid (ARA) fatty acids and are used in infant foods. Egg phospholipid isolates with greater than 95% PC content are used in clinical nutrition as emulsifiers in parenteral fat emulsions for vein injection and enteral fat emulsions, which are important in foods.

Soy phosphatidylcholine fractions (linoleic acid, C18:2; linolenic acid, C18:3) are used in infant nutrition, dietary supplements, and health-enhancing food.

Marine lecithins such as fish roe (DHA C22:6, EPA C20:5) and krill oil (DHA, EPA) are new suppliers of these omega fatty acids. Clinical studies are ongoing, trying to answer the question of whether fatty acids in the polar phospholipid structure are absorbed more efficiently than the same fatty acids bound in the triglyceride molecule structure.

Biological Effects

According to the Dietary Supplement and Health and Education Act (DSHEA, 1994) lecithin is an essential bio-available nutrient that improves heart health and reduces cholesterol, boosts memory and physical performance, is important in reproduction and growth, and acts as a choline supplement. The listed properties justify the continued search for clinical evidence regarding the detailed effects of phospholipids on human health.

Milk lecithin contains phosphatidylserine, a phospholipid with positive signalled effects on cognitive performance and gut health. Further clinical research data are required for substantiating biological effects.

Health Claim Regulation

Documented dossiers with results of state-of-the-art clinical trials are required for the application of approved health claims of specific lecithin and phospholipid fractions in foods. National and international regulatory safety agencies investigate the dossiers in detail before approval by the regulatory authorities will be given.

These requirements are strict and are widened to the dietary and health supplement sectors. In Europe, the European Food Safety Authority (EFSA) has not approved submitted dossiers yet. In the United States, the FDA generally recognized as safe (GRAS) status of various lecithins and choline may support the marketing of health food supplements.

Analytical Aspects

Chemical analyses of lecithins control the quality during processing, trade specification, and supplier–user agreements. The methods were developed for soy, canola, and sunflower lecithins and may not always be applicable for other lecithin sources. Table 9.D lists the most used and validated analytical methods approved by AOCS and DGF (German Society for Fat Science).

Chemical Analysis

A number of methods are routinely used for determining the specifications during lecithin production, trading, and incoming quality control.

- *Acetone Insoluble (AI).* The amount of acetone-insoluble matter (%AI) is the approximate indication for the amount of phospholipids, glycolipids, and carbohydrates present, while the oil and fatty acids dissolve in acetone. It is a rough indication for the "active substance."
- *Hexane Insoluble (HI) or Toluene Insoluble (TI).* This method measures the content of impurities (residual fiber, protein, and dirt). The level of HI/TI matter in crude lecithin should not exceed 0.3%, and today it rarely exceeds 0.1%. In

Table 9.D Survey of Approved Analytical Methods

Method	AOCS	DGF	Method	AOCS	DGF
Acetone Insoluble	Ja 4-46	F-I 5	Viscosity (rotation)	Ja 10-87	F-I 2a
Toluene Insoluble	—	F-I 4b	Viscosity (bubble)	Ja 11-87	—
Hexane Insoluble	Ja 3-87	—	TLC	Ja 7-86	—
Moisture KF	Ja 2b-87	F-I 4	HPLC-UV	Ja 7b-91	—
Acid value	Ja 6-55	F-I 3	HPLC-LSD	Ja 7c-07	F-I-6a
Peroxide value	Ja 8-87	F-I 3b	HPTLC	—	F-I-6
Gardner color	Ja 9-87	C-IV 4a	Iron	Ca 15-75	F-I 4a

international trade, the AOCS method using hexane as an analytical solvent is usually applied, whereas in Europe, regulations advise using the toluene as solvent (DGF method). The porosity of the used glass filters is of larger importance on the result than the type of solvent used.

- *Acid value (AV).* The AV expresses the acidity in mg KOH/g of the sample. The AV represents the acidity contributed by phospholipids (in vegetable lecithin it is often 18–24 mg KOH/g) and (added) free fatty acids.
- *Moisture.* The water content of lecithin products is usually less than 1.0%. Moisture is determined by the Karl Fisher method. A less accurate moisture level can also be determined by azeotropic toluene distillation or drying in an oven at 105 °C.
- *Color.* By convention, the amber color tones of lecithin are measured on the Gardner color scale or the Iodine colour scale. The color range of most clear lecithins is generally in the range of Gardner values 11–17 in the undiluted products. If the lecithin is turbid, color can be measured in a filtered dilution in hexane or toluene, usually in a 1:10 ratio, a condition that should be reported in the analysis certificate.
- *Peroxide value (POV).* The POV of lecithin, produced from fresh or optimal stored oilseeds, is usually below 2 meq/kg. The POV in lecithin is mostly a result of residual hydrogen peroxide from the bleaching process step. This in contrast to POV values in oil as a result of oxidation.
- *Consistency.* Vegetable lecithins are available in fluid, paste-like, and plastic (solid) forms. Liquid lecithins show non-Newtonian flow characteristics. The viscosity is not a constant ratio of shear stress to shear rate. The viscosity profile of lecithins is a complex function of acetone-insoluble content, moisture, mineral content, and acid value. High AI and moisture content yield higher viscosity, whereas an increased acid value decreases viscosity. The sample preparation for the rotation viscosity method eliminates some pseudo-plastic effects. The measured result is a useful indication for flow properties of lecithin during pumping and in dosing equipment.

- *Clarity.* High levels of toluene-insoluble (TI) or hexane-insoluble matter (HI) may partition with the lecithin gums on separation from the oil. This lipid-insoluble material can cause haziness in fluid lecithins. Intensive oil filtration prior to degumming will improve clarity and transparency.
- *Microbiology.* Lecithins with low microbiological counts are required for food and pharmaceutical uses. The lecithin production should be carried out in closed equipment, complying with state-of-the-art quality insurance (QA) and Hazard Analysis and Critical Control Points (HACCP) standards. With GMP production, it is possible to produce food-grade lecithins. A low addition (e.g., 0.1%) of hydrogen peroxide 35% solution to the gums may further reduce the total microbiological count.

Phospholipid Composition

Determination of the various phospholipids of a lecithin is an important tool for supporting effective degumming in the oil mills and refineries, understanding emulsifying functions of lecithin, and recording nutritional phospholipid composition.

- *TLC.* Qualitative thin layer chromatography (TLC) is a useful, simple, and fast method for determining the four main phospholipids PC, PE, PI, PA and their lysophospholipid derivates LPC, LPE, LPI, LPA. The intensity and size of the colored spots on the chromatogram can be used for semiquantitative determination.
- *HPTLC.* More exact data are obtained using high performance thin layer chromatography (HPTLC) methods with good repeatability within qualified laboratories. New analytical apparatus allow precise HPTLC determination of all phospholipid and glycolipid classes in vegetable lecithins in a reproducible manner.
- *HPLC-LSD.* High performance liquid chromatography with a light scattering detector (HPLC-LSD) is a common method in measuring six phospholipid classes in standard and fractionated lecithins. This method with good reproducibility is preferred over HPLC with a UV detector. The validated method has been published as the official unified DGF method and the AOCS approved method.
- *31P-NMR.* The ultimate, most sophisticated method for PL analysis is the 31P-Nucleic Magnetic Resonance (31P-NMR) method, which quantitatively analyzes 15 lysophospholipid classes with phosphorus as the absolute reference (Diehl, 2001). Typical 31P-NMR analyses of commercially produced lecithins from canola, soybeans, and sunflower kernels show slight differences in phospholipid composition, which may be caused by seed origin and variations in oilseed processing. PL profiles of new sources of lecithins from milk, fish, krill, and algae have recently been assessed or are under development, and novel food and health claim regulations require state-of-the-art validated procedures of analysis.

Regulatory Specification

According to most legislative definitions for use as food ingredients, lecithins are mixtures of phosphatides (phospholipids) derived from vegetable and animal origins. In the United States, lecithins are classified in the Food Chemical Codex (FCC III). Under Title 21, part 184 of the Code of Federal Regulations, the FDA has regulated the use of lecithin as generally recognized as safe (GRAS) food substances. Standard lecithin, fractions, and acetylated lecithin are listed in §184.1400, enzyme-modified lecithin in §184.1063, and hydroxylated lecithin in §172.814. Table 9.E lists the various purity specifications of food-grade lecithin.

The EU-approved food additive number E322 covers all standard, physical fractionated, and enzymatically hydrolyzed lecithins for use in the food industry. Hydroxylated and acetylated lecithins do not have an EU-approved food additive status.

EU legal specifications are similar or close to the Codex Alimentarius recommended specifications and methods of analysis. The Codex numbers are recommendations to nations for worldwide use; however, national laws are binding.

Regulations for use in health foods, supplements, or pharmaceutical products may require further specific purity specifications.

Table 9.E Legal Purity Specifications of Food-Grade Lecithin

Purity	FAO/WHO Codex Alimentarius	European Union E322	Food Chemical Codex
Acetone insoluble %	>60	>60 Hydrolyzed > 56	>50
Hexane insoluble %	—	—	<0.3
Toluene insoluble %	<0.3	<0.3	<0.3
Moisture %	—	—	<1.5
Drying loss %	<2.0	<2.0	<1.5
Acid value mgKOH/g	<36	<35 Hydrolyzed < 45	<36
Peroxide value meq/Kg	<10	<10	<100
Arsenic ppm	<3	<3	—
Lead ppm	<10	<5	<1
Mercury ppm	—	<1	—
Heavy metals (as Pb) ppm	<40	<10	—

Processing Technologies

Canola Lecithin Production

Lecithin recovery from canola seeds starts with seed treatment, followed by expelling (pressing) to squeeze out seed oil to lower the content from 42% down to 15–20% in the press cake. Then, the grinded cake is usually hexane-extracted for further removing the oil to below 0.5% (Figure 9.3). The press oil and extraction oil streams are combined for degumming, in which the lecithin gums are separated in continuous decanters. Well-filtered lecithins with a transparent glossy appearance have an HI value far below 0.1% and iron content less than 50 ppm. However, in a large-scale plant operation, it is not possible to effectively filter high viscous turbid lecithin to achieve a visually clear lecithin.

Canola contains chlorophyll components, giving canola lecithin a greenish-brown tone. The natural tocopherol content is about 400 ppm, consisting of 100 ppm of α-tocopherols and 200–300 ppm of γ-tocopherols. Canola lecithin yields are about 0.2–0.3% on seed basis, compared to 0.5% soy lecithin on a bean basis.

Nowadays, canola lecithin is produced in a number of Canadian, European, and Chinese crushing plants. In principle, it is possible to process lecithin with good phospholipid composition from expelled seed oils only, but their availability is small. Lecithin processing is performed in four subsequent unit operations that are extensively described in a review publication (van Nieuwenhuyzen and Tomás, 2008).

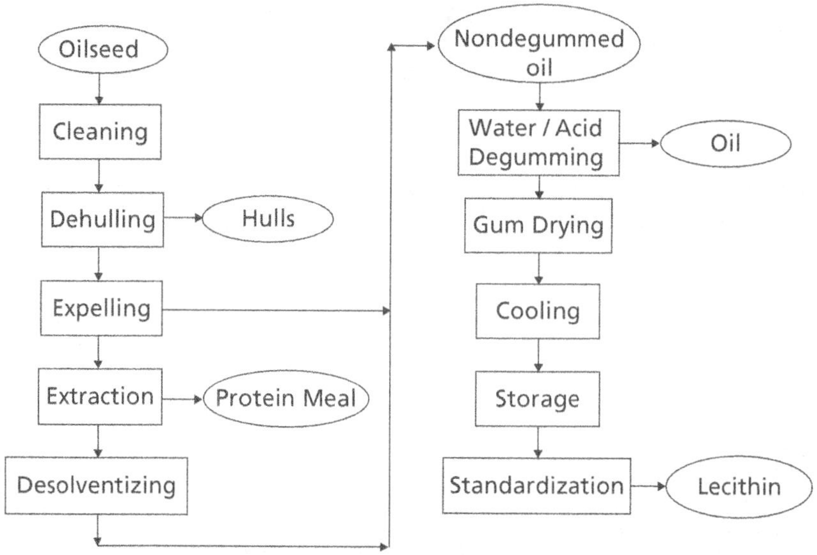

Figure 9.3 Canola lecithin recovery and processing.

Hydration of Phospholipids

The easily hydratable phospholipids are PC, PI, and LPC. PE and PA have low hydrating properties, and are therefore marked as nonhydratable phospholipids (NHP). In practice, a mix of these phospholipids is separated in the lecithin gum. In the degumming process, the polar phospholipids hydrate within 1 hour from a gum with a higher specific density than the oil.

Separation of the Lecithin Gums

Gums often contain a maximum of 50% water, a minimum of 33% acetone insolubles, and a maximum of 17% oil. The gums are separated from the oil in continuously operating centrifuges at 50–70 °C. If gums are not dried to lecithins, they are often sprayed on the meal in the oil mill.

Drying

To achieve long shelf life and fluidity, lecithin is dried to a content of less than 1% moisture. Film evaporators have the advantage of a high-performance capacity per unit drying surface and a short drying time of 1–2 minutes, which is adequate for achieving a good color quality. Therefore, continuous drying film evaporators are used for the drying of the lecithin gums batch and semi-batch. Long batch drying times cause severe darkening due to the Maillard reaction of the adherent sugars and the Amadori reaction between sugars and PE.

Cooling

Rapid cooling of the lecithin to below 50 °C is effective to prevent post-darkening by Maillard reactions. Bulk storage for lecithin at around 40 °C under dry conditions is recommended, using adequate tanks with stirring facilities for homogeneity.

Canola lecithin can be standardized by adjusting the AI content to 62–64% by oil and fatty acid addition, giving a liquid consistency at room temperature.

Research in the 1960s and 1970s with rapeseed lecithin gums showed the good potential for modification of rapeseed lecithin by hydrolysis and alcohol fractionation as a functional emulsifier in food (Pardun, 1988). Crude canola lecithin is an attractive material for acetone-deoiling, alcohol fractionation, and purification by super critical carbon dioxide extraction (Temelli and Dunford, 1995). Full acetylation of the ethanol-insoluble fraction gives products with improved emulsifying properties (Sosada et al, 2003).

Table 9.F gives the typical total composition of canola lecithin in comparison to soy lecithin and sunflower lecithin. The phospholipid compositions do not differ very much, which is also due to the quite similar oil extraction processes. Hence, the fatty acid pattern largely follows the fatty acid composition of the triglycerides as is shown in Table 9.G. This fatty acid division is measured in the liquid lecithin, consisting of phospholipids and triglycerides.

Table 9.F Phospholipid Composition of Vegetable Lecithins

	Canola Lecithin (%)	Soy Lecithin (%)	Sunflower Lecithin (%)
Acetone-insoluble matter			
Phospholipids			
Phosphatidylcholine	17	15	16
Phosphatidylethanolamine	9	11	8
Phosphatidylinositol	10	10	14
Phosphatidic acid	4	4	3
Other phospholipids	6	7	6
Subtotal: All phospholipids	46	47	47
Glycolipids	11	11	11
Complex carbohydrates	4	4	4
Total: Acetone-insoluble matter	62	62	63
Acetone-soluble matter			
Oil + added fatty acid	37	37	36
Moisture	<1	<1	<1
Total	100	100	100

Table 9.G Fatty Acid Composition of Vegetable Lecithins

Fatty Acid	Canola Lecithin%	Soy Lecithin%	Sunflower Lecithin%
C16:0	7	16	11
C18:0	1	4	4
C18:1	56	17	18
C18:2	25	55	63
C18:3	6	7	0
Others	5	1	4

Sunflower lecithin is produced in a similar way as canola lecithin, combining the pressed oil and extracted oil streams for the degumming. In addition to the monograph by Pardun (1988), production and modification processes are also reported by Hollo et al. (1993). High quality sunflower lecithins are on the market.

Soy lecithin is recovered from the hexane-extracted oil because the 18% oil in the bean does not justify economic oil expelling prior to solvent extraction in large scale plants.

Modification and Isolation of Vegetable Phospholipids

Modified lecithins have dedicated emulsifying properties due to the increased hydrophilicity. The process options are: (1) enzymatic and chemical adaptation of the phospholipid molecules, (2) physical fractionation for separating oil from the phospholipids, and (3) fractionation of phospholipids. This knowledge from decades (1960s–1990s) of research and product development has been updated by new experiments and analytical procedures at Unilever Research Vlaardingen, which was one of the birth places for patented lecithin innovations (Doing and Diks, 2003). In the past, most commercial production units sourced soy lecithin as raw material. Hence, canola lecithin and sunflower lecithin are sound alternative sources. The modification principles can be applied for all three sources of lecithin.

Modification by Enzymes

The availability of pure phospholipase A_2, A_1, and lipase enzymes enables precise degrees of hydrolysis of the various PLs. Enzymatically hydrolyzed lecithins have technological and commercial benefits because they are excellent O/W emulsifiers. The enzyme, phospholipase A_2, specifically hydrolyzes the fatty acid at the β position of the PL molecule. Lecithins with varying degrees of hydrolysis of lysophospholipids and dedicated emulsifying properties are processed. The modifications are essential for achieving and adjusting optimal ratios between hydrophilic and lipophilic properties and for ensuring optimal surfactant properties.

This knowledge of enzymatic hydrolysis was also the incentive for starting enzymatic degumming of crude oil, turning the nonhydratable PLs into hydratable lysophospholipids. First, enzymatic degumming plants started with canola oil (Buchold et al., 1994). In addition to the liquid hydrolyzed lecithins, oil-free hydrolyzed lecithin powder and special spray-dried products used as carriers are also available.

The phospholipase enzyme technologies can also be used in the synthesis of triglycerides and PLs with special structure fatty acid compositions (Vikbjerg et al., 2007; Xu et al., 2008).

Modification by Chemicals

Acetylation

The amino group in PE reacts with acetic anhydride, resulting in acetyl-PE that enhances lecithin O/W emulsifying properties. The principle of the process is that the zwitterionic group of the PE is blocked.

$$-NH2 \rightarrow NH\text{–}COCH_3$$

Acetylation can be performed starting from lecithin gums or with dried lecithin. Under plant-scale conditions, the acetylating reaction with crude lecithin often gives partially acetylated PE. To overcome this, the lecithin is treated with 1.5–5% acetic anhydride at 50–60 °C. Acetylated lecithin has better resistance to browning when heated because fewer amine groups in the phosphatidylethanolamine are available.

Hydroxylation

Hydroxylation of the double bonds in the unsaturated fatty acids of the PLs is done in the presence of peroxide and organic acids, resulting in the highest possible lecithin hydrophilicity.

$$-CH{=}CH- \rightarrow -CH(OOH)-CH(OOH)- \rightarrow -CH(OH)-CH(OH)-$$

In this process, the crude lecithin is mixed with 2–14% hydrogen peroxide and 1% lactic acid at 50–60 °C. The lactic acid reduces the pH and facilitates the reaction. During the reaction, the colored components are bleached, which yields a light-colored lecithin. Hydroxylated lecithin is superior for O/W emulsions. It easily disperses in cold water as a whitish emulsion. This is a good indication that the HLB value increases significantly.

Acetylation and Hydroxylation

The amino group of PE has a negative influence on the hydroxylation process. Therefore, it may be appropriate to use acetylated lecithin as a raw material for hydroxylation. These combined modifications yield a lecithin with very hydrophilic properties. The HLB value is reported to reach 11 (Bonekamp, 2008).

Hydrolysis and Hydroxylation

Enzymatically hydrolyzed lecithin is also used as a raw material for producing hydroxylated lecithins. The hydroxylation process conditions are quite similar to using standard lecithin. This results in lecithin with slightly more hydrophilic properties.

Hydrogenation

Vegetable lecithins contain unsaturated fatty acids in the phospholipid molecules. Hydrogenation of the unsaturated fatty acids into saturated fatty acids is done by using catalysts such as palladium.

$$-CH{=}CH- \rightarrow -CH_2-CH_2-$$

The hydrogenated products have higher melting points and are stable against oxidation, but they are less soluble in oils and fats. The process is preferably carried out with PC fractions that are quite pure. The catalysts are poisoned by the impurities if crude lecithin is used and the reaction is not very well controlled.

Fractionation for Oil Removal

Phospholipids possess polar hydrophilic groups by which they can be separated from the apolar triglycerides. Membrane filtration of lecithin-in-hexane miscella is feasible by separating the small triacylglycerol molecules (permeate) from the AI phospholipids (retentate). The miscella should be well filtered and without impurities, which may foul the membranes. Super critical carbon dioxide extraction is also an attractive technology for the production of lecithin powders. The up-scaling of this technique for large continuous production needs further improvement to justify investments.

Acetone Extraction

Use of acetone in the solvent extraction process is still the robust technology for producing oil-free lecithin powders and granules. Triglycerides dissolve in acetone in contrast to the other more polar components of standard lecithin. Deoiling with acetone is executed as an efficient continuous process, in which the crude lecithin is mixed and agitated with acetone. The phospholipids, glycolipids, and adherent carbohydrates precipitate as sediments, which are centrifuged and/or filtered. A careful drying process is required to eliminate the residual acetone. Deoiled lecithins have only 2–5% remaining oil and are marketed with an AI matter of minimum 95%. Low oil content is beneficial for excellent free-flowing properties of the powders or granules. The classification between powders and granules is often made by sieving. The process can also be used for the deoiling of enzyme and chemically modified lecithins and lecithin fractions.

Fractionation of Phospholipids

Phospholipids have different loading and solubility properties in solvents, so aqueous alcoholic solvents can be used alone or in conjunction with chromatography for separation.

Alcohol Fractionation

PC dissolves better and faster in alcohols in comparision to other PLs present in crude lecithin. Therefore, the phospholipid mixture in crude lecithin can be fractionated into the alcohol-soluble and alcohol-insoluble fractions. Ethanol-soluble fractions contain a high PC:PE ratio of often 5:1, which is close to the PC:PE ratio of egg lecithin. Consequently, the insoluble fraction has a low PC:PE ratio. Prior to the alcohol evaporation, mixtures of refined oil and monodiglycerides or alkali may be added for producing stable liquid PC-enriched fractions.

Chromatographic Isolation

PC products can be produced in pure form by using column chromatography with an adsorbent, such as aluminium oxide or silica gel. On a commercial plant scale, oil-free lecithin or its ethanol-soluble PC fraction is subjected onto a chromatographic column with aluminium oxide or silica gel adsorbents (Günther, 1984). The PC from soy lecithin is isolated up to 70–95% purity. Canola lecithin is not yet used as a raw material.

Recovery of Phospholipids from Fish, Krill, Milk, and Egg

Egg Lecithin

Egg yolk is a complex combination of phospholipid–protein interactions. Fresh egg yolk and dried egg yolk powder are used as emulsifiers and stabilizers in many food products. Phospholipase A_2 treated egg yolk forms lysophospholipids, which support the heat stability of mayonnaise and salad dressing (Dutilh, 1988). Although these foods have high oil contents, the products are structured O/W emulsions.

Extraction of the phospholipids from liquid egg yolk or egg yolk powder is carried out with help of solvents, usually ethanol and/or acetone. The block diagram in Figure 9.4 shows the processing alternatives. One may extract at first the oil with acetone, followed by PL fractionation with ethanol. Or it can be done in the ranking

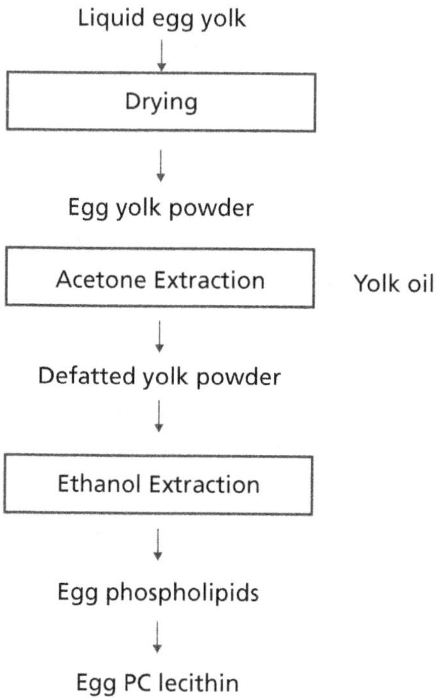

Figure 9.4 Egg lecithin processing.

ethanol, followed by acetone deoiling. Solvent fractionation of egg lecithin gives a product with a minimum of 95% AI matter. The phospholipid part consists of 74% PC, 17% PE, and 9% various minor phospholipids (Palacio and Wang, 2005). For use in parenteral fat emulsions, the absence of any residual protein is a must and the target may be greater than 95% PC, which requires a PC concentration step.

Milk

Dry ultra-filtered butter serum contains 20% phospholipids. Further purification of the PL concentrates requires ethanol and acetone extraction. Membrane filtration techniques are investigated to separate the phospholipids from the milk proteins (Rombaut and Dewettinck, 2006).

Fish

Fish roe was often considered a waste product of the fish processing industry. In the last decade, research on the phospholipid extraction from fish roe shows interesting products with high PC and DHA/EPA content. Solvent extraction with ethanol and/or acetone is required for the production of these novel high-value lecithins.

Krill

After fishing, krill is immediately pressed to krill meal and frozen. The krill oil is recovered by alcohol extraction. Krill oil is a natural mixture of triglycerides and phospholipids, predominantly PC, with a high content of EPA, DHA, and astaxanthin.

Indicative phospholipid composition of these lecithins is given in Table 9.H, whereby different methods of PL analysis may have been used.

Table 9.H Phospholipid Composition of Animal and Marine Lecithins

Phospholipids	Egg[a] (%)	Milk[b] (%)	Krill[c] (%)	Cod Roe[d] (%)
Phosphatidylcholine	74	26	89	45
Phosphatidylethanolamine	17	32	3	35
Phosphatidylinositol	1	5	<2	16
Phosphatidylserine	—	—	<1	4
Sphingomyelin	—	23	—	—
Other phospholipids	3	14	7	—

[a]Palacios and Wang (2005).

[b]Rombaut and Dewettinck (2006).

[c]Winther et al. (2011).

[d]Lovaas (2000).

Applications

Soy lecithin has been tested in many applications in the last century. The fast growing soybean crops in the United States prompted further research, product development, and applications beginning in the 1940s. Lecithin was and is relatively cheap and is potentially available in large quantities. The functional properties of soy and egg lecithins in foods, pharmaceuticals, and cosmetics are reviewed by Hernandez and Quezada (2008).

Food

The uses of lecithins in food are many; the origin of use was often egg lecithin in the form of egg or egg yolk. Egg yolks were often expensive or not always available in the world market, which was the incentive to investigate and develop soy lecithins with effective performance. This was followed by lecithins from canola and sunflower seed. In principle, the surface-active lecithins may have technological functions in all types of emulsions and suspensions, but the effective use also depends on the price:performance ratio in comparison to other emulsifiers and surfactants.

Margarine

Margarines are W/O emulsions with 20% water or milk homogenized in the continuous 80% fat phase. Lecithin acts as a co-emulsifier with unique antispattering properties. In salted margarine, the shallow frying properties without any spattering of fat are obtained through the use of 0.3–0.8% standard lecithin. A film of phospholipids surrounds the water droplets and prevents undesired coalescence. Antispattering means that the stabilized water droplets are transported gently to the surface, whereas a slow evaporation of the water takes place without fat spattering. In some European countries, consumers prefer salt-free or low-salt margarine. These recipes require 0.1–0.3% PC-enriched fractions with PC:PE ratios over 4.2 or enzymatic hydrolyzed lecithin as excellent performing antispattering agents (Hooft et al., 2005).

For convenience and reduction of fat consumption, numerous low-fat spreads are offered to consumers for table use. Although yellow fat spreads with less than 40% fat are not used for frying, the lecithin and its fractions can be used as co-emulsifiers together with monoglycerides for stabilization of the spread with a pleasant melting profile in the mouth.

Baked Goods

Baked goods are a tremendous food segment worldwide. Lecithin, or lecithin in combination with other emulsifiers, is used in many bakery ingredients and recipes. Some examples are given in this section.

Bread

Bread improvers containing mixes of lecithins, synthetic emulsifiers, ascorbic acid, and enzymes are added to wheat and grain flowers in bakeries. Volume and freshness are important quality parameters in yeast-leavened products such as bread. Lecithins, particularly hydrolyzed lecithin, enhance both properties. The phospholipids are physically linked with the wheat gluten by H-bridges as a sort of lipoprotein, by which elasticity and baking volume are enlarged and the fermentation tolerance is improved. Good complexation of hydrolyzed phospholipids into the amylose a-helix retards recrystallization of amylose and thus enhances anti-staling, crumb freshness, and shelf life properties. The hydrolyzed, hydroxylated, and acetylated lecithins may match or even exceed the functional properties of esters of monoglycerides such as diacetyl tartaric acid (DATA), glycerol lacto palmitate (GLP), and polysorbates.

Pretzels

In pretzel recipes, lecithin acts as lubricating agent and improves the processing speed. The addition of lecithin in soft pretzel recipes reduces stickiness and improves the subsequent machineability. The quality of pretzels is improved with a glossy product surface, resulting in higher consumer approval.

Cookies

Lecithin in cookie recipes supports the good dispersion of the shortening during dough preparation. For production of light cookies, a 30% fat reduction is possible with lecithin. The reduced-fat recipe without lecithin results in sticky dough, but the addition of lecithin up to 0.5% reduces the stickiness, improves dough smoothness, and ensures optimal dough handling.

Release Agents

Special refined liquid lecithins in combination with waxes and vegetable oils are used in release agents for frying. Pan-release sprays for kitchen use are established products on the supermarket shelf. The hydrophilic acetylated, hydroxylated, and fractionated lecithins with improved heat resistance properties are often used in industrial food production. Lecithin can be added to oil to prevent food from sticking to oven belts, molds, meat casings, and cooking surfaces. A typical pump spray formulation consists of 25% medium-chain triglyceride (MCT) oil, 65% vegetable oil, and 10% hydrophilic lecithin. In the case of meat casings, for example, an aqueous lecithin solution is used in the belt container.

Chocolate

In chocolate, lecithin facilitates the lubrication of the sugar and cocoa particles, and possibly fat crystals, thereby making the flow of the chocolate mass more liquid. The lecithin functions are viscosity reduction, good flow during tempering, and good flow performance during molding or enrobing. Technically, chocolate is a suspension with hydrophilic sugar and lipophilic cocoa particles dispersed in the continuous fat phase of 30–34% cocoa butter. After mixing and refining (i.e., grinding to particle size less than 25 m), the flaked ingredients are conched (rolled) into a paste-like chocolate mass. At the end of the conch process, 0.3–0.5% lecithin is added. Flow properties of chocolate are measured with rotary viscometers, and the plastic viscosity and yield value are calculated with the standard Casson equation according to the official International Cocoa Organization (ICCO) analytical procedure. Plastic viscosity expresses the flow behavior when the chocolate mass is pumped through a pipeline. The yield value of the non-Newtonian chocolate is an indication of the spontaneous spread of the mass during enrobing and molding. In the smooth chocolate production process, a good standard lecithin with constant phospholipid composition is desired. In chocolate enrobing recipes with an extremely low yield value requirement, mixtures of lecithin with polyglycerolpolyricineolate (PGPR) are used.

Agglomerated Instant Powders

Lecithins are used as wetting and dispersing agents in instant and infant foods. First, the PLs reduce the surface tension in the free fat–water interface, and second, the interactions between the lecithin and the proteins are formed. When dispersed in water or milk, lecithin promotes controlled hydration of the powders. Therefore, in the production of agglomerated whole milk powder (WMP) and fat-filled milk powder (FFMP), low amounts (<0.05–0.2%) of lecithin are applied. After spray drying, the warm powder is lecithinated by spraying lecithin dispersion on the powder during the agglomeration stage in a fluid bed unit. Depending on the legal and technological opportunities, deoiled lecithin/butter fat, liquid lecithins with adjusted PL composition, or hydrophilic lecithin/water dispersions are applied.

Agglomerated particles have larger sizes than the original powder and improved free-flowing properties, dispersion, and shelf life properties. Lecithins also improve the heat stability of reconstituted milk.

Egg lecithin containing ARA and DHA fatty acids is applied in infant milk as a replacement for human breast milk. Egg phospholipids show good bioavailability of ARA and DHA, are stable against oxidation, and protect the mucosa of the intestinal tract against the adhesion of pathogenic bacteria, reducing the risk of severe diarrhea.

Lecithin improves the wetting performance of the particles in instant cocoa powders, promoting convenient dispersion in water or milk. The lipophilic fatty acid chains of the PLs are dissolved in the cocoa butter, and the hydrophilic part has affinity to the water or milk. The PLs prevent fat exudation to the surface of the capillaries in the cocoa powder agglomerate. Technologically, it is possible to incorporate up to 5% lecithin in the hot cocoa cake just after pressing. Later on, these powders are mixed with sugars and other ingredients to produce ready-mixed instant cocoa drinks. In other processes, standard cocoa powder, sugar, and other ingredients are agglomerated with spraying up to 1.5% lecithin onto the mix.

Liposome Encapsulation

Liposomes are unilamellar and multilamellar vesicles with internal water cores suitable for the inclusion of hydrophilic components. The liposome technique can be applied for flavor encapsulation. There are at least two reasons why baked goods, especially microwave-baked low-fat products, release flavor. First, fat contributes to good flavor release, so fat reduction might give sensory discomfort. Second, volatile flavors disappear during microwave treatment. The use of encapsulated flavors in liposomes made with PC may contribute to good sensory quality in these types of foods. The technology can also be applied in the joint encapsulation of hydrophilic and lipophilic antioxidant mixtures, demonstrating improved efficiency and stability against oxidation (Ghyczy, 1994). Table 9.I summarizes the important food uses of lecithin.

Table 9.I Survey of Lecithin Application in Selected Foods

Application	Functionality	Types of Lecithin[a]
Baked goods	Volume improvement, fat dispersion, anti-staling, firmness, freshness	Standard, hydrolyzed, deoiled
Chocolate	Rheology, yield value, viscosity modification	Standard
Chewing gum base	Rheology, tackiness, brittleness	Standard, hydrolyzed
Instant drinks dairy/cocoa	Agglomeration, wetting, dispersibility	Standard, hydrolyzed, PC fraction, deoiled
Milk protein and replacer	Emulsification, emulsion stabilization, wetting, anti-dusting	Standard, hydrolyzed, PC fraction, deoiled
Margarine	Anti-spattering in frying, emulsification, mouth feel	Standard, hydrolyzed, PC fraction
Flavor	Liposome encapsulation	PC fraction
Pan release agent	Wetting, separation, machineability	Standard, hydrolyzed, acetylated

Note: Standard: standard liquid lecithin; PC fraction: phosphatidylcholine enriched fraction.

[a]Choice of lecithin is based on European Food additive legislation. Acetylated and hydroxylated lecithins may be applicable in other continents.

Feed

Lecithins are used in animal feed recipes as economic emulsifiers, instantizing agents, choline supplements, and essential fatty acid sources.

Milk Replacers

Lecithins are used for stabilization of milk replacers, either alone or in combination with other emulsifiers, for feeding calves, piglets, and other young animals. The technologies that are used are quite identical to the spray-drying processes the of the dairy industry. Milk-replacing powders consist of vegetable fat mixtures and whey and soy proteins, and they replace the more expensive milk ingredients. The artificial milk is prepared daily and should be stable without fat creaming or protein sedimentation until feeding. The emulsion stability is sometimes affected, giving undesired protein sedimentation and fat creaming, negatively influencing feed intake, resorption, and growth. Lecithins may reconstitute the emulsion performance.

Fish Feed

Fish farming, or aquaculture, is a fast developing agricultural industry worldwide. PLs are essential nutrients both for larvae and for shrimps. In particular, feed pellets for cultured shrimps (tiger prawn, kuruma prawn, Chinese white shrimp) require lecithin for pellet floating properties and nutritional benefits. PLs are actively involved in the absorption of dietary fat and fat-soluble vitamins, and are essential in the intermediary metabolism and fatty acid metabolism. Therefore, the deoiled lecithin powders and liquid lecithins are used in recipes, improving the settling and density of the feed granules.

Poultry Feed

Vegetable lecithins containing choline and high amounts of essential fatty acids are used in poultry feed. PLs act as metal complexing agents, enhancing antioxidant properties. Also, the organic bound choline optimizes feed quality. Vegetable lecithins contain choline and high amounts of essential fatty acids with a metabolic energy content of about 8 kcal (equivalent to 34 joule per gram). Linoleic acid (C18:2) has a vital function in preventing adipose deposits in broiler chickens, preventing fatty liver syndrome.

Pharmaceuticals

Soy lecithin and synthetic PLs are used as excipients for targeted drug delivery, encapsulation of drug molecules, and cell membrane passage.

Liposome encapsulation of high-value functional ingredients is an innovative tool for achieving optimal targeted performance at the lowest concentrations. Liposomes enable the encapsulation of combined oil-soluble plus water-soluble drugs and controlled drug transfer and release (Fricker et al., 2010). Liposomes made with PLs vary in size (50 nm to 300 nm) and are unilamellar and multilamellar vesicles. The liposome vesicles have an internal water core suitable for the inclusion of hydrophilic components. PC can form the desired structured bilayers. Sometimes the vesicles leak the encapsulated agent due to the configuration of the polyunsaturated fatty acids of the PC molecules. To overcome this, more stable liposomes can be obtained by using PC molecules with hydrogenated saturated fatty acids.

Cosmetics

Cosmetics with advertising slogans that mention "liposomes" and "nanosomes" have been launched on the market. If the cosmetic recipe requires a surfactant with a high HLB value, lecithins should be used in combination with other surfactants. The PC fractions contribute to excellent skin humidity and emollient properties (Hoogevest et al., 2014). It is a technological challenge to achieve the optimal performance of cosmetics ingredients encapsulated in liposomes.

Lecithins containing various PLs, including polyunsaturated fatty acid, have skin softening properties in cosmetic formulations such as creams, lotions, foundations and cleansing creams, sunscreens, soaps, bath oils, shampoos, and hair conditioners. Lecithin can also reduce the undesirable oily feeling in cosmetics containing oils, and it supports a moisturizing effect. PLs penetrate the skin and facilitate the penetration of other essential cosmetic compounds.

Technical Products

Lecithins are evaluated and used in a number of industrial nonfood uses. The surface-active and emulsifying properties will compete with synthetic surfactants on performance, price, and sustainability. Hence, under practical circumstances, synergy can often be reached by using lecithins in combination with surfactants.

Release and Lubrication Agents

Lecithins are used in lubricant formulations in metal and other technical industries. The fatty acid tails of polar PLs stick on the metal surface and form a thin film, while the surface-active heads form a hydrophilic layer. Whereas hydrophilic lecithins are used as dispersing and wetting agents in lipid-based products, the more hydrophobic standard lecithin and PC-depleted lecithin fractions with low HLB values are used for reducing the hydrophilic properties of powdered plastic intermediates. This leads to more convenient handling and fewer dusting properties.

Leather

The production of leather from animal skins involves a series of processes, from the slating of fresh skins to tanning and fatting. The tanning is performed with various vegetable, synthetic, and mineral tanning agents. After the tanning, the skin has to be greased to produce supple leather suitable for manufacturing goods. Fat liquors consist of sulfonated and sulphated oils with polar groups in the triglycerides. Natural lecithins can be added to the fat blend as raw material sources for the sulfonation and sulphating processes. The hydroxylated lecithins are also used because these products fulfill the requirements for a good fatting agent: (1) deep penetration into the skin, (2) fixation in the leather, (3) good lightfastness (resistance), and (4) low residual fat content in the used liquor. In particular, the very hydrophilic combined hydrolyzed and hydroxylated lecithins give deep penetration and good fixation, resulting in leathers with firm grain, good tear, and tensile strength.

Paper Coating

Paper and paperboard are often coated with lecithins to improve qualities such as printability and appearance. A coating with superior uniformity, flow properties, stability, and UV brightness is obtained with a blend of fatty acids and lecithin. The rheological properties of aqueous coatings for rapidly moving webs are enhanced. Inclusion of hydroxylated or acetylated lecithin in the coating recipe supports the requirements for high production speed in paper manufacturing. The coating facilitates excellent properties of tensile strength and UV light absorbency.

Paints

Lecithins function as interfacial agents in paints, lacquers, and printing inks, influencing the wetting, dispersing, suspending, and stabilizing agents in both oil-based and latex/resin emulsion paints. The broad range of functional properties of lecithin makes it highly suitable for many different coating formulations such as paints, waxes, polishes, and wood preservatives. In paints and lacquers, the choice of wetting agents depends on the nature of the pigment, the vehicle, and the processing procedure. As a rule, natural grades of lecithin have been recommended up to 1% on a pigment weight basis. Lecithin products facilitate a pigment dispersion and redispersion and also reduce thixotropic viscosity properties. In water-based paint systems, water-dispersible lecithins are recommended, either use alone or in combination with other surfactants.

Plant Crop Protection

The use of PLs facilitates the transport of agrochemicals in the leaves and stems of plants or in the organs of insects. The modified vegetable lecithins with enhanced hydrophilicity are preferred. PL vesicles have been successfully used as model membrane systems to study permeability, ion transport, fluidity, and other properties of biological membranes. The surfactant function supports the formation of small particle droplet sizes of the pesticide suspension, making efficient transfer through the cell membranes possible. Lower dosages of agrochemicals per spraying period are feasible.

Soil Bioremediation

Soils in different parts of the world are polluted by chemicals of varying persistence and toxicity. Such wastes may spread into the ground and disperse into the ground water. The soil contaminants can be cleaned by enhancing their degradation through the use of micro-organisms. PLs facilitate the cleanup of contaminated soils through a combined effect of enhancing bacterial growth by supplying essential nutrients while modifying the physical states of oil or wastes. Tests were conducted on soils artificially contaminated with polychlorinated biphenyls (PCBs) using deoiled lecithin as a natural surfactant. The PCBs were significantly enhanced at lecithin concentrations in slight excess of the critical micelle concentration (Fava and Gioia, 2001).

These are only a few examples of lecithin use in the nonfood industrial sectors. There are also many other specific uses. Some other examples are antidusting agents on sandy roads, viscosity enhancers in brick production, mosquito control systems, and asphalt rheology.

Future Trends

Lecithin is a co-product of the oilseed milling industry. The availability of large volumes of lecithin gums from seed oil degumming and refining challenges further economic valorization of lecithins as sustainable surface-active agents in many applications in food, feed, cosmetics, pharmaceutical, and technical industries.

European sourcing requirements for identity preserved (IP) non-GMO soy lecithin for the European food market drastically changed the lecithin world market supply system beginning in 1996. Traditional non-GMO soybean availability will become more scarce, which presents a market opportunity for high-quality IP non-GMO canola, soy, and sunflower lecithins.

PC fractions with specific essential omega fatty acid profiles, sourced from milk, fish, krill oil, fungi, and algae, are in the research and development pipeline or have reached the market stage for use as nutritional ingredients and health supplements. Extensive clinical studies are required for proving and substantiating health claims, and dossiers have to be approved by the regulatory safety authorities before placing these dedicated lecithins on the market.

Conclusion

Lecithins are primarily obtained from soybean, canola, and sunflower kernels as co-products from seed oil crushing. The phospholipid modification by enzymatic hydrolysis, solvent fractionation, acetylating, and hydroxylation processes produce lecithins with specific enhanced hydrophilicity and O/W emulsifying properties. The present knowledge of phospholipid chemistry, biology, and technological advancement, combined with precise analytical methods to characterize the composition, emulsion particle sizes, and emulsion stability, support the understanding of the functionality of the various phospholipids. Natural and modified lecithins are used alone or in combination with other surfactants in a large range of technical applications. The products are renewable and have good biodegradability.

Egg yolk was and is an efficient food emulsifier and a nutritional source of choline, phosphatidylcholine, and essential long-chain fatty acids for infant food, parenteral nutrition, and health food supplements. Phosphatidylcholine fractions from soy lecithin fulfill these nutritional requirements to a certain extent.

Milk phospholipids, fish phospholipids, and krill oil phospholipids have been launched on the world market.

References

Bergenståhl, B.; Claesson, P. M. Surface Forces in Emulsions. In *Food Emulsions*; Larsson, K., Friberg, S. E., Eds.; Marcel Dekker: New York, 1990; pp 41–96.

Bonekamp, A. Chemical Modification. In *Phospholipid Technology and Applications*; Gunstone, F. D., Ed.; The Oily Press: Bridgewater, UK, 2008; pp 141–151.

Buchold, H.; Boensch, R.; Schroeppel, J. Process for Enzymatically Degumming Vegetable Oil. European Patent 0654527, 1994.

Cho, E.; Zeisel, S. H. Dietary Choline and Betaine Assessed by Food-frequency Questionnaire in Relation to Plasma Total Homocysteine Concentration in the Framingham Offspring Study. *Am. J. Clin. Nutr.* **2006**, *83,* 905–911.

Diehl, B. K. W. High Resolution NMR Spectroscopy. *Eur. J. Lipid Sci. Technol.* **2001**, *103,* 16–20.

Doig, S. M.; Diks, R. M. M. Toolbox for Modification of the Lecithin Head Group. *Eur. J. Lipid Sci. Technol.* **2003**, *105,* 359–376.

Dutilh, C. E. Heat-sterilizable Water and Oil Emulsion. European Patent 0328789B1, 1988.

Fava, F.; Gioia, D. D. Soya Lecithin Effects on the Aerobic Biodegradation of Polychlorinated Biphenyls in an Artificially Contaminated Soil. *Biotechnology and Bioengineering* **2001**, *72* (2), 177–184.

Fricker, G.; Kromp, T.; Wendel, A.; Blume, A.; Zirkel, J.; Rebmann, H.; Setzer, C.; Quikert, R.-J.; Martin, F.; Müller-Goymann, C. Phospholipids and Lipid-Based Formulations in Oral Drug Delivery. Published online Apr 22, 2010. DOI: 10.1007/s1 1095-010-0130-x.

Ghyczy, M. Liposomes for the Food Industry. *Food Tech Europe* **1994,** *1* (5), 44–46.

Griffin, W. C. Classification of Surface-active Agents by "HLB." *J. Soc. Cosm., Chem.* **1949**, *1,* 311–326.

Günther, B. R. Process for the Separation of Oil and/or Phosphatidylethanolamine from Alcohol Soluble Phosphatidylcholine Products, Containing the Same. U.S. Patent 4,425,276, 1984.

Hernandez, E.; Quezada, N. Use of Phospholipids as Functional Ingredients. In *Phospholipid Technology and Applications*; Gunstone, F. D., Ed.; The Oily Press: Bridgewater, UK, 2008; pp 83–94.

Hollo, J.; Peredi, J.; Ruzics, A.; Jeranek, M.; Erdelyi, A. Sunflower Lecithin and Possibilities for Utilisation. *J. Am. Oil Chem. Soc.* **1993**, *70,* 997–1001.

Hooft, C.; van 't Kommer, M.; van den Segers, J. C. Food Composition Suitable for Shallow Frying Comprising Sunflower Lecithin. European Patent EP1607003, 2005.

Hoogevest, P.; Van Prusseit, B.; Wajda, R. Phospholipids: Natural Functional Ingredients and Actives for Cosmetic Products. *Inform* **2014,** *25* (3), 182–188.

Institute of Medicine, National Academy of Science. Choline. In *Dietary Reference Intakes for Folate, Thiamin, Riboflavin, Niacin, Vitamin B12, Panthothenic Acid, Biotin and Choline;* National Academy Press: Washington, DC, 1998; pp 390–422.

Judde, A.; Villeneuve, P.; Rossignol-Castera, A.; Le Guillou, A. Antioxidant Effect of Soy Lecithin on Vegetable Oil Stability and Their Synergism with Tocopherols. *J. Am. Oil Chem. Soc.* **2003**, *80* (12), 1209–2013.

Lovaas, E. Marine Phospholipids. Presented at AOCS Annual Meeting, 2000.

Palacios, L. E.; Wang, T. Egg/yolk Fractionation and Lecithin Characterization. *J. Am. Oil Chem. Soc.* **2005**, *82,* 571–578.

Pan, L. C.; Tomás, M. C.; Añón, M. C. Oil in Water Emulsions (O/W) Formulated with Sunflower Lecithins: Vesicle Formation and Stability. *J. Am. Oil Chem. Soc.* **2004**, *81,* 241–244.

Pardun, H. *Die Pflanzenlecithine, Verlag für chem.* Industrie H. Ziolkowsky: Augsburg, Germany, 1988; pp 247, 271–272, 286, 296.

Rombaut, R.; Dewettinck, K. Properties, Analysis and Purification of Milk Polar Lipids. *Int. Dairy J.* **2006**, *16,* 1362–1373.

Schneider, M. Major Sources, Composition and Processing. In *Phospholipid Technology and Applications*; Gunstone, F. D., Ed.; The Oily Press: Bridgewater, UK, 2008; pp 21–39.

Scuriatti, P.; Wagner, J. R.; Tomás, M. C. Influence of Soybean Proteins-phosphatidylcholine (PC) Interaction on the Stability of Oil in Water Emulsions (O/W). *J. Am. Oil Chem. Soc.* **2003**, *80,* 1093–1100.

Sosada, M.; Pasker, B.; Gabzdyl, R. Optimization by Full Factorial Design of the Emulsifying Properties of Ethanol Insoluble Fraction from Rapeseed Lecithin. *Eur. J. Lipid Sci. Tech.* **2003**, *105,* 672–676.

Temelli, F.; Dunford, N. T. Modification of Crude Canola Lecithin for Food Use. *J. Food Sci.* **1995**, *60* (1), 160–163.

van Nieuwenhuyzen, W. Lecithin and Other Phospholipids. In *Surfactants from Renewable Resources*; Kjellin, M., Johansson, I., Eds.; Wiley, 2010; pp 191–212.

van Nieuwenhuyzen, W.; Tomás, M. C. Update of Vegetable Lecithin and Phospholipid Technologies. *Eur. J. Lipid Sci. Tech.* **2008**, *5,* 472–486.

van Ruijven, M.; van Dalen, G.; Nijsse, J.; Regismond, S. CSLM and SEM: Imaging of Plant Material for Food Emulsion Structuring: A Cryo-SEM and CSLM Study, 2009, G.I.T.

Imaging & Microscopy 4/2009, 32-34. Published online Oct 30, 2009. DOI: 10.1002/imic.200990083.

Vikbjerg, A. F.; Mu, H.; Xu, X. Synthesis of Structured Phospholipids by Immobilized Phospholipase A2 Catalyzed Acidolysis. *J. Biotechnol.* **2007**, *128,* 545–554.

Winther, B.; Hoem, N.; Berge, K.; Reubsaet, L. Elucidation of Phosphatidylcholine Composition in Krill Oil Extracted from Euphausia Superb. *Lipids* **2011,** *46* (1), 25–36.

Xu, X.; Vikbjerg, A. F.; Guo, Z.; Zhang, L.; Acharya, A. K. Enzymatic Modification of Phospholipids and Related Polar Lipids. In *Phospholipid Technology and Applications*; Gunstone, F. D., Ed.; The Oily Press: Bridgewater, UK, 2008; pp 41–82.

10

Autoxidation of Plasma Lipids, Generation of Bioactive Products, and Their Biological Relevance

Arnis Kuksis and Waldemar Pruzanski ■ *University of Toronto, Canada*

Introduction

Plasma contains cholesteryl, glyceryl, and phosphoglyceryl esters of fatty acids as major lipid components. These compounds are carried in plasma and provide a source of metabolic fuel that drives all biochemical reactions required for normal cellular activity, and they provide structural material for cell membranes and lipoproteins. These compounds also include molecules that signal receptor activation on the cell surface, triggering subsequent signaling cascades through various protein kinases and phospholipases.

The unsaturated fatty acid–containing lipid molecules are subject to autoxidation and free radical attack depending on their structure and location in the cellular architecture. Excessive recent consumption of polyunsaturated fatty acids has stimulated much interest in the chemistry and biochemistry of lipid autoxidation, the toxicity of the oxidation products, and the role of antioxidants in the prevention and modulation of their toxicity. Due to the extreme complexity of the oxidation process, a multitude of oxo-lipid species arise from plasma lipids, but only a few species have been isolated and tested for their biological activity. The metabolic effects have been attributed to such ill-defined oxo-lipid entities as thiobarbituric acid–reactive substances (TBARS), minimally modified LDL, oxo-LDL, oxo-phospholipids, oxo-PtdCho, and oxo-palmitoyl/arachidonoyl-GroPCho, which may include both primary and secondary oxidation products of the component glycerolipids and cholesteryl esters (ChE).

It was recognized early that a determination of the biological activity would require isolation and testing of individual molecular species of the oxo-lipids, which could also provide support for plausible mechanisms of their action. The initial work was performed with the volatile aldehydes, such as malonaldehyde, nonenal, and 4-hydroxy-nonenal, which led to documentation of specific biochemical and biological effects. More recent work has resulted in isolation and identification of the core aldehydes of cholesteryl esters, glycerophospholipids, and triacylglycerols (TAGs), which are produced in a mole/mole ratio to the volatile products, and in determination of their metabolic activity and biological effects.

This chapter provides a brief summary of plasma lipid composition and of the main chemical and biochemical methods of detecting and quantifying the major primary and secondary products of oxidation of plasma lipoproteins, followed by a discussion of the physicochemical, chemical, and metabolic activities of the major individual molecular species or small groups thereof. The chapter concludes by calling attention to possible involvement of oxo-lipids in metabolic syndromes and disease conditions, as well as prevention of oxidation.

Plasma and Lipoprotein Lipid Composition

The nature and extent of plasma lipid autoxidation depends on the fatty acid composition of glycerolipids, glycerophospholipids, and cholesteryl esters of plasma lipoproteins. The fatty acid composition of the lipoproteins varies with the diet and to a lesser extent with the age, sex, and general health of the individual.

Total Lipids

Quehenberger et al. (2010) recently reported the plasma total lipid composition utilizing a lipidomics approach based on normal phase liquid chromatography-mass spectrometry/mass spectrometry (LC-MS/MS). In a quantitative assessment, over 500 distinct molecular species were recognized that were distributed among the main lipid catagories: fatty acids, glycerolipids, glycerophospholipids, sphingolipids, sterols, and prenols. The general findings agreed with previous less-elaborate analyses by Kuksis et al. (1969, 1981), Uran et al. (2001), and Pang et al. (2008). Individual molecular species, however, were not always determined. Thus, the analysis of triacylglycerols (TAGs) did not allow the definition of molecular species of each subset, which previous studies had revealed to be made up of complex enantiomeric and multiple isobaric entities (McAnoy et al., 2005). The cholesteryl esters were found to make up the single most-abundant lipid class in human plasma with ChE18:2 comprising about 50% of the total. Over 200 species of glycerophospholipids were detected in the plasma reference material. Due to the presence of isobaric species, only 158 PtdOH, PtdCho, PtdEtn, PtdGro, PtdIns, and PtdSer species were quantified. Sphingomyelin (SM) accounted for the largest fraction of plasma sphingolipids and permitted estimation of about 100 subspecies, which was about twice the number arrived at previously on the basis of GC analyses of the ceramides released by phospholipase C (Kuksis et al., 1969). Oxidized species of neutral or polar lipids were not reported or detected.

Lipoprotein Lipids

Dashti et al. (2011) recently published a phospholipidomic analysis of all defined human plasma lipoproteins: LC/ESI-MS, LC-ESI-MS/MS, and HPTLC analysis of different lipoprotein fractions collected from pooled plasma. It revealed the presence of PtdEtn, PtdIns, and SM only on lipoproteins, whereas PtdCho and lysoPtdCho were present on both lipoproteins and plasma lipoprotein free fractions. Surprisingly, cardiolipin, PtdGro, and PtdSer were observed neither in the lipoprotein fractions nor in lipoprotein free fraction.

Detailed molecular species composition of plasma HDL, HDL_3, APHDL, and LDL was reported for SM (Pruzanski et al., 2000), PtdCho (Pruzanski et al., 1998, 2005), and PtdEtn, including plasmalogens, PtdIns, PtdGro, and PtdSer (Pruzanski et al., 2007). Up to 20 molecular species for each lipid class in each lipoprotein fraction were identified and quantified using LC/ESI-MS. The molecular species composition of the major plasma lipoproteins was in good qualitative agreement with that of whole plasma reported by Uran et al. (2001), but significant quantitative variation was found among the lipoprotein classes.

Autoxidation

The exact mechanism of oxo-lipid formation is not known with certainty, but plausible hypotheses have been advanced for both autoxidation and chemical oxidation to account for many of the final products that have been isolated and identified beyond doubt by chromatographic (GC and HPLC), mass spectrometric (MS/MS), and nuclear magnetic resonance (NMR) methods. It is assumed that these mechanisms apply also to the autoxidation of plasma and lipoprotein lipids.

The fatty acid hydroperoxides, hydroxides, epoxides, isoprostanes, and core aldehydes generated by autoxidation or chemical oxidation with hydroperoxide (e.g., *tert*-butyl hydroperoxide) are believed to be similar to those formed by free radical oxidation in vivo (Halliwell and Gutteridge, 1989; Porter et al., 1995). Because in vivo peroxidation takes place in the presence of proteins, labile peroxidation intermediates are likely to be trapped by the protein (Kaur et al., 1997), which would lead to differences from peroxidation products generated by chemical oxidation in protein absence.

Primary Products

The primary oxidation products are characterized by retention of an intact fatty acid chain. The reaction of oxygen with unsaturated lipids, as normally presented, involves free radical initiation, propagation, and termination processes (Frankel, 1998). The fundamental issues of lipid peroxidation have been recently reviewed with special emphasis on routes to 4-hydroxy 2-nonenal (Schneider et al., 2008; Spickett, 2013). Yin et al. (2011) discussed chemical mechanisms for achieving a stereochemical control of product formation.

Another important way in which unsaturated lipids can be oxidized involves exposure to light and a sensitizer such as methylene blue (Frankel, 1998). Through this non-free-radical process, oxygen becomes activated to the singlet state by transfer of energy from the photosensitizer. Two forms of excitation have been recognized: the singlet and the triplet state (Frankel, 1998). The triplet state has a longer lifetime and is believed to initiate all photosensitized reactions. It can proceed along two different pathways depending on the photosensitizer. Pathway 1 leads to a hydrogen or electron transfer to an unsaturated lipid to yield a conjugated lipid radical, which reacts further with oxygen. Hydroperoxides formed by Pathway 1 and by normal free radical oxidation are identical, and they differ from hydroperoxides produced by the alternate Pathway 2 reaction (Garscha et al., 2008). According to Pathway 2, unsaturated lipids are attacked on either side of a double bond by electrophilic single oxygen according to a concentrated "ene" addition mechanism. Figure 10.1 shows

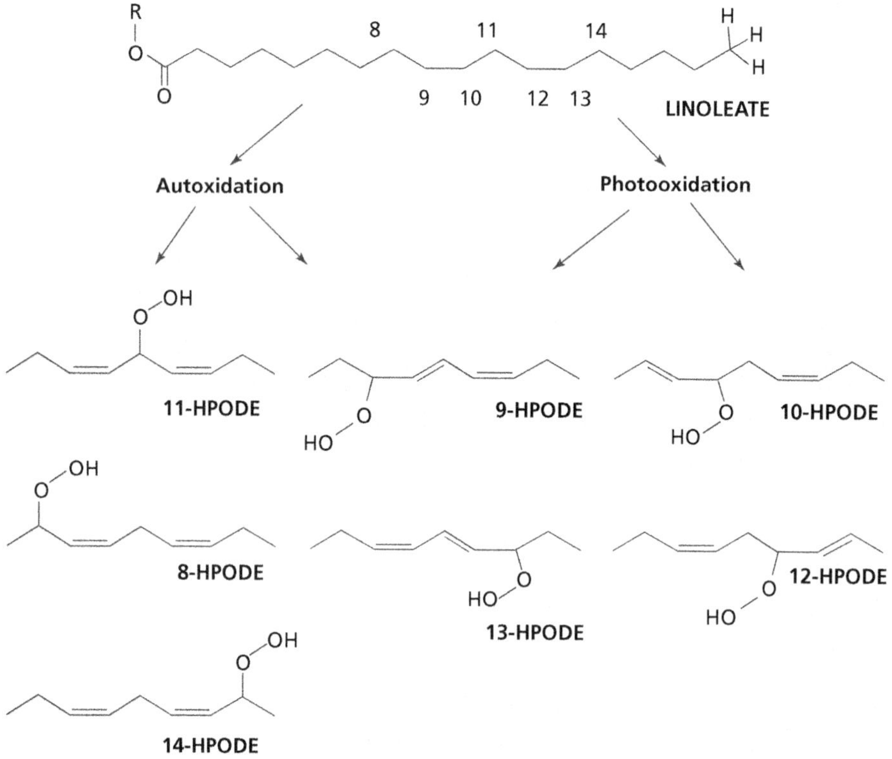

Figure 10.1 Summary of hydroperoxy-octadecadienoic acids (HPODEs) generated by autoxidation and photooxidation (modified from Garscha et al., 2008).

the hydroperoxy-octadecadienoic acid (HPODE) generation by autoxidation and photooxidation.

The resulting singlet oxygen ($^1O_2$) produced by this process is extremely reactive. Linoleate is reported to react at least 1500 times faster with $^1O_2$ than with normal oxygen in the triplet ground state ($^3O_2$). This hydroperoxidation reaction is so rapid that it has been postulated to initiate free radical autoxidation. In direct photooxidation, free radicals formed by UV radiation decompose to oxygen-containing lipids following the free radical chain reaction.

The isoprostanes (IsoPs) and isofurans (IsoFs) are also oxidation products generated from the non-enzymatic oxidation of arachidonic acid and its esters, which retain the intact oxo–fatty acid chain. IsoFs are preferentially formed under increased oxygen tension (Milne et al., 2013). Figure 10.2 shows the mechanism of formation of F_2-IsoPs and IsoFs.

Adachi et al. (2006) reported the presence of the 9- and 13-epoxides of 16:0/18:2-GroPCho and 18:/18:2-GroPChos in human plasma, and Morisseau et al. (2010) reported the natural occurrence of the monoepoxides of arachidonic (ARA),

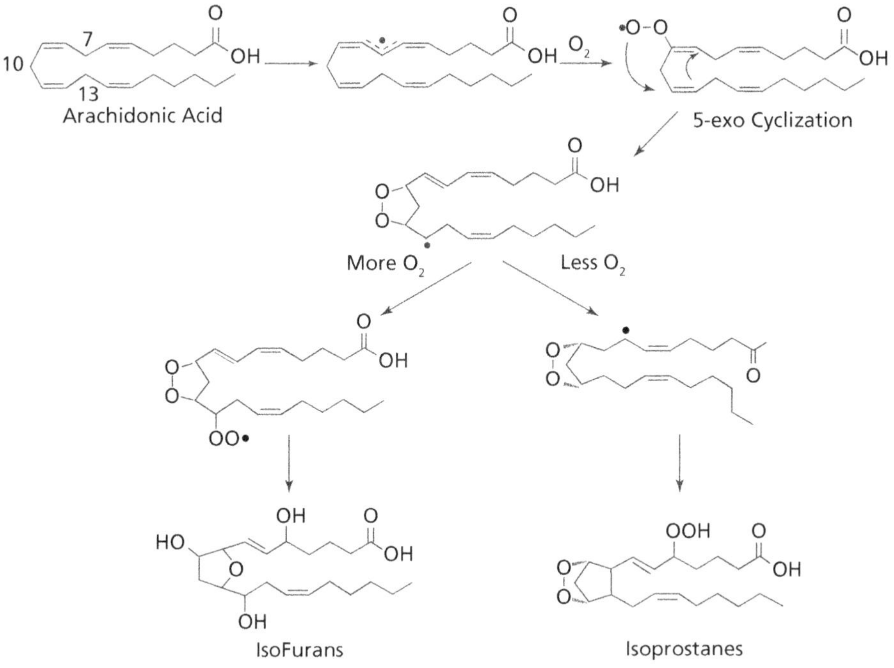

Figure 10.2 The mechanism of formation of isofurans compared to the mechanism of formation of F2-isoprostanes (redrawn from Milne et al., 2013).

eicosapentaenoic (EPA), and docosahexaenoic (DHA) acids, which are believed to be products of CYP enzymes (Fer et al., 2008).

Secondary Products

All peroxides are subject to decomposition by homolytic cleavage or to form alkoxy radicals and their reduction products, including aldehydes, ketones, alcohols, hydrocarbons, esters, furans, and lactones. PtdCho γ-hydroxyalkenals (PC-HAS) are formed by β-scission of an alkoxyl radical derived from dihydroperoxide, which produces two γ-hydroxy-α,β-unsaturated aldehydes, that is, a methyl-terminal hydroxynonenal (HNE) molecule and a mirror image of HNE, still esterified to PtdCho (namely 9-hydroxy-[12-oxo]-10-dodecenoic acid [HODA], its GroPCho ester from linoleate and 5-hydroxy-[8-oxo]-6-octenoic acid [HOOA], or its GroPCho ester from arachidonate) (Kaur et al., 1997).

Figure 10.3 shows the free radical–induced cleavage of palmitoyl/arachidonoyl-GroPCho to generate HOOA-GroPCho, a hydroxyalkenal phospholipid analog of HNE, as well as a saturated core aldehyde and core acid.

In earlier studies (Kamido et al., 1995), the HOOA-GroPCho was not isolated along with the saturated core aldehydes presumably because of irreversible protein binding. Its intermediate formation, however, has been demonstrated during per-

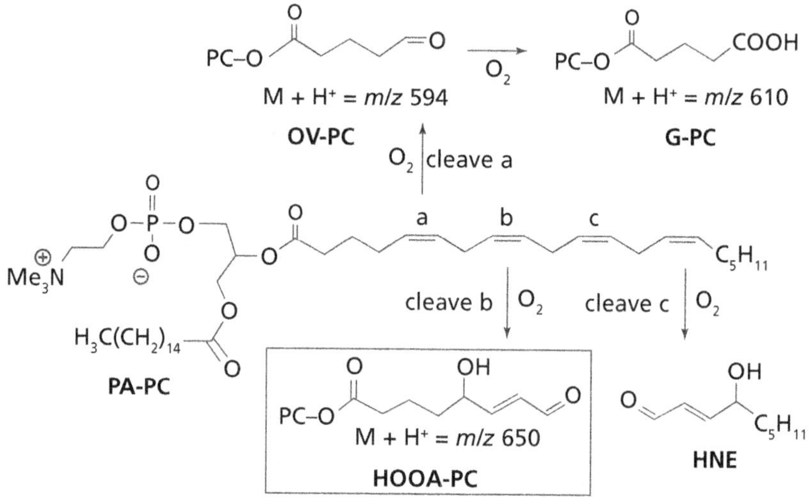

Figure 10.3 Free radical–induced oxidative cleavage of palmitoyl/arachidonoyl GroPCho to generate HOOA-PtdCho, a hydroxyalkenal phospholipid analogue of HNE (redrawn from Salomon et al., 2011).

oxidation of arachidonoyl-GroPCho in presence of albumin, which serves to trap an oxo-lipid in form of a Michael addition product (see below). The trapped oxo-lipid intermediate could be recognized following hydrolysis of the oxo-lipid-albumin complex. Because they possess a γ-hydroxy-α,β-unsaturated terminal aldehyde like HNE, GroPCho-hydroxyalkenals form Michael adducts with primary amino groups of lysyl residues and thiol groups of cysteinyl residues, as well as pentylpyrrole adducts, incorporating the ε-amino groups of lysyl residues (Salomon et al., 2011).

Schaich (2013) discussed the chemistry of alternate reactions to show that lipid oxidation is much more complex than the simplistic radical chain reactions normally presented. For the purpose of the present review, only the final products of lipid oxidation are considered, regardless of the exact mechanism of formation. Figure 10.4

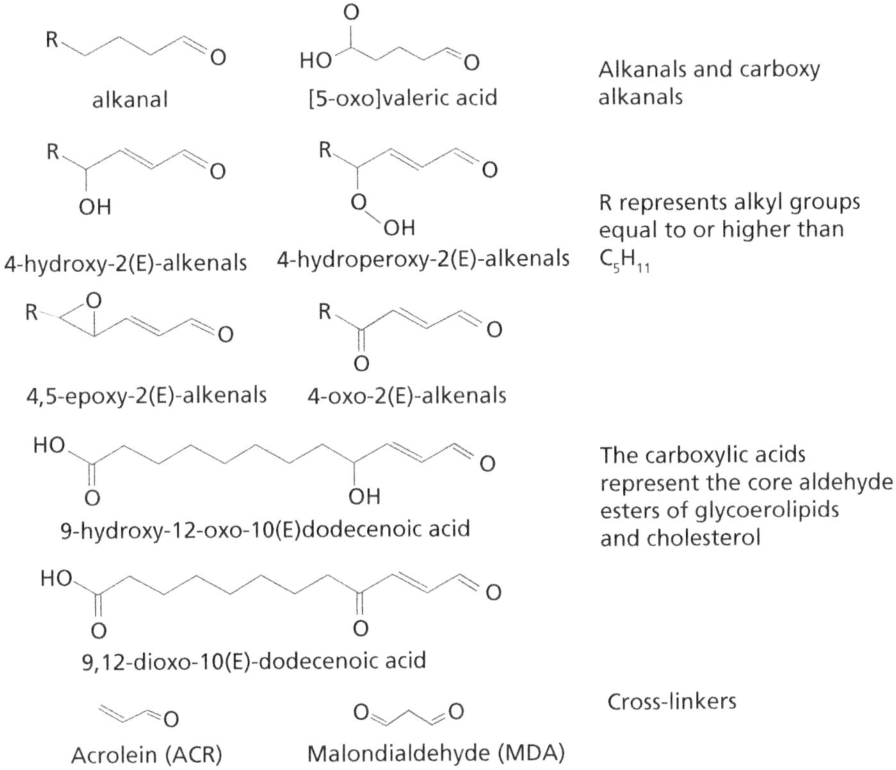

Figure 10.4 Structures of selected bioactive bifunctional alkenals and core aldehydes. Representative saturated aldehydes derived from both methyl and carboxy terminals of the fatty acids are included (from Kuksis, 2010).

summarizes the structures of the low molecular weight aldehydes formed by free radical oxidation of interest to the present discussion.

Biological Oxidation

Lipoxygenases and Cyclooxygenases

Enzymatic oxidation of unsaturated lipids is catalyzed by different lipoxygenases (LOXs), which are widely distributed in human tissues. These enzymes are non-heme iron-containing proteins, which among other activities, catalyze dioxygenation of the 1,4-*cis,cis*-pentadiene moiety of unsaturated fatty acids, such as linoleic (L), linolenic (Ln), ARA, and EPA acids, and yield hydroperoxides with one pair of conjugated double bonds (Gardner, 1995; Marchand et al., 2002). The oxidation process consists of removal of a hydrogen atom to form a free radical, conjugation of the double bonds, rearrangement of the radical electron, and insertion of di-oxygen. Enzymatically catalyzed processes are regio- and stereo-specific, producing a variety of positional, geometric, and optical isomers (Gardner, 1995; Marchand et al., 2002; Schneider et al., 2007). The reactions with DHA, docosapentaenoic-ω-3, and docosapentaenoic-ω-6 with 5-, 12-, and 15-LOXs produce oxylipins, which also have been identified and characterized by LC-ESI-MS/MS (Serhan et al., 2000). The structure and stereochemistry of many of the synthetic products have been unambiguously obtained on the basis of their total synthesis from chiral starting materials of known stereochemistry. Serhan and Petussis (2011) have provided references to the chemical and biochemical synthesis of these lipid mediators, now referred to as resolvins and protectins, and have discussed the role of PUFA in inflammation and resolution. Cyclooxygenases (COX) catalyze the reaction of achiral PUFA with oxygen to form a chiral peroxide of high regio- and stereochemical purity. These enzymes also employ free radical chemistry, but execute efficient control during catalysis to form a specific product over a multitude of isomers found in the non-enzymatic reaction (Steenhorst-Slikkerveer et al., 2000). Four mechanistic models have been presented that could account for the specific reactions of molecular oxygen with a fatty acid in the LOX as well as in the COX active site. The puzzling issue in understanding how the LOX and COX enzymes control the regio- and stereochemistry of their catalytic reactions is the uncontrolled access of oxygen to the reactive lipid intermediate (Furse et al., 2006).

Cytochrome P450 Epoxygenases

Whereas the prostaglandins and leukotrienes are best known as products of arachidonic acid metabolism by cyclo-oxygenases and lipoxygenases (Gronert, 2008), arachidonic acid is also a substrate for the cytochrome P450 (CYP) epoxygenases CYP2C8 and CYP2J2, which convert it to 4-regioisomeric epoxyeicosatrienoic (EET) acids (5,6-EET; 8,9-EET; 11,12-EET; and 14,15-EET). The bioactive EETs are produced predominantly in the endothelium and are metabolized by soluble epoxide hydrolase to less-active dihydroxyeicosatrienoic acids (DHETs) (Campbell and Falck, 2007; Fleming, 2008).

CYP enzymes, CYP2J and CYP2C in particular, produce a number of bioactive epoxy-fatty acids (Spector and Norris, 2006). Hepatic and renal cytochrome P-450 epoxygenases react equally well with EPA, DHA, and ARA to yield multiple epoxyeicosatetraenoic, epoxydocosapentaenoic, and epoxyeicosatrienoic acid regioisomers, respectively.

Prevention of in Vitro Peroxidation

Because unsaturated fatty acids and their esters are readily peroxidized, exposure of plasma and lipoprotein samples to air/oxygen must be avoided or limited. Sample handling during extraction and work-up requires elaborate means to prevent peroxidation of lipid samples. Although it has been reported (Choi et al., 2008; Rodenburg et al., 2006; Tsimikas, 2006) that various biomarkers are stable to prolonged freezing and transport to processing sites on dry ice, there is evidence that lipoproteins become peroxidized during ultracentrifugation as well as during storage, even at dry-ice temperature (–78 °C) (Engstrom et al., 2009). Napoli et al. (1997) demonstrated peroxidation of lipoproteins during classical ultracentrifugation compared to short-run centrifugation (Chung et al., 1980). Kuksis and Pruzanski (2014) observed peroxidation of lipoprotein lipids during transport to processing sites at dry-ice temperature. According to Arneson and Roberts (2007), neuroprostane (NP) peroxidation can be minimized or prevented by not leaving biological fluids at room temperature or at –20 °C. NP oxzidation also occurred when stored at –20 °C because tissue fluid is not ice-solid at –20 °C. There were several steps one could take to prevent extraneous lipid peroxidation from occurring. Thus, ex vivo formation of F_4-NPs was minimal if the biological fluids or tissues were frozen immediately after procurement and if butylated hydroxytoluene (BHT, a free radical scavenger) and/or triphenylphosphine (a reducing agent) were added to the organic solvents during extraction and hydrolysis of phospholipids. Addition of BHT alone to a Folch solution used to extract lipids from plasma did not suppress ex vivo lipid peroxidation entirely. Addition of triphenylphosphine (50 mg/100 ml Folch; 0.005% BHT) along with BHT was necessary (Arneson and Roberts, 2007).

Gruber et al. (2012) reported attempts to minimize lipid peroxidation by purging samples and solvents with argon and maintaining an argon atmosphere throughout sample handling and storage. Nitrogen blanketing was not adequate.

Analysis of Lipid Oxidation Products

The lipid oxidation products range from low molecular weight chain cleavage products to intact lipid ester molecules containing oxo–fatty acid residues. The low molecular weight products are usually analyzed by GC/MS, but LC/MS is also utilized following preparation of suitable derivatives. Not all of the chemical reaction products have yet been demonstrated in vivo, but they may be found in the future.

Identification of Intact Oxo-Lipid Esters

Although recent advances in tandem mass spectrometry (MS/MS) technology have permitted detailed analyses of molecular species without prior lipolysis or chromatographic resolution of lipid molecules (Han and Gross, 2003), MS analyses of intact oxo-lipids from total lipid extracts have usually been performed with at least a partial prior chromatographic resolution (LC/MS) (Pruzanski et al., 1998, 2000). Kuksis and colleagues (Kuksis, 2010; Kuksis et al., 2009) published LC/ESI-MS protocols for lipidomic analysis of intact molecular species of neutral and polar glycerolipids and their peroxidation products.

Yin et al. (2009) described a definitive identification of intact oxidized products of glycerophospholipids including PtdCho, PtdEtn, and PtdSer in vitro and in vivo using ion trap MS. For these analyses, the negative ions of the oxidation products of phospholipids are fragmented by MS^n and unequivocal structural characterization is obtained based on collision-induced dissociation (CID) of the *sn*-2-carboxy ion. The oxo-phospholipids were analyzed by ultra-performance liquid chromatography/ESI-Ion Trap-MS with Phenomenex Luna 3-μ C_8 reversed-phase column. The method used synthetic standards, products of free radical in vitro oxidation, and phospholipids from rat liver.

Hydroperoxides, Hydroxides, and Epoxides

The determination of hydroperoxides, hydroxides, and epoxides in plasma and plasma lipoproteins is fraught with uncertainty and contradiction because of the presence of various endogenous reducing systems, which lead to oxo-lipid interconversions. Definitive identification of hydroperoxides, hydroxides, and epoxides from human plasma was first reported by Adachi et al. (2006), who employed LC/ESI-TOF-MS. The authors observed that incubation of 16:0/18:2 GroPCho-OOH were 18:0/20:4 GroPCho-OOH aerobically in human plasma gave 16:0/18:2 GroPCho-OH and 18:0/20:4 GroPCho-OH derivatives. The PtdCho-OOHs were reduced to PtdCho-OHs by apolipoproteins A-I, A-II, and B-100 in human plasma. Adachi et al. (2006) detected epoxyhydroxy and trihydroxy derivatives of 16:0/18:2 GroPCho and 18:0/18:2 GroPCho.

Subsequently, Reis et al. (2007) used liquid chromatography coupled with electrospray tandem mass spectrometry (LC-MS/MS) to identify palmitoyl/linoleoyl GroPCho oxidation products, which were separated using a gradient of aqueous ammonium acetate (5 mM)/acetonitrile (90:10, v/v) (eluent A) and acetonitrile/aqueous ammonium acetate (5 mM) (90:10, v/v) (eluant B) programmed as follows: 60% B for 30 minutes followed by a linear increase to 100% B at 33 minutes held for 5 minutes. Flow rate was 0.8 ml/min. No arachidonoyl GroPChos were detected. Hui et al. (2010) identified PtdCho hydroperoxides in human plasma, including the molecular species of 16:0/18:2-OOH and 18:0/18:2-OOH, which were estimated at 89 nM and 32 nM, respectively.

Gruber et al. (2012) reported a simplified procedure for semi-targeted lipidomic analysis of oxidized PtdChos, including identification of 16:0/20:4, 18:0/20:4, 16:0/18:2, and 18:0/18:2 GroPCho oxidation products by exposure of dry lipids to air for about 48 hours. The oxo-lipids were analyzed by reversed-phase LC/ESI-MS/MS and various molecular species of hydroxyl, hydroperoxy, and epoxy derivatives, including di- and trisubstituted derivatives, were identified. The methodology included a preliminary separation of neutral lipids and fatty acids from phospholipids by solvent-solvent extraction, which was accomplished on-column when using normal phase LC. In attempts to minimize lipid peroxidation, the samples and solvents were purged with argon.

Hui et al. (2013) reported qualitative analyses of TAG-OOH molecular species in human lipoproteins by use of reversed phase LC with LTQ Orbitrap XL mass spectrometer. No TAG-OOH species were detected in normal LDL or HDL, whereas 11 species of TAG-OOH molecules were detected in oxo-LDL and oxo-HDL. The species were not identified beyond the presence of the hydroperoxide group.

Based on the reports of Adachi et al. (2006) and Hui et al. (2010), Kuksis and Pruzanski (2013) reported the presence of the major mono-and di-hydroxy and hydroperoxy derivatives of 16:0(18:0)/18:2 GroPCho and 16:0(18:0)/20:4 GroPCho as well as the trihydroxy 16:0(18:2) GroPCho and 16:0(18:0)/20:4 GroPCho derivatives in human plasma LDL, HDL, HDL_3, and acute phase HDL (APHDL) following a 4-hour incubation at 37 °C. The oxo-PtdCho levels in fresh samples were extremely variable and ranged from undetectable to high picomol and low nanomol/mg protein.

Figure 10.5 shows the total ion current profiles and the mass chromatograms of the hydroperoxide (A) and hydroxide (B) derivatives of PtdCho of LDL subjected to incubation at 37 °C for 4 hours.

The peak at m/z 790 in Figure 10.5A was attributed to the 16:0/18:2 monohydroperoxide, which could have overlapped with its isobaric companions 16:0/18:2 GroPCho dihydroxide and 16:0/18:2 GroPCho monohydroxy epoxide. The peak at m/z 818 was attributed to 18:0/18:2 GroPCho monohydroperoxide and its isobaric

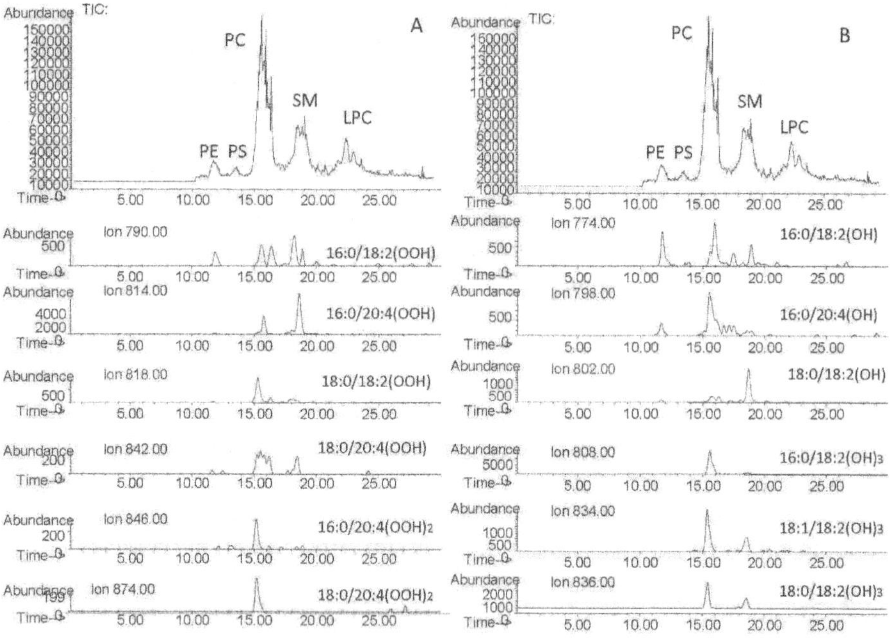

Figure 10.5 Total ion current profiles and mass chromatograms of the hydroperoxide (A) and hydroxide (B) derivatives of LDL PtdCho subjected to incubation at 37 °C for 4 hours (Kuksis and Pruzanski, 2014, unpublished). LC/ESI-MS methodology was as described by Pruzanski et al. (1998).

dihydroxy and hydroxyepoxy isobaric companions. The peak at m/z 814 was attributed to 16:0/20:4 GroPCho monohydroperoxide and its isobaric dihydroxy and hydroxyl epoxy isobaric companions, whereas the peak at m/z 842 was attributed to the 18:0/20:4 GroPCho monohydroperoxide and its isobaric dihydroxy and hydroxyepoxy companions, if present. Furthermore, the peak at m/z 846 was attributed to 16:0/20:4 GroPCho dihydroperoxide and its hydroperoxy dihydroxide, and the peak at m/z 874 was attributed to 18:0/20:4 GroPCho dihydroperoxide.

The peak at m/z 774 in Figure 10.5B was attributed to 16:0/18:2 GroPCho monohydroxide and its isobaric epoxide. The peak at m/z 798 was attributed to 16:0/20:4 GroPCho monohydroxide and its isobaric epoxide. Likewise, the peak at m/z 802 was attributed to 18:0/18:2 GroPCho monohydroxide and its isobaric epoxide, while the peak at m/z 826 was attributed to 18:0/20:4 GroPCho monohydroxide and its isobaric epoxide. On the other hand, the peak at 808 was attributed to 16:0/18:2 trihydroxide and its isobaric dihydroxyepoxide, if present, and the peak at m/z 836 was attributed to 16:0/20:4 trihydroxide. The quantitative amounts varied widely and ranged from high picomoles to low nanomoles/mg protein. Comparable amounts of hydroperoxides and hydroxides were also found in the APHDL and HDL_3 lipoprotein classes.

Adachi et al. (2006) identified the 9- and 13-monoepoxides of 16:0(18:0)/18:2-GroPCho in human plasma. Naturally occurring monoepoxides of ARA, EPA, and DHA have been recently reported in the rat central nervous system (Morisseau et al., 2010).

Hui et al. (2012) reported detection and characterization of cholesteryl ester hydroperoxides (CE-OOH) and the unique time-dependent changes for PtdCho monohydroperoxide in oxo-LDL and oxo-HDL. No CE-OOH molecules were detected in normal LDL and normal HDL, whereas six CE-OOH molecules were detected in the oxo-LDL and oxo-HDL. CE-OOH has been detected in oxo-LDL and oxo-HDL by LC/ESI-MS (Ahmed et al., 2003) and TAG-OOH has been reported in dietary TAGs (Sjovall et al., 2001). Oxo-CE is the most abundant class of oxidized lipids in minimally oxidized LDL (Harkewicz et al., 2008).

Isoprostanes and Neuroprostanes

The F_2-isoprostanes are a family of prostaglandin-like compounds produced in vivo primarily by a non-enzymatic, free radical–induced oxidation of arachidonic acid (Morrow et al., 1990a). Unlike cyclo-oxygenase-derived prostanoids, the F_2-IsoPs are initially formed in situ on glycerophospholipids, from which they are subsequently released by PLA_2s (Morrow et al., 1992a). These compounds possess a 1,3-dihydroxycyclopentane ring (PGF ring) with hydroxyls mainly in the *syn* configuration. Up to 64 isomers divided in four structural classes could be generated, depending on which of the labile hydrogen atoms are first abstracted by free radical attack. The D_2 and E_2-ISOPs are PGD and PGE ring compounds and arise from a rearrangement of the bicylic endoperoxid PGH_2-like intermediates (Morrow et al., 1994).

Lynch et al. (1994) determined the F_2-IsoP content of plasma LDL following exposure to copper oxidation. In isolated LDL exposed to aqueous peroxyl radicals or Cu^{++}, consumption of endogenous ubiquinol-10 and α-tocopherol was followed by rapid formation and subsequent breakdown of lipid hydroperoxides and esterified F_2-IsoPs, and a continuous increase in LDL's electronegativity. The F_2-IsoPs were quantified by GC/MS. Moore et al. (1995) incubated LDL with peroxynitrite (0.125 to 1 mmol/L) or the peroxynitrite donor, SIN-1 (0.5 and 1 mmol/L) and induced a concentration-dependent increase in the formation of F_2-IsoPs, reaching a maximum of 5.5-fold and 18-fold above control values, respectively. Incubation of plasma with peroxynitrite or SIN-1 yielded similar results. Ahmed et al. (2003) observed formation of PtdCho F_2-IsoPs during oxidation of HDL with peroxinitrite.

Pruzanski et al. (2000) reported the presence of PtdCho-IsoPs in plasma lipoproteins based on positive total ion current profiles. Figure 10.6 shows the total ion current profiles and the mass chromatograms of the PtdCho-IsoPs of LDL (Figure 10.6A) and HDL (Figure 10.6B) subjected to a 4-hour incubation at 37 °C. The peak at m/z 828

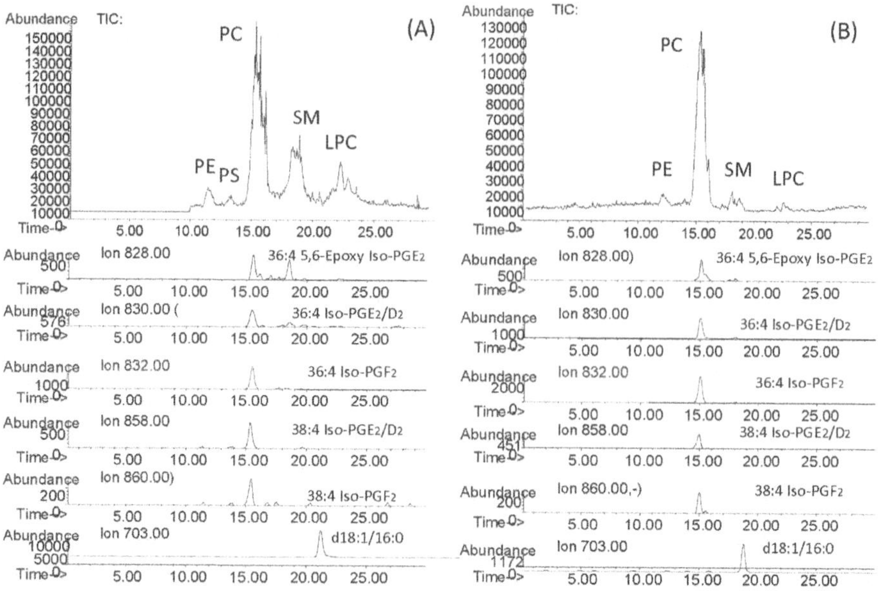

Figure 10.6 Total ion current profiles and mass chromatograms of the isoprostane derivatives of PtdCho of LDL (A) and HDL_3 (B) subjected to incubation at 37 °C for 4 hours (Kuksis and Pruzanski, 2014, unpublished). Internal standard (d18:1/16:0PCho) at 7 nanomoles/mg HDL_3 protein. Other legends as in Figure 10.5.

was attributed to 16:0/5,6-Epoxy $IsoPGE_2$GroPCho, which may overlap with 16:0/20:4 GroPCho hydroperoxy ketone. The peak at m/z 830 was attributed to 16:0/Iso-PGE_2/D_2GroPCho, which may overlap with its isobaric isoLG (E_2,D_2)-PPC (PAPC) and 16:0/20:2 GroPCho hydroperoxy hydroxide. The peak at 832 was attributed to 16:0/$IsoPGF_2$GroPCho, which may overlap with its isobaric analogue 18:0/18:2 GroPCho hydroperoxide ketone. The peak at m/z 858 was attributed to 18:0/$IsoPGE_2$/D_2GroP-Cho, which may overlap with isoLG (E_2,D_2)-SPC (SAPC) and 18:0/20:4 GroPCho hydroperoxyhydroxide, and the peak at 860 was attributed to 18:0/$IsoPGF_2$GroPCho, which may overlap with 16:0/20:4 hydroperoxyhydroxy epoxide, if present. Identical peaks were found in the total lipid extracts of HDL_3, APHDL, and NHDL. The total amounts varied widely and ranged from high picomoles to low nanomoles/mg protein.

Previous studies comparing enzyme-linked immunosorbent (ELISA) and more specific GC/MS assays had already indicated that three different ELISAs may overestimate 15-F(2t)-IsoP concentrations in human plasma. Klawitter et al. (2011) have now shown that the three ELISAs measured substantially higher 15-F(2t)-IsoP concentrations (2.1- to 182.2-fold higher in plasma) than LC/LC-MS/MS, which was assumed to be the more physiological level. Solid phase extraction maintained the difference. LC/MS, however, is two to three orders of magnitude less sensitive than GC/MS (Lawson et al., 1999). Both methods have been used as indices of lipid peroxidation in vivo (Halliwell et al., 2010). The study of arachidonoyl GroPCho peroxidation and the formation of F_2-IsoPs has allowed this research to be extended to other PUFAs and their lipid peroxidation products. F-ring IsoPs have been shown to be generated also from peroxidation of linolenic acid (C18:3, omega-3, F_1-IsoPs), EPA (C20:5, omega-3, F_3-IsoPs) and DHA (C22:6, omega-3, F_4-NPs) (Gao et al., 2006; Roberts et al., 1998). Besides the F_4-NPs, the oxidation of DHA also yields other IP-like compounds including D_4/E_4-NPs (Reich et al., 2000) and J_4/A_4-NPs (Fam et al., 2002).

Oxidation of EPA in vitro yielded a series of compounds that were structurally established to be F_3-IsoPs using a number of chemical and mass spectrometric approaches. The amounts formed were extremely large (up to 8.7±1.0 µg/mg EPA) and greater than levels of F_2-IsoPs generated from ARA. Isoprostane-like compounds (i.e., F_3, A_3, and J_3-IsoPs), the last two with cyclopentenone rings, are formed from the oxidation of EPA in the heart muscle of mice in vivo (Gao et al., 2006; Brooks et al., 2008).

Brain tissue contains relatively high proportions of DHA and this gives rise to isoprostane-like compounds that have been characterized and termed *neuroprostanes* (Roberts et al., 1998). DHA is preferentially enriched in some tissues, such as the brain, retina, and testes. Also, cellular signaling is affected by DHA (Stillwell and Wassall, 2003). Musiek et al. (2004) described stable isotope dilution mass spectrometric assays for the quantification of F-ring isoprostane-like compounds (F_4-neuroprostanes).

Bernoud-Hubac et al. (2001) described the formation of highly reactive γ-ketoaldehydes (neuroketals) as products of the neuroprostane pathway.

Isofurans and Neurofurans

Isofurans (IsoFs) are oxidation products of arachidonate that contain substituted tetrahydrofuran rings (isomers with tetrahydropyran rings are also known) (Fessel et al., 2002). The IsoFs are stable compounds that are present in readily detectable amounts in normal tissues and body fluids, including plasma. IsoFs are generated by a mechanism similar to that of IsoPs and they share a common carbon-centered radical as an intermediate. The mechanism of formation F_2-IsoPs and IsoFs (Milne et al., 2013) is reproduced in Figure 10.2. IsoFs esterified to glycerophospholipids are also found in plasma, and Milne et al. (2013) provided a protocol for their determination by GC/MS following saponification of the esters. Due to a softer mode of ionization utilized, LC/MS methodologies for the measurement of F_2-IsoPs provide a less destructive alternative, along with eliminating the need for derivatization (Taylor and Traber, 2010).

Recent studies have focused on identifying and characterizing the DHA-derived, IsoF-like compounds, termed *neurofurans* (NFs). NFs are formed by peroxidation of

DHA and they can be purified and quantified. The NFs co-purify with the F_4-NPs using the methodology described by Arneson and Roberts (2007), which is an advantage because it avoids additional sample preparation. The NPs in tissues are formed in situ on phospholipids: probably on PtdSer and PtdEtn in brain, but on PtdCho in plasma (Kuksis and Pruzanski, 2014). Arneson and Roberts (2007) presented neurofuranstructures of ARA and DHA lipid peroxidation along with the structures of isoprostanes and isofurans (NFs) derived from ARA, and of neuroprostanes and neurofurans derived from DHA. These NFs have also been identified by GC/MS using an established protocol (Milne et al., 2013).

Identification of Chain Cleavage Products

Among the chain cleavage products can be recognized both low and high molecular weight products. The volatile low molecular weight products are generated from the methyl terminal of the fatty acid or ester, whereas the nonvolatile high molecular weight products are generated from the carboxyl or ester end of the molecule. The volatile products are noted here only in passing, and the volatile chain cleavage products of lipid peroxidation (HNE and HHE) are discussed here only to the extent to which they have provided critical information about the mode of production of certain phospholipid core aldehydes, which, in the presence of protein molecules, are irreversibly trapped as adducts. Figure 10.7 shows representative structures of isoprostane, neuroprostane, isofuran, and neurofuran products of ARA and DHA peroxidation.

Malonaldehyde and Acrolein

Malondialdehyde (MDA) is a major end-product of oxidation of omega-3 and omega-6 PUFAs, and is frequently measured as a thiobarbituric acid reactive sub-

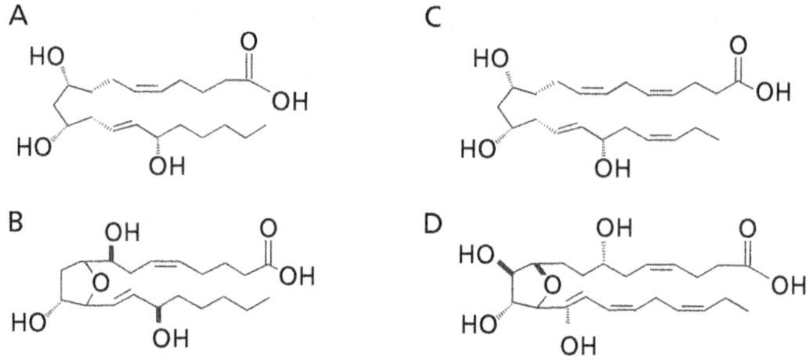

Figure 10.7 Lipid peroxidation products of ARA and DHA: (A) F_2-IsoP (15-F_2-IsoP); (B) Iso F (ent-8-epi-AT-Δ^{13}-9-IsoF); (C) F_4-NPs (17_{4c}-NP); (D) NF (7-epi-AC-Δ^{8}-10-NF) (from Arneson and Roberts, 2007).

stance (TBARS). There are methodological problems with the TBA assay, including lack of specificity and generation of artifactual TBARS under various assay conditions (Janero, 1990). MDA is derived only from certain lipid peroxidation products and is neither the sole end product nor one of lipid peroxidation only (Halliwell and Whiteman, 2004). MDA forms Schiff-base adducts with lysine residues and cross-links proteins in vitro. Miyatake and Shibamoto (1996) developed an effective method for simultaneous determination of acrylamide (ACR), MAD, and HNE from lipids peroxidized with Fenton's Reagent (H_2O_2/Fe^{++}). The aldehydes are derivatized to N-methylpyrazoline with N-methylpyrazole and 5-(1'-hydroxyhexyl)-1-methyl-2-pyrazolin with N-methylhydrazine, respectively, and determined by capillary GLC with nitrogen-phosphorus detector. The maximum amounts of ACR were 9.7±2.1 nmol/ml and of MDA 61±2 nmol/ml of cod liver oil.

Requena et al. (1997) developed methods, utilizing $NaBH_4$ reduction, to stabilize these adducts to conditions used for acid hydrolysis of protein and prepared reduced forms of lysine-MDA (3[N^{ε}-lysino]propan-1-ol) (LM), the lysine-MDA-lysine iminopropene cross-link (1,3-[di N^{ε}-lysino]propane) (LML), and lysine-HNE (3-[N^{ε}-lysino]-4-hydroxynonan-1-ol) (LHNE). The products were analyzed by quantitative GC/MS and an excellent agreement was found between measurement of MDA bound to RNAse as LM and LML and as thiobarbituric acid-MDA adducts measured by HPLC. LM and LML (0.002–0.12 mmol/mol of lysine) were also found in freshly isolated LDL from healthy human subjects. LHNE was not detected in native LDL.

The biologically active carbonyl compounds derived from lipid peroxidation include acrolein (ACR) (Uchida et al., 1998). The major precursor of ACR is glycerol (Umano et al., 1988). ACR is the strongest electrophile in the α,β-unsaturated aldehyde series; its reaction with the thiol group of cysteine is about 100–150 times faster than that of HNE (Esterbauer et al., 1991). The toxicity of ACR is related to its ability to deplete glutathione and to form DNA and protein adducts.

Hydroxy-2-nonenal and 4-Hydroxy-2-hexanal

4-Hydroxy-2-nonenal (HNE) is the most investigated species of the products of lipid peroxidation. It possesses high biological activity. Due to its strong hydrophobicity, HNE is associated mostly with the membrane in which it is produced, but it can also diffuse to different cellular compartments. HNE forms adducts with three different types of amino acids (Cys, His, and Lys residues) via Michael addition either to thiol or to amino groups, but it can also modify protein structure through Schiff base formation with lysyl residues, leading to pyrrole formation (Sayre et al., 2006). Rauniyar et al. (2010) have identified the carbonylation sites in apomyoglobin by LC/ESI-MS/MS after exposure to 4-hydroxy-2-nonenal.

Only some of these adducts, such as the HNE and ONE-derived Michael adducts on Cys and His residues, were found to survive the conditions of proteolysis and

LC-MS. Figure 10.8 shows the formation of Michael addition products of amino, imino, and thiol groups with the double bond of oxygenated α,β-unsaturated aldehydes.

The 4-hydroxy-2-hexenal (HHE) has reactivity comparable to that of HNE. The hydroxyalkenals are easily detectable and they reflect separately the peroxidation of all omega-6 (4-HNE as marker) and all omega-3 (4-HHE as marker) fatty acids. A low molecular weight dicarbonyl compound, glyoxal, has also been detected among the products formed during oxidation of PUFA in vitro (Fu et al., 1996).

Core Aldehydes and Acids

Prior to the discovery of carboxyethylpyrroles and the pyrrole type of core aldehydes (Salomon et al., 2011), much effort was expanded in pursuing the formation and biological activity of the saturated glycerophospholipid core aldehydes derived from arachidonoyl and linoleoyl PtdChos: 1-palmitoyl(stearoyl)/2-[5-oxo]valeroyl and 1-palmitoyl(stearoyl)/2-[9-oxo]nonanoyl-GroPChos (C_5 and C_9 core aldehydes, respectively). Similar C_5 and C_9 aldehydes derived from Ch-arachidonate were also isolated from lipid peroxidation reactions.

Kamido et al. (1992a) were the first to prepare and characterize the cholesteryl ester and glycerophospholipid core aldehydes from egg yolk lipoproteins. Reference compounds were first obtained by subjecting the lipid esters to hydroxylation with

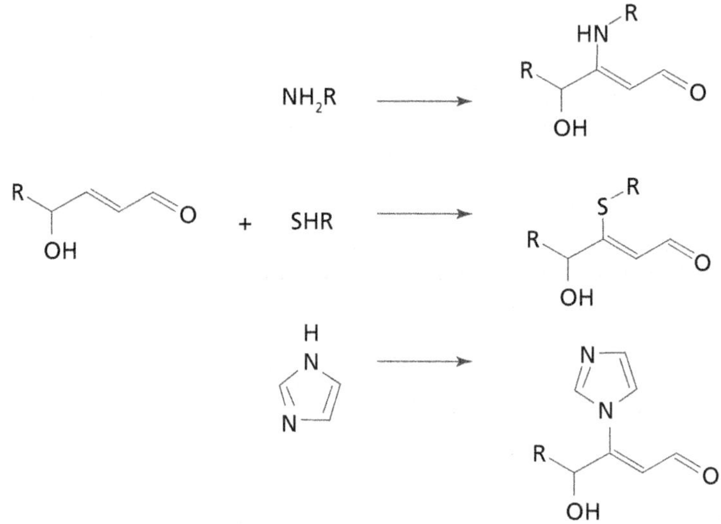

Figure 10.8 Formation of Michael reaction compounds by addition of amino, imino, or thiol groups to the double bond of oxygenated α,β-unsaturated aldehydes (modified from Sayre et al., 2006).

osmium tetroxide followed by carbon–carbon bond cleavage with periodic acid. The resulting core aldehyde esters (mainly [5-oxo]valerate and [9-oxo]nonanoate esters of cholesterol and the choline and ethanolamine glycerophospholipids) were converted to DNPH derivatives, isolated and purified by TLC and reversed phase HPLC, and identified by FAB-MS. The aldehyde-containing diacylglycerol moieties following dephosphorylation with phospholipase C and preparation of the monomethoximes and TMS ethers were also separated and identified by GC/MS. The formation of PtdCho core aldehydes is accompanied by the formation of the dicarboxylic acid esters of the GroPCho (glutaric and azelaic). The C_5 and C_9 core aldehydes of PtdCho are readily synthesized via ozonization of the 16:0(18:0)/20:4- and 16:0(18:0)/18:2-GroPCho (Kamido et al., 1992b; Ravandi et al., 1995; Subbanagounder et al., 2000).

Besides the C_9 core aldehydes, Kamido et al. (1993) later isolated and identified other core aldehydes produced by *tert*-butyl hydroperoxide (TBHP) oxidation from Ch-linoleate as smaller amounts of the [8-oxo]octenoates, [10-oxo]decenoates, [11-oxo]undecenotes, and [12-oxo]dodecenoates. Peroxidation of Ch-arachidonate yielded [5-oxo]valerates of cholesterol and the oxidized-cholesterols as the main products with smaller amounts of the [4-oxo]butyrates, [6-oxo]hexenoates, [7-oxo]heptenoates, [8-oxo]octenoates, [9-oxo]nonenoates, [9-oxo]nonadienenoates and [10-oxo] decadienoes. The oxocholesterols resulting from the peroxidation of the steroid ring were identified as mainly 7-keto-, 7α-hydroxy-, and 7β-hydroxy-cholesterols and 5α,6β- and 5β,6β-epoxy-cholestanols.

Subsequently, Kamido et al. (1995) presented detailed analyses of the lipid ester bound aldehydes among the copper-catalyzed peroxidation products of human plasma lipoproteins. The aldehydes were isolated by extraction with acidified $CHCl_3$-MeOH containing 2-4-dinitrophenylhyrazine (DNPH). The DNPH derivatives were resolved by HPLC and identified by on-line LC/MS. The major PtdCho core aldehydes from oxidized LDL and HDL were identified as 1-palmitoyl(stearoyl)-2-[9-oxo]nonanoyl-1-palmitoyl(stearoyl)-2-[8-oxo]octanoyl, and 1-palmitoyl(stearoyl) 2-[5-oxo]valeroyl-sn-glycerols after phospholipase C digestion of the DNPH derivatives of the glycerophospholipids.

Figure 10.9 shows the composition of PtdCho core aldehydes recovered from HDL as the DNPH hydrazones (Kamido et al., 1995).

The triacylglycerol core aldehydes have been isolated by Sjovall et al. (1997) from oxidized herring oil.

Nakanishi et al. (2009) used a myocardial ischemia-reperfusion model in mouse to investigate the formation of oxo-PtdChos under the stress of the system. The results showed that in the 18:2- or 20:4-containing oxidized PtdCho series, hydroxide, hydroperoxide, and aldehyde forms were detected at significant levels. In contrast, 22:6-containing oxidized PtdCho series consisted mainly of aldehyde forms followed

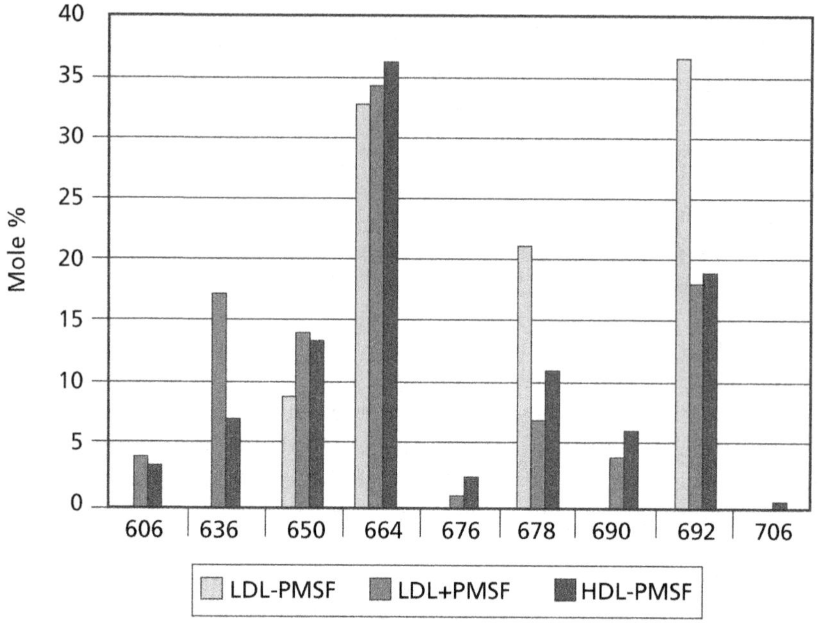

Figure 10.9 PtdCho core aldehydes recovered as DNPH derivatives from peroxidized LDL (from Kamido et al., 1995).
PMSF: phenylmethanesufonyl fluoride, a PAF-AH inhibitor.

by hydroxide or hydroperoxide forms. The aldehyde forms seemed to be significantly higher in 22:6-derived oxidized PtdChos than in 18:2- or 20:4-derived oxidized PtdChos. DHA-containing PtdChos were abundant in heart tissue.

These core aldehydes must be distinguished from the carboxyalkylpyrroles that, although generated and complexed with amino acids and proteins, still retain the ester linkage and are recovered as carboxyalkyl pyrroles following alkaline hydrolysis of the adduct (Kaur et al., 1997; Salomon et al., 2011; Zhang et al., 2011).

Oxidation of either arachidonic or linoleic acid in the presence of human serum albumin (HAS) produced an HNE-derived 2-pentylpyrrole (CPP) epitope. However, only oxidation of linoleic acid formed HAS-bound carboxyheptyl pyrrole (CHP), while only oxidation of arachidonic acid generated HAS-bound CPP. The epitope was detected by antibodies, which exhibit high structural selectivity in competitive binding inhibition assays with the corresponding HAS-bound pyrroles. No cross-activity was detected for HAS-bound 2-pentylpyrrole, an epitope that is generated by reaction of 4-hydroxy-2-nonenal (HNE) with protein lysyl residues. Analogous reactions of polyunsaturated fatty acyl (PUFA)-derived γ-hydroxyalkenal oxo-PLs were then shown to produce ω-carboxyalkylpyrrole modifications of proteins after lipolysis of intermediate phospholipid (PL) adducts (Kaur et al., 1997; Salomon et al., 2011). For example, oxidation of 1-palmitoyl-2-linoleoyl-sn-glycerophosphocholine (PL-PC) in LDL delivers carboxyheptylpyrrole (CHP)-protein modification and oxidation of 1-palmitoyl-2-arachidonoyl-sn-glycerophosphocholine (PA-PC) and gives carboxypropylpyrrole (CCP)-protein modifications.

Isolevuglandins (isoLGs) are a family of reactive γketoaldehydes, analogous to the levuglandins, which are generated from isoprostanes by opening of the cyclopentane ring. They are distinguished from the levuglandins on the basis of their variable geometry. Isolevuglandins are highly reactive and were overlooked in biological samples for many years until discovered as protein adducts by an immunological approach. Zhang et al. (2011) recently provided an updated and comprehensive overview of the generation and detection of LGs and isoLGs in vitro and in vivo. The γ-keto functionality of the LGs and isoLGs renders them extraordinarily reactive toward primary amino groups in proteins. LGs and isoLGs initially react in seconds to form Schiff base adducts with the primary amino groups, which are transformed to pyrrole products in minutes. However, these highly alkylated pyrroles are chemically sensitive compounds in the presence of oxygen and are further oxidized in a few hours to stable lactam and hydroxylactam end products (Zhang et al., 2011).

Physicochemical and Chemical Reactivity of Oxo-Lipids

The physicochemical effects of oxolipids on cellular and lipoprotein lipid membranes have been recently reviewed (Catala, 2012; Huang et al., 2011), as has been chemical reactivity of the oxo-lipids with the amino lipids, amino acids, and peptides of plasma lipoproteins (Kuksis, 2010). For the present purposes, only the effects of the glycerolipid and cholesteryl ester core aldehydes have been summarized.

Lipid Membrane Targets

Autoxidation of phospholipids yields hydroperoxy-, hydroxy-, epoxy-, and isoprostane-containing glycerophospholipids as well as phospholipid core aldehydes, which, upon incorporation into plasma membranes and lipoprotein lipid monolayers, tend to disorganize their structures due to the larger surface area of the oxo-lipid molecules.

Van den Berg et al. (1993) were the first to suggest changes in membrane configuration and fluidity to result from lipid peroxidation, and Greenberg et al. (2008) elaborated upon a lipid whisker model as a working hypothesis for binding and microdomain formation by chain-shortened and oxygenated PtdCho derivatives, including core aldehydes. There was evidence that addition of a polar oxygen atom on numerous peroxidized fatty acids reorients the acyl chain, whereby it no longer remains buried within the lipid membrane interior but rather protrudes into the aqueous compartment. However, simply adding an oxygen to the terminal end of the *sn*-2 chain of an oxo-PL is not always sufficient to buoy the acyl group into the aqueous phase.

Leidy et al. (2006) have recently shown that group IIA $sPLA_2$ activation by PtdOH could be mapped onto a multicomponent phase diagram in which the ionic lipid content of the fluid-phase domains, for example, can be determined based on the phase lines. In this way, the anionic lipid content could be clearly related to the anionic lipid threshhold of the enzyme.

In addition to a mere dissolution of the lipid monolayer, the phospholipid core aldehydes may become chemically bound to the aminophospholipids in the lipid monolayer of the lipoprotein particle. Kuksis (2010) recently reviewed the Schiff base and Michael adduct formation of lipid aldehydes with aminophospholipids, amino acids, and peptides. Early work with pure Ch-[5-oxo]valerate showed that it tends to dimerize upon storage giving a m/z 969, while mixtures of Ch-[5-oxo]valerate and Ch-[9-oxo]nonanoate yielded m/z values corresponding to C_5–C_5 and C_9–C_9, as well as mixed C_5–C_9 combinations (Kamido et al., 1992b; 1995). Ravandi et al. (1995, 1997) demonstrated that the core aldehydes of PtdCho formed a Schiff base with the PtdEtn and PtdSer. In vitro, the Schiff base complex could be stabilized by chemical reduction with $NaBH_3CN$, while in vivo the stabilization may have been brought about by an endogenous reducing system.

TAG and MAG core aldehydes also undergo self-condensation and react with aminophospholipids such as PtdEtn and PtdSer (Kurvinen et al., 1999). The volatile short-chain aldehydes, such as malonyl dialdehyde (MDA), form Schiff bases with the amino groups of phospholipids (Ishii et al., 2008), but the present discussion is limited to the lipid core aldehydes.

Amino Acid and Protein Targets

PtdCho core aldehydes were shown to form Schiff bases with the amino acids, including the N^{ε}-amino group of lysine (Ravandi et al., 1995, 1997). The binding of lipid core aldehydes to the apoproteins of both low- and high-density lipoproteins has been reported. An early demonstration of PtdCho core aldehyde binding to a myoglobin, which is a protein of low molecular weight (16,000), showed that two moles of PtdCho C_5 aldehyde were bound per mole of myoglobin (Ravandi et al., 1997).

Another approach to core aldehyde binding to plasma lipoproteins has been taken by Ahmed et al. (2003), who demonstrated it on the basis of recovery of appropriate phosphorus/protein ratios for oxidized HDL and for apo A-I exposed to C_5 and C_9 core aldehydes. Core aldehydes can form Schiff base adducts with the ε-amino groups of lysine of the apolipoproteins. Following conversion of the adduct to a stable compound, by reduction of the adduct with $NaBH_3CN$, native HDL contained about 1 mole phosphorus per mole of HDL protein. After 20 hours of oxidation, HDL contained 8.5 mol phosphorus per mol of HDL protein. A similar increase in the phosphorus/protein ratio was observed when apoA-I was incubated with C_5 and C_9 PtdCho core aldehydes. Figure 10.10 shows binding of oxo-PtdCho (A) to oxidized HDL and native HDL (open squares) following 0–20 hours of incubation; (B) shows binding of C_5 (solid bars) and C_9 (open bars) core aldehydes to apo A following a 6-hour incubation (Ahmed et al., 2003).

Gugiu et al. (2008) reported protein binding of biotinylated core aldehydes of glycerophospholipids. At least 20 different biotinylated human aortic endothelial cell (HAEC) proteins were recognized to which 1-palmitoyl-2-[5-oxo]valeroyl GroPEtn-N-biotin (Oxo-PAPE-N-biotin) was covalently bound. Such adducts were not detected during treatment with unoxidized 16:0/20:4 GroPEtn-N-biotin. The biotin derivatives permitted a selective isolation of the adducted proteins using avidin binding beads and SDS-PAGE. The 16:0/[5-oxo]valeroylGroPEtn-N-biotin bound peptides were characterized by MS/MS following trypsin digestion, reduction, and alkylation. Basic research on protein modification by 4-hydroxy-2-nonenal, a γ-hydroxyalkenal product of phospholipid oxidation, led to the finding that it forms covalent adducts in vivo that incorporate the ε-amino group of protein lysyl residues in pentylpyrrole modifications. Figure 10.11 shows the covalent adduction of γ-hydroxyaldehydes with proteins to generate alkyl and carboxyalkyl pyrrole modifications (Salomon et al., 2011). Analogous reactions take place with the polyunsaturated fatty acyl chain–derived γ-hydroxyalkenals of oxo-GroPCho (Kaur et al., 1997; Salomon et al., 1999, 2011).

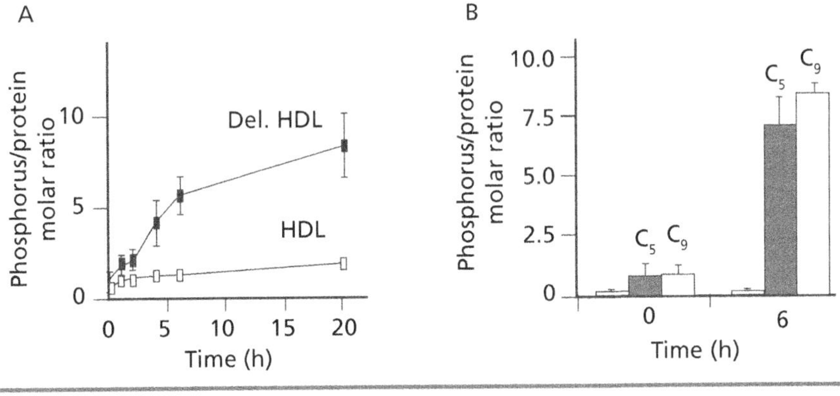

Figure 10.10 Binding of oxo-PtdCho (A) to delipidated oxidized HDL and native HDL (open squares) following 0–20 hours of incubation; (B) binding of C_5 (solid bars) and C_9 (open bar) core aldehydes to apo A following a 6-hour incubation (from Ahmed et al., 2003).

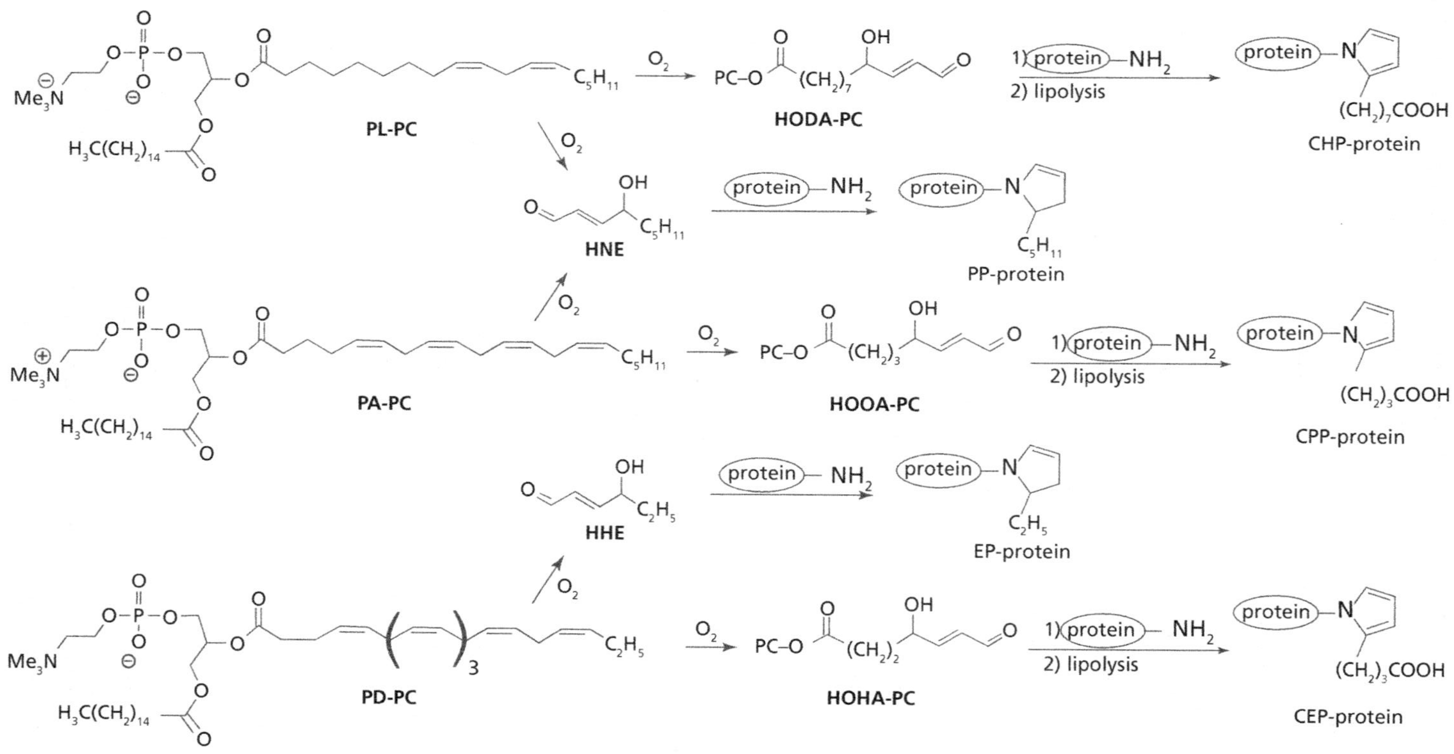

Figure 10.11 Structures and reactions of phosphatidylcholine γ-hydroxyalkanals (PC-Has) (reprinted from Salomon et al., 2011).

Binding of ChE core aldehydes to amino acids, peptides, and aminophospholipids has also been reported, as has been the binding of radiolabeled ChE core aldehydes to mouse peritoneal macrophages and serum proteins (Karten et al., 1999). A monoclonal antibody prepared to lysine-bound oxidized ChE recognized exclusively protein bound ChE core aldehydes, including Ch-[9-oxo]nonanoyl-lysine and Ch-[5-oxo]valeroyl-lysine, in atherosclerotic plaques (Hoppe et al., 1997). Likewise, 7-keto-ChE core aldehydes were demonstrated to bind to lysine and to LDL apo B, as confirmed by a monoclonal antibody.

Kurvinen et al. (1999) showed that MAG core aldehydes also reacted with amino acids and peptides, as demonstrated by reversed-phase LC/ESI-MS of the $NaBH_3CN$ reduced Schiff's base of 2-[9-oxo]nonanoyl Gro and glycine-histidine-lysine (GHK). Shiff's base adducts of triacylglycerol core aldehydes with aminophospholipids and amino acids were reported by Sjovall et al. (1998).

Low molecular weight aldehyde adducts of LDL have been related to conversion of LDL to an atherogenic form (Steinberg, 1995). Requena et al. (1997) described a GC/MS method for the specific quantification of MDA-lysine Schiff base adducts (LM) and cross-links (LML), as well as the Michael addition adducts of lysine to HNE (LHNE), in proteins. Both LM and LML were detected in native LDL and increased in concert with conjugated dienes during the course of metal-catalyzed oxidation. The maximal yields of LM, LML, and LHNE in oxidized LDL accounted for about 0.5% (about 5 mmol/mol) of lysine residues in the protein. Pizzimenti et al. (2013) recently discussed the protein oxo-lipid adducts identified along with the mechanism of their formation.

Antibodies are available for measuring MDA and HNE epitopes on plasma and tissue proteins, but these assays provide only qualitative information on the extent of protein modification by lipid peroxidation products. The measurement of changes in circulating and tissue isoprostanes (Morrow and Roberts, 1991) provides an alternative means of detecting products of non-enzymic lipid peroxidation reactions, but does not measure direct protein modification. Milne et al. (2008) demonstrated that cyclopentenone IsoPs derived from EPA possess potent bioactivity because they readily form Michael adducts with proteins and alter protein structure and function. Specifically, J_3-IsoPs were produced in large amounts in tissues of mice supplemented with EPA. These compounds were capable of modulating nuclear translocation of the transcription factor NrF_2. NrF_2 is a major regulator of the antioxidant response in cells (Gao et al., 2007), which likely occurs because of the ability of J_3-IsoPs to modify critical cysteine residues in the cytosolic NrF_2-Keap1 domain that subsequently destabilizes the complex. Milne et al. (2008) suggested that part of the mechanism by which EPA may be beneficial in some pathological states is by its ability to decrease F_2-IsoP generation. Therefore, a supplementation with fish oil may be of benefit in populations associated with increased levels of F_2-IsoPs.

The earlier findings with hydroxyl-2-nonenal would now have to be expanded in view of the discovery of the formation of carboxyethylpyrroles upon interaction of amino groups of proteins (Salomon and Gu, 2011; Salomon et al., 2011). Fu et al. (1996) reported that Ne-(carboxymethyl)lysine (CML) is also formed during oxidation of LDL, and this accounted for modification of another 0.1% of lysine residues. N^{ε}-(carboxymethyl)lysine CML levels reached about the same level as the MDA and HNE adducts.

The dicarbonyl compound glyoxal has also been detected among the products formed during oxidation of PUFA in vitro (Lloid-Stahlhoffen and Spiteller, 1994; Mlakar and Spiteller, 1994, 1996) and is known to be a precursor of CML (Wells-Knecht et al., 1994).

Nitrogenous Base and Nucleic Acid Adducts

All four bases and nucleosides are known to undergo modification as a result of exposure to reactive oxygen species (ROS) (Evans et al., 2004; Kuksis, 2010). Most of the low molecular weight aldehydes were found to react with the exocyclic amino groups of DNA. Thus, products of DNA treated with 4-hydroxy-2-nonenal have been isolated and identified by HPLC-MS/MS methods, as have the reaction products of 4-hydroperoxy-2-noneal and 2′-deoxyguanosine (Blair, 2008; Lee and Blair, 2001). Figure 10.12 shows the formation of etheno DNA and heptanone-etheno DNA adducts through homolytic decomposition of lipid hydroperoxides and formation of urinary excretion products through base exchange repair.

Thum et al. (2008) have reported that receptor blockade abrogates oxo-LDL induced oxidation DNA damage.

Biological Activity of Oxo-Lipids

Oxidized lipoproteins and oxidized palmitoyl/arachidonoyl GroPCho are known to yield upon further oxidation complex mixtures of oxo-lipids, including both full-chain and chain-shortened products as well as nonesterified products. Nevertheless, such preparations have been used to investigate their biological activity, including receptor binding. A much smaller number of studies have investigated the effects of individual oxidized phospholipids, such as the truncated products, C_5 and C_9 core aldehydes and C_5 and C_9 core acids, and full-chain-length phospholipids, palmitoyl/(5,6-epoxy)isoprostane E_2-GroPCho, and palmitoyl/epoxycyclopentenone GroPCho (Ashraf and Srivastava, 2012; Stemmer and Hermetter, 2012). The discovery of the formation of carboxyethylpyrroles as intermediates in lipid ester oxidation and protein complexing may lead to more rapid progress in the understanding of the biological activity of oxo-lipids (Salomon and Gu, 2011; Salomon et al., 2011).

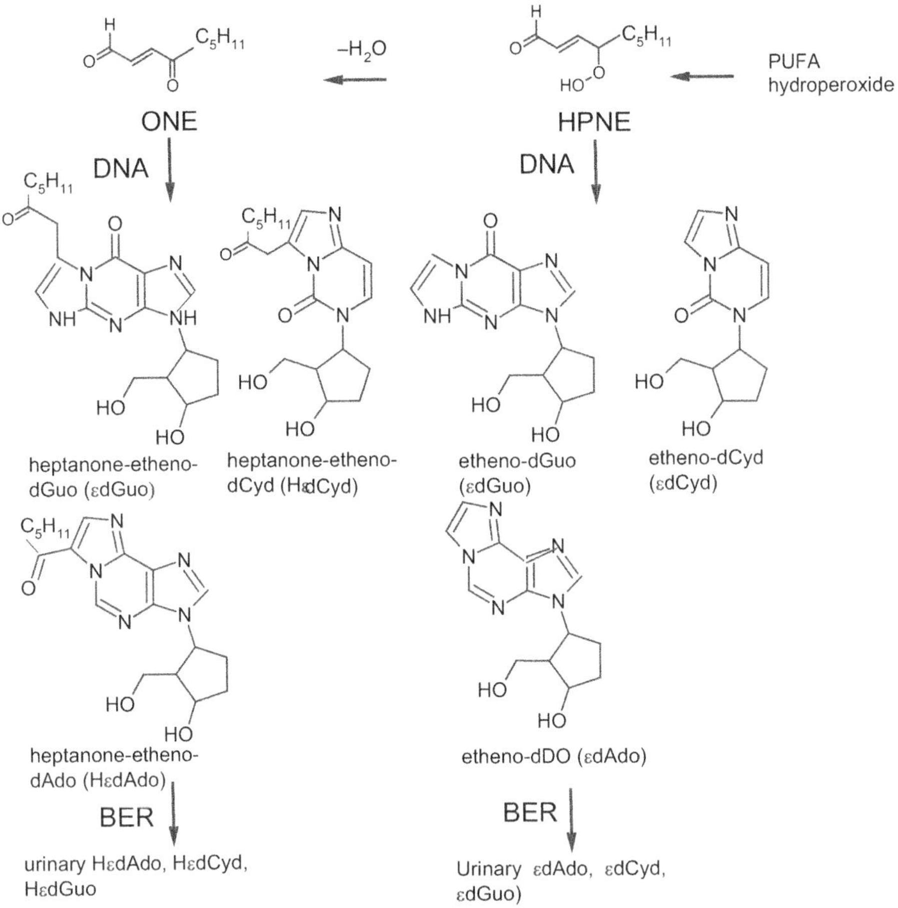

Figure 10.12 Formation of etheno DNA and heptanone-etheno DNA adducts through hemolytic decomposition of lipid hydroperoxides, urinary excretion products through base exchange repair (redrawn from Blair, 2008).

Receptor Binding

Receptors are macromolecules involved in chemical signaling between and within cells. The receptors may be located on the cell surface membrane or within cytoplasm. Molecules that bind to a receptor are called *ligands*. A ligand may activate or inactivate a receptor. A large number of receptors have been described for various products of lipid peroxidation. However, in very few instances, the chemical nature of the receptor has been characterized or the mechanism of binding identified. Only the better characterized receptors have been referred to here.

PAF Receptors

The PAF receptor was one of the earliest suggested targets of oxidized phospholipids, owing to the structural similarities between PAF and truncated products from oxidation of phospholipids (Marathe et al., 2001). The activation of the PAF receptor by specific oxidized lipids (chain-abridged derivatives), also termed *PAF-like lipids,* was demonstrated in vascular cells and macrophages (Marathe et al., 2002; Pegorier et al., 2006), although the presence of PAF itself in the oxo-LDL cannot be ruled out (Stafforini et al., 2006). One of the actions of oxo-phospholipids on the PAF receptor is the activation of platelets, inducing platelet aggregation (Chen et al., 2009).

The importance of the PAF receptor in phospholipid signaling is further seen from the fact that two competitive antagonists at the PAF receptor inhibited vascular smooth muscle cell (VSMC) proliferation induced by oxo-LDL (Heery et al., 1995). Similarly, monocyte binding to HAECs induced by oxo-LDL is also blocked by PAF receptor antagonists. However, PAF alone did not induce monocyte binding (Leitinger et al., 1997), which suggests that oxo-PAPC might be eliciting its effects through a receptor distinct from the PAF receptor. PG(glutaric)PC can activate human neutrophils, and this was also blocked by the action of PAF receptor antagonists (Smiley et al., 1991).

Recent research has moved away from the PAF receptor being the primary mechanism of modified phospholipid signaling, since some of the actions of modified phospholipids cannot be mimicked by PAF alone. Even in platelet activation by SAz(azelaic)PC, SOV([5-oxo]valeric]PC), and SG(glutaric)PC, an increase in intracellular Ca^{++} was not observed, suggesting that the PAF receptor is not responsible for this effect (Gopfert et al., 2005).

Prostaglandin Receptors

Takahashi et al. (1992) reported evidence for interaction of 8-epi-prostaglandin $F_{2\alpha}$ with thromboxane A_2 receptor, whereas Morrow et al. (1992a) reported that F_2-isoprostane, that is 8-epi-prostaglandin $F_{2\alpha}$, was a platelet thromboxane/endoperoxide receptor antagonist. Early research on the possible role of G-protein-coupled receptors (GPCRs) focused on the effects of mmLDL and a putative Gs-coupled receptor (Parhami et al., 1995), although it was subsequently suggested that free oxo-fatty acids were in fact responsible for effects following hydrolysis of oxidized phospholipids (Obinata et al., 2005). More recently it has been found that other GPCRs, specifically prostaglandin E_2 and prostaglandin D_2 receptors, can be activated by oxo-PAPC and PEIPC, but not POVPC. These effects do not seem to involve direct binding to the receptors, although it was unclear whether oxidized phospholipids or free oxidized fatty acids were responsible (Li et al., 2006). The PAF receptor is also a GPCR and was one of the earliest suggested possibilities (Honda et al., 2002).

Audoly et al. (2000) reported that cardiovascular responses to the isoprostanes iso-$PF_{2\alpha}$-III and iso-PE_2-III are mediated via the thromboxane A_2 receptor in vivo. Khasawneh et al. (2008) reported a unique coordination of 8-iso-$PGF_{2\alpha}$ with the thromboxane A_2 receptor and activation of a separate cAMP-dependent inhibitory pathway in human platelets. These results provide evidence for a novel isoprostane function in platelets that is mediated through a cAMP-coupled receptor.

As previously noted, it has been proposed that isoprostanes exert their biological effects on VSMC via thromboxane receptors. However, Fukunaga et al. (1997) have shown that VSMC cells, but not glomerular mesangial cells, have distinct F_2-IsoP binding sites, although these putative receptors bear homology to the thromboxane A_2 (TXA_2) receptor. At pharmacological concentrations, TXA_2 receptor antagonists, such as SQ29,548, can reverse the biological actions of F_2-isoprostanes in vivo (Takahashi et al., 1992) and in vitro (Fukunaga et al., 1993). Yura et al. (1999) examined the effects of 8-iso-$PGF_{2\alpha}$ on cell proliferation and gene expression and production of endothelin-12 (FT-1) in cultured bovine aortic endothelial cells as well as its signal transduction mechanism.

In vascular smooth muscle cells, 8-iso-$PGF_{2\alpha}$ exerts its biological actions through enhancement of PtdIns turnover as a second-messenger system, with subsequent stimulation of cellular mitogenesis, as does TXA_2 (Fukunaga et al., 1993; Hanasaki et al., 1990). In subsequent work with bovine aortic endothelial cells, it was demonstrated that 8-iso-$PGF_{2\alpha}$ provoked an increase in [^{3}H]-thymidine incorporation and subsequent enhancement of cellular proliferation via a mechanism involving the release of IP_3 and mobilization of intracellular Ca^{++} (Yura et al., 1999).

Scavenger Receptors

The scavenger receptors were originally identified by their ability to recognize modified lipoproteins; however, it is now emerging that they may have also a variety of other functions (Canton et al., 2013). The scavenger receptor family was originally known as membrane receptors that recognize oxidized and acetylated LDL. Previously, oxidized HDL had been shown to induce scavenger receptors in macrophages (Tontonoz et al., 1998). However, it is now known that they may have a variety of other functions. Modified lipids have been suggested as possible ligands for scavenger receptor families such as cluster differentiation 36 (CD36), scavenger receptor class A types I and II (SR-AI/II), and the class B scavenger receptor. The CD36 receptor is a platelet-integral membrane glycoprotein and is highly conserved between humans and rodents (Febbraio et al., 2001). SRA I/II and CD36 have been shown to be primary receptors involved in the uptake of oxidized phospholipids by macrophages. Genetic inactivation of both of these receptors caused a significant decrease in atherosclerotic plaques in mice (Kunjathoor et al., 2002). Furthermore, some truncated oxidized phospholipids containing hydroxyalkenal moieties (levuglandins) have been found to bind with high affinity to the CD36 receptor (Podrez et al., 2002), although other oxidized phospholipids such as palmitoyl/[5-oxo]valeroyl GroPCho and oxo-PtdSer have also been shown to be ligands at the receptor, which in turn was found to play an important role in macrophage phagocytosis of apoptotic cells (Greenberg et al., 2006). The CD36 receptor has been found to play a critical role in oxo-LDL-induced adhesion of macrophages to endothelial cells (Kopprasch et al., 2004).

Han et al. (2001) demonstrated that macrophage expression of SR-BI is inhibited in response to oxo-LDL, resulting in reduced HDL-cholesteryl ester uptake and HDL-mediated cholesterol efflux. In contrast, Hirano et al. (1999) demonstrated that incubation of human monocyte-derived macrophages with either oxo-LDL or acetylated LDL increased SR-RI expression by four times. The reason for this discrepancy demonstrating an inverse relationship between cellular cholesterol content and SR-BI expression is unknown (Hirano et al., 1999; Sun et al., 1999).

Peroxisome Proliferator Activated Receptors

Peroxisome proliferator activated receptors (PPARs) are intracellular ligand-activated transcription factors and part of the nuclear receptor superfamily. PPARs are divided into three subtypes, which can give rise to pro- or anti-inflammatory outcomes. It has been observed that incubation of mmLDL and the oxidation products of palmitoyl/arachidonoyl GroPCho transfected HAECs activated PPARs (Namgaladze et al., 2010). Oxidatively fragmented alkyl phospholipids, present in LDL, were found to have high affinity for PPAR (Davies et al., 2001). In contrast, LDL modified by PLA_2 activated PPAR in macrophages and changed gene expression in relation to lipid metabolism, suggesting an anti-inflammatory response (Namgaladze et al., 2010). Sethi et al. (2002) reported that oxidized omega-3 fatty acids in fish oil inhibited leukocyte-enthothelial interaction through activation of PPARs.

Toll-Like Receptors

Another family of cell surface and intracellular receptors that has been investigated for its potential role in signaling proinflammatory effects of oxidized phospholipids is the toll-like receptor (TLR) family. These are innate immune receptors that recognize pathogen-associated molecular patterns or respond to products of host tissue damage. Erridge et al. (2008) examined the mechanism by which oxo-PAPC inhibits TLR signaling induced by diverse ligands in macrophages, smooth muscle cells, and epithelial cells. The study established that, contrary to expectation, oxo-PAPC is an inhibitor of only TLR2- and TLR4-dependent signaling that is mediated largely via interaction with accessory molecules including CD14, LBP, and MD2, and that this inhibition does not extend to other TLRs. Palmitoyl/[5-oxo]valeroyl GroPCho and palmitoyl/glutaroyl GroPCho significantly inhibited Pam3CSK4-induced TLR2 signaling, and LPS induced TLR4 signaling at concentrations of 30 nM. Fuentes-Antras et al. (2014) reviewed the activation of TLRs and inflammatory complexes and have suggested that they may be key inducers for inflammation through necrosis factor κB(NF-κB) activation and reactive oxygen species overproduction.

Cellular Signaling

Adduction of HNE occurs on proteins involved in cellular signaling, including LKB1, a serine/threonine kinase, altering their activity and function. LKB1 is a key signaling kinase that regulates cellular growth. Studies investigating HNE-mediated changes in LKB1 signaling used an HNE concentration of 40 μM (Dolinsky et al., 2009), which resulted in significant degradation of LKB1 protein within 1 hour. Pathophysiological responses to HNE treatment in cardiac myocytes with concentrations of 400 μM have been reported to include calcium overload and reactive oxygen species generation (Nakamura et al., 2009). Calamaras et al. (2012) tested the hypothesis that HNE inhibits LKB1 activity through adduct formation on a specific reactive residue of the protein. They also reported that the reactive aldehyde 4-hydroxynonenal (HNE) forms adducts on LKB1 that directly inhibit kinase activity. It was concluded that oxidative modification of Lys-97 by HNE is sufficient for inhibition of LKB1. The study describes how a key growth suppressor protein is inactivated during states of oxidant stress. An important aspect of this study was the use of low concentration of HNE (10 μM). The concentration of HNE in human plasma ranges from 0.3 to 0.7 μM (Strohmaier et al., 1995), but can reach levels as high as 10 μM under conditions of oxidative stress (Uchida, 2003).

Choi et al. (2013) reported that ChE hydroperoxide activates TLR3- and SYK-dependent signaling in macrophages. Using LC/MS/MS and biological assays, the authors identified an oxidized cholesteryl arachidonate with bicyclic endoperoxide and hydroperoxide groups (BEP-ChE) as a specific OxoChE that activates macrophages in a TLR4/MD-2 dependent manner. BEP-ChE induced TLR4/MD-2 binding and TLR4 dimerization, phosphorylation of SYK, ERK1/2, JNK and C-Jun, as well as cell spreading and uptake of dextran and native LDL by macrophages. It was suggested that BEP-ChE is an endogenous ligand that activates the TLR4/SYK signaling system. Kawai et al. (2003) previously demonstrated a covalent binding of oxo-ChE aldehydes to LDL.

Enzyme Activation

Several groups have reported activation of phospholipases by autoxidized glycerophospholipids, including the core aldehydes, but the mechanism of this activation has remained unclear. Furthermore, reactive low molecular weight aldehyde (e.g., 4-HNE) binds covalently to AMP-activated kinase reducing its activity, rather than activating it (Shearn et al., 2014).

Snake Venom sPLA$_2$

Sevanian and Kim (1985) and Sevanian et al. (1988) had shown oxo-PtdCho to be a preferred substrate for low molecular weight PLA$_2$s. Van den Berg et al. (1993) reported an accelerated hydrolysis of oxo-phospholipids by snake and bee venom PLA$_2$s and suggested that oxo-phospholipids exhibit an altered conformation, which may activate the enzyme by facilitating access to the *sn*-2 bonds. The small amounts of these oxo-lipids could not account for the observed 30-80% destruction of the PtdCho by group IIA sPLA$_2$.

Group IIA sPLA$_2$

Eckey et al. (1997) had reported that oxo-PtdCho of LDL served as an activator of group IIA sPLA$_2$. The stimulation of the activity of group IIA sPLA$_2$ by oxo-PtdCho was attributed to a deformation of the lipid monolayer of the lipoprotein particle, but specific oxo-PtdChos involved in the process were not identified. It was therefore of interest to relate the content of PtdCho-hydroperoxides, PtdCho-hydroxides, and PtdCho isoprostanes, the major oxidation products of the lipoproteins, to their susceptibility to hydrolysis by group IIA sPLA$_2$, which normally has low activity towards zwitterionic PtdChos (Pruzanski et al., 1998).

Figure 10.13 shows the time course of hydrolysis of the 16:0/20:4 and 18:0/20:4 GroPCho, the precursors of PtdCho-IPs, and of the iso-PGF$_2$ ester of 16:0(18:0) GroPCho of LDL (A) and APHDL (B) by group IIA sPLA2 (2.5 ug/ml) over a period of 1–24 hours (Kuksis and Pruzanski, 2014).

There are about 7 nanomoles /mg protein of isoPGF$_2$ ester in the LDL sample at 0 time, but it increases to 18 nmoles/mg protein after 2 hours of incubation, after which it decreases to 10, 5, and 0 nmoles/mg protein at 4, 8, and 24 hours. In B, a closely similar trend of hydrolysis for the group IIA sPLA$_2$ is seen for the APHDL, except that the effect is reproduced at a nearly 10 times lower concentration of PtdCho-IsoPs. In the meantime, the hydrolysis of the PtdCho-IsoP precursors (native PtdCho) proceeds at a nearly linear rate, not responding to the changes in the isoprostane production of hydrolysis.

On the basis of model building, Morrow et al. (1990) suggested that the kinked structures of isoprostanes would affect cell membrane structure and function. The minimal presence and the rather slow hydrolysis of the oxo-PtdCho would appear to

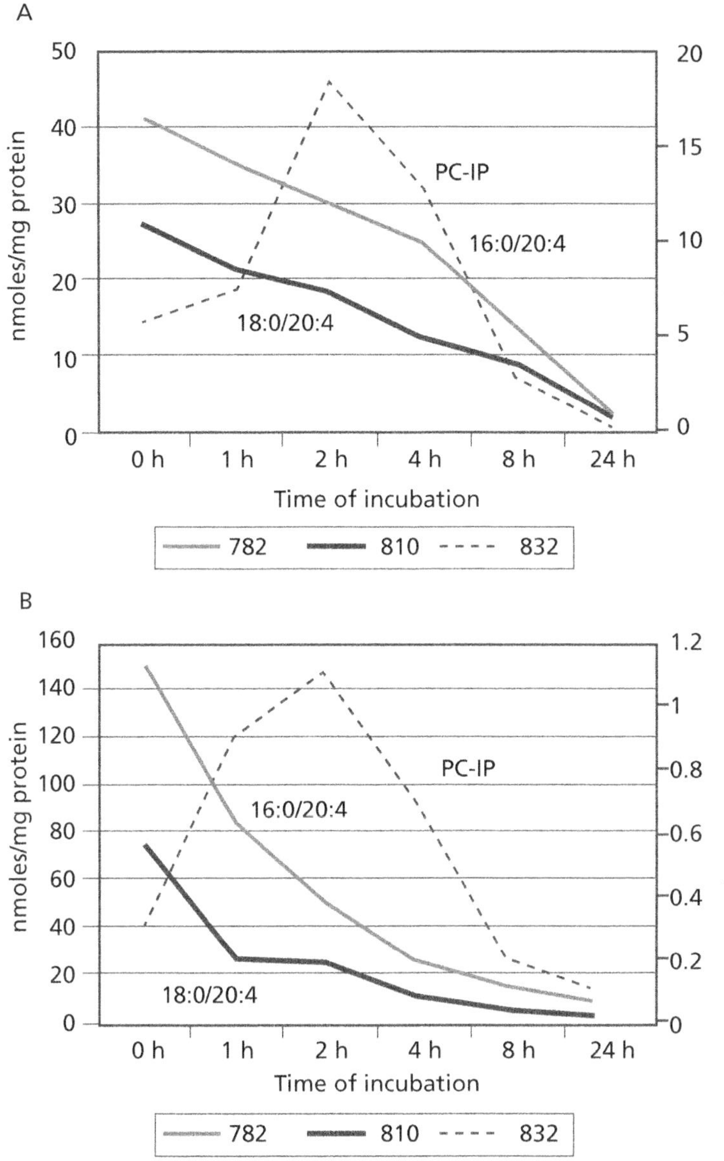

Figure 10.13 Time course of hydrolysis of the 16:0/20:4 and 18:0/20:4 GroPCho, the precursors of PtdCho-IPs, and of the iso-PGF2 ester of 16:0(18:0) GroPCho (PC-IP) of LDL (A) and APHDL (B) by group IIA sPLA2 (2.5 ug/ml) over a period of 1–24 hours (from Kuksis and Pruzanski, 2014, unpublished).

exclude the PtdCho hydroperoxides and hydroxides as likely candidates for the activation of group IIA $sPLA_2$ by oxo-PtdCho (Kuksis and Pruzanski, 2014). It was calculated that for a demonstrable activation of the enzyme in the present study, about 1 nanomole of PtdCho isoprostane/mg protein was necessary, which worked out to less than 1 molecule of isoprostane per 500 molecules of native PtdCho species.

Alternatively, the activation of group IIA $sPLA_2$ by oxo-PtdCho could have resulted from the presence of PtCho core aldehydes, which are also generated during PtdCho oxidation as shown by Kamido et al. (1992b). This possibility finds support in recent reports by Korotaeva et al. (2010) and Code et al. (2010). Using [^{14}C] arachidonoyl GroPEtn as substrate, Korotaeva et al. (2010) demonstrated a two-fold stimulation of group IIA $sPLA_2$ activity by oxidied palmitoyl/arachidonoyl GroPCho incorporated into liposomes or LDL. Furthermore, Korotaeva et al. (2010) have now reported that fresh plasma lipoproteins inhibit the activity of group IIA $sPLA_2$, but that the inhibition can be relieved by PtdCho autoxidation. One mechanism of relieving the enzyme inhibition might be a covalent binding of the PtdCho core aldehyde. Covalent binding of the core aldehydes leading to enzyme activation would avoid the need for lipid monolayer penetration by the $sPLA_2$, which has now been demonstrated not to take place (Huang et al., 2011). Code et al. (2010) have proposed that the core aldehyde forms a transient Schiff base with PLA_2, inducing aggregates that trap PLA_2 into transient highly reactive protofibril intermediates, increasing the overall activity.

In view of the demonstration of carboxyalkylpyrrole formation (Salomon et al., 2011) during oxidation of arachidonoyl PtdCho in presence of protein, including enzyme proteins, it is possible that Michael adduct formation with an ε-amino group of lysine is responsible for the core aldehyde binding and group IIA $sPLA_2$ activation.

Bee Venom PLA$_2$

Code et al. (2010) recently demonstrated the activation of bee venom phospholipase A_2, which is believed to be closely similar to group IIA $sPLA_2$ in structure and activity, by incubating 1-palmitoyl-2-[9-oxo]nonanoyl GroPCho in a dipalmitoyl GroPCho matrix. The presence of the core aldehyde shortened and abolished the lag time in action of the enzyme and concomitantly induced enhanced hydrolysis of a fluorescent phospholipid analogue (C_{28}-OPHPM, 1-octacosanyl-2-[pyren-1-yl]hexanoyl-sn-glycero-3-phosphatidylmonomethyl ester). The effect of the core aldehyde was abolished by the aldehyde scavenger methoxyamine. The authors proposed that PtdCho core aldehyde forms a transient Schiff base with PLA_2, inducing aggregates that trap PLA_2 into transient highly reactive profibril intermediates increasing the overall activity.

Phospholipase D

Natarajan et al. (1995) reported that oxo-LDL mediated activation of PLD in smooth muscle cells. The activation was specific for copper oxidized, TBAR reactive product;

native LDL or acetylated LDL had no effect. The activation of PLD by oxo-LDL generated second-messenger-like PtdOH and lysoPtdOH derivatives, which modulate mitogenesis.

Medical Relevance

Oxo-phospholipids have profound effect on gene expression. Oxo-PAPCs have been shown to modulate the expression of approximately 1000 genes in human aortic ECs, which include both up-regulated and down-regulated mRNAs (Gargalovic et al., 2006). Although some autoxidation products of polyunsaturated PtdCho have been shown to exert specific biochemical and physiological effects, the chemical diversity and relative abundance of the oxo-PtdCho products has made it difficult to relate and quantify any disease condition to a specific component of the oxo-PtdCho products.

The molecular properties of oxo-phospholipids related to disease have been reviewed (Fruwirth et al., 2007) as have been the pathological aspects of lipid peroxidation (Negre-Salyre et al., 2010). The recent discovery of the carboxylalkylpyrrole modification of proteins is now considered of great pathogenic importance in age-related macular degeneration, autism, and cancer, and in promotion of wound healing (Salomon and Gu, 2011; Salomon et al., 2011).

Inflammation and Vascular Effects

Inflammation is characterized by a massive production of ROS. The oxo-PLs production in response to inflammation is induced by different cell types, including leukocytes. Phorbol ester-stimulated neutrophils and monocytes incubated with PUFA-PCs produced mono- and bis-hydroperoxides of PtdCho as well as PtdCho-isoPs, thus suggesting that activated phagocytes can oxidize lipids in the surrounding medium (Jerlich et al., 2003).

Basu (2010) presented a comprehensive review of prostaglandin $F_{2\alpha}$ and F_2-IsoPs in inflammation and oxidative stress–related pathology and has concluded that the emerging roles of isoprostanes in oxidative stress and inflammation will open a new frontier of oxidative stress research, both preclinically and clinically, in the coming decades.

Araki et al. (2011) reported that peroxidation of n-3 polyunsaturated fatty acids (EPA and DHA) inhibit the induction of iNOS gene expression in proinflammatory cytokine-stimulated hepatocytes. It was suggested that peroxidized products, but not malonic acid, suppressed the induction of iNOS gene expression through both the transcriptional and posttranslational steps, leading to the prevention of hepatic inflammation. The peroxidation products responsible for this effect were not identified or discussed, although the formation of neuroprostanes, for example, from the long chain n-3 fatty acids has been long known (Roberts et al., 1998), as has been the formation of hydroperoxides, hydroxides, and aldehydes other than malonic.

Greig et al. (2012) reviewed the physiological effects of oxidized phospholipids and their cellular signaling mechanisms in inflammation. The authors point out that oxo-phospholipids were viewed as culprits, in line with observations that they have proinflammatory effects and enhance inflammatory cytokine production, cell adhesion, migration, proliferation, apoptosis, and necrosis, especially in vascular endothelial cells, macrophages, and smooth muscle cells. However, evidence has emerged that these compounds also have protective effects in some situations and cell types; a notable example is their ability to interfere with signaling by certain TLRs induced by microbial products that normally lead to inflammation. The various and sometimes contradictory effects that have been observed for oxidized phospholipids depend on their concentration, their specific structure, and the cell type investigated. For example, enzymatically oxygenated derivatives of omega-3 fatty acids of DHA and EPA, known as *resolvins,* have potent inflammation resolution activity (Dangi et al., 2009). Rosenson et al. (2012) have discussed the potential benefits of modulating the peroxisome proliferator–activated receptors in clinical practice, along with its future prospects.

Galano et al. (2013) described, total synthesis of isoprostanes and neuroprostanes and discussed their usefulness as biomarkers of oxidative stress in humans. Chemically synthesized 5-hydroxy-[8-oxo]-6-octenoyl-GroPCho (HOOA-PC) exhibited properties of a chemical mediator of chronic inflammation. HOOA-PC was found unbound and in pyrrole adducts in lipid extracts of oxo-LDL and human atheroma (Hoff et al., 2003; Podrez et al., 2002).

Recently, Pruzanski and Kuksis (2014) demonstrated the effects of $sPLA_2$ hydrolysis of proinflammatory products of oxo-PtdCho on mitogenesis of VSMC. The oxo-PtdChos were generated during incubation of HDL and LDL with the $sPLA_2$s at 37 °C with maximum formation of oxo-PtdCho between 2 and 4 hours. The major molecular species of oxo-PtdCho were identified and quantified by LC-ESI-MS at 0, 4, and 24 hours of incubation of LDL and HDL with VSMC and with group IIA, V, and X $sPLA_2$s in the presence of normal and oxo-HDL and oxo-LDL additives. The mechanism responsible for the apparent increase in the mitogenic activity of the VSMC cells following incubation with the hydrolysis products of oxo-PtdCho-isoprostanes is not known, as is not known the special role of group V $sPLA_2$, although it has been shown to hydrolyze the PtdCho-linoleoyl-hydroperoxides and hydroxides in preference to those of the PtdCho-arachidonoyl-hydroperoxides and hydroxides. Group X $sPLA_2$ showed slight preference for hydrolysis of the oxo-arachidonoyl GroPChos. Earlier, Kogure et al. (2003) reported temporary membrane distortion of VSMCs to be responsible for the apoptosis by oxo-PtdCho, while Moumtzi et al. (2007) documented the fate of fluorescent analogs of oxo-PtdCho following import into VSMCs. Palmitoyl/glutaroyl–GroPCho derivatives were translocated to lysosomes, but palmitoyl/oxo-valeroyl-GroPCho analogs were initially captured in the plasma membrane. The effects of 4-hydroxynonenal on vascular endothelial and smooth muscle cell redox signaling and function in health and disease have been reviewed by Chapple et al. (2013). The production of reactive aldehydes, including 4-hydroxy-2-nonenal (4-HNE), is a key component of the pathogenesis in a spectrum of chronic inflammatory hepatic diseases including alcoholic liver disease (ALD). One consequence of ALD is increased oxidative stress and altered β-oxidation in hepatocytes. Shearn et al. (2014) identified AMPK as a direct target of 4-HNE adduction resulting in inhibition of both H_2O_2 and 5-aminoimiadazole-4-carboxyamide ribonucleoside (AICAR)-induced downstream signaling. Carbonylation of proteins contributes to increased hepatocellular damage during alcoholic liver disease. Mass spectrometry identified Michael addition adducts of 4-hydroxynoneal (4-HNE) on Cys130, 174,227, and 304 on recombinant 5′ AMP protein kinase (AMPK) as a direct target of 4-HNE adduction resulting in inhibition of both H_2O_2 and downstream signaling (Shearn et al., 2014). Molecular modeling analysis of identified 4-HNE adducts on AMPK suggested that inhibition of AMPK occurs by steric hindrance of the active site pocket and by inhibition of hydrogen peroxide induced oxidation.

Epoxyeicosatrienoic acids (EETs) are small molecules produced by cytochrome P450 epoxygenases. They are lipid mediators that act as autocrine or paracrine factors to regulate inflammation and vascular tone (Fleming, 2008). The contribution of autoxidation to the formation of resolvins and protectins in inflammation-resolution does not appear to have been assessed (Serhan and Petassis, 2011).

Atherosclerosis

Originally considered purely a lipid storage disease, atherosclerosis is now recognized as a disease linked to inflammation, particularly to involvement of innate and adaptive immunity (Jafraim et al., 2012). Haberland et al. (1988) and Palinski et al. (1989) demonstrated that monoclonal antibodies directed against protein adducts, such as MDA-modified LDL, bind epitopes in atherosclerotic lesions. They proposed the hypothesis that the formation of MDA and modification of the lysine residues of LDL apoB occur in vivo as a prerequisite to the formation of arterial foam cells and contribute to the development of atherosclerosis. The participation of reactive aldehydes in atherosclerosis is also suggested by a series of immune-histochemical analyses of atherosclerotic lesions from the human aorta using various antibodies against short-chain aldehyde-amino acid adducts, such as 4-hydroxy-2-nonenal-histidine (Uchida et al., 1995), MDA-lysine (Uchida et al., 1997; Yamada et al., 2001), and acrolein-lysine (Uchida et al., 1998). These studies provide evidence for the presence of oxidized LDL (or at least of antigens closely related to it) in arterial lesions at significant concentrations and suggest that reactive aldehydes may play a role in the formation of arterial foam cells.

The presence of oxo-PtdCho-apoprotein B adducts in vivo is supported by the detection with monoclonal antibodies specific for 1-palmitoyl/2-[9-oxo]nonanoyl-GroPCho, their elevated plasma levels in acute coronary syndromes (Ehara et al., 2001), and their immunological detection in atherosclerotic lesions (Haberland et al., 1988; Horkko et al., 1997; Palinski et al., 1996). Hedrick et al. (2000) provided in vivo evidence for the presence of oxidized phospholipids in HDL. The data of Ahmed et al. (2003) indicate that this modification is functionally significant and could result in the uptake of oxidized HDL by macrophages.

Ox-PtdCho/apoB content was measured by chemiluminescence enzyme-linked immunosorbent assay, as described previously in detail, using the murine monoclonal antibody E06, which binds to the phosphocholine head group of oxidized but not native PtdCho (Tsimikas et al., 2004, 2005). The oxo-PtdCho/apoB assay does not measure the total oxo-PtdCho present in plasma or ox-PtdCho not detected by E06, and E06 does not cross-react with apolipoprotein A [Lp(a)]. Tsimikas et al. (2010) recently simplified such attempts by relying on oxidation-specific biomarkers to establish pathophysiological links to coronary artery disease (CAD). A prospective case-control study based on the European Prospective Investigation of Cancer–Norfolk cohort of 45- to 79-year-old apparently healthy men and women followed for up to 6 years was designed. Baseline levels of oxo-phospholipids on apoB particles and Lp a were measured in 763 cases and 1397 controls. Their relationship to group IIA $sPLA_2$ mass and activity, myeloperoxidase mass, and lipoprotein-associated PLA_2 activity and association with CAD events were determined. The oxidation-specific biomarkers provided cumulative predictive value when added to traditional cardiovascular risk factors.

Ascribing specific biological effects of oxo-LDL to defined chemical components of oxo-LDL has been accomplished in a number of laboratories. Steinbrecher (1987) reported oxidized phospholipid adducts with the lysine group of apoB. Antibody-based studies revealed the presence of carboxyethoxypyrroles (HPs) and carboxypropylopyrroles (CPPs) in oxo-LDL (Kaur et al., 1997). Also, the CHP immunoreactivity, reflecting the presence of protein adducts of 9-hydroxy-[12-oxo]-10-dodecenoic acid (HODA) or its PtdCho ester in human plasma, was significantly higher in the plasma of patients with atherosclerosis and end-stage renal disease than in healthy controls (Kaur et al., 1997). HODSA-protein adducts were produced in vivo from 9-hydroxy-12-oxo-10-dodecenoyl-PtdCho (HODA-PC), one of the oxidized lipids derived from 1-palmitoyl-2-arachidonoyl-sn-glycerol-3-phosphocholine (PA-PC), altogether referred to as oxo-PA-PC. Chemically synthesized 5-hydroxy-[8-oxo]-6-octenoyl-phosphatidylcholine (HOOA-PC) exhibited properties of a chemical mediator of chronic inflammation. HOOA-PC was found unbound and in pyrrole adducts in lipid extracts of oxo-LDL and human atheroma (Hoff et al., 2003; Podrez et al., 2002).

The ability of peroxidized lipids to covalently react with proteins is not limited only to low molecular weight short-chain aldehydes. It has been demonstrated that, after oxidation, radiolabeled phospholipids that are incorporated into LDL become associated with apoB (Steinbrecher, 1987; Tertov et al., 1995). Itabe et al. (1994) also provided evidence that oxidized phospholipids form complexes with lysine residues on proteins due to the presence of [9-oxo]nonanoyl GroPCho. In addition, the Schiff base adducts of phospholipid core aldehydes with lysine residues of myoglobin (Ravandi et al., 1997) and thyroglobin (Wang and Tai, 1990) have been identified. Furthermore, the binding of phospholipids to apo B during LDL oxidation was independently shown by measuring the phosphorus incorporated in the protein (Gilotte et al., 2000), as was the incorporation of oxo-PtdCho in to HDL protein, and of PtdCho core aldehydes into apoA1 (Ahmed et al., 2003).

Ahmed et al. (2003) demonstrated that PtdCho core aldehyde binding by HDL and apoAI results in their uptake by THP-1 macrophages in a manner indistinguishable from the uptake of oxidized LDL. The uptake of DMPC proteosomes by THP-1 macrophages was much lower compared to the uptake of PtdCho core aldehyde proteosomes. These data are consistent with the requirement for specific modifications of the apolipoproteins for recognition by receptors expressed on macrophages, rather than selectivity for a specific apolipoprotein (Steinberg and Witztum, 1999). Furthermore, these results are consistent with the immunohistochemical identification of apoAI in atherosclerotic lesions (Mackness et al., 1997; Vollmer et al., 1991). The intracellular location of apoAI and apoAII in the intima of early stage human coronary and aortic lesions (Vollmer et al., 1991) is consistent with the binding and uptake of oxidized HDL by THP-1 cells. In view of the results of Ahmed et al. (2003), the modification of apoAI following HDL oxidation and its uptake by THP-1 macrophages suggests a specificity for oxidized PtdCho. In addition, regulation of nuclear signaling pathway via PPAR γ-ligands activators has been demonstrated using oxidatively fragmented alkyl phospholipid (2-azelaoyl-GroPCho) and 9-HODE, 3-HODE from oxo-LDL (Davies et al., 2001; Nagy et al., 1998). Oxidized phospholipid HOOA-PC increases production of both IL-8 and MCP-1 from endothelial cells (Subbagounder et al., 2002). These are but a few examples of the biological effects associated with oxidized LDL.

Kamido et al. (2002) identified the same major core aldehydes of alkylglycerophosphocholines in atheroma and observed their induction of platelet aggregation and inhibition of endothelium-dependent arterial relaxation. The nature of the oxolipids present in various fractions of atherosclerotic plaques have been identified and quantified by LC-ESI-MS by Ravandi et al. (2004), Kamido et al. (2002), and Hoppe et al. (1997). Along this line, LDL glycation has been suggested to be responsible for the increased susceptibility to atherogenesis of diabetic subjects. Thus, Ravandi et al. (1999) reported that PtdEtn glycation of LDL is accounted for completely by the known effect of LDL glycation on macrophage uptake, ChE, and TAG accumulation.

Kawai et al. (2003) summarized the evidence that aldehyde molecules generated from lipid peroxidation play a role in the pathogenesis of atherosclerosis. The authors call attention to the following major points: (1) increasing plasma level of reactive aldehydes in relation to extensive aortic atherosclerosis; (2) high concentration of aldehydes generated during LDL lipid oxidation; (3) structural and functional changes in LDL upon interaction with aldehydes; and (4) reaction of aldehydes with critical lysine residues of apoB, that produce internalization by scavenger receptor of human monocyte-macrophages, and subsequent intracellular accumulation of the lipoprotein derived ChE.

Kawai et al. (2003) speculated that the modification by cholesteryl ester core aldehydes renders proteins relatively resistant to intercellular proteolytic degradation, which, together with the demonstrated resistance of cholesteryl ester core aldehydes to hydrolysis by cholesteryl ester hydrolase (Hoppe et al., 1997), suggests an accumulation of the epitopes in the macrophages. In line with the potential pathophysiological roles of these compounds, the cholesteryl ester and 7-ketocholesteryl ester core aldehydes of varying chain lengths were identified in human atheromas (Hoppe et al., 1997; Karten et al., 1998; 1999).

Both 9-HODE and cholesteryl ester 9-HODE produced during LDL oxidation can induce IL-1 beta from human monocyte derived macrophages (Ku et al., 1992; Thomas et al., 1994). Oxo-sterols 7-ketosterol and 7-hydroxycholesterol, which are formed during LDL oxidation, have been shown to be responsible for oxo-LDL cytotoxicity on porcine aortic small smooth muscle cells (Hughes et al., 1994).

Alzheimer's Disease and Parkinson's Disease

Alzheimer's disease (AD) is a neurodegenerative disorder that is clinically characterized by progressive cognitive decline and dementia. The major neurophysiological hallmarks of this age-related disease are the extracellular accumulation of amyloid β-peptide (Aβ), the intracellular presence of neurofibrilliary tangles composed of hyperphosphorylated tau, and the loss of cholinergic neurons in the brain (Jafraim et al., 2012). Alzheimer's subjects are believed to be in an heightened state of oxidative stress, which is characterized by high levels of lipid peroxidation products in blood (acrolein, 4-HNE, and MDA) (Arlt et al., 2002; Butterfield et al., 2005, 2010; Schrag et al., 2013).

The lipid peroxidation product 4-hydroxy-2-nonenal (HNE) is proposed to be a toxic factor in the pathogenesis of Alzheimer's disease. HNE was generated from one of the phospholipid hydroperoxides, 1-palmitoyl-2-(13-hydroperoxy-cis-9,trans-11-octadecadienoyl) GroPCho, by free Cu^{2+} in the presence of ascorbic acid through Cu^{2+} reduction and degradation of PLPC-OOH. HNE generation was markedly inhibited by equimolar concentrations of Aβ. However, Aβ binding 2 or 3 molar equivalents of Cu^{2+} acted as a pro-oxidant to form HNE from PLPC-OOH. These findings suggested that at moderate concentrations of copper, Aβ acts primarily as an antioxidant to prevent Cu^{2+}-catalyzed oxidation of biomolecules, but that in the presence of excess copper, pro-oxidant complexes of Aβ with Cu^{2+} are formed (Butterfield et al., 2010).

In comparative stages of Alzheimer's disease, there are significantly higher CSF concentrations of 24S-hydroxycholesterol, suggesting increased cholesterol turnover in the CNS during degeneration (Leoni and Caccia, 2011). However, cholesteryl arachidonate is also readily oxidized to core aldehydes, the effect of which has not been examined in parallel to that of HNE.

Pizzimenti et al. (2013) recently reviewed the role of lipid peroxide-derived aldehyde adducts of proteins whose oxidative modifications might be related to the neuronal dysfunctions observed in Alzheimer's disease. The authors have tabulated the HNE-protein adducts detected in Alzheimer's disease in relation with the progress of the disease. HNE can react with the Aβ peptide. HNE modified the three histidyl residues of Aβ, so the HNE-modified Aβ molecules had increased affinities for membrane lipids and adopted a similar conformation as mature amyloid fibrils (Liu et al., 2008; Murray et al., 2007). Sultana et al. (2011) reviewed the evidence for other low molecular weight aldehyde involvement in protein adduct formation in Alzheimer's and have summarized the results in a redox model of Alzheimer's disease pathogenesis.

Starting from the early observation that protein modified by 2-pentylpyrrole incorporation of lysyl ε-amino groups, upon covalent addition of HNE, accumulates in the brain neurons of patients with Alzheimer's disease (Sayre et al., 1997), it is obvious that γ-hydroxyalkenal phospholipids and their ω-carboxyalkylpyrrole derivatives contribute strongly to the pathogenesis of Alzheimer's disease (Salomon and Gu, 2011; Stemmer and Hermeter, 2012).

In contrast, Parkinson's disease (PD) is a neurodegenerative disease involving the dopaminergic neurons in the *substantia nigra.* It is characterized clinically by bradykinesia, resting tremor, and progressive rigidity (Lowe et al., 1997). Oxidative stress has been implicated as a key event in PD neurodegeneration. Products of oxidative damage have been shown to be elevated in the *substantia nigra* from PD patients, including the lipid peroxidation product 4-hydroxy-nonenal and protein carbonyl oxidation products (Alam et al., 1997). Alam et al. (1997) measured protein carbonyls in postmortem brain tissues from patients with PD compared to controls. Increased carbonyl levels were also found in areas of the brain not thought to be affected in PD. The data showed that L-DOPA treatment probably contributes to protein oxidation.

Fessel et al. (2003) reported that isofurans, but not F_2-isoprostanes, are increased in the *substantia nigra* of patients with PD and with dementia with Lewy body disease. The preferential increase in IsoFs in the *substantia nigra* of patients with PD or Lewy body disease presented a unique model of oxidant injury in these diseases, but also suggested different underlying mechanisms of dopaminergic neurodegeneration in PD and Lewy body disease from those of degenerative conditions. Fessel et al. (2003) attributed the preferential isofuran formation to an increased local oxygen tension in the nigral tissue in these two diseases, although the mechanisms are not fully understood.

Seet et al. (2010) recently attempted to reassess the oxidative damage in Parkinson's disease by measuring the levels of accurate biomarkers. Using products of lipid and DNA oxidation measured by accurate methods, the authors assessed the extent of oxidative damage in PD patients. The levels of plasma F_2-isoprostanes (F_2-IsoPs), hydroxyeicosatetraenoic acid products (HETEs), cholesterol oxidation products, neuroprostane (F_4-NPs), phospholipase A_2 (PLA_2) and PAF-AH activities, urinary 8-hydroxy-2'-deoxyguanosine (8-OHdG), and serum C-reactive protein were compared in 61 PD patients and 61 age-matched controls. The levels of plasma F_2-IsoPs, HETEs, 7β- and 27-hydroxycholesterol, 7-ketocholesterol, F_4-NPs, and urinary 8-OHdG were elevated, whereas the levels of plasma PLA_2 and PAF-AH activities were lower in PD patents compared to controls. It was concluded that oxidative damage markers are systematically elevated in PD, which may give clues to the relation of oxidative damage to the onset and progression of PD.

Liu et al. (2014) recently pointed out that lipid hydroperoxides could react with the primary amino group of dopamine to form amide-linkage dopamine adducts, which could affect the etiology of Parkinson's disease.

A comparative review (Sayre et al., 2008) has highlighted the role of oxidative stress in Alzheimer's, Parkinson's, and Huntington's diseases, as well as amyotrophic lateral sclerosis and multiple sclerosis, which are recognized as neurodegenerative neuro-inflammatory disorders, where there is evidence for a primary contribution of oxidative stress in neuronal death.

Carcinogenesis

There is a suggestion that dietary fat is a risk factor for cancer development. The presence of peroxides and their decomposition products could contribute to this by promoting mutation, activation of carcinogens, and abnormal cell proliferation (Halliwell et al., 2000). In contrast, epidemiological and preclinical evidence supports the idea that omega-3 dietary fatty acids (fish oil) reduce the risk of cancer (Abeywardena and Patten, 2011). The mechanisms by which these omega-3 lipids inhibit tumorigenesis, however, remain to be explained.

Epoxyeicosatrienoic acids (EETs) are small molecules produced by cytochrome P450 epoxygenases (Panigrahy et al. 2010). EET levels are directly influenced by nutrients and inflammatory processes (Node et al., 1999). Panigrahy et al. (2012) found in several tumor models that EETs stimulate multiorgan metastases and escape from tumor dormancy in mice. The systemic metastasis depended on endothelium-derived

EETs at the site of metastasis. Administration of synthetic EETs recapitulated these results, while EET antagonists suppressed tumor growth and metastasis, demonstrating in vivo that pharmacological modulation of EETs can affect cancer growth.

Zhang et al. (2013) have shown that epoxydocosapentaenoic acids (EDPs), which are lipid mediators produced by cytochrome P450 epoxygenases from omega-3 fatty acid docosahexaenoic acid (Arnold et al., 2010), inhibit vascular endothelial growth factor (VEGF)-induced and fibroblast growth factor 2-induced angiogenesis in vivo and suppress endothelial cell migration and protease production in vitro via a VEGF receptor 2-dependent mechanism. By shutting off a tumor's blood supply, these compounds can act to dramatically slow tumor growth and prevent spread. Contrary to the effects of EDPs, the corresponding metabolites derived from omega-6 arachidonic acid, such as epoxyeicosatrienoic acids, increase angiogenesis and tumor progression (Panigrahy et al., 2012; Zhang et al., 2013).

The possibility that ROS provide a a link between chronic inflammation and cancer has been suggested by Salman and Ashraf (2013).

Diabetes

Although glycation of proteins has been investigated extensively, little attention has been paid to lipid glycation. In vivo, lipid glycation is likely to induce changes in the biosynthesis and turnover of membranes, the activity of membrane-bound enzymes, and the susceptibility to oxidative stress. These changes may contribute to the pathology of the chronic disease, atherosclerosis, and aging.

Protein glycation progresses during normal aging and at an accelerated rate in patients with diabetes mellitus (Krolewski et al., 1991). Oxidative stress on lipids also plays a role in the pathophysiology of diabetes (Baynes, 1991). The presence of a free amino group in aminophospholipids could be modified by glycation (Bucala et al., 1993). There was a four-fold increase of Maillard Reaction Products (MRP) lipids in LDL from diabetic patients versus healthy subjects that reported using enzyme-linked immunoadsorbent assay (ELISA), suggesting a greater increase in Maillard-type modifications of aminophospholipids by diabetes (Bucala et al., 1993). Ravandi et al. (1995) prepared the Schiff base adducts/Amadori products of glucosylated aminophospholipids. Using LC/MS, it was possible to demonstrate the presence of Amadori-PtdEtn adduct in diabetic blood plasma, red blood cells, and human atherosclerotic plaques (Ravandi et al., 1996, 2000). Subsequently, using a QTRAPLC/MS method, Nakagawa et al. (2005) confirmed that Amadori-PtdEtn was present in both normal and diabetic human plasma. Furthermore, they showed that Amadori-PtdEtn product was present in higher amounts in plasma of diabetic patients than in healthy controls. Recently, Sookwong et al. (2011) proposed Amadori-glycated PtdEtn as a potential marker for hyperglycemia in streptozotocin-induced diabetic rats.

Aging

It was postulated some 50 years ago that complications arising as individuals age could be due to lipid peroxidation and an accumulation of molecular and cellular damage induced by free radicals (Harman, 1956). A decrease in mitochondrial function has repeatedly been advanced as a primary key event, especially on the basis of analyses of skeletal muscle mitochondria (Liu and Ames, 2005). Under normal physiological conditions, about 1–5% of oxygen consumed by mitochondria is converted into reactive oxygen species. Several sites of the mitochondrial respiratory chain (MRC) are involved in the generation of such oxygen species, which have been claimed to be maintained at a relatively high steady state level in mitochondrial matrix (Lee et al., 1997). Miro et al. (2000) found a progressive, significant increase of heart membrane lipid peroxidation with aging. The study included subjects 8–86 years old, of which 75% were males and 29% were smokers. Lipid oxidation was measured by spectrophotometric and fluorometric methods. MRC enzymes, however, remained preserved in the heart with aging and could not be considered the main cause of the increased oxidative damage with aging. Prashant et al. (2007) reported significant age-related increases in malondialdehyde and decreases in antioxidants in normal elderly people. Highly significant increases in MDA and decreases in antioxidants were observed in elderly people when complicated by diabetes and hypertension. The changes in lipid peroxidation and antioxidants in elderly people (60–75 years of age, of both sexes) were determined. Malondialdehyde was measured in red blood cells by means of the TBA reaction, while the antioxidants catalase and glutathione were determined by spectrometric methods. Kinoshita et al. (2000) have reported age related increases in plasma PtdCho hydroperoxide concentrations in control subjects and patients with hyperlipidemia.

Although aging is regularly mentioned as a side effect of lipid peroxidation, no specific oxo-lipid has been identified as the culprit. Most of the tests have been indirect and of insufficient duration. An exception would appear to be the Maillard reaction products (MRPs) of lipids. Like proteins, Maillard reaction products of lipids are of interest to aging due to their potential contribution to the physiology of the aging process and age-associated diseases such as diabetes and atherosclerosis. There is little evidence, however, that demonstrates MRP-lipid accumulation in biological membranes during aging (Naudi et al., 2013).

The reaction of the cholesteryl ester and PtdCho core aldehydes with primary amines of the proteins may represent a process common to the formation of carbonyl-modified proteins during aging and related diseases (Kawai et al., 2003).

Detoxification of Reactive Oxo-Lipids and Reversal of Autoxidation

Normally, free radicals are neutralized by various free radical defense mechanisms. However, when the endogenous defense mechanisms are overwhelmed, the peroxidation products accumulate and both reversible and irreversible damage result, which may lead to aging and disease (Greig et al., 2012; Salomon et al., 2011).

Lumenal Defense Mechanisms

The gastrointestinal GSH peroxidase isozymes represent the first line of defense against ingested hydroperoxides. Other protective mechanisms rely on antioxidants and lipolytic enzymes, as well as limitation of ingestion of already peroxidized or easily peroxidized foods.

There is evidence that the lipid peroxide absorption from the lumen depends on the availability of cellular reductants (GSH and NADPH) (Awe et al., 1992; LeGrand and Aw, 2001) that largely dictate the kinetics and extent of intracellular peroxide detoxification by the intestine.

GSH peroxidase and GSSG reductase act as an enzyme pair in the reduction of peroxides, with concomitant oxidation of GSH and the regeneration of GSH with reducing equivalents donated from NADPH (LeGrand and Aw, 2001). The enzyme pair is responsible for much of the protection against intestinal absorption of hydroperoxides. GSH reduces the amount of hydroperoxides transported from the human gut into lymph.

It has been shown that TAG-OOH appearance in lymph and chylomicrons is very low or undetectable as long as ascorbate and/or ubiquinols are present (Mohr et al., 1999). Toxicity of aldehydes can be prevented by replenishing the exhausted thiol scavengers (Fenaille et al., 2002). Mitochondrial aldehyde dehydrogenase 2 (ALDH2) detoxifies HNE by oxidizing its aldehyde group (Flohe-Brigelius, 1999).

Other antioxidant systems based on ascorbate, ubiquinol, and α-tocopherol have also been found to provide an efficient defense mechanism against formation of lipid peroxides in mesenteric lymph (Wingler et al., 2000). However, α-tocopherol (Upston et al., 1999) has been occasionally shown to be ineffective as an antioxidant, and ascorbic acid (Lee et al., 2001) has been found to actively decompose linoleic hydroperoxides via one electron reduction, yielding a variety of reactive aldehyde and ketone products. Aldo-keto reductase (AQKR) family member 10 (AKR1B10) is a monomeric enzyme that efficiently catalyzes the reduction to corresponding alcohols of a range of aromatic and aliphatic aldehydes and ketones, including highly electrophilic α,β-unsaturated carbonyls and antitumor drugs containing carbonyl groups with NADPH as a co-enzyme. AKR1B10 is primarily expressed in the normal human colon and small intestine. AKR1B10 silencing increases the levels of α,β-unsaturated carbonyls, leading to a two- to three-fold increase of cellular lipid peroxides. Other protective mechanisms consist of lipolytic activity of pancreatic lipase, cholesterol ester hydrolase, and pancreatic phospholipase A_2.

Vascular Defense Mechanisms

The common antioxidants (tocopherols, vitamin C, ubiquinols, etc.) are usually thought of first as the preventers of plasma lipid peroxidation. α-Tocopherol rapidly reacts with lipid peroxyl radicals and is generally regarded as the major lipid-soluble antioxidant in human tissues and lipoproteins (Esterbauer et al., 1992).

Of interest, however, is the coexistence of oxidized lipids and α-tocopherol in all lipoprotein density fractions isolated from advanced human atherosclerotic plaques (Niu et al., 1999). Low and very low density fractions contained most of the lesion lipids and α-tocopherol. Two to five percent of lesion Ch18:2 was present as Ch18:2-O(O)H and distributed more or less equally among all density fractions, yet the content of α-tocopherol per unit of Ch18:2 was higher than that in corresponding plasma lipoproteins. These results demonstrate that α-tocopherol and oxidized lipids coexist in all lesion density fractions, further supporting the notion that large proportions of lipids in lipoproteins of advanced stages of atherosclerosis are oxidized.

Niu et al. (1999) point out that lipoprotein lipid peroxidation resembles emulsion polymerization and that 1-electron oxidants can cause the peroxidation of a substantial proportion of lipids in isolated lipoproteins despite the presence of vitamin E (for a review see Witting et al., 1998). The results of Niu et al. (1999) cast doubt on the rationale of dietary supplementation with vitamin E alone as a strategy to prevent intimal lipoprotein oxidation. This seems to contradict the commonly held view of action of vitamin E in LDL and other lipoproteins as a chain-breaking antioxidants (Esterbauer et al., 1992).

It has been reported that HDL has a protective effect against oxidative modification of LDL (Klimov et al., 1989). The effect was believed to be due to the presence of lecithin-cholesterol acyltransferase in the HDL subfraction, but Mackness et al. (1991) suggested that an enzyme purified from HDL and named *paraoxonase* could be responsible for decreasing the accumulation of lipid peroxidation products from oxo-LDL. The suggestion of Mackness et al. (1991) was later confirmed by Ahmed et al. (2003), who demonstrated that a preparation of paraoxonase obtained from Mackness's laboratory was capable of hydrolyzing PtdCho-hydroperoxides, hydroxides, and isoprostanes as well as PtdCho core aldehydes. However, further investigation proved that this activity was likely due to contamination of the paraoxonase preparation with PAF-AH (Connelly et al., 2005).

Yoshikawa et al. (1997) suggest that HDL_3 exerts more powerful antioxidative effect against LDL oxidation than HDL_2 although the mechanism is not known. It is possible that the difference in distribution of antioxidants between HDL_3 and HDL_2 causes the difference in the antioxidative, protective effects against copper-catalyzed LDL oxidation. Other defense mechanisms available in plasma consist of various hydroperoxide reductases, endogenous lipases, and *trans* esterases. In a review of the chemistry and biochemistry of lipi peroxidation, Gueraud et al (2010) have emphasized 4-hydroxynonenal regulated apoptosis and autophagy as other oxo-lipid detoxification mechanisms.

A specific hydrolysis of oxidized chains at the *sn*-2 position is thought to contribute to detoxification, which could represent a feedback mechanism of limiting proinflammatory effects. PAF acetyl hydrolase activity is well known to attack chain-shortened and oxygenated fatty acids at the *sn*-2 position of PtdCho (Stafforini et al., 2006). Stafforini et al. (2006) reported that PAF acetylhydrolases were solely responsible for hydrolysis of plasma PtdCho-IPs. Kuksis and Pruzanski (2013) have recently shown that the secretory PLA_2s, especially the group V and X enzymes, also attack the PtdCho-IPs, as well as hydroperoxides and hydroxides. Hydrolysis of the PtdCho-IP, PtdCho-OOH, and PtdCho-OHs has also been demonstrated for group IIA $sPLA_2$ when using increased enzyme levels and extended times of enzyme digestion.

Figure 10.14 shows the time course of hydrolysis of PtdCho-IPs by group V (A) and X (B) $sPLA_2$s 1 µg/ug lipoprotein protein. The isoprostanes are identified by their masses (m/z) as indicated in Figure 10.6 earlier in the chapter. The LC/ESI-MS conditions were as given in Figure 10.6.

Figure 10.15 shows the hydrolysis of the HDL_3 PtdCho-OOHs and PtdCho-OHs by group V and X $sPLA_2$s using 0.1 to 2.5 µg enzyme and 4-hour incubation times (Kuksis and Pruzanski, 2013). The hydroperoxides and hydroxides are identified by their masses (m/z) as explained in Figure 10.5, which also gives LC/ESI-MS conditions. There appears to be significant variation in the rates of hydrolysis of the different molecular species of the oxo-PtdChos, which is difficult to evaluate in view of a continued biosynthesis of the oxo-PtdChos especially during incubations with minimal amounts of the enzyme.

There is evidence that the group IIA $sPLA_2$ enzyme, which is slow to digest zwitterionic PtdCho, may be activated by oxo-PtdChos, including the core aldehydes, to hydrolyze both PtdCho and oxo-PtdCho (Eckey et al., 1997). Other endogenous PLA_2 activities that are believed to attack PtdCho-IPs, hydroperoxides, and hydroxides include the lipoprotein-associated PLA_2 and paraoxonase 1, both activities having been recognized as those of PAF acetylhydrolase contamination (Marathe et al., 2003; Stafforini et al., 2006). The substrates and products of of the lipoprotein lipase-associated PLA_2 in oxo-LDL have been identified by ESI-MS (Davis et al., 2008). Still other protective activities include PtdCho-cholesterol acyl transferase, which has been demonstrated to remove oxo–fatty acids from the *sn*-2 position of oxo-PtdCho and transfer them to cholesterol for transport out of plasma as cholesteryl esters. There are reports (Bielicki and Forte, 1999), however, that claim that oxo-PtdCho inhibits PtdCho-cholesterol acyltransferase (LCAT).

Dietary and Therapeutic Intervention

The production of oxo-lipids requires a source of unsaturated lipids, which is inadvertently provided by the diet. Feeding of DHA increases hydroperoxide levels in rat plasma (191–192% of control), liver (170–230%), and kidney (250–340%) when the α-tocopherol level is reduced (21–73%) concomitantly (Song et al., 2000). Consistent with these results, the DHA-oil-fed rats had more TBA reactive substances in

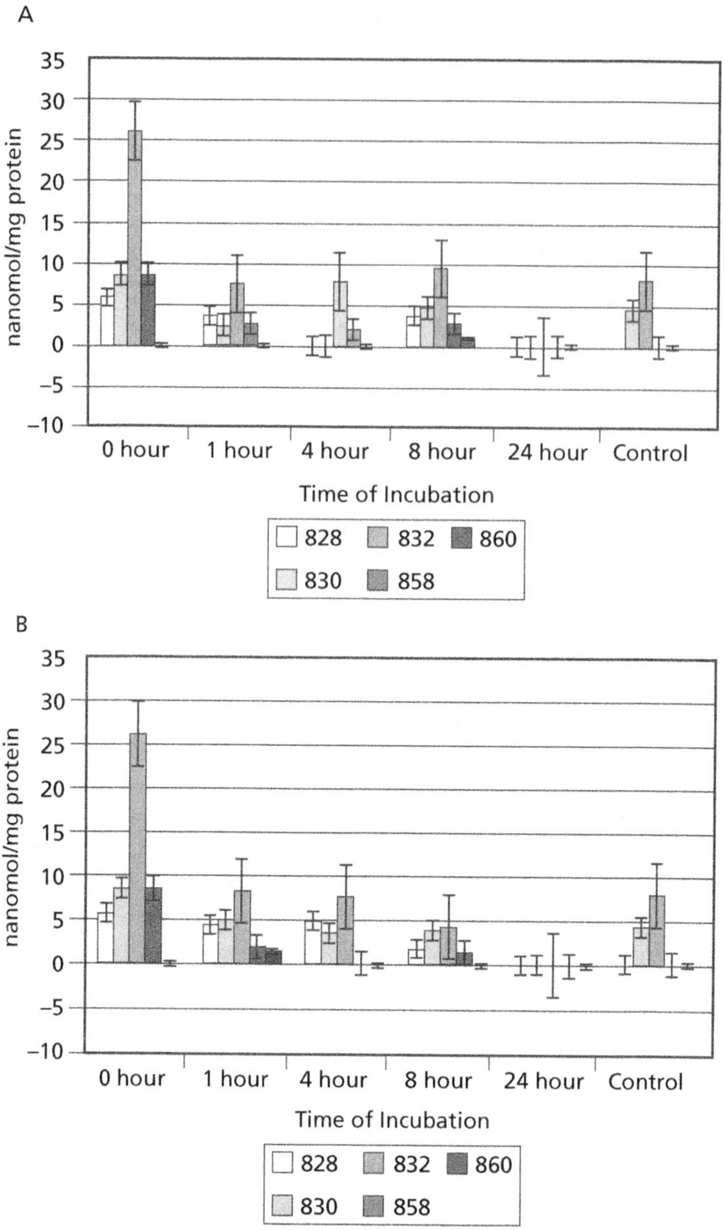

Figure 10.14 Time course of hydrolysis of LDL PtdCho-IPs by group V (A) and X (B) $sPLA_2s$ (1μg/mg protein) (Kuksis and Pruzanski, 2013, unpublished). S.E.M. (*n* = 4–6). Control, 24 hour control. Other legends as given in Figure 10.6.

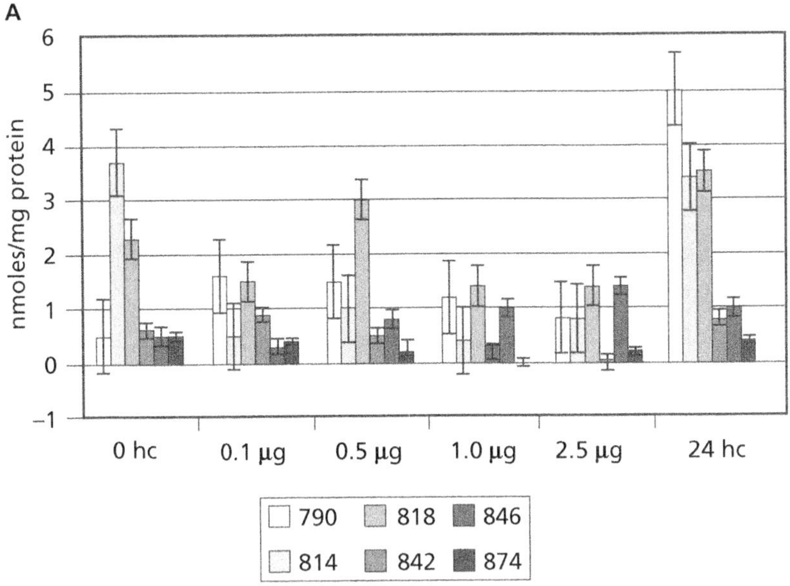

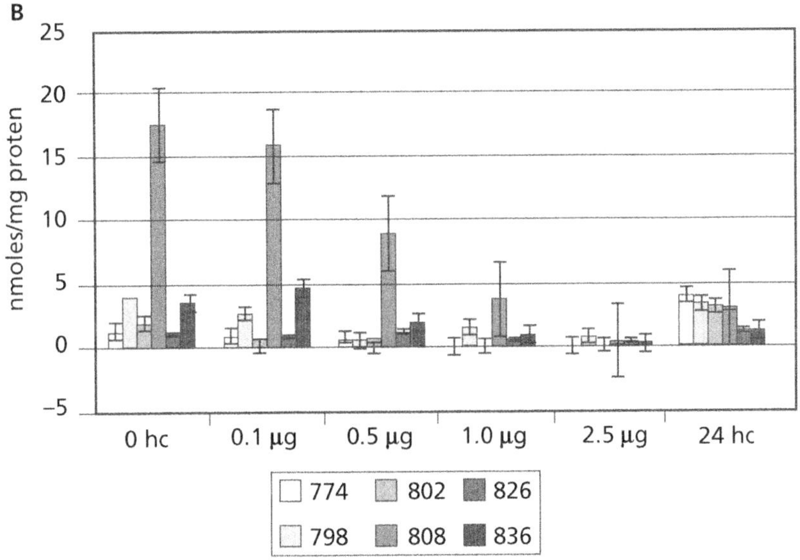

Figure 10.15 Hydrolysis of HDL_3 PtdCho-OOHs (A) and PtdCho-OHs (B) by group V sPLA2s and HDL_3 PtdCho-OOHs (C) and HDL_3 PtdCho-OHs by group X sPLA2 (D) (from Kuksis and Pruzanski, 2013). S.E.M. (n = 3–6). 24 hc: 24 hour control. LC/ESI-MS and ion identification as in Figure 10.5.

(continued)

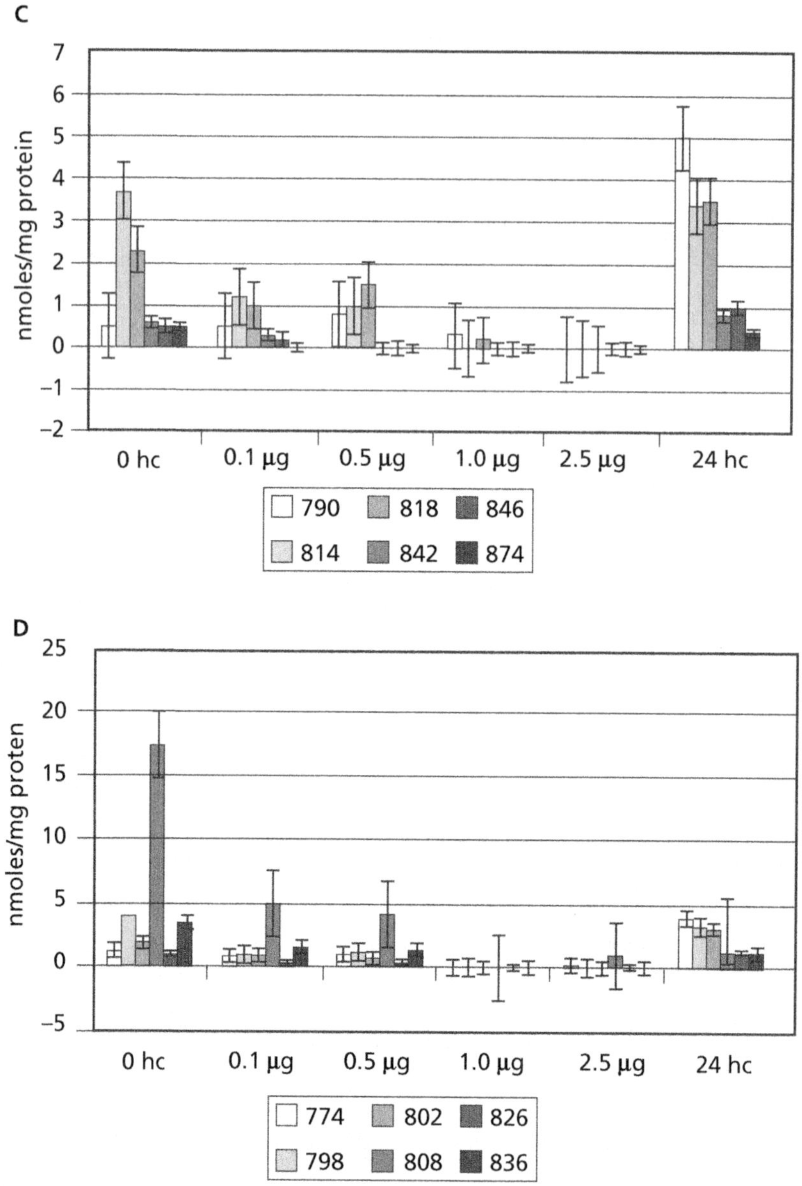

Figure 10.15 *Continued.*

these organs. Various cardiac societies (Von Schacky and Harris, 2007) have recommended an intake of 1 g/day of the omega-3 fatty acids EPA and DHA for cardiac disease prevention; this may also be within the protection level provided by various endogenous antioxidation systems. However, popular press and open market advise many times larger quantities of omega-3 oils, which are likely to exceed the endogenous defense mechanisms resulting in oxo-lipid accumulation. Interestingly, Spiteller (2005) has raised the possibility that furan fatty acids, which are minor components of fish oil, may be responsible for the cardiac protective effects of a fish diet.

Dietary intervention by antioxidant supplementation is widely advocated (Bouayad and Bohn, 2010). Vitamin E supplementation from food and supplement sources has been recently observed to significantly increase plasma α- and γ-tocopherol in adult populations (Zhao et al., 2014), However, none of the tocopherols could prevent the formation of oxo-lipids or promote their degradation in a gastric digestion model (Tarvainen et al., 2012). Previously, Raghavamenon et al. (2009) had suggested that α-tocopherol may promote accumulation of toxic aldehydes, while Kontush et al. (1996) had shown both pro-oxidant and anti-oxidant activity of α-tocopherol. Tsimikas et al. (2010) pointed out that most of the oxidative biomarkers can be modified by therapeutic interventions such as statins and other CVD therapies (Tsimikas and Witztum, 2008). In addition, $sPLA_2$ and $Lp\text{-}PLA_2$ inhibitors have been undergoing evaluation in clinical trials to assess whether they reduce the incidence of CVD events (Rosenson et al., 2009; Serruys et al., 2008).

Leonardi et al. (2007) have shown in C_6 glioma cells that low doses (25 μM) of DHA strengthened the cellular antioxidant defense system, as shown by a rise in glutathione peroxidase and catalase activity and decreased levels of lipid peroxidation. The opposite effect was observed with high doses (50 and 75 μM) of DHA and increased time exposure (72 hours or longer). Recently, Nagai et al. (2012) and Frizzell and Baynes (2014) proposed chelation therapy for management of diabetic complications arising from advanced glycation end product and oxo-lipid accumulation. Earlier, Moreau et al (2005) had reported reversal of age related increases of glycoxidation and lipoxidation by aminoguanidine in the cardiovascular system of Fisher 344 rats.

There is also evidence that suggests that the formation of oxidized phospholipids may sometimes act as a feedback loop, limiting the extent to which inflammation can progress. This has led to interest in the potential therapeutic value of these compounds and the synthesis of analogues such as lecinoxoids with enhanced biological stability (Feige et al., 2010).

Conclusions

The study of plasma lipid peroxidation has made significant progress. Many oxidation products with high biological activity have been recognized, and their protein and nucleic acid adducts have been identified. Among the core aldehydes, the lipid ester hydroxyalkenals have attracted special attention; these have been identified as the carboxyhydroxy alkenals following alkaline hydrolysis of their protein adducts. These aldehydes are highly reactive and have been trapped during peroxidation by albumin as the Michael's adduction products.

The accumulated data show that oxidized neutral lipids and phospholipids are formed in vivo in a variety of pathological conditions. Assays performed with test materials such as oxo-LDL, oxo-PtdCho, or even oxo-palmitoyl/arachidonoyl glycerophosphocholine are not adequate for the identification of specific interactions or specific mechanisms of biological activity. The various markers suggested for lipid peroxidation in plasma or tissues are also inadequate for this purpose because they may have little resemblance to the compounds actually involved in biological activity. There remains a clear need for testing individual molecular species of specific oxo-lipids and for exact identification of their interaction products with specific sites of proteins and nucleic acids. Modern methods of chromatography and mass spectrometry would appear adequate for the isolation and identification of the products involved. Development of antibodies of high reactivity and specificity has been shown to facilitate the development of meaningful test protocols.

While the overall level of plasma lipid peroxidation remains low even in severe disease conditions, the concentration can be greatly increased by microdomain formation, which leads to an enrichment of bioactive oxo-lipids at specific sites. As a result, the concentration of the active lipids may far exceed the need for one or a few molecules to inactivate a protein or a nucleic acid derivative. Nevertheless, there is evidence that lipid peroxidation even in the absence of disease may result in exhaustion of reducing equivalents provided by the various defense systems. Overdosing with polyunsaturated lipids is a major contributor to this exhaustion. The detrimental effects of lipid peroxidation appear to be long term and cumulative, involving cell loss and tissue waste, which we may have to accept at our own peril.

References

Abeywardena, M. Y.; Patten, G. S. Role of ω-3 Long Chain Polyunsaturated Fatty Acids in Reducing Cardio-metabolic Risk Factors. *Endocrine, Metabolic & Immune Disorders—Drug Targets* **2011,** *11*, 232–246.

Adachi, J.; Asano, M.; Yoshioka, N.; Nushida, H.; Ueno, O. Analysis of Phosphatidylcholine Oxidation Products in Human Plasma Using Quadrupole-time-of-flight Mass Spectrometry. *Kobe J. Med. Sci.* **2006,** *52,* 127–140.

Ahmed, Z.; Ravandi, A.; Maguire, G. F.; Kuksis, A.; Connelly, P. W. Formation of Apolipoprotein Ai-phosphatidylcholine Core Aldehyde Schiff Base Adducts Promotes Uptake by THP-1 macrophages. *Cardio. Res.* **2003,** *58,* 712–720.

Alam, Z. I.; Daniel, S. E.; Lees, A. J.; Marsden, D. C.; Jenner, P.; Halliwell, B. A. General Increase in Protein Carbonyls in the Brain in Parkinson's but not Incidental Lewy Body Disease. *J. Neurochem.* **1997,** *69,* 1326–1329.

Araki, Y.; Matsumiya, M.; Matsuura, T.; Oishi, M.; Kaibori, M.; Okumura, T.; Nishizawa, M.; Takada, H.; Kwon, A. H. Peroxidation of n-3-polyunsaturated Fatty Acids Inhibits the Induction of iNOS Gene Expression in Proinflammatory Cytokine-stimulated Hepatocytes. *J. Nutr. Metab.* **2011,** doi:1155/2011/374542.

Arlt, S.; Beisiegel, U.; Kontush, A. Lipid Peroxidation in Neurodegeneration: New Insights into Alzheimer's Disease. *Curr. Opin. Lipidol.* **2002a**, *13,* 289–294.

Arlt, S.; Kontush, A.; Zerr, I.; Buhmann, C.; Jacobi, C.; Schroter, A.; Poser, S.; Beisiegel, U. Increased Lipid Peroxidation in Cerebrospinal Fluid and Plasma from Patients with Creutzfeld-Jakob Disease. *Neurobiol. Dis.* **2002b**, *10,* 150–156.

Arneson, K. O.; Roberts, L. J., II. Measurement of Products of Docosahexaenoic Acid Peroxidation, Neuroprostanes, and Neurofurans. *Methods Enzymol.* **2007,** *433,* 127–143.

Arnold, C.; Markovic, M.; Blossey, K.; Wallukat, G.; Fischer, R.; Dechend, R.; Konkel, A.; Von Schacky, C.; Luft, F. C.; Muller, D. N.; et al. Arachidonic Acid-metabolizing Cytochrome P450 Enzymes Are Targets of ω-3 Fatty Acids. *J. Biol. Chem.* **2010,** *285,* 32720–32733.

Ashraf, M. Z.; Srivastava, S. Oxidized Phospholipids: Introduction and Biological Significance in Lipoproteins—Role in Health and Diseases. *Intech* **2012,** 409–430.

Audoly, L. L.; Rocca, B.; Fabre, J. E.; Koller, B. H.; Thomas, D.; Loeb, A. L.; Coffman, T. M.; FitzGerald, G. A. Cardiovascular Responses to the Isoprostanes $iPF_{2\alpha}$-III and iPF_2-III Are Mediated via the Thromboxane A_2 Receptor in Vivo. *Circulation* **2000,** *101,* 2833–2840.

Aw, T. Y., Williams, M. W.; Gray, L. Absorption and Lymphatic Transport of Peroxidized Lipids by Rat Small Intestine in Vivo: Role of Mucosal GSH. *Am. J. Physiol.* **1992,** *262,* G99–G106.

Basu, S. Bioactive Eicosanoids: Role of Prostaglandin $F_{2\alpha}$ and F_2-isoprostanes in Inflammation and Oxidative Stress Related Pathology. *Mol. Cells* **2010,** *30,* 383–391.

Bernoud-Hubac, N.; Davies, S. S.; Boutaud, O.; Montine, T. J.; Roberts, L. J., II. Formation of Highly Reactive γ-Ketoaldehydes (Neuroketals) as Products of the Neuroprostane Pathway. *J. Biol. Chem.* **2001**, *176,* 30964–30970.

Bielicki, J. K.; Forte, T. M. Evidence That Lipid Hydroperoxides Inhibit Plasma Lecithin: Cholesterol Acyltransferase Activity. *J. Lipid Res.* **1999,** *40,* 948–954.

Blair, I. DNA Adducts with Lipid Peroxidation Products. *J. Biol. Chem.* **2008,** *283,* 15545–15549.

Bouayad, J.; Bohn, T. Exogenous Antioxidants—Double-edged Swords in Cellular Redox State: Health Beneficial Effects at Physiological Doses versus Deleterious Effects at High Doses. *Oxid Med. Cell. Longev.* **2010,** *3* (4), 228–237.

Brooks, J. D.; Milne, G. L.; Yin, H.; Sanchez, S. C.; Porter, N. A.; Morrow, J. D. Formation of Highly Reactive Cyclopentenone Isoprostane Compounds (A_3/J_3-isoprostanes) in Vivo from Eicosapentaenoic Acid. *J. Biol. Chem.* **2008,** *283,* 12043–12055.

Bucala, R.; Makita, Z.; Koschinsky, T.; Cerami, A.; Vlasara, H. Lipid Advanced Glycosylation: Pathway for Lipid Oxidation in Vivo. *Proc. Natl. Acad. Sci. USA* **1993,** *90,* 6434–6438.

Butterfield, D. A.; Boud-Kimball, D. The Critical Role of Methionine 35 in Alzheimer's Amyloid β Peptide (1–42) Induced Oxidative Stress and Neurotoxicity. *Biochim. Biophys. Acta* **2005,** *1703,* 149–156.

Butterfield, D. A.; Bader Lange, M. L.; Sultana, R. Involvement of the Lipid Peroxidation Product, HNE, in the Pathogenesis and Progression of Alzheimer's Disease. *Biochim. Biophys. Acta* **2010,** *1801,* 924–929.

Calamaras, T. D.; Lee, C.; Lan, F.; Ido, Y.; Siwik, D. A.; Colucci, W. S. Post-translational Modification of Serine/Threonine Kinase LKB1 via Adduction of the Reactive Lipid Species 4-Hydroxy-trans-2-nonenal (HNE) at Lysine Residue 97 Directly Inhibits Kinase Activity. *J. Biol. Chem.* **2012,** *287,* 42400–42406.

Campbell, W. B.; Falck, J. R. Arachidonic Acid Metabolites as Endothelium-derived Hyperpolarizing Factors. *Hypertension* **2007,** *49,* 590–596.

Canton, J.; Neculai, D.; Grinstein, S. Scavenger Receptor Homeostasis and Immunity. *Nature Revs. Immunol.* **2013,** *13,* 621–634.

Catala, A. Lipid Peroxidation Modifies the Picture of Membranes from the "Fluid Mosaic Model" to the "Lipid Whisker Model." *Biochimie* **2012,** *94,* 101–109.

Chapple, S. J.; Cheng, X.; Mann, G. E. Effects of 4-Hydroxynonenal on Vascular Endothelial and Smooth Muscle Cell Redox Signaling and Function in Health and Disease. *Redox Biology* **2013,** *1,* 319–331.

Chen, R.; Chen, X.; Salomon, R. G.; McIntyre, T. M. Platelet Activation by Low Concentrations of Intact Oxidized LDL Particles Involves the PAF Receptor. *Arterioscler. Thromb. Vasc. Biol.* **2009,** *29,* 363–371.

Choi, S.-H.; Chae, A.; Miller, E.; Messig, M.; Nitanios, F.; DeMaria, A. N.; Nissen, S. E.; Witztum, J. L.; Tsimikas, S.. Relationship between Biomarkers of Oxidized Low-density Lipoprotein, Statin Therapy, Quantitative Coronary Angiography, and Atheroma: Volume Observations from the REVERSAL (Reversal of Atherosclerosis with Aggressive Lipid Lowering) study. *J. Am.Coll. Cardiol.* **2008,** *52,* 24–32.

Choi, S.-H.; Yin, H.; Ravandi., A.; Armando, A.; Dumlao, D.; Kim, J.; Almazan, F.; Taylor, A. M.; McNamara, C. A.; Tsimikas, S.; et al. Polyoxygenated Cholesteryl Ester Hydroperoxide Activates TLR4 and SYK Dependent Signaling in Macrophages. *PLOS ONE* **2013,** *8* (12), e83145.

Chung, B. H.; Wilkinson, T.; Greer, J. C.; Segrest, J. P. Preparative and Quantitative Isolation of Plasma Lipoproteins: Rapid, Single Discontinuous Density Gradient Ultracentrifugation in a Vertical Rotor. *J. Lipid Res.* **1980,** *21,* 284–291.

Code, C.; Mahalka, A. K.; Bry, K.; Kinnunen, P. K. Activation of Phospholipase A_2 by 1-Palmitoyl-2-[9-oxo]-nonanoyl-sn-glycero-3-phosphocholine in Vitro. *Biochim. Biophys. Acta* **2010,** *1798,* 1593–1600.

Connelly, P. W.; Draganov, D.; Maguire, G. F. Paraoxonase-1 Does Not Reduce or Modify Oxidation of Phospholipids by Peroxynitrite. *Free Rad. Biol. Med.* **2005,** *38,* 164–174.

Dangi, B.; Obeng, M.; Nauroth, J. M.; Teymourlouei, M.; Needham, M.; Raman, K.; Arterburn, L. M. Biogenic Synthesis, Purification, and Chemical Characterization of Anti-inflammatory Resolvins Derived from Docosapentaenoic Acid (DPA-n6). *J. Biol. Chem.* **2009,** *284,* 14744–14759.

Dashti, M.; Kulik, W.; Hoek, F.; A Phospholipidomic Analysis of All Defined Human Plasma Lipoproteins. *Sci. Rep.* **2011,** *1,* 139.

Davies, S. S.; Pontsler, A. V.; Marathe, G. K.; Harrison, K. A.; Murphy, R. C.; Hinshaw, J. C.; Prestwich, G. D.; Hilaire, A. S.; Prescott, S. M.; Zimmerman, G. A.; McIntyre, T. M. Oxidized Alkyl Phospholipids Are Specific High Affinity Peroxisome Proliferator-activated Receptor γ Ligands and Activators. *J. Biol. Chem.* **2001,** *276,* 16015–16023.

Davis, B.; Koster, G.; Douet, L. J.; Scigelova, M.; Woffendin, G.; Ward, J. M.; Smith, A.; Humphries, J.; Burnand, K. G.; Macphee, C. H.; et al. Electrospray Ionization Mass Spectrometry Identifies Substrates and Products of Lipoprotein-associated Phospholipase A_2 in Oxidized Human Low Density Lipoprotein. *J. Biol. Chem.* **2008,** *283,* 6428–6437.

Dolinsky, V. W.; Chan, A.Y.; Robillard, F.; Robillard Frayne, I.; Light, P. E.; DesRosiers, C.; Dyck, J. R. Resveratrol Prevents the Prohypertrophic Effects of Oxidative Stress on LKB1. *Circulation* **2009,** *119,* 1643–1652.

Eckey, R.; Menschikowski, M.; Lattke, O.; Jaross, W. Minimal Oxidation and Storage of Low Density Lipoproteins Result in an Increased Susceptibility to Phospholipid Hydrolysis by Phospholipase A_2. *Atherosclerosis* **1997,** *132,* 165–176.

Ehara, S.; Ueda, M.; Naruko, T.; Haze, K.; Itoh, A.; Otsuka, M.; Komatsu, R.; Matsuo, T.; Itabe, H.; Takano, T.; et al. Elevated Levels of Oxidized Low-density Lipoprotein Show a Positive Relationship with the Severity of Acute Coronary Syndromes. *Circulation* **2001,** *103,* 1955–1960.

Engstrom, K.; Saldeen, A. S.; Yang, B.; Mehta, J. L.; Saldeen, T.. Effect of Fish Oils Containing Different Amounts of EPA, DHA, and Antioxidants on Plasma and Brain Fatty Acids and Brain Nitric Oxide Synthetase Activity in Rats. *Uppsala J. Med. Sci.* **2009,** *114,* 206–213.

Erridge, C.; Kennedy, S.; Sprickett, C. M.; Webb, D. T. Oxidized Phospholipid Inhibition of Toll-like Receptor (TLR) Signaling Is Restricted to TLR2 and TLR4: Roles for CD14, LPS-Binding Protein, and MD2 as Targets for Specificity of Inhibition. *J. Biol. Chem.* **2008,** *283,* 24748–24759.

Esterbauer, H.; Schaur, R.; Zoellner, H. Chemistry and Biochemistry of 4-Hydroxynonenal, Malondialdehyde, and Related Aldehydes. *Free Radic. Biol. Med.* **1991**, *11,* 81–128.

Esterbauer, H.; Gebicki, J.; Puhl, H.; Jurgens, G. The Role of Lipid Peroxidation and Antioxidants in Oxidative Modification of LDL. *Free Radic. Biol Med.* **1992,** *13,* 341–390.

Evans, M. D.; Dizdaroglu, M.; Cooke, M. S. Oxidative DNA Damage and Disease: Induction, Repair and Significance. *Mutat. Res. Rev.* **2004,** *567,* 1–61.

Fam, S. S.; Murphey, L. J.; Terry, E. S.; Zackert, W. E.; Chen, Y.; Gao, L.; Pandalai, S.; Milne, G. L.; Roberts, L. J.; Porter, N. A.; et al. Formation of Highly Reactive A-ring and J-ring Isoprostane-like Compounds (A_4/J_4-neuroprostanes) in Vivo from Docosahexaenoic Acid. *J. Biol. Chem.* **2002,** *277,* 36076–36084.

Febbraio, M.; Hajjar, D. P.; Silverstein, R. L. CD36: A Class B Scavenger Receptor Involved in Angiogenesis, Atherosclerosis, Inflammation, and Lipid Metabolism. *J. Clin. Invest.* **2001,** *108,* 785–791.

Feige, E.; Mendel, I.; George, J.; Yacov, N.; Harats, D. Modified Phospholipids as Anti-inflammatory Compounds. *Curr. Opin. Lipidol.* **2010,** *21,* 525–529.

Fenaille, F.; Tabet, J. C.; Guy, P. A. Immunoaffinity Purification and Characterization of 4-Hydroxy-2-nonenal and Malondialdehyde Modified Proteins by Electrospray Ionization Tandem Mass Spectrometry. *Anal. Chem.* **2002,** *74,* 6298–6304.

Fer, M.; Dreano, Y.; Lucas, D.; Corcos, L.; Salaun, J. P.; Berthou, F.; Amet, Y. Metabolism of Eicosapentaenoic and Docosahexaenoic Acids by Recombinant Human Cytochrome P450. *Arch. Biochem. Biophys.* **2008,** *471,* 116–125.

Fessel, J. P.; Porter, N. A.; Moore, K. P.; Sheller, J. R.; Roberts, L. J., II. Discovery of Lipid Peroxidation Products Formed in Vivo with a Substituted Tetrahydrofuran Ring (Isofurans) That Are Favored by Increasing Oxygen Tension. *Proc. Natl. Acad. Sci. USA* **2002,** *99,* 16713–16718.

Fessel, J. P.; Hulette, C.; Powell, S.; Roberts, L. J., II. Zhang, J. Isofurans, but not F_2-isoprostanes, Are Increased in the *Substantia Nigra* of Patients with Parkinson's Disease and with Dementia with Lewy Body Disease. *J. Neurochem.* **2003,** *85,* 645–650.

Fleming, I. Vascular Cytochrome P450 Enzymes: Physiology and Pathophysiology. *Trends Cardiovasc. Med.* **2008,** *18,* 20–25.

Flohe-Brigelius, R. Tissue Specific Functions of Individual Glutathione Peroxidases. *Free Radical Biol. Med.* **1999**, *27,* 951–965.

Frankel, E. N. *Lipid Oxidation.* The Oily Press: Dundee, Scotland, 1998; pp 55–77.

Frizzell, N.; Baynes, J. W. Chelation Therapy for the Management of Diabetic Complications: A Hypothesis and a Proposal for Clinical Laboratory Assessment of Metal Ion Homeostasis in Plasma. *Clin. Chem. Lab. Med.* **2014,** *52,* 69–75.

Fruhwirth, G. O.; Loidl, A.; Hermetter, A. Oxidized Phospholipids: From Molecular Properties to Disease. *Biochim. Biophys. Acta* **2007,** *1772,* 718–736.

Fu, M.-X.; Requena, J. R.; Jenkins, A. J.; Lyons, J. J.; Baynes, J. W.; Thorpe, S. R. The Advanced Glycation End Product, N-(carboxymethyl)lysine, Is a Product of Both Lipid Peroxidation and Glycoxidation Reactions. *J. Biol. Chem.* **1996,** *271,* 9982–9986.

Fuentes-Antras, J.; Ioan, A. M.; Tunon, J.; Egido, J.; Lorenzo, O. Activation of Toll-like Receptors and Inflammasome Complexes in the Diabetic Cardiomyopathy-associated Inflammation. *Int. J. Endocrinol.* **2014.** http://dx.doi.org/10.1155/2014/847827.

Fukunaga, M.; Makita, N.; Roberts, L. J.; Morrow, J. D.; Takahashi, K.; Badr, K. F. Evidence for the Existence of F_2-isoprostane Receptors on Rat Vascular Smooth Muscle Cells. *Am. J. Physiol* **1993**, *264,* C1619–C1624.

Fukunaga, M.; Yura, T.; Grygorczyk, R.; Badr, K. F. Evidence for Distinct Nature of F_2-isoprostane Receptors from Those of Thromboxane A_2. *Am. J. Physiol.* **1997,** *272,* F477–F483.

Furse, K. E.; Pratt, D. A.; Schneider, C.; Brash, A. R.; Porter, N. A.; Lybrand, T. P. Molecular Dynamics Simulation of Arachidonic Acid-derived Pentadienyl Radical Intermediate Complexes with COX-1 and COX-2: Insight into Oxygenation Regio and Stereospecificity. *Biochemistry* **2006,** *45,* 3206–3218.

Galano, J.-M.; Mas, E.; Barden, A.; Mori, T. A.; Signorini, C.; DeFelice, C.; Barrett, A.; Opere, C.; Pinot, E.; Schwedhelm, E.; et al. Isoprostanes and Neuroprostanes: Total Synthesis, Biological Activity and Biomarkers of Oxidative Stress in Humans. *Prostaglandins and Other Lipid Mediators.* **2013,** *107,* 95–102.

Gao, L.; Yin, H.; Milne, G. L.; Porter, N. A.; Morrow, J. D. Formation of F-ring Isoprostane-like Compounds (F_3-isoprostanes) in Vivo from Eicosapentaenoic Acid. *J. Biol. Chem.* **2006,** *281,* 14092–14099.

Gao, L.; Wang, J.; Sekhar, K. R.; Yin, H.; Yared, N. F.; Schneider, S. N.; Sasi, S.; Dalton, T. P.; Anderson, M. E.; Chan, J. Y. Novel n-3 Fatty Acid Oxidation Products Activate Nrf2 by Destabilizing the Association between Keap1 and Cullin3. *J. Biol. Chem.* **2007,** *282,* 2529–2537.

Gardner, H. W. Recent Investigations into the Lipoxygenase Pathway of Plants (a Review). *Biochim. Biophys Acta* **1995,** *1084,* 221–239.

Gargalovic, P. S.; Imura, B.; Zhang, N. M.; Gharavi, N. M.; Clark, M. J.; Pagnon, J.; Yang, W. P.; He, A.; Truong, A.; Patel, S. Identification of Inflammatory Gene Modules Based on Variations of Human Endothelial Cell Responses to Oxidized Lipids. *Proc. Natl. Acad. Sci. USA* **2006,** *103,* 12741–12746.

Garscha, U.; Nilsson, T.; Oliw, E. H. Enantiomeric Separation and Analysis of Unsaturated Hydroperoxy Fatty Acids by Chiral Column Chromatography-Mass Spectrometry. *J. Chromatogr B.* **2008,** *872,* 90–98.

Gilotte, K. L.; Horkko, S.; Witztum, J. L.; Steinberg, D. Oxidized Phospholipids, Linked to Apoprotein B of Oxidized LDL Are Ligands for Macrophage Scavenger Receptors. *J. Lipid Res.* **2000,** *41,* 824–833.

Gopfert, M. S.; Siedler, F.; Siess, W.; Sellmayer, A. Structural Identification of Oxidized Acyl-phosphatidylcholines That Induce Platelet Activation. *J. Vasc. Res.* **2005,** *42,* 120–132.

Greenberg, M. E.; Sun, M.; Zhang, R.; Febbraio, M.; Silverstein, R.; Hazen, S. L. Oxidized Phosphatidylserine-CD36 Interactions Play an Essential Role in Macrophage-dependent Phagocytosis of Apoprotic Cells. *JEM* **2006,** *203,* 2613–2625.

Greenberg, M. E.; Li, X.-M.; Gugiu, B. GGu, X.; Qin, J.; Salomon, R. G.; Hazen, S. L. The Lipid Whisker Model of the Structure of Oxidized Cell Membranes. *J. Biol. Chem.* **2008,** *283,* 2385–2396.

Greig, F. H.; Kennedy, S.; Spickett, C. M. Physiological Effects of Oxidized Phospholipids and Their Cellular Signaling Mechanisms in Inflammation. *Free Rad. Biol. Med.* **2012,** *52,* 266–280.

Gronert, K. Lipid Autacoids in Inflammation and Injury Responses: A Matter of Privilege. *Mol. Interv.* **2008,** *8,* 28–35.

Gruber, F.; Bicker, W.; Oskolkova, O. V.; Tschachler, E.; Bochkov, V. N. A Simplified Procedure for Semi-targeted Lipidomic Analysis of Oxidized Phosphatidylcholines Induced by UV Irradiation. *J. Lipid Res.* **2012,** *53,* 1232–1242.

Gueraud, F.; Atalay, M.; Bresgen, N.; Cipak, A.; Ecki, P. M.; Huc, L.; Jouanin, I; Siems, W.; Uchida, K. Chemistry and Biochemistry of Lipid Peroxidation Products. *Free Radic, Res.* **2010,** *44,* 1098–1124.

Gugiu, B. G.; Mouillesseaux, V.; Duong, T.; Herzog, T.; Hekimian, A.; Koroniak, L.; Vondriska, T. M.; Watson, A. D. Protein Targets of Oxidized Phospholipids in Endothelial Cells. *J. Lipid Res.* **2008,** *49,* 510–520.

Haberland, M. E.; Fong, D.; Cheng, L. Malondialdehyde-altered Protein Occurs in Atheroma of Watanabe Heritable Hyperlipemic Rabbits. *Science* **1988,** *241,* 215–218.

Halliwell, B.; Gutteridge, J. M. C. *Free Radicals in Biology and Medicine.* Clarandon Publishing: Oxford, UK, 1989.

Halliwell, B.; Lee, C. Y. J. Using Isoprostanes as Biomarkers of Oxidative Stress: Some Rarely Considered Issues. *Antioxidants Redox Signaling* **2010,** *13,* 145–156.

Halliwell, B.; Whiteman, M. Measuring Reactive Species and Oxidative Damage in Vivo and in Cell Culture: How Should You Do It and What Do the Results Mean? *Brit. J. Pharmacol.* **2004,** *142,* 231–255.

Halliwell, B.; Zhao, K.; Whiteman, M. The Gastrointestinal Tract: A Major Site of Antioxidant Action? *Free Rad. Res.* **2000,** *33,* 819–830.

Han, J.; Nicholson, A. C.; Zhou, X.; Feng, J.; Gotto, A. M. Jr; Hajjar, D. P. Oxidized Low Density Lipoprotein Decreases Macrophage Expression of Scavenger Receptor B-I. *J. Biol. Chem.* **2001,** *276,* 16567–16572.

Han, X.; Gross, R. W. Global Analyses of Cellular Lipidomes Directly from Extracts of Biological Samples by ESI Mass Spectrometry. *J. Lipid Res.* **2003,** *44,* 1071–1079.

Hanasaki, K.; Nakano, T.; Kasai, H.; Arita, H. Receptor-mediated Mitogenic Effect of Thromboxane A_2 in Vascular Smooth Muscle Cells. *Biochem. Pharmacol.* **1990,** *40,* 2535–2542.

Harkewicz, R.; Hartvigsen, K.; Almazan, F.; Dennis, E. A.; Witztum, J. L.; Miller, Y. L. Cholesteryl Ester Hydroperoxides Are Biologically Active Components of Minimally Oxidized Low Density Lipoprotein. *J. Biol. Chem.* **2008,** *283,* 10241–10251.

Harman, D. Ageing: A Theory Based on Free-radical and Radiation Chemistry. *J. Gerontol.* **1956**, *11,* 298–300.

Hedrick, C. C.; Hassan, K.; Hough, G. P.; Yoo, J. H.; Simzar, S.; Quinto, C. R.; Kim, S. M.; Dooley, A.; Langi, S.; Hama, S. Y. Short-term Feeding of Atherogenic Diet to Mice Results in Reduction of HDL and Paraoxonase That May Be Mediated by an Immune Mechanism. *Arterioscler. Thromb. Vasc. Biol.* **2000,** *20,* 1946–1952.

Heery, J. M.; Kozak, M.; Stafforini, D. M.; Jones, D. A.; Zimmerman, G. A.; McIntyre, T. M.; Prescott, S. M. Oxidatively Modified LDL Contains Phospholipids with Platelet-activating Factor-like Activity and Stimulates the Growth of Smooth Muscle Cells. *J. Clin. Invest.* **1995**, *96,* 2322–2330.

Hirano, K.; Yamashita, S.; Nakagawa, Y.; Ohya, T.; Matsuura, F.; Tsukamoto, K.; Okamoto, Y.; Matsuyama, A.; Matsumoto, K.; Miyagawa, J.; Matsuzawa, Y. Expression of Human Scavenger Receptor Class B Type 1 in Cultured Human Monocyte-derived Macrophages and Atherosclerotic Lesions. *Circ. Res.* **1999**, *85,* 108–116.

Hoff, H. F.; O'Neil, J.; Wu, Z.; Hoppe, G.; Salomon, R. L. Phospholipid Hydroxyalkenals: Biologhical and Chemical Properties of Specific Oxidized Lipids Present in Atherosclerotic Lesions. *Arterioscler. Thromb. Vasc. Biol.* **2003,** *23,* 275–282.

Honda, Z.; Ishiii, S.; Shimizu, T. Platelet Activating Factor Receptor. *J. Biochem.* **2002,** *131,* 773–779.

Hoppe, G.; Ravandi, A.; Herrera, D.; Kuksis, A.; Hoff, H. F. Oxidation Products of Cholesteryl Linoleate Are Resistant to Hydrolysis in Macrophages, Form Complexes with Proteins and Are Present in Human Atherosclerotic Lesions. *J. Lipid Res.* **1997,** *38,* 1347–1360.

Horkko, S.; Miller, E.; Branch, D. W.; Palinski, W.; Witztum, J. L. The Epitopes for Some Antiphospholipid Antibodies Are Adducts of Oxidized Phospholipid and β2-Glycoprotein 1 (and Other Proteins). *Proc. Natl. Acad Sci USA* **1997,** *94,* 10256–10361.

Huang, W.-N.; Chen, Y.-H.; Chen, C.-L.; Wu, W.-W. Surface Pressure-dependent Interactions of Secretory Phospholipase A_2 with Zwitterionic Phospholipid Membranes. *Langmuir* **2011,** *27,* 7034–7041.

Hughes, H.; Mathews, B.; Lenz, A. M.; McGuyton, J. R. Toxicity of Oxidized LDL to Porcine Aortic Smooth Muscle Cells Is Associated with Oxysterols 7-Keto-cholesterol and 7-Hydroxycholesterol. *Atheroscl Thromb Vasc. Biol.* **1994,** *14,* 1177–1185.

Hui, S.-P.; Chiba, J.; Jin, S.; Nagasaka, H.; Kurosawa, T. Analyses for Phosphatidylcholine Hydroperoxides by LC/MS. *J. Chromatogr. B.* **2010,** *878,* 1677–1682.

Hui, S.-P.; Sakurai, T.; Ohkawa, F.; Furumaki, H.; Jin, S.; Fuda, H.; Takeda, S.; Kurosawa, T.; Chiba, H. Detection and Characterization of Cholesteryl Ester Hydroperoxides in Oxidized LDL and Oxidized HDL by use of an Orbitrap Mass Spectrometer. *Anal. Bioanal. Chem.* **2012,** *404,* 101–112.

Hui, S.-P.; Sakurai, T.; Takeda, S.; Jin, S.; Fuda, H.; Kurosawa, T.; Chiba, H.. Analysis of Triacylglycerol Hydroperoxides in Human Lipoproteins by Orbitrap Mass Spectrometer. *Anal Bioanal. Chem.* **2013,** *405,* 4981–4987.

Ishii, T.; Ito, S.; Kumazawa, S.; Sakurai, T.; Yamaguchi, S.; Mori, T.; Nakayama, T.; Uchida, K. Site-specific Modification of Positively Charged Surface on Human Serum Albumin by Malondialdehyde. *Biochem. Biophys Res. Commun.* **2008,** *371,* 28–32.

Itabe, H.; Takeshima, E.; Iwasaki, H.; Kimura, J.; Yoshida, Y.; Imanaka, T.; Takano, T. A Monoclonal Antibody against Oxidized Lipoprotein Recognizes Foam Cells in Atherosclerotic Lesions. Complex Formation of Oxidized Phosphatidylcholines and Polypeptides. *J. Biol. Chem.* **1994,** *269,* 15274–15279.

Jafraim, V.; Uchida, K.; Narayanaswami, V. Pathophysiology of Lipoprotein Oxidation. In *Lipoproteins—Role in Health and Disease*; Frank, S., Kostner, G., Eds.; Intech, 2012; pp 383–408. http://dx.doi.org/10.5772/50622.

Janero, D. R. Malondialdehyde and Thiobarbituric—Reactivity as Diagnostic Indices of Peroxidative Tissue Injury. *Free Radic Biol Med.* **1990,** *9,* 515–540.

Jerlich, A.; Schaur, R. J.; Pitt, A. R.; Spickett, C. M. The Formation of Phosphatidylcholine Oxidation Products by Stimulating Phagocytes. *Free Radical Res.* **2003,** *37,* 645–653.

Kamido, H.; Kuksis, A.; Marai, L; Myher, J. J.; Pang, H. Preparation, Chromatography and Mass Spectrometry of Cholesteryl Ester and Glycerolipid Bound Aldehydes. *Lipids* **1992a,** *27,* 645–650.

Kamido, H.; Kuksis, A.; Marai, L.; Myher, J. J. Identification of Cholesterol-bound Aldehydes in Copper-oxidized Low Density Lipoprotein. *FEBS Letters* **1992b,** *304,* 269–272.

Kamido, H.; Kuksis, A.; Marai, L.; Myher, J. J. Identification of Core Aldehydes Among in Vitro Peroxidation Products of Cholesteryl Esters. *Lipids* **1993,** *28,* 331–336.

Kamido, H.; Kuksis, A.; Marai, L.; Myher, J. J. Lipid Ester-bound Aldehydes among Copper-catalyzed Peroxidation Products of Human Plasma Lipoproteins. *J. Lipid Res.* **1995,** *36,* 1876–1886.

Kamido, H.; Eguchi, H.; Ikeda, H.; Ikeda, H.; Imaizumi, T.; Yamana, K.; Hartvigsen, K.; Ravandi, A.; Kuksis, A. Core Aldehydes of Alkyl Glycerophosphocholines in Atheroma Induce Platelet Aggregation and Inhibit Endothelium-dependent Arterial Relaxation. *J. Lipid Res.* **2002,** *43,* 158–166.

Karten, B.; Boechzelt, H.; Abuja, P. M.; Mittelbach, M.; Oettl, K.; Sattler, W. Femtomole Analysis of 9-Oxononanoyl Cholesterol by High Performance Liquid Chromatography. *J. Lipid Res.* **1998,** *39,* 1508–1519.

Karten, B.; Boechzelt, H.; Abuja, P. M.; Mittelbach, M.; Sattler, W. Macrophage-enhanced formation of cholesteryl ester-core aldehydes during oxidation of low density lipoprotein. *J. Lipid Res.* **1999**, *40,* 1240–1253.

Kaur, K.; Salomon, R. G.; O'Neil, J.; Hoff, H. F. (Carboxyalkyl) Pyrroles in Human Plasma and Oxidized Low-density Lipoproteins. *Chem. Res. Technol.* **1997,** *10,* 1387–1396.

Kawai, Y.; Saito, A.; Shibata, N.; Kobayashi, M.; Yamada, S.; Osawa, T.; Uchida, K. Covalent Binding of Oxidized Cholesteryl Esters to Protein. *J. Biol. Chem.* **2003,** *278,* 21040–21049.

Khasawneh, F. T.; Huang, J.-S.; Mir, F.; Srinivasan, S.; Tiruppathi, C.; LeBreton, G. C. Characterization of Isoprostane Signaling: Evidence for a Unique Coordination Profile of 8-iso-PGF$_{2\alpha}$ with the Thromboxane A$_2$ Receptor, and Activation of a Separate cAMP-dependent Inhibitory Pathway in Human Platelets. *Biochem. Pharmacol.* **2008,** *75,* 2301–2315.

Kinoshita, M.; Oikawa, S.; Hayasaka, K. Age-related Increases in Plasma Phosphatidylcholine Hydroperoxide Concentrations in Control Subjects and Patients with Hyperlipidemia. *Clin. Chem.* **2000,** *46,* 822–828.

Klawitter, J.; Haschke, M.; Shokati, T; Klawitter, J.; Christians, U. Quantification of 15-F$_{2t}$-isoprostane in Human Plasma and Urine: Results from Enzyme-linked Immunoassay and Liquid Chromatography/Tandem Mass Spectrometry Cannot Be Compared. *Rapid Commun. Mass Spectrom.* **2011,** *25,* P463–468.

Klimov, A. N.; Nikiforova, A. A.; Pleskov, V. M.; Kalashnikova, N. N. The Protective Action of High Density Lipoproteins, Their Subfractions, and Lecithin-cholesterol Acyltransferase in the Peroxide Modification of Low Density Lipoproteins. *Biokhimiia* **1989,** *54,* 118–123.

Kogure, K.; Nakashima, S.; Tsuchie, A.; Tokumura, A.; Fukuzawa, K. Temporary Membrane Distortion of Vascular Smooth Muscle Cells Is Responsible for Their Apoptosis Induced by Platelet-activating Factor-like Oxidized Phospholipids and Their Degradation Product, Lysophosphjatidylcholine. *Chem. Phys. Lipids* **2003,** *126,* 29–38.

Kontush, A.; Finck, B.; Karten, B.; Kohlschutter, A.; Beisiegel, U. Antioxidant and Prooxidant Activity of α-Tocopherol in Human Plasma and Low Density Lipoprotein. *J. Lipid Res.* **1996,** *37,* 1436–1448.

Kopprasch, S., Pietzsch, J.; Westendorf, T.; Kruse, H. J.; Grassler, J. The Pivotal Role of Scavenger Receptor CD36 and Phagocyte-derived Oxidants in Oxidized Low Density Lipoprotein-induced Adhesion to Endothelial Cells. *Int. J. Biochem. Cell. Biol.* **2004,** *36,* 460–471.

Korotaeva, A. A.; Samoilova, E. V.; Piksina, G. F.; Prokazova, N. V. Oxidized phosphatidylcholine Stimulates Activity of Secretory Phospholipase A$_2$ Group IIA and Abolishes Sphingomyelin-induced Inhibition of the Enzyme. *Prostaglandins and Other Lipid Mediators* **2010,** *91,* 38–41.

Krolewski, A. S.; Warram, J. H.; Valsania, P.; Martin, B. C.; Laffel, L. M.; Christlieb, A. R. Evolving Natural History of Coronary Artery Disease in Diabetes Mellitus. *Am. J. Med.* **1991,** *90,* 56–61.

Ku, G.; Thomas, C. E.; Akesson, A. L; Jackson, R. L. Induction of Interleukin-1 Beta from Human Peripheral Blood-derived Macrophages by 9-Hydroxyoctadecenoic Acid. *J. Biol. Chem.* **1992,** *267,* 14183–14188.

Kuksis, A. Lipidomics and Metabolomics of Dietary Lipid Peroxidation. In *Mass Spectrometry and Nutrition Research*; *RSC Food Analysis Monographs,* No. 9; Fay, L. B., Kussmann, M., Eds.; The Royal Society of Chemistry: Cambridge, UK; 2010; pp 102–162.

Kuksis, A.; Pruzanski, W. Release of Fatty Acid Hydroperoxides and Hydroxides from Lipoprotein Phospholipids by Group IIA, V and X Human Secretory Phospholipases (sPLA$_2$).

In *Abstracts,* 104th AOCS Annual Meeting and Expo, April 28–May 1, 2013, Montreal, Quebec, Canada.

Kuksis, A.; Pruzanski, W. Generation of Phosphatidylcholine Hydroperoxides and Phosphatidylcholine Isoprostanes during Ultracentrifugation and Storage of Lipoproteins. *Lipid Technology* **2014,** *26,* 11–14.

Kuksis, A.; Stachnyk, O.; Holub, B. J. Improved Quantitation of Plasma Lipids by Direct Gas-liquid Chromatography. *J. Lipid Res.* **1969,** *10,* 660–667.

Kuksis, A.; Myher, J. J.; Geher, K.; Breckenridge, W. C.; Jones, G. J.; Little, J. A. Lipid Class and Molecular Species Interrelationships among Plasma Lipoproteins of Normolipemic Subjects. *J. Chromatogr. Biomedical Applications* **1981,** *224,* 1–23.

Kuksis, A.; Suomela, J.-P.; Tarvainen, M.; Kallio, H. Lipidomic Analysis of Glycerolipid and Cholesteryl Ester Autooxidation Products. *Mol. Biotechnol.* **2009,** *42,* 224–268.

Kunjathoor, V. V.; Febbraio, M.; Podrez, E. A.; Moore, K. J.; Andersson, L.; Koehn, S.; Rhee, J. S.; Silverstein, R.; Hoff, H. F.; Freeman, M. N. Scavenger Receptors Class A-1/I and CD36 Are the Principal Receptors Responsible for the Uptake of Modified Low Density Lipoprotein Leading to Lipid Loading in Macrophages. *J. Biol Chem.* **2002,** *277,* 49982–49988.

Kurvinen, J. P.; Kuksis, A.; Ravandi, A; Sjovall, O.; Kallio, H. Rapid Complexing of Oxoacylglycerols with Amino Acids, Peptides and Aminophospholipids. *Lipids* **1999,** *34,* 299–305.

Lawson, J. A.; Rokach, J.; FitzGerald, G. A. Isoprostanes: Formation, Analysis and Use as Indices of Lipid Peroxidation in Vivo. *J. Biol. Chem.* **1999,** *274,* 24441–24444.

Lee, H. C.; Wei, Y. H. Role of Mitochondria in Human Ageing. *J. Biomed. Sci.* **1997,** *4,* 319–326.

Lee, S. H.; Blair, J. A. Oxidative DNA Damage and Cardiovascular Disease? *Trends Cardiovasc. Med.* **2001,** *11,* 148–155.

LeGrand, T. S.; Aw, T. Y. In *Intestinal Lipid Metabolism;* Mansbach, C. H., III, Tso, P., Kuksis, A., Eds.; Kluwer Academic/Plenum Publishers: New York, 2001; pp 351–366.

Leidy, C. L.; Linderoth, L.; Andresen, T. L.; Mouritsen, O. G.; Jorgensen, K.; Peters, G. H. Domain-induced Activation of Human Phospholipase A_2 Type IIA: Local versus Global Lipid Composition. *Biophys. J.* **2006,** *90,* 3165–3175.

Leitinger, N.; Watson, A. D; Faull, K. F.; Fogelman, A. M.; Berliner, J. A. Monocyte Binding to Endothelial Cells Induced by Oxidized Phospholipids Present in Minimally Oxidized Low Density Lipoprotein Is Inhibited by a Platelet Activating Factor Receptor Antagonist. *Adv. Exp. Med. Biol.* **1997,** *433,* 379–382.

Leonardi, F.; Attorri, L.; Di Benedetto, R.; Biase, A. D.; Sanchez, M.; Tregno, F. P.; Nardini, M.; Salvati, S. Docosahexaenoic Acid Supplementation Induces Dose and Time Dependent Oxidative Changes in C6 Glioma Cells. *Free Radical Res.* **2007,** *41,* 748–756.

Leoni, V.; Caccia, C. Oxysterols as Biomarkers in Neurodegenerative Diseases. *Chem. Phys. Lipids* **2011,** *164,* 515–524.

Li, R.; Mouillesseaux, K. P; Montoya, D.; Cruz, D.; Gharavi, N.; Dun, M.; Koroniak, L.; Berliner, J. A. Identification of Prostaglandin E_2 Receptor Subtype 2 as a Receptor Activated by OxPAPC. *Circ. Res.* **2006,** *98,* 642–650.

Liu, J.; Ames, B. B. Reducing Mitochondrial Decay with Mitochondrial Nutrients to Delay and Treat Cognitive Dysfunction, Alzheimer's Disease and Parkinson's Disease. *Nutr. Neurosci.* **2005,** *8,* 67–89.

Liu, X.; Yamada, N.; Maruyama, W.; Osawa, T. Formation of Dopamine Adducts Derived from Brain Polyunsaturated Fatty Acids: Mechanism for Parkinson Disease. *J. Biol. Chem.* **2008,** *283,* 34887–34895.

Liu, X.; Yamada, N.; Osawa, T. Amide-type Adduct of Dopamine—Plausible Cause of Parkinson Diseases. *Subcell Biochem.* **2014**, *77,* 49–60.

Lloidl-Stahlhofen, A.; Spiteller, G. α-Hydroxyaldehydes, Products of Lipid Peroxidation. *Biochim. Biophys. Acta* **1994,** *1211,* 156–160.

Lowe, J.; Graham, L.; Leight, P. N. Disorders Of movement and System Degeneration. In *Greenfeld's Neurophysiology,* 6th ed.; Graham, D. I., Lantos, P. L., Eds.; Arnold: London, 1997; Vol. II, pp 281–366.

Lynch, S. M.; Morrow, J. D.; Roberts, L. J.; Baiz, F. Formation of Non-cyclooxygenase-derived Prostanoids (F_2-isoprostanes) in Plasma and Low Density Lipoprotein Exposed to Oxidative Stress in Vitro. *J. Clin. Invest.* **1994,** *93,* 998–1004.

Mackness, B.; Arrol, S.; Dorrington, P. N. Paraoxonase Prevents Accumulation of Lipoperoxides in Low Density Lipoprotein. *FEBS Lett.* **1991,** *286,* 152–154.

Mackness, B.; Hunt, R.; Durrington, P. N.; Mackness, M. I. Increased Immunolocalisation of Paraoxonase, Clusterin, and Apolipoprrotein A-I in the Human Artery Wall with the Progression of Atherosclerosis. *Arterioscler Thromb. Vasc. Biol.* **1997,** *17,* 1233–1238.

Marathe, G. K.; Prescott, S. M.; Zimmerman, G. A.; McIntyre, T. M Oxidized LDL Contains Inflammatory PAF-like Phospholipids. *Trends Cardiovasc. Med.* **2001,** *11,* 139–142.

Marathe, G. K.; Zimmerman, G. A.; Prescott, S. M.; McIntyre, T. M. Activation of Vascular Cells by PAF-like Lipids in Oxidized LDL. *Vascul Pharmacol.* **2002,** *38,* 193–200.

Marathe, G. K.; Zimmerman, G. A.; McIntyre, T. M. Platelet Activating Factor Acetylhydrolase, and Not Paraoxonase-1, Is the Oxidized Phospholipid Hydrolase of High Density Lipoprotein Particles. *J. Biol. Chem.* **2003,** *278,* 3937–3947.

Marchand, D.; Grossi, V.; Hirsschler-Rea, A.; Ronfani, J.-F. Regiospecific Enzymatic Oxygenation of *Cis*-vaccenic Acid during Aerobic Senescence of the Halophilic Purple Sulfur Bacterium *Thiohalocapsa halophila. Lipids* **2002,** *37,* 541–548.

McAnoy, A. M.; Wu, C. C.; Murphy, R. C. Direct Qualitative Analysis of Triacylglycerols by Eletrospray Mass Spectrometry Using a Linear Ion Trap. *J. Am. Soc. Mass. Spectrom.* **2005,** *16,* 1498–1509.

Milne, G. L.; Yin, H.; Morrow, J. D. Human Biochemistry of the Isoprostane Pathway. *J. Biol Chem.* **2008,** *283,* 15533–15537.

Milne, G. L.; Yin, H.; Hardy, K. D.; Davies, S. S.; Roberts, L. J., II. Isoprostane Generation and Function. *Chem Revs.* **2011,** *111,* 5973–5996.

Milne, G. L.; Gao, B.; Terry, E. S.; Davies, S. S.; Roberts, L. J., II. Measurement of F_2-isoprostanes and Isofurans Using Gas Chromatography-Mass Spectrometry. *Free Radical Biol. Med.* **2013,** *59,* 36–44.

Miro, O.; Casademont, J.; Casals, E.; Perea, M.; Urbano-Marquez, A.; Rustin, P.; Cardellach, F. Ageing Is Associated with Increased Peroxidation in Human Hearts, but Not with Mitochondrial Respiratory Chain Enzyme Defects. *Cardiovasc. Res.* **2000,** *47,* 624–631.

Miyake, T.; Shibamoto, T. Simultaneous Determination of Acrolein, Malondialdehyde and 4-Hydroxy-2-nonenal Produced from Lipids Oxidized with Fenton's Reagent. *Food Chem. Toxicol.* **1996,** *34,* 1009–1011.

Mlakar, A.; Spiteller, G. Reinvestigation of Lipid Peroxidation of Linolenic Acid. *Biochim Biophys Acta* **1994,** *1214*, 209–220.
Mlakar, A.; Spiteller, G. Previously Unknown Aldehydic Lipid Peroxidation Compounds of Arachidonic Acid. *Chemistry and Physics of Lipids* **1996,** *79*, 47–53.
Mohr, D.; Umeda, Y.; Redgrave, T. G.; Stocker, R. Antioxidant Defenses in Rat Intestine and Mesenteric Lymph. *Redox Rep* **1999,** *4*, 79–87.
Moore, K. P.; Darley-Usmar, V.; Morrow, J.; Roberts, L. J., II. Formation of F_2-isoprostanes During Oxidation of Human Low-density Lipoproteins and Plasma by Peroxynitrite. *Circulation Research* **1995,** *77*, 335–341.
Moreau, R.; Nguyen, B. Y.; Doneanu, C. E.; Hagen, T. M. Reversal by Aminoguanidine of the Age-related Increase in Glycoxidation and Lipoxidation in the Cardiovascular System of Fischer 344 Rats. *Biochem. Pharmacol.* **2005,** *69*, 29–40.
Morisseau, C.; Inceoglu, B.; Schmelzer, K.; Tsai, H. J.; Jinks, J. L.; Hegedus, C. M.; Hammock, B. D. Naturally Occurring Monoepoxides of Eicosapentaenoic Acid and Docosahexaenoic Acid Are Bioactive Antihyperalgesic Lipids. *J. Lipid Res.* **2010,** *51*, 3481–3490.
Morrow, J. D.; Roberts, L. J., II. Quantification of Noncyclooxygenase Derived Prostanoids as a Marker of Oxidative Stress. *Free Rad Biol Med.* **1991,** *10*, 195–200.
Morrow, J. D.; Harris, T. M.; Roberts, L. J. Non-cyclooxygenase Oxidative Formation of a Series of Novel Prostaglandins: Analytical Ramifications for Measurement of Eicosanoids. *Anal Biochem.* **1990a,** *184*, 1–10.
Morrow, J. D.; Hill, K. E.; Burk, R. F.; Nammour, T. M.; Badr, K. F.; Roberts, L. J., II. A Series of Prostaglandin F_2-like Compounds Are Produced in vivo in Humans by a Non-cyclooxygenase, Free Radical-catalyzed Mechanism. *Proc. Natl. Acad. Sci. USA* **1990b,** *87*, 9383–9387.
Morrow, J. D.; Awad, J. A.; Boss, H. J.; Blair, I. A.; Roberts, L. J., II. Non-cyclooxygenase-derived Prostanoids (F_2-isoprostanes) Are Formed in situ on Phospholipids. *Proc Natl Acad Sci. USA* **1992a**, *89*, 10721–10725.
Morrow, J. D.; Minton, T. A.; Roberts, L. J., II. The F_2-isoprostane, 8-epi-prostaglandin $F_{2\alpha}$, a Potent Agonist of the Vascular Thromboxane/Endoperoxide Receptor, Is a Platelet Thromboxane/Endoperoxide Receptor Antagonist. *Prostaglandins* **1992b,** *44*, 155–163.
Morrow, J. D.; Minton, T. A.; Mukundan, C. R.; Campbell, M. D.; Zackert, W. E.; Daniel, V. C.; Badr, K. F.; Blair, I. A.; Roberts, L. J., II. Free Radical-induced Generation of Isoprostanes in Vivo. Evidence for the Formation of D-ring and E-ring Isoprostanes. *J. Biol. Chem.* **1994,** *269*, 4317–4326.
Moumtzi, A.; Trenker, M.; Zenzmaier, E.; Saf, R.; Hermetter, A. Import and Fate of Fluorescent Analogs of Oxidized Phospholipids in Vascular Smooth Muscle Cells. *J. Lipid Res.* **2007,** *48*, 565–582.
Murray, L. V. J.; Liu, L.; Komatsu, H.; Uryu, K.; Xiao, G.; Lawson, J. A.; Axelsen, P. H. Membrane-mediated Amyloidogenesis and the Promotion of Oxidative Lipid Damage by Amyloid β Proteins. *J. Biol. Chem.* **2007,** *282*, 9335–9345.
Musiek, E. S.; Cha, J. K.; Yin, H.; Zackert, W. E.; Terry, E. S.; Porter, N. A.; Montine, T. J.; Morrow, J. D. Quantification of F-ring Isoprostane-like Compounds (F_4-neuroprostanes) Derived from Docosahexaenoic Acid in Vivo in Humans by a Stable Isotope Dilution Mass Spectrometric Assay. *J. Chromatogr. B. Analyt. Technol. Biomed. Life Sci.* **2004,** *799*, 95–102.

Nagai, R.; Murray, D. B.; Metz, T. O.; Baynes, J. W. Chelation: A Fundamental Mechanism of Action of AGE Inhibitors, AGE Breakers, and Other Inhibitors of Diabetes Complications. *Diabetes* **2012,** *61,* 549–559.

Nagy, L.; Tontonoz, P.; Alvarez, J. G.; Chen, H.; Evans, R. M. Oxidized LDL Regulates Macrophage Gene Expression through Activation of PPAR-γ. *Cell* **1998,** *93,* 229–240.

Nakagawa, K.; Oak, J.-H.; Higuchi, O.; Tsuzuki, T.; Oikawa, S.; Otoni, H.; Mune, N.; Cai, H.; Miyazawa, T. Ion-trap Mass Spectrometric Analysis of Amadori Glycated Phosphatidylethanolamine in Human Plasma with or without Diabetes. *J. Lipid Res.* **2005,** *46,* 2514–2524.

Nakamura, K.; Miura, D.; Kusano, K. F.; Fujimoto, Y.; Sumita-Yoshikawa, N.; Fuke, S.; Nishii, N.; Nagase, S.; Hata, Y.; Morita, H.; et al. 4-Hydroxy-2-nonenal Induces Calcium Overload via the Generation of Reactive Oxygen Species in Isolated Rat Cardiac Myocytes. *J. Card. Fail.* **2009,** *15,* 709–716.

Nakanishi, H.; Iida, Y.; Shimizu, T.; Taguchi, R. Analysis of Oxidized Phosphatidylcholines as Markers for Oxidative Stress, Using Multiple Reaction Monitoring with Theoretically Expanded Data Sets with Reversed-phase Liquid Chromatography/Tandem Mass Spectrometry. *J. Chromatogr. B.* **2009,** *877,* 1366–1374.

Namgaladze, D.; Morbitzer, D.; von Knethen, A.; Brune, B. Phospholipase A_2 Modified Low Density Lipoprotein Activates Macrophage Peroxisome Proliferator-activated Receptors. *Arterioscler. Thromb. Vasc. Biol.* **2010,** *30,* 313–320.

Napoli, C.; Mancini, F. P.; Corso, G.; Malomi, A.; Crescenzi, E.; Postiglione, A.; Palumbo, G. A Simple and Rapid Purification Procedure Minimizes Spontaneous Oxidative Modifications of Low Density Lipoprotein and Lipoprotein(a). *J. Biochem.* **1997,** *121,* 1096–1101.

Natarajan, V.; Scribner, W. M.; Hart, C. M.; Parthasarathy, S. Oxidized Low Density Lipoprotein-mediated Activation of Phospholipase D in Smooth Muscle Cells: A Possible Role in Cell Proliferation and Atherogenesis. *J. Lipid Res.* **1995,** *36,* 2005–2016.

Naudi, A.; Jove, M.; Ayala, V.; Cabre, R.; Portero-Otin, M.; Pamplona, R. Non-enzymatic Modification of Aminophospholipids by Carbonyl-amine Reactions. *Int. J. Mol. Sci.* **2013,** *14,* 3285–3313.

Negre-Salvayre, A.; Auge, N.; Ayala, V.; Basaga, H.; Boada, J.; Brenke, R.; Chapple, S.; Cohen, G.; Feher, J.; Grune, T.; et al. Pathological Aspects of Lipid Peroxidation. *Free Radical Res.* **2010,** *44,* 1125–1171.

Niu, X.; Zammit, V.; Upston, J. M.; Dean, R. T.; Stocker, R. Coexistence of Oxidized Lipids and α-Tocopherol in All Lipoprotein Density Fractions Isolated from Advanced Human Atherosclerotic Plaques. *Arterioscler Thromb Vasc Biol* **1999,** *19,* 1708–1718.

Node, K.; Huo, Y.; Ruan, Y.; Yang, B.; Speicker, M.; Ley, K; Zeldin, D. C.; Liao, J. K. Anti-inflammatory Properties of Cytochrome P450 Epoxygenase-derived Eicosanoids. *Science* **1999,** *285,* 1276–1279.

Obinata, H.; Hattori, T.; Nakane, S.; Tatei, K.; Izumi, T. Identification of 9-Hydroxyoctadecadienoic Acid and Other Oxidized Free Fatty Acids as Ligands of the G-protein-coupled Receptor G2A. *J. Biol. Chem.* **2005,** *280,* 40676–40683.

Palinski, W.; Rosenfeld, M. E.; Yla-Herttuala, S.; Gurtner, G. C.; Socher, S. S.; Butler, S. N.; Parthasarathy, S.; Carew, T. E.; Steinberg, D.; Witztum, J. L. Low-density Lipoprotein Undergoes Oxidative Modification in Vivo. *Proc. Natl Acad Sci. USA* **1989,** *86,* 1372–1376.

Palinski, W.; Horkko, S.; Miller, E.; Steinbrecher, U. P.; Powell, H. C.; Curtiss, C. K.; Witztum, J. L. Cloning of Monoclonal Autoantibodies to Epitopes of Oxidized Lipoproteins from Apolipoprotein E-deficient Mice. Demonstration of Epitopes of Oxidized Low-density Lipoprotein in Human Plasma. *J. Clin. Invest.* **1996,** *98,* 800–814.

Pang, L. Q.; Liang, Q. L.; Wang, Y.. M.; Ping, L.; Luo, G. A. Simultaneous Determination and Quantification of Seven Major Phospholipid Classes in Human Blood Using Normal-phase Liquid Chromatography Coupled with Electrospray Mass Spectrometry and the Application in Diabetes Nephropathy. *J. Chromatogr. B Analyt. Technol. Biomed. Life Sci.* **2008,** *869,* 118–125.

Panigrahy, D.; Kaipainen, A.; Greene, E.; Huang, S. Cytochrome 450-Derived Eicosanoids: The Neglected Pathway in Cancer. *Cancer Metastasis Rev.* **2010,** *29,* 723–735.

Panigrahy, D.; Edin, M. L.; Lee, C. R.; Huang, S.; Bielenberg, D. R.; Barnes, C. N.; Mammoto, A.; Mammoto, T.; Luria, A.; Benny, O. Epoxyeicosanoids Stimulate Multiorgan Metastasis and Tumor Dormancy Escape in Mice. *J. Clin. Invest.* **2012,** *122,* 178–191.

Parhami, F.; Fang, Z. T.; Yang, B.; Fogelman, A. M.; Berliner, J. A. Stimulation of Gs and Inhibitions of Gi Protein Functions by Minimally Oxidized LDL. *Arterioscler. Thromb. Vasc. Biol.* **1995,** *15,* 2019–2024.

Pegorier, S.; Stengel, D.; Durand, H.; Croset, M.; Ninio, E. Oxidized Phospholipid: POVPC Binds to Platelet-activating-factor Receptor on Human Macrophages: Implications in Atherosclerosis. *Atherosclerosis* **2006,** *188,* 433–443.

Pizzimenti, S.; Ciamporcero, E.; Daga, M.; Pettazzoni, R.; Arcara, A.; Cetrangolo, G.; Minelli, R.; Dianzani, C.; Lepore, A.; Gentile, F.; Barrera, G. Interactions of Aldehydes Derived from Lipid Peroxidation and Membrane Proteins. *Frontiers in Physiology* **2013,** *4,* 1–17.

Podrez, E. A.; Poliakov, E.; Shen, Z.; Zhang, R.; Deng, Y.; Sun, M.; Finton, P. G., Shan, L.; Febbraio, M.; Hajjar, D. P.; et al. A Novel Family of Atherogenic Oxidized Phospholipids Promotes Macrophage Foam Cell Formation via the Scavenger Receptor CD36 and Is Enriched in Atheromatous Lesions. *J. Biol. Chem.* **2002,** *277,* 38517–38523.

Porter, N. A.; Caldwell, S. E.; Mills, K. A. Mechanisms of Free Radical Oxidation of Unsaturated Lipids. *Lipids* **1995,** *30,* 277–290.

Prashant, A. V.; Harishchandra, H.; D'souza, V.; D'souza, B. Age Related Changes in Peroxidation and Antioxidants in Elderly People. *Indian J. Clin. Biochem.* **2007,** *22,* 131–134.

Pruzanski, W.; Kuksis, A. Secretory Phosholipases as Mediators of Inflammatory and Proatheromatous Changes in Rheumatic Diseases. The European League Against Rheumatism (EULAR), 2014 Report, Paris, France, June 11–14, 2014.

Pruzanski, W.; Stefanski, E.; deBeer, F. C.; deBeer, M. C.; Vadas, P.; Ravandi, A.; Kuksis, A. Lipoproteins Are Substrates for Human Secretory Group IIA Phospholipase A_2: Preferential Hydrolysis of Acute Phase HDL. *J. Lipid Res.* **1998,** *39,* 2150–2160.

Pruzanski, W.; Stefanski, E.; deBeer, F. C.; deBeer, M. C.; Ravandi, A.; Kuksis, A. Comparative Analysis of Lipid Composition of Normal and Acute-phase High Density Lipoproteins. *J. Lipid Res.* **2000,** *41,* 1035–1047.

Pruzanski, W.; Lambeau, L.; Lazdunsky, M.; Cho, W.; Kopilov, J.; Kuksis, A. Differential Hydrolysis of Molecular Species of Lipoprotein Phosphatidyl Choline by groups IIA, V and X Secretory Phospholipases A_2. *Biochim. Biophys. Acta* **2005,** *1736,* 38–50.

Pruzanski, W.; Lambeau, G.; Lazdunski, M.; Cho, W.; Kopilov, J.; Kuksis, A. Hydrolysis of Minor Glycerophospholipids of Plasma Lipoproteins by Human Group IIA, V and X Secretory Phospholipases A_2. *Biochim. Biophys. Acta* **2007,** *1771,* 5–19.

Quehenberger, O.; Armando, A. M.; Brown, A. H.; Milne, S. B.; Myers, D. S.; Merrill, A. H.; Bandyopadhyay, S.; Jones, K. N.; Shaner, R. L.; et al. Lipidomics Reveals a Remarkable Diversity of Lipids in Human Plasma. *J. Lipid Res.* **2010,** *51,* 3299–3305.

Raghavamenon, A. C.; Garelnabi, M.; Babu, S.; Aldrich, A.; Litvinov, D.; Parthasarathy, S. α-Tocopherol Is Ineffective in Preventing the Decomposition of Pre-formed Lipid Peroxides and May Promote the Accumulation of Toxic Aldehydes: A Potential Explanation for the Failure of Antioxidants to Affect Human Atherosclerosis. *Antioxid. Redox. Signal* **2009,** *11,* 1237–1248.

Rauniyar, N.; Prokai-Tatrai, K.; Prokai, L. Identification of Carbonylation Sites in Apomyoglobin after Exposure to 4-Hydroxy-2-nonenal by Solid-phase Enrichment and Liquid Chromatography-electropsray Ionization Tandem Mass Spectrometry. *J. Mass Spectrom.* **2010,** *45,* 398–410.

Ravandi, A.; Kuksis, A.; Myher, J. J.; Marai, L. Determination of Lipid Ester Ozonides and Core Aldehydes by High-performance Liquid Chromatography with On-line Mass Spectrometry. *J. Biochem. Biophys. Methods* **1995,** *30,* 271–285.

Ravandi, A.; Kuksis, A.; Marai, L.; Myher, J. J.; Steiner, J.; Lewisa, G.; Kamido, H. Isolation and Identification of Glycated Aminophospholipids from Red Cells and Plasma of Diabetic Blood. *FEBS Lett.* **1996,** *381,* 77–81.

Ravandi, A.; Kuksis, A.; Shaikh, N.; Jackowski, G. Preparation of Schiff Base Adducts of Phosphatidylcholine Core Aldehydes and Aminophospholipids, Amino Acids, and Myoglobin. *Lipids* **1997,** *32,* 989–1001.

Ravandi, A.; Kuksis, A; Shaikh, N. Glycated Phosphatidylethanolamine Promotes Macrophage Uptake of Low Density Lipoprotein and Accumulation of Cholesteryl Esters and Triacylglycerols. *J. Biol. Chem.* **1999,** *274,* 16494–16500.

Ravandi, A.; Kuksis, A.; Shaikh, N. Glucosylated Glycerphosphoethanolamines Are the Major LDL Glycation Products and Increase LDL Susceptibility to Oxidation: Evidence of Their Presence in Atherosclerotic Lesions. *Arterioscler. Thromb. Vasc. Biol.* **2000,** *20,* 467–477.

Ravandi, A.; Babaei, S., Leung, R.; Monge, J. C., Hoppe, G.; Hoff, H.; Kamido, H.; Kuksis, A. Phospholipids and Oxophospholipids in Atherosclerotic Plaques at Different Stages of Plaque Development. *Lipids* **2004,** *39,* 97–109.

Reich, E. E.; Zackert, W. E.; Brame, C. J.; Chen, Y.; Roberts, L. J., II. Hachey, D. L.; Montine, T. J.; Morrow, J. D. Formation of Novel D-ring and E-ring Isoprostane-like Compounds (D_4/E_4-neuroprostanes) in Vivo from Docosahexaenoic Acid. *Biochemistry* **2000,** *39,* 2376–2383.

Reis, A.; Domingues, M. R.; Amado, F. M.; Ferrer-Correia, A. J.; Domingues, P. Radical Peroxidation of Palmitoyol-linoleoyl-glycerophosphocholine Liposomes: Identification of Long-chain Oxidized Products by Liquid Chromatography-Tandem Mass Spectrometry. *J. Chromatogr. B Analyt. Technol. Biomed. Life Sci.* **2007,** *855,* 186–199.

Requena, J. R.; Fu, M. X.; Ahmed, M. U.; Jenkins, A. J.; Lyons, T. J.; Baynes, J. W.; Thorpe, S. R. Quantification of Malondialdehyde and 4-Hydroxynonenal Adducts to Lysine Residues in Native and Oxidized Human Low-density Lipoprotein. *Biochem. J.* **1997,** *322,* 317–325.

Roberts, L. J., II; Montine, T. J.; Marksbery, W. R.; Tapper, A. R.; Hardy, P.; Chemtob, S.; Dettbarn, W. D.; Morrow, J. D. Formation of Isoprostane-like Compounds (Neuroprostanes) in Vivo from Docosahexaenoic Acid. *J. Biol. Chem.* **1998,** *273,* 13605–13612.

Rodenburg, J.; Vissers, M. N.; Wiegman, A.; Miller, E. R.; Ridker, P. M.; Witztum, J. L.; Kastelein, J. J.; Tsimikas, S. Oxidized Low Density Lipoprotein in Children with Familial Hypercholesterolemia and Unaffected Siblings: Effect of Provastatin. *J. Am. Coll. Cardiol.* **2006,** *47,* 1803–1810.

Rosenson, R.S.; Hislop, C.; McConnell, D.; Elliot, M; Stasir, Y.; Wang, N.; Waters, D. D. Effects of 1-H-indole-3-glyoxamide (A-002) on Concentration of Secretory Phospholipase A_2 (PLASMA stud): A Phase II Double-blind, Randomized, Placebo Controlled Trial. *Lancet* **2009,** *373,* 649–658.

Rosenson, R. S.; Wright, R. S.; Farkouh, M.; Plutzky, J. Modulating Peroxisome Proliferator–activated Receptors for Therapeutic Benefit? Biology, Clinical Experience, and Future Prospects. *Am. Heart J.* **2012,** 164, 672-680.

Salman, K. A.; Ashraf, S. Reactive Oxygen Species: A Link between Chronic Inflammation and Cancer. *AsPac J. Mol. Biol. Biotechnol.* **2013,** *21,* 42–49.

Salomon, R. G.; Gu, X. Critical Insights into Cardiovascular Disease from Basic Research on the Oxidation of Phospholipids: The γ-Hydroxyalkenal Phospholipid Hypothesis. *Chem. Res. Technol.* **2011,** *24,* 1791–1802.

Salomon, R. G.; Sha, W.; Brame, C.; Kaur, K.; Subbanagounder, G.; O'Neil, J.; Hoff, H. F.; Roberts, L. J., II. Protein Adducts of Iso-levulglandin E_2, a Product of the Isoprostane Pathway, in Oxidized Low Density Lipoprotein. *J. Biol. Chem.* **1999,** *274,* 20271–20280.

Salomon, R. G.; Kaur, K.; Podrez, E.; Hoff, H. F.; Krushinsky, A. V.; Sayre, L. M. HNE-derived 2-Pentylpyrroles Are Generated during Oxidation of LDL, Are More Prevalent in Blood Plasma from Patients with Renal Disease or Atherosclerosis, and Are Present in Atherosclerotic Plaques. *Chem. Res. Toxicol.* **2000,** *13,* 557–564.

Salomon, R. G.; Hong, L.; Hollyfield, J. G. The Discovery of Carboxyethylpyrroles (CEPs): Critical Insights into AMD, Autism, Cancer, and Would Healing from Basic Research on the Chemistry of Oxidized Phospholipids. *Chem. Res. Technol.* **2011,** *24,* 1803–1816.

Sayre, L. M.; Zelasko, D. A.; Harris, P. L.; Perry, G.; Salomon, R. G.; Smith, M. A. 4-Hydroxynonenal-derived Advanced Lipid Peroxidation End Products Are Increased in Alzheimer's Disease. *J. Neurochem.* **1997,** *68,* 2092–2097.

Sayre, L. M.; Lin, D.; Yuan, Q.; Zhu, X.; Tang, X. Protein Adducts Generated from Products of Lipid Oxidation: Focus on HNE and ONE. *Drug. Metab. Revs.* **2006,** *38,* 651–675.

Sayre, L. M.; Perry, G.; Smith, M. A. Oxidative Stress and Neurotoxicity. *Chem. Res. Toxicol.* **2008,** *21,* 172–188.

Schaich, K. M. Challenges in Elucidating Lipid Oxidation Mechanisms: When, Where, and How Do Products Arise? In *Lipid Oxidation*; Logan, A., Nieabar, U., Pan, X., Eds.; AOCS Press: Urbana, IL, 2013; pp 1–52.

Schneider, C.; Pratt, D. A.; Porter, N. A.; Brash, A. R. Control of Oxygenation in Lipoxygenase and Cyclooxygenase Catalysis. *Chem. Biol.* **2007,** *14,* 473–488.

Schneider, C.; Porter, N.; Brash, A. Routes to 4-Hydroxynonenal: Fundamental Issues in the Mechanism of Lipid Peroxidation. *J. Biol. Chem.* **2008,** *283,* 15539–15543.

Schrag, M.; Mueller, C.; Zabel, M.; Crofton, A.; Kirsch, W. M.; Ghribi, O.; Squitti, R.; Perry, G. Oxidative Stress in Blood in Alzheimer's Disease and Mild Cognitive Impairment: A Meta-analysis. *Neurobiol. Dis.* **2013,** *59,* 100–110.

Seet, R. C.; Lee, C. Y.; Lim, E. C.; Tan, J. J.; Quek, A. M.; Chong, W. L.; Looi, W. F.; Huang, S. H.; Wang, H.; Chan, Y. H.; Halliwell, B. Oxidative Damage in Parkinson Disease: Measurement Using Accurate Biomarkers. *Free Rad. Biol. Med.* **2010,** *48,* 560–566.

Serhan, C. N.; Petassis, N. A. Resolvins and Protectins in Inflammation Resolution. *Chem. Rev.* **2011,** *111,* 5922–5943.

Serhan, C. N.; Clish, C. B.; Brannon, J.; Colgan, S. P.; Chiang, N.; Gronert, K. Novel Functional Sets of Lipid-derived Mediators with Anti-inflammatory Actions Generated from ω-3 Fatty Acids via Cyclooxygenase 2-Nonsteroidal Anti-inflammatory Drugs and Transcellular Processing. *J. Exp. Med.* **2000**, *192,* 1197–1204.

Serruys, P. W.; Garcia-Garcia, H. M.; Buszman, P.; Erne, P.; Berheye, S.; Aschermann, M.; Duckers, H.; Bleie, O.; Dudek, D.; Bøtker, H. E.; et al. Effects of the Direct Lipoprotein-associated Phospholipase A_2 Inhibitor Darapladib on Human Coronary Atherosclerotic Plaque. *Circulation* **2008,** *118,* 1172–1182.

Sethi, S.; Ziouzenkova, O.; Ni, H.; Wagner, D. D.; Plutzky, J.; Mayadas, T. N. Oxidized ω-3-fatty Acids in Fish Oil Inhibits Leukocyte-endothelial Interactions through Activation of PPARα. *Blood* **2002,** *100,* 1340–1346.

Sevanian, A.; Kim, E. Phospholipase A_2 Dependent Release of Fatty Acids from Peroxidized Membranes. *Free Radic. Biol. Med.* **1985,** *1,* 263–271.

Sevanian, A.; Wratten, M. L.; McLeod, L. L.; Kim, E. Lipid Peroxidation and Phospholipase A_2 Activity in Liposomes Composed of Unsaturated Phospholipids: A Structural Basis for Enzyme Activation. *Biochim. Biophys. Acta* **1988,** *961,* 316–327.

Shearn, C. T.; Backos, D. S.; Orlicky, D. J.; Smathers-McCullough, R. L.; Petersen, D. R. Identification of 5'-Activated Kinase as a Target of Reactive Aldehydes during Chronic Ingestion of High Concentrations of Ethanol. *J. Biol. Chem.* **2014,** *289,* 15449–15462.

Sjovall, O.; Kurvinen, J.-P.; Kallio, H. Lipid Ester Hydroperoxides and Core Aldehydes in Oxidized Triacylglycerols of Baltic Herring Oil. *Proceedings of the 19th Nordic Lipid Symposium,* Scandinavian Forum for Lipid Research and Technology, Ronneby, Sweden, 1997; p 57.

Sjovall, O.; Koivusalo, M.; Kallio, H. Hydrolysis of Core Aldehyde and Epoxy Triacylglycerol Regioisomers by Pancreatic Lipase and Preparation of Schiff Base Adducts of Triacylglycerol Core Aldehydes with Aminophospholipids and Amino Acids. In *Abstracts,* 89th AOCS Annual Meeting & Expo, Chicago, IL, 1998.

Sjovall, O.; Kuksis, A.; Kallio, H. Reversed Phase High-performance Liquid Chromatographic Separation of *Tert*-butyl Hydroperoxide Oxidation Products of Unsaturated Triacylglycerols. *J. Chromatogr. A* **2001,** *905,* 119–132.

Smiley, P. L; Stremler, K. E.; Prescott, S. M.; Zimmerman, G. A.; McIntyre, T. M. Oxidatively Fragmented Phosphatidylcholines Activate Human Neutrophils through the Receptor for Platelet-activating Factor. *J. Biol. Chem.* **1991,** *266,* 11104–11110.

Song, J. H.; Fujimoto, K.; Miyazawa, T. Polyunsaturated (n-3) Fatty Acids Susceptible to Peroxidation Are Increased in Plasma and Tissue Lipids of Rats Fed Docosahexaenoic Acid-containing Oils. *J. Nutr.* **2000,** *130,* 3028–3033.

Sookwong, P.; Nakagawa, K.; Fujita, I.; Shoji, N.; Miyazawa, T. Amadori-glycated Phosophatidylethanolamine. A Potential Marker for Hyperglycemia in Streptozotocin-induced Diabetic Rats. *Lipids* **2011,** *46,* 943–952.

Spector, A. A.; Norris, A. W. Action of Epoxyeicosatrienoic Acids on Cellular Function. *Am. J. Physiol. Cell Physiol.* **2006,** *292,* C996–C1012.

Spickett, C. M. The Lipid Peroxidation Product 4-Hydroxy-2-nonenal: Advances in Chemistry and Analysis. *Redox Biology* **2013,** *1,* 145–152.

Spiteller, G. Furan Fatty Acids: Occurrence, Synthesis, and Reactions. Are Furan Fatty Acids Responsible for the Cardiac-protective Effects of a Fish Diet. *Lipids* **2005,** *40,* 755–771.

Stafforini, D. M.; Sheller, J. R.; Blackwell, T. S. Sapirstein, M. A.; Yull, F. E.; McIntyre, T. M.; Bonventre, J. V.; Prescott, S. M.; Roberts, L. J., II. Release of Free F_2-isoprostanes from Esterified Phospholipids Is Catalyzed by Intracellular and Plasma Platelet-activating Factor Acetylhydrolases. *J. Biol. Chem.* **2006,** *281,* 4616–4623.

Steenhorst-Slikkerveer, L.; Loufer, A.; Janssen, H.-G.; Bauer-Plank, C. Analysis of Nonvolatile Lipid Oxidation Products in Vegetable Oils by Normal-phase High-performance Liquid Chromatography with Mass Spectrometric Detection. *J. Am. Oil Chem. Soc.* **2000,** *27,* 837–845.

Steinberg, D. Role of Oxidized LDL and Antioxidants in Atherosclerosis. *Adv. Exp. Med. Biol.* **1995,** *369,* 39–48.

Steinberg, D.; Witztum, J. L. Lipoproteins, Lipoprotein Oxidation and Atherogenesis. In *Molecular Basis of Cardiovascular Disease;* Chien, K. R., Ed.; W. B. Sanders: Philadelphia, PA, 1999.

Steinberg, D.; Witztum, J. L. Oxidized Low-density Lipoprotein and Atherosclerosis. *Arterioscler. Thromb. Vasc. Biol.* **2010,** *30,* 2311–2316.

Steinbrecher, U. P. Oxidation of Human Low-density Lipoprotein Results in Derivatization of Lysine Residues of Apolipoproein B by Lipid Peroxide Decomposition Products. *J. Biol. Chem.* **1987,** *262,* 3603–3608.

Stillwell, W.; Wassall, S. R. Docosahexaenoic Acid. Membrane Properties of a Unique Fatty Acid. *Chem. Phys. Lipids* **2003,** *126,* 1–27.

Stremmer, U.; Hermetter, A. Protein Modification by Aldehydophospholipids and Its Functional Consequences. *Biochim. Biophys. Acta* **2012,** *1818,* 2436–2445.

Strohmaier, H.; Hinghoifer-Szalkay, H.; Schaur, R. I. Detection of 4-Hydroxynonenal (HNE) as Physiological Component in Human Plasma. *J. Lipid Mediat. Cell Signal.* **1995,** *11,* 51–61.

Subbanagounder, G.; Watson, A. D.; Berliner, J. A. Bioactive Products of Phospholipid Oxidation: Isolation, Identification, Measurement and Activities. *Free Radical Biology & Medicine* **2000,** *28,* 1751–1761.

Subbanagounder, G.; Wong, J. W.; Lee, H.; Faull, K. F.; Miller, E.; Witztum, J. L.; Berliner, J. A. Epoxyisoprostane and Epoxycyclopentanone Phospholipids Regulate Monocyte Chemotactic Protein-1 and Interlukin-8 Synthesis. Formation of These Oxidized Phospholipids in Response to Interleukin-1β. *J. Biol. Chem.* **2002,** *277,* 7271–7281.

Sultana, R.; Perluigi, M.; Butterfield, D. A. Lipid Peroxidation Triggers Neurodegeneration, Redox Proteomics View into the Alzheimer Disease Brain. *Free Radic. Biol. Med.* **2011,** *62,* 157–169.

Sun, Y.; Wang, N.; Tall, A. R. Regulation of Adrenal Scavenger Receptor-B1 Expression by ACTH and Cellular Cholesterol Pools. *J. Lipid Res.* **1999,** *40,* 1799–1805.

Takahashi, K.; Nammour, T. M., Fukunaga, M.; Ebert, J.; Morrow, J. D.; Roberts, L. J., II; Hoover, R. L.; Badr, K. F. Glomerular Actions of a Free Radical-generated Novel Prostaglandin, 8-Epi-prostaglandin $F_{2\alpha}$, in the Rat. Evidence for Interaction with Thromboxane A_2 Receptor. *J. Clin. Invest.* **1992,** *90,* 136–141.

Tarvainen, M.; Phuphusit, A.; Suomela, J.-P.; Kuksis, A.; Kallio, H. Effects of Antioxidants on Rapeseed Oil Oxidation in an Artificial Digestion Model Analyzed by UHPLC-ESI-MS. *J. Agric. Food Chem.* **2012,** *60,* 3564–3579.

Taylor, A. W.; Traber, M. G. Quantitation of Plasma Total 15-series F_2-isoprostanes by Sequential Solid Phase and Liquid-Liquid Extraction. *Anal. Biochem.* **2010,** *396,* 310–321.

Tertov, V. V.; Kaplun, V. V.; Dvoryantsev, S. N.; Orekhov, A. N. Apoprotein B-bound Lipids as a Marker for Evaluation of Low Density Lipoprotein Oxidation in Vivo. *Biochem. Biophys. Research Commun.* **1995,** *214,* 608–613.

Thomas, C. E.; Jackson, R. L.; Ohlloweiler, D. F.; Ku, G. Multiple Lipid Oxidation Products in Low Density Lipoproteins Induce Interleukin-1 β Release from Human Blood Mononuclear Cells. *J. Lipid Res.* **1994,** *35,* 417–427.

Thum, T.; Borlak, J. LOX-1 Receptor Blockade Abrogates oxLDL-induced Oxidative DNA Damage and Prevents Activation of the Transcriptional Repressor Oct-1 in Human Coronary Arterial Endothelium. *J. Biol. Chem.* **2008,** *283,* 19456–19464.

Tontonoz, P.; Nagy, L.; Alvarez, J. G.; Thomazy, V. A.; Evans, R. M. PPARγ Promotes Monocyte/Macrophage Differentiation and Uptake of Oxidized LDL. *Cell* **1998,** *93,* 241–252.

Tsimikas, S. Oxidative Biomarkers in the Diagnosis and Prognosis of Cardiovascular Disease. *Am. J. Cardiool.* **2006,** *98,* 9P–17P.

Tsimikas, S.; Miller, Y. I. Oxidative Modification of Lipoproteins: Mechanisms, Role in Inflammation and Potential Clinical Applications in Cardiovascular Disease. *Curr. Pharm. Des.* **2011,** *17,* 27–37.

Tsimikas, S.; Witztum, J. L. The Role of Oxidized Phospholipids in Mediating Lipoprotein(a) Atherogenicity. *Curr. Opin. Lipidology* **2008,** *19,* 369–377.

Tsimikas, S.; Lau, H. K.; Han, K. R.; Shortal, B.; Miller, E. R.; Segev, A.; Curtiss, L. K.; Witztum, J. L.; Strauss, B. H. Percutaneous Coronary Intervention Results in Acute Increases in Oxidized Phospholipids and Lipoprotein(a): Short-term and Long-term Immunologic Responses to Oxidized Low Density Lipoprotein. *Circulation* **2004,** *109,* 3164–3170.

Tsimikas, S.; Brilakis, E. S.; Miller, E. R.; McConnell, J. P.; Lennon, R. J.; Komman, K. S.; Witztum, J. L.; Berger, P. B. Oxidized Phospholipids, Lp(a) Lipoprotein, and Coronary Artery Disease. *N. Engl. J. Med.* **2005,** *353,* 46–57.

Tsimikas, S.; Mallat, Z.; Talmud, P. J.; Kastelein, J. J.; Waveham, N. J.; Sandhu, M. S.; Miller, E. R.; Benessiano, J.; Tedqui, A.; Witztum, J. L.; et al. Oxidation-specific Biomarkers, Lipoprotein(a), and Risk of Fatal and Non-fatal Coronary Events. *J. Am. College of Cardiol.* **2010,** *56,* 946–955.

Uchida, K. 4-Hydroxy-2-nonenal: A Product and Mediator of Oxidative Stress. *Progr. Lipid Res.* **2003,** *42,* 318–343.

Uchida, K.; Itakura, K.; Kawakishi, S.; Hiai, H.; Toyokuni, S.; Stadtman, E. R. Characterization of Epitopes Recognized by 4-Hydroxy-2-nonenal Specific Antibodies. *Arch Biochem. Biophys.* **1995,** *324,* 241–248.

Uchida, K.; Sakai, K.; Itakura, K.; Osawa, T.; Toyokuni, S. Protein Modification by Lipid Peroxidation Products: Formation of Malondialdehyde-derived N-(2-propenal)lysine in Proteins. *Arch. Biochem. Biophys.* **1997,** *346,* 45–52.

Uchida, K.; Kanematsu, M.; Morimitsu, Y.; Osawa, T.; Noguchi, N.; Niki, E. Acrolein Is a Product of Lipid Peroxidation Reaction: Formation of Acrolein and Its Conjugate with Lysine Residues in Oxidized Low Density Lipoprotein. *J. Biol. Chem.* **1998,** *273,* 16058–16066.

Umano, K.; Dennis, K. J.; Shibamoto, T. Analysis of Free Malondialdehyde in Photo-irradiated Corn Oil and Beef Fat via a Pyrazole Derivative. *Lipids* **1988,** *23,* 811–814.

Upston, J. M.; Neuzil, J.; Socker, R. Oxidation of LDL by Recombinant Human 15-Lipoxygenase: Evidence for α-Tocopherol Dependent Oxidation of Esterified Core and Surface Lipids. *J. Lipid Res.* **1996,** *37,* 2650–2661.

Upston, J. M.; Terentis, A. C.; Stocker, R. Tocopherol-mediated Peroxidation of Lipoproteins: Implications for Vitamin E as Potential Antiatherogenic Supplement. *FASEB J.* **1999,** *13,* 977–994.

Uran, S.; Larsen, A.; Jacobsen, P. B.; Skotland, T. Analysis of Phospholipid Species in Human Blood Using Normal Phase Liquid Chromatography Coupled with Electrospray Ionization Ion-trap Mass Spectrometry. *J. Chromatogr. B* **2001,** *758,* 265–275.

Vollmer, E.; Brust, J.; Roessner, A.; Bosse, A.; Burwikel, F.; Kaesberg, B.; Harrach, B.; Robenek, H.; Bocker, W. Distribution Patterns of Apoproteins A I, A II, and B in the Wall of Atheriosclerotic Vessels. *Virchows Arch. Anat. Histopathol.* **1991,** *419,* 79–88.

Van den Berg, J. J.; Op den Kamp, J. A.; Lubin, B. H.; Kuypers, F. A. Conformational Changes in Oxidized Phospholipids and Their Preferential Hydrolysis by Phospholipase A_2: A Monolayer Study. *Biochemistry* **1993,** *32,* 4962–4967.

Von Schacky, C.; Harris, W. S. Cardiovascular Benefits of Omega-3 Fatty Acids. *Cardiovasc. Res.* **2007,** *73,* 310–315.

Wang, C. J.; Tai, H. H. A Facile Synthesis of an Aldehydic Analog of Platelet Activating Factor and Its Use in the Production of Specific Antibodies. *Chem. Phys. Lipids* **1990,** *55,* 265–273.

Watson, A. D.; Leitinger, N.; Navab, M.; Faull, K. F.; Horkko, S.; Witztum, J. L.; Palinski, W.; Schwenke, D.; Salomon, R. G.; Sha, W. Structural Identification by Mass Spectrometry of Oxidized Phospholipids in Minimally Oxidized Low Density Lipoprotein That Induce Monocyte/Endothelial Interactions and Evidence for Their Presence in Vivo. *J. Biol. Chem.* **1997,** *272,* 13597–13607.

Wells-Knecht, K. J.; Zyzak, D. V.; Litchfield, J. E.; Thorpe, S. R.; Baynes, J. W. Mechanism of Autooxidative Glycosyation: Identification of Glyoxal and Arabinose as Intermediates in the Autooxidative Modification of Proteins by Glucose. *Biochemistry* **1994,** *34,* 3702–3709.

Wingler, K.; Muller, C.; Schmehl, K. Gastrointestinal Glutathione Peroxidase Prevents Transport of Lipid Hydroperoxides in Caco-2 Cells. *Gastroenterology* **2000,** *119,* 420–430.

Witztum, J. L.; Steinberg, D. Role of Oxidized Low Density Lipoprotein in Atherogenesis. *J. Clin. Invest.* **1991**, 88, 1785–1792.

Witting, P. K.; Upston, J. M.; Stocker, R. The molecular Action of α-Tocopherol in Lipoprotein Lipid Peroxidation: Pro and Antioxidant Activity of Vitamin E in Complex Heterogeneous Lipid Emulsions. In *Subcellular Biochemistry: Fat Soluble Vitamins*; Quinn, P., Kagan, V., Eds; Plenum: London, 1998; pp 345–390.

Yamada, S.; Kumazawa, S.; Ishii, T.; Nakayama, T.; Itakura, K.; Shibata, N.; Kobayashi, M.; Sakai, K.; Osawa, T.; Uchida, K. Immunochemical Detection of a Lipofuscin-like Fluorescence from Malondialdehyde-lysine. *J. Lipid Res.* **2001,** *42,* 1187–1196.

Yin, H.; Cox, B. E.; Liu, W.; Porter, N. A.; Morrow, J. D.; Milne, G. L. Identification of Intact Oxidation Products of Glycerophospholipids in Vitro and in Vivo Using Negative Ion Electrospray Ion-trap Mass Spectrometry. *J. Mass Spectrom.* **2009,** *44,* 672–680.

Yin, H.; Yu, L.; Porter, N. A. Free Radical Lipid Peroxidation: Mechanism and Analysis. *Chem. Revs.* **2011,** *111,* 5944–5972.

Yoshikawa, M.; Sakuma, N.; Hibino, T.; Sato, T.; Fujinami, T. HDL_3 Exerts More Powerful Anti-oxidative Protective Effects against Copper-catalyzed LDL Oxidation Than HDL_2. *Clin. Biochem.* **1997,** *30,* 221–225.

Yura, T.; Fukunaga, M.; Khan, R.; Nassar, G. N.; Badr, K. F.; Montero, A. Free-radical-generated F_2-isoprostane Stimulates Cell Proliferation and Endothelin-1 Expression on Endothelial Cells. *Kidney International* **1999,** *56,* 471–478.

Zhao, Y.; Monahan, F. J.; McNulty, B. A.; Gibnev, M.; Gibnev, E. R. Effect of Vitamin E Intake from Food and Supplement Sources on Plasma α- and γ-Tocopherol Concentrations in a Healthy Irish Adult Population. *Br. J. Nutr.* **2014,** 112, 1575–1585.

Zhang, G.; Panigrahy, D.; Mahakian, L. M.; Yang, J.; Lin, J. Y.; Stephen Lee, K. S.; Wettersten, H. I.; Ulu, A.; Hu, Y.; et al. Epoxy Metabolites of Docosahexaenoic Acid (DHA) Inhibit Angiogenesis, Tumor Growth, and Metastasis. *PNAS* **2013,** *110,* 6530–6535.

Zhang, M.; Li, W.; Li, T. Generation and Detection of Levuglandins and Isolevuglandins in Vitro and in Vivo. *Molecules* **2011,** *16,* 5333–5348.

11

Lysophospholipids: Advances in Synthesis and Biological Significance

Moghis U. Ahmad, Shoukath M. Ali, Ateeq Ahmad, Saifuddin Sheikh, and Imran Ahmad ▪ *Jina Pharmaceuticals Inc., Libertyville, Illinois, USA*

Introduction

A lysophospholipid (LP) is a phospholipid that is missing one of its two O-acyl chains (free hydroxyl group in either the *sn-1* or *sn-2* position). Lysophospholipids are good emulsifying and solubilizing agents as well as synthons for many synthetic phospholipids used in food, cosmetics, agrochemicals, and pharmaceuticals (Aoi, 1990; Dennis et al., 2006; Fujita and Suzuki, 1990; Reblova and Pokorny, 1995). Also, LPs play an important role in phospholipid metabolism and influence a variety of biological processes (Birgbauer and Chun, 2006). LPs are membrane-derived signaling molecules produced by phospholipases that exhibit a wide range of biological activities (Moolenaar, 2000) and are also involved in a wide range of cellular processes including membrane–protein or membrane–membrane interactions (Torkhovskaya et al., 2007). The two LPs, lysophosphatidic acid (LPA) and sphingosine 1-phosphate (S1P) (Figure 11.1), regulate diverse biological processes. The LPA (Ishii et al., 2004; Mills and Moolenaar, 2003; Moolenaar, 1995; Tigyi and Parrill, 2003; Tokumura, 2002) and S1P (Ishii et al., 2004; Spiegel et al., 1994) have created interest for their extracellular signaling properties. S1P has recently been implicated in the etiology of cancer due to its involvement in tumor growth, angiogenesis, and metastatic potential (Shida et al., 2008b). LPs are mediated by specific G protein-coupled receptors (GPCRs) and implicated in a growing number of disorders such as inflammation, autoimmune diseases, neuropathic pain, atherosclerosis, cancer, and obesity (Chun and Rosen, 2006; Dorbnik et al., 2003; Kim et al., 2006; Mills and Moolenaar, 2003; Raynal et al., 2005; Sutphen et al., 2004; Xu et al., 1998; Zhao et al., 2007).

LPs are generally produced by phospholipase A-type enzymatic activity of phospholipids such as phosphotidylcholine or phosphatidic acid. They can also be generated by the acylation of glycerophospholipids or the phosphorylation of monoacylglycerols. In this chapter, we will mainly focus on the chemical and enzymatic synthesis of LPs and the biological significance of three LPs, namely 2-lysophosphatidylcholine (2-LPC), 2-lysophosphatidic acid (2-LPA), and S1P. Alkyl-lysophospholipids (ALPs), a family of promising anticancer drugs, will also be discussed.

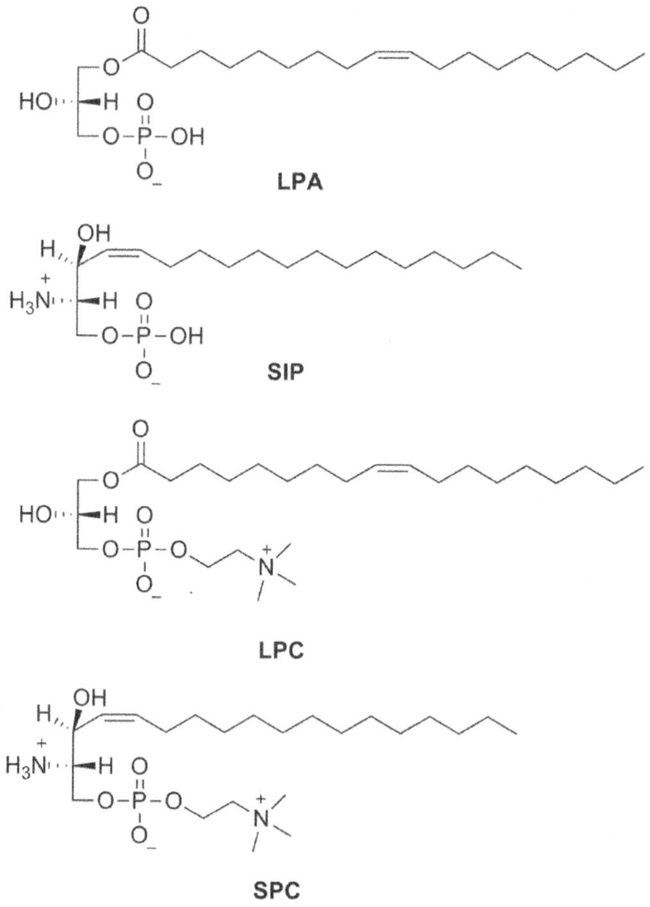

Figure 11.1 Chemical structures of some of the bioactive phospholipids. LPA: lysophosphatidic acid; SIP: sphingosine 1-phosphate; LPC: lysophosphatidylcholine; SPC: sphingosylphosphorylcholine.

Synthetic Lysophospholipids

Development of new synthetic methods for the preparation of LPs is an extremely important as well as a challenging problem of membrane chemistry and biochemistry today. Elucidation of the mechanistic details involved in the enzymological, cell-biological, and membrane-biophysical roles of LPs depends on the availability of structurally variable LP derivatives. By definition, an LP is a phospholipid that is missing one of its two O-acyl chains (i.e., a phospholipid having a free alcohol in either the *sn-1* or the *sn-2* position). 1-Lysophospholipids (1-LPs) maintain the acyl chain in *sn*-2-position, and 2-lysophospholipids (2-LPs) are only acylated at *sn*-1 position (Figure 11.2).

The importance of LPs and their receptors' roles in the development, activation, and regulation of the immune system and cancerous cell proliferation makes them attractive targets for developing drugs for various diseases, especially for the treatment of cancers. In fact, LPs and their receptors have been found in different tissues and cell types that are important in many physiological processes, including vascular development (Karliner, 2004), nervous system (Chun, 2005; Gardell et al., 2006), and reproduction (Parrill, 2008). In the last decade, it has become clear that LPs' activities are mediated by specific GPCRs and are responsible for a growing number of disorders such as inflammation, autoimmune diseases, neuropathic pain, atherosclerosis, cancer, and obesity (Chun and Rosen, 2006; Drobnik et al., 2003; Kim et al., 2006; Mills and Moolenaar, 2003; Raynal et al., 2005; Sutphen et al., 2004; Xu et al., 1998; Zhao et al., 2007). Recently, LPs become the focus of special attention since it was discovered that in addition to LPs' role in phospholipid metabolism, they also function as second messengers, thereby exhibiting a broad range of biological activities. The simplicity and flexibility of these molecules result in high activity toward biomolecular targets. However, unlike PLs, the LPs are found only in small amounts in biological cell membranes (Birgbauer and Chun, 2006). The structural and dynamic studies of biomembranes to establish the structure-activity relationships, PLs-protein interactions, and mechanisms of action of PLs metabolizing enzymes require the preparation of PLs derivatives as the

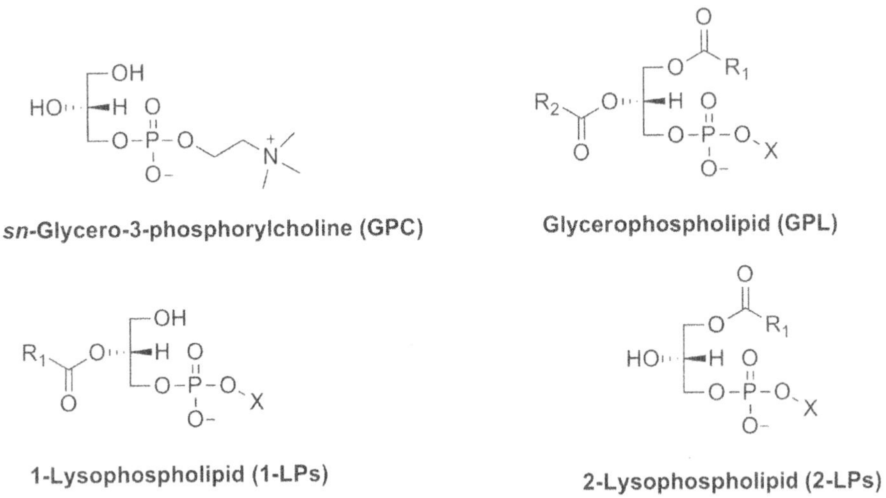

Figure 11.2 Phospholipids and derivatives, R_1, R_2 = alkyl chain.

first important step in advancing membrane chemistry and biochemistry. Moreover, PL analogues and LPs incorporating spectroscopically active groups have been shown to be structural probes to study the aggregation properties of these molecules and the fate of products generated by PL metabolizing enzymes.

Chemical Synthesis

A variety of chemical and enzymatic routes have been studied in different laboratories, including ours (Ahmad, 2004). New synthetic methods for the preparation of LPs and PLs have been reported and recently reviewed (D'Arrigo and Scotti, 2013; D'Arrigo and Servi, 2010). However, only a few chemical synthetic methods have been developed for large-scale preparation of LPs. The challenges in the chemical synthesis of LPs are the construction of the chiral structure and the maintenance of the configuration in further chemical processing. The chemical synthesis of optically active LPs involves the extensive use of protecting groups. Acyl migration is a challenge to synthetic chemists and is often noticed in selective synthesis of LPs regioisomers. Intramolecular acyl group migration results in a mixture of 1-acyl and 2-acyl glycerophosphates derivatives. Acyl migration (i.e., intramolecular transfer of one fatty acid moiety from one –OH group to an adjacent one) is often encountered in selective synthesis of LP regioisomers. The acyl migration is generally controlled either by using a special reagent, such as the addition of borate (Vaz et al., 2000) or by controlling the pH of the reaction mixture. 2-LPC, the regioisomer deprived of the acyl chain in position 2, is much more stable than the 1-LPC isomer, and the equilibrium between the two isomeric compounds at 25 °C is reported approximately as 9:1 (Adlercreutz et al., 2002).

Similar to the synthesis of PLs, the optically pure glycerol derivatives as C-3 building blocks are powerful starting materials for the synthesis of LPs. The general procedure includes the preparation of a stereospecific acyl or ether-substituted glycerol backbone and further phosphorylation of glycerol derivatives. A few important chemical syntheses are described next.

Synthesis from Glycidol

Use of enantiomerically pure glycidol is an alternative to chiral glycerol derivatives and is generally obtained from the same starting meterials, such as carbohydrates. One of the syntheses starting from C-3 synthon relies on a nucleophilic ring opening of phosphorylated glycidol derivatives, yielding a small amount of the migrated *sn*-2-acyl byproduct and providing an efficient sequence to LPs (Lindberg et al., 2002). In this process, phosphorylation of commercially available (*S*)-glycidol (1) was performed with di-*tert*-butyl-*N*,*N*-diisopropyl phosphoroamidite (2) and subsequent oxidation using *meta*-chloroperbenzoic acid (*m*-CPBA) to produce (*R*)-di-*tert*-butylphosphorylglycidol (3) in high yield (Figure 11.3). Regioselective opening of the ep-

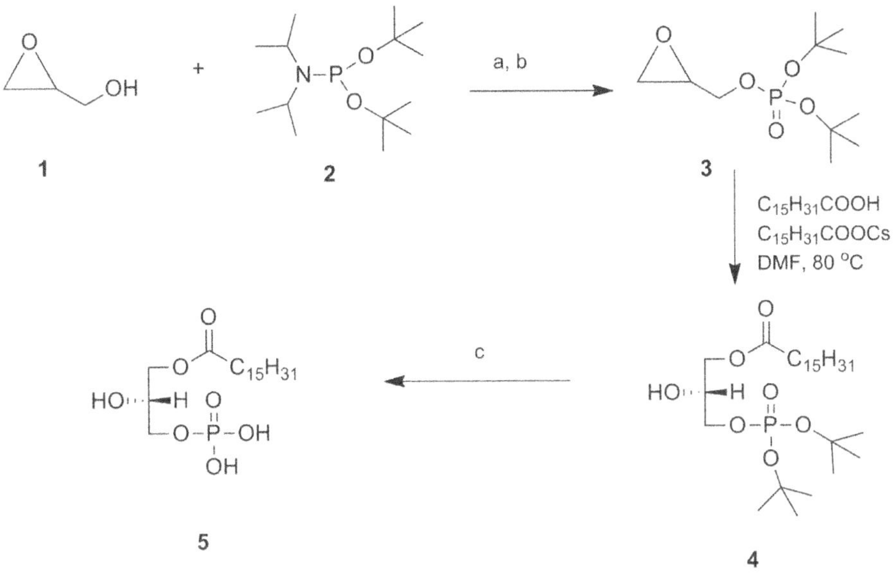

Figure 11.3 Synthesis of 2-LPA (5) from (S)-glycidol (1). Reagents: (a) 1H-tetrazole, (b) m-CPBA, (c) CH_2Cl_2, TFA 4:1.

oxide with cesium palmitate gives the intermediate (4), which on further treatment with triflouro acetic acid (TFA) in methylene chloride produced the deprotected 2-LPA (5). The overall yield was reported to be 46%. During the reaction process, minor amounts of regioisomeric 2-*O*-palmitoyl derivative were also produced by acyl migration from the primary position.

In an alternate procedure, the phosphorylation of (*S*)-glycidol (1) with phosphorous oxychloride (6) and subsequent treatment with choline tosylate and pyridine gives choline derivative (7) in high yield. The nucleophilic opening of the epoxide with cesium palmitate produces 2-LPC (8) (Figure 11.4).

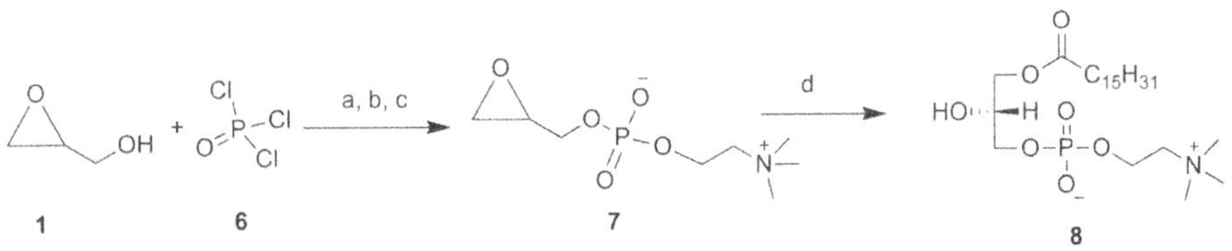

Figure 11.4 Synthesis of LPC (8) from (S)-glycidol (1). Reagents: (a) DIPEA, $CHCl_3$, (b) choline OTS, (c) H_2O, (d) $C_{15}H_{31}COOH$, $C_{15}H_{31}COOCs$, DMF, 80 °C.

The enantiomeric forms of LPA and LPC can analogously be obtained from (*R*)-glycidol. Both natural and unnatural enantiomers of a broad range of biologically important LPs can be obtained with this method in high optical purity and yields.

Synthesis from *p*-Nitrophenyl Glycerate

The starting material *p*-nitrophenylester of (*R*)-glyceric acid is first prepared from the commercially available (*R*)-2, 3-*O*-isopropylideneglyceric acid methyl ester (9) by reaction with sodium hydroxide in aqueous dioxane, followed by dicyclohexylcarbodiimide (DCC)-promoted condensation with *p*-nitrophenyl and subsequent acidolytic cleavage of the isopropylidene function to give the active ester of the D-glyceric acid (10) in high yield (Rossetto et al., 2004). Regiospecific monoacylation of compound 10 at the primary alcohol function can be accomplished by treating with two-fold

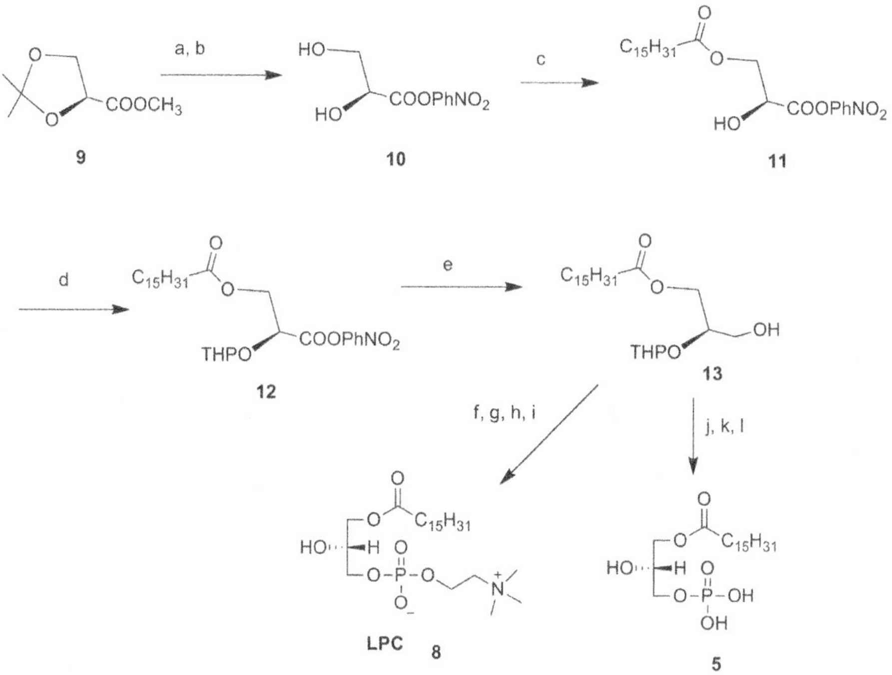

Figure 11.5 Synthesis of LPC (8) and LPA (5) from p-nitrophenyl glycerate (10). Reagents: (a) NaOH, dioxane/H_2O; (b) p-$NO_2C_6H_4OH$, DCC; (c) $C_{15}H_{31}COCl$; (d) 3,4-dihydropyran, PPTS; (e) $NaBH_4$, glyme; (f) $iPr_2NP(OCH_2CH_2CN)_2$; (g) H_2O_2, aq.; (h) DBU; (i) HCl 0.15N dioxane-H_2O; (j) ethylene chlorophosphate; (k) $N(CH_3)_3$; (l) HCl 0.15N dioxane-H_2O.

molar excess of palmitoyl chloride in acetonitrile/benzene (1:1) at room temperature. The column purification on Sephadex LH-20 gives pure *sn*-1-palmitoyl ester (11). Tetrahydropyranylation at the *sn*-2 glycerol position with 3, 4-dihydropyran in chloroform in the presence of pyridinium *p*-toluensolfonate (PPTS) as catalyst at room temperature gives the compound (12). The compound 12 was then reduced with an excess of sodium borohydrate ($NaBH_4$) in 1, 2-dimethoxyethane at room temperature to give 1-palmitoyl-2-tetrahydropyranyl-*sn*-glycerol (13). The glycerol derivative 13 on phosphorylation using a reliable method and cleavage of *sn*-2 tetrahydropyranyl group produces LPA (5) and LPC (8) (Figure 11.5). Under these conditions, no acyl migration is reported.

Synthesis from *p*-Toulenesulfonate-*sn*-Glycerol

A new stereoselective synthesis of LPC was recently reported (Rosseto et al., 2007), which is based on: (1) the use of 3-*p*-toluenesulfonyl-*sn*-glycerol (14) (Figure 11.6) to provide the stereocenter for construction of the optically active LPs molecule (8); (2) tetrahydropyranylation of the secondary alcohol function to achieve orthogonal protection of the *sn*-2 and *sn*-3 glycerol positions (15); and (3) elaboration of the phosphodiester head group using a cyclic phosphochloridate, 2-chloro-1,3,2-dioxaphospholane-2-oxide, and trimethylamine. In the course of developing the synthesis, it has been discovered that the conversion of the tosyl group into the hydroxyl in compound (16) can be achieved by methoxyacetate displacement of the *sn*-3-*p*-toluenesulfonate cleaving the reactive methoxyacetyl ester with methanol/*tert*-butylamine, while the ester group at the *sn*-1 position remains unaffected. Because of the order in which the functional groups are introduced, this synthetic strategy requires

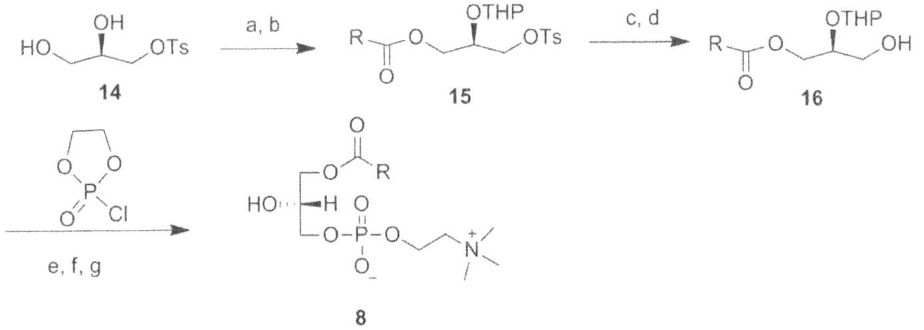

Figure 11.6 Synthesis of LPC (8) from p-toluenesulfonyl-sn-glycerol. Reagents: (a) RCOOH, DCC, DMF; (b) 3,4-dihydropyran, PPTS; (c) $CH_3OCH_2COO\text{-}ET_4N^+$, CH_3CN; (d) $(CH_3)_3CNH_2$, CH_3OH; (e) TEA, benzene; (f) $N(CH_3)_3$, CH_3CN; (g) HCl 0.15N dioxane-H_2O.

minimal use of protecting groups, and the sequence is suitable for preparation of spectroscopically labelled LPCs.

Synthesis from *sn*-Glycero-3-Phosphorylcholine

The *sn*-Glycero-3-phosphorylcholine (GPC) is soluble in water or methanol and less soluble in other organic solvents, and this characteristic has hampered its use in synthesis. Obviously, alcohols cannot be used as solvents for esterification of the hydroxyl group of GPC. However, GPC is readily available from alcoholysis of natural lecithins (17) (Figure 11.7) and is a suitable starting material for the preparation of different type of PLs because it has the same chirality that PLs have in nature (Kim et al., 2003). Most of the methods for the GPC acylation employ the cadmium chloride complex of GPC, which is much more soluble in dimethylformamide (DMF) or any other polar organic solvent (Chada, 1970). However, the use of a heavy metal like cadmium is not suitable for the large-scale synthesis of GPC derivatives and is not suitable for foods or environmental uses because heavy metals can be carcinogenic. To overcome this hurdle, mono- and regioselective acylation of GPC can be readily achieved by tin technology (Fasoli et al., 2006). Monoacylation at position 1 of GPC can be performed by a tin-mediated acylation, exploiting the high reactivity and selectivity of stannylene ketals formed in situ. The method consists of the

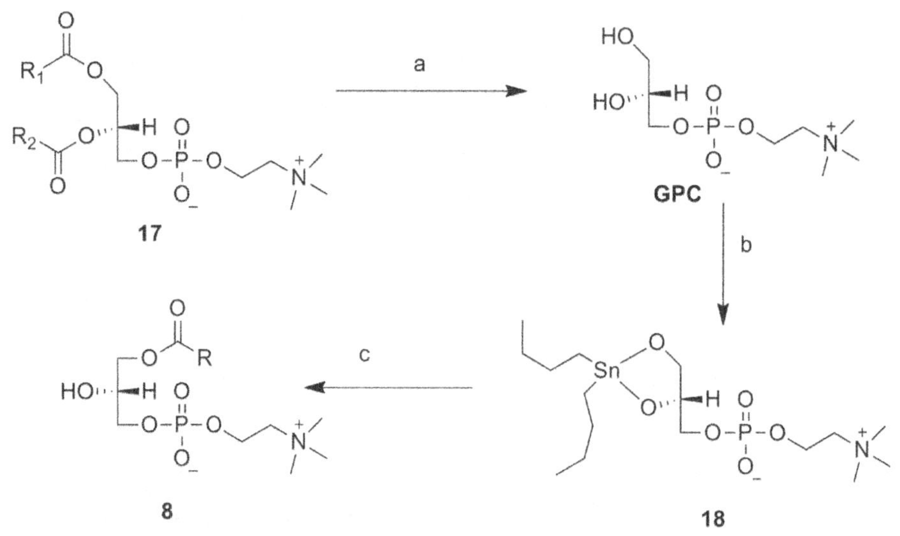

Figure 11.7 Synthesis of LPC (8) from natural lecithins. Reagents: (a) hydrolysis, (b) DBTO, 2-propanol, (c) RCOCl, base.

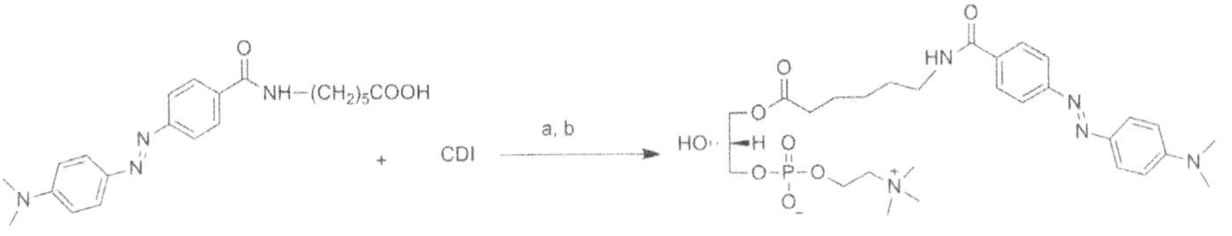

Figure 11.8 Synthesis of fluorogenic GPC (19). Reagents: (a) CDI, $CHCl_3$, DBU, DMF, (b) GPC.

formation of a stannylene derivative (18) in 2-propanol and subsequent acylation with a stoichiometric amount of acylating agent, resulting in the exclusive formation of LPC (8). The advantages of the tin-mediated mono-functionalization of the GPC are the high regio- and chemoselectivity and the completion in a short time, compared with conventional chemical acylation. When compared to other methods of mono-functionalization of GPC, this process gives mono-acyl derivatives free from the regioisomer usually produced by reaction using the acidic conditions during the reactions or in the isolation steps. In general, this method gives comparable yields with saturated and unsaturated fatty acid derivatives of different chain lengths (D'Arrigo et al., 2007). This method is also applicable to the acylation of glycerol-3-phosphate (GPA).

A similar synthesis of LPC-fluorogenic compound starting from GPC has been developed to probe biological phenomena in living cells because fluorescent sensors are advantageous due to their high sensitivity. Rose and Prestwich (2006) reported the synthesis of 1-*O*-(6-(*p*-methyl red)-aminohexanoyl)-*sn*-glycerylphosphatidylcholine (19) (Figure 11.8). The process involves the reaction of carbonyldiimidazole (CDI) with (*p*-methyl red) aminohexanoic acid in appropriate solvent. The solvent is then removed from the reaction mixture and the residue obtained is added to 1, 8-diazabicyclo [5.4.0] undec-7-ene (DBU) and GPC in appropriate organic solvent. The reaction after completion is purified by column chromatography to give the desired fluorescent product (19).

Synthesis from Glyceric Acid

Bibak and Hajdu (2003) reported the stereospecific synthesis of LPCs based on orthogonal protection of a hydroxyl group of glyceric acid using fluorenylmethylcarbonate (FMOC). The starting material, 1-palmitoyl-*sn*-glycerol (22), prepared by reduction of (*S*)-2,3-*O*-isopropylidenglyceric acid methyl ester (20) with lithium borohydrate ($LiBH_4$) in an appropriate organic solvent, followed by acylation with palmitic acid in the presence of DCC and dimethylaminopyridine (DMAP).

Compound (21), after purification, acid-catalyzed deprotection, and chromatography, gives (22). Regioselective monoacylation of compound (22) was carried out using FMOC-chloroformate and DMAP. Tetrahydropyranylation of the *sn*-2 hydroxyl group and subsequent cleavage of the FMOC-carbonate function gives compound (24). Phosphorylation of compound (24) and removal of the protecting group gives the desired LPC, 1-palmitoyl-*sn*- glycerophosphocholine (8) (Figure 11.9).

Synthesis from D-Mannitol

D-mannitol [(2R, 3R, 4R, 5R)-hexan-1, 2, 3, 4, 5, 6-hexol] (25) is an optically active compound and can be transformed into LPs in a multistep reaction (Xia and Hui, 1997). In this process, D-mannitol is first transformed in 1, 2, 3, 4, 5, 6-triisopropylidene-D-mannitol with acetone-sulfuric acid, then ditritylated with dry trityl chloride in dry pyridine. The resulting product, monotrityl derivative, is then benzylated to give compound (26). The trityl group is selectively removed from compound (26) in methanol/2-propanol/sulfuric acid to obtain 2,5,6-tribenzyl-3,4-isopropylidene-D-

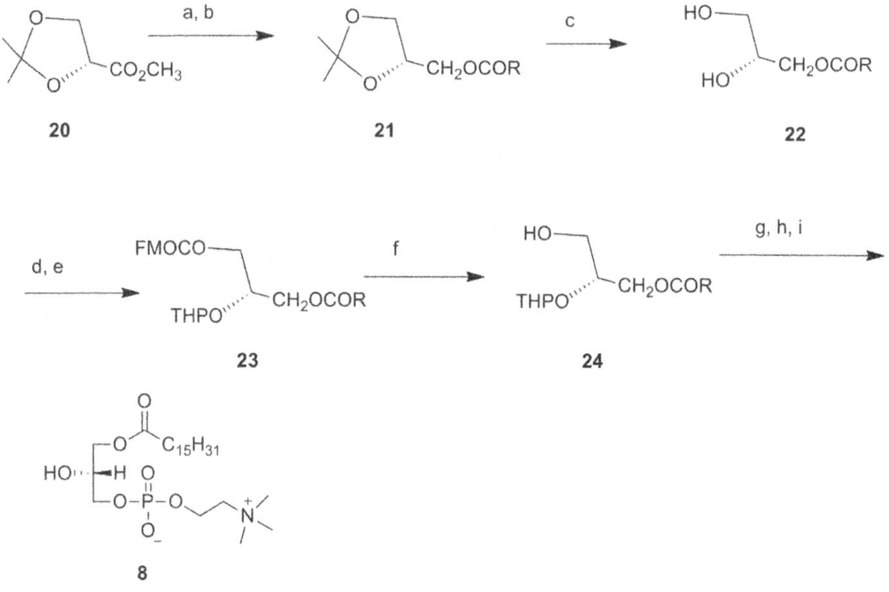

Figure 11.9 A stereospecific synthesis of LPC (8) from glyceric acid. Reagents: (a) $LiBH_4$, Et_2O; (b) RCO_2H, DCC, DMAP, CH_2Cl_2; (c) 0.4N HCl dioxane, H_2O; (d) Fmoc-Cl, DMAP, CH_2Cl_2; (e) THP, PPTS, CH_2Cl_2; (f) Pyr., CH_2Cl_2; (g) ethylene glycol chlorophosphate, NEt_3, C_6H_6; (h) $N(CH_3)_3$; (i) HCl 0.15N dioxane-H_2O.

mannitol, which is subsequently esterified with saturated fatty acid using DCC and a catalytic amount of DMAP in dry methylene chloride to produce 2,5,6-tribenzyl-1-acyl-3,4-isopropylidene-D-mannitol (27). Compound (27), after treatment with trifluroacetic acid (TFA)/perchloric acid (70% $HClO_4$) in methylene chloride, is then treated with lead tetraacetate in dry ethyl acetate. The resulting aldehyde is then reduced with $NaBH_4$ to give products (28a) and (28b). The product (28b) is phosphorylated with chloroethylphosphoric acid dichloride in dry methylene chloride and anhydrous triethylamine to give product (29) with natural configuration. The product (29) is aminated with trimethylamine in chloroform-ethanol to give compound (30), followed by hydrogenation in the presence of 10% Pd/C to give the final LPC (8) (Figure 11.10). This process is not suitable for large-scale synthesis because of multiple steps and low yields.

Enzymatic Synthesis

LPs can be synthesized through enzymatic catalysis with fairly good yields (Hass et al., 1994; Kim et al., 1997; Mustranta et al., 1995; Sarney et al., 1994). Enzymatic synthesis of LPs has many advantages over chemical synthesis because of the selectivity and specificity of enzymes, less exposure of the chemical reagents, and less or no

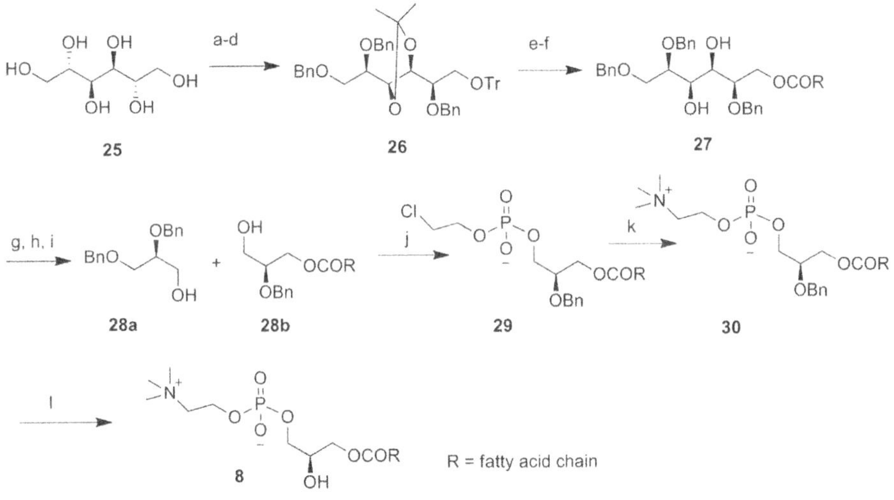

Figure 11.10 Synthesis of LPC (8) from D-mannitol. Reagents: (a) acetone, H_2SO_4; (b) AcOH; (c) Tr-Cl, Pyr.; (d) $PhCH_2Cl$, KOH; (e) MeOH, IPA, H_2SO_4; (f) RCO_2H, DCC, DMAP; (g) CF_3CO_2H, $HClO_4$; (h) $Pb(OAc)_4$; (i) $NaBH_4$, MeOH; (j) $ClCH_2CH_2OP(O)Cl_2$, Et_3N, CH_2Cl_2; (k) $N(CH_3)_3$, EtOH, $CHCl_3$; (l) Pd/C, H_2, $CHCl_3$.

contamination of trace amounts of toxic chemicals used in the synthetic process, the mild reaction conditions, and the easier purification procedure. Another important advantage in enzymatic synthesis and purification is that in the case of an incomplete reaction, the possible impurities in the final product will be much more acceptable if starting materials are GRAS (generally recognized as safe), like natural PLs and enzymes are, than if they are residual chemical reagents. The potential of PLs modifying enzymes and enzymatic synthesis of LPs is therefore of great interest, particularly on an industrial scale.

Phospholipases are the enzymes that selectively catalyze the transformation of PLs (D'Arrigo and Servi, 1997). They play crucial roles in cellular regulation, metabolism, and biosynthesis of PLs. Each of the four major phospholipases selectively recognizes one of the four ester bonds, two carboxylic ester bonds, and two phosphates ester bonds. According to the type of the cleaved bond, phospholipases are classified in three main groups: phospholipase A (PLA), phospholipase C (PLC), and phospholipase D (PLD). PLAs are members of the carboxylester hydrolyase family. They catalyze the hydrolysis of one of the two carboxylic ester bonds in phospholipids either at the *sn*-1 (PLA_1) or *sn*-2 (PLA_2) position. PLCs and PLDs are phosphoric diester hydrolases. PLCs catalyze the hydrolysis of the phosphate ester bond at the glycerol end of the phosphate group, releasing diacylglycerol and phosphoalcohol moieties, while PLDs catalyze the hydrolysis of the phosphate ester bond at the alcohol end, releasing phosphatidic acid and alcohol moieties (Figure 11.11).

Besides phospholipases, there are many other enzymes that can be used for the modification of PLs. Such enzymes are constituted by lipases, nonspecific enzymes with broad substrate specificity. Lipases have been more studied than phospholipases for theoretical and practical understanding. Lipase catalyzes the esterification of glycerophosphatidyl derivatives with free fatty acid to produce LPs (Han and Rahee, 1998; Kim and Kim, 1998; Virto et al., 1999). Lipase catalyzes the esterification

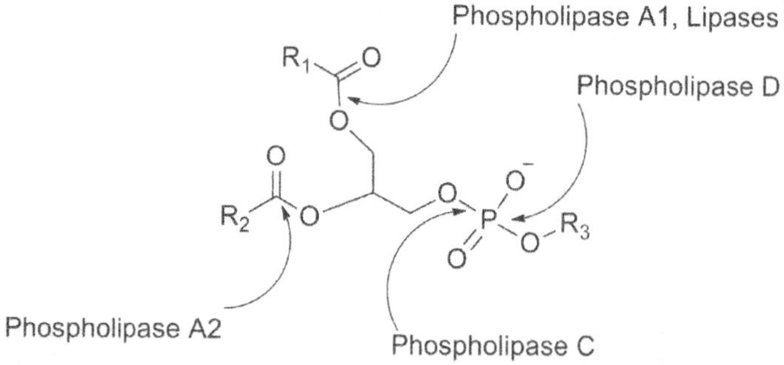

Figure 11.11 Enzymes' specific sites of action on phospholipids.

mainly at the *sn*-1 position of glycerophosphatidyl moiety. This is a one-step esterification process in which only two substrates, a free fatty acid and a molecule with a glycerophosphatidyl group, and an enzyme are required for the reaction. However, this process has a few drawbacks, such as use of excess of fatty acid and low yield.

LPCs were synthesized from GPC by lipase-catalyzed esterification in a solvent-free system. To improve the yield, factors affecting the esterification of GPC in a solvent-free system have been reported (Ha and Rhee, 1998; Kim and Kim, 1998). Lipase-catalyzed GPC esterification is a water-producing reaction in a nonaqueous medium, in which the free fatty acid condenses with a hydroxyl group of GPC. Therefore, water activity control is important to accomplish nonaqueous enzymatic reactions successfully. Typically, high vacuum is used to remove water from the reaction mixture. Ha and Rhee (1998) demonstrated more sophisticated water activity control with salt hydrate pairs for the production of LPs in a solvent-free system. Addition of a small amount of hydrophilic solvent, such as DMF, and controlling the water content of the enzyme has also been reported to enhance the reaction rate and maximize the yield of LPC (Kim and Kim, 2000). Use of DMF also reduces the formation of byproducts.

Phospholipases have been used for PL transforming reactions both in the laboratory and on an industrial scale (Adlercreutz et al., 2003; Mustranta et al., 1994; Servi, 1999; Ulbrich-Hofman, 2000). The enzymatic catalysis provides routes for preparation of PLs and LPs with the natural configuration at the *sn*-2 center. As described in the Introduction, LPs have numerous applications in the food, cosmetics, and pharmaceutical industries because of their superior emulsification properties. However, their properties depend on the fatty acid component and on the polar head bound to glycerol backbone. By changing the hydrophilic/lipophilic balance in these compounds using enzymes, such as lipases or phospholipases, it is possible to produce varieties of derivatives for specific applications. The enzymatic production of LPs using phospholipase A_1 (PLA_1), phospholipase A_2 (PLA_2), and lipases is summarized in Figure 11.12.

All these enzymes can be used in hydrolysis reaction of PLs, alcoholysis of PLs, or esterification of GPC (Figure 11.12). A hydrolysis reaction of natural PA or PC with PLA_2 gives 2-LPA and 2-LPC. The enzymatic alcoholysis of PLs in the presence of lipases can simultaneously produce 1-LPs and fatty acid esters. The starting alcohol can be glycerol, methanol, or ethanol, and the direction of reaction depends on the equilibrium constant and the reactant concentrations (Adlercreutz et al., 2003; Sarney et al., 1994). The enzymatic synthesis of LPC is carried out by esterification of GPC with fatty acid anhydrides or with free fatty acids in a solvent free system (Mazur et al.,1991; Kim and Kim, 1998).

Alternatively, in a consecutive two-step chemoenzymatic synthesis of 2-LPC from GPC, a fatty acid derivative and an appropriate enzyme in a microaqueous system is used. PLA_2 can also be used in a chemoenzymatic process starting from GPC, obtained

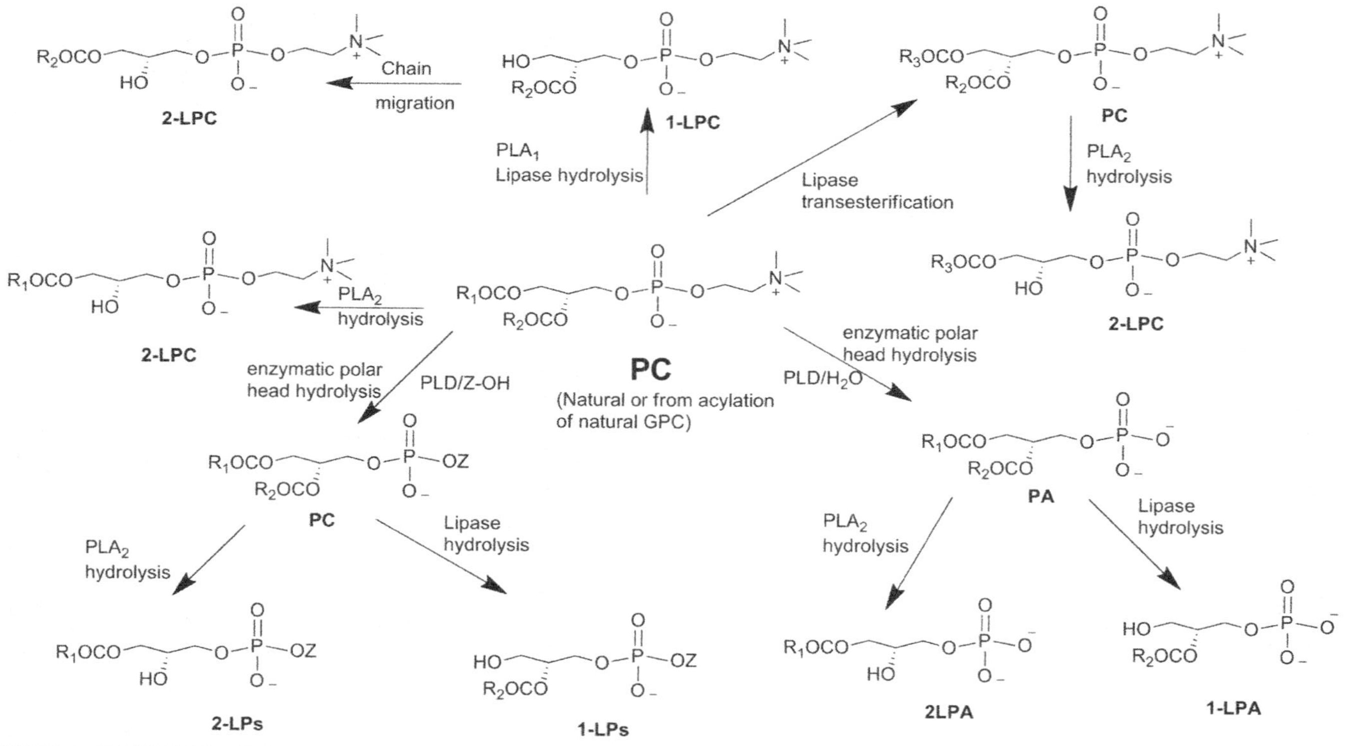

Figure 11.12 Enzymatic synthesis of lysophospholipids.

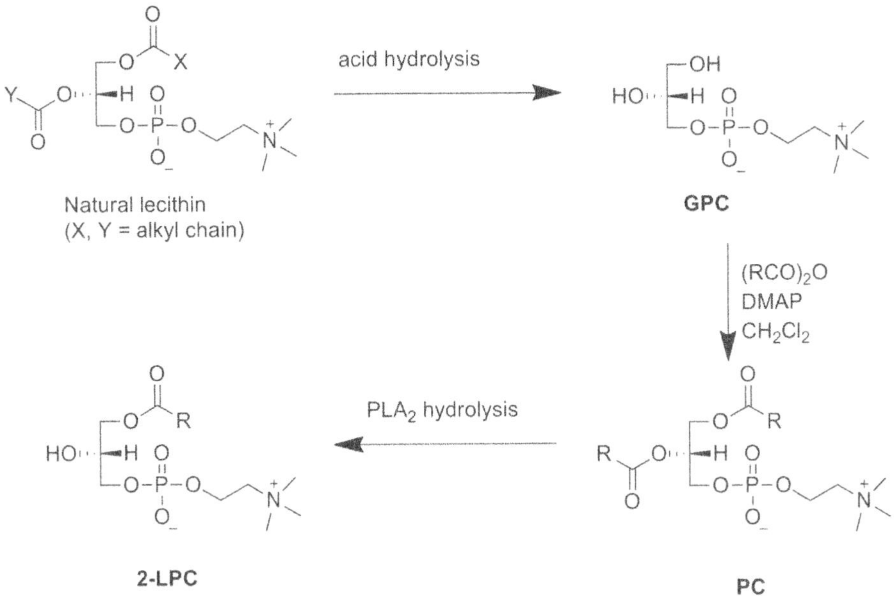

Figure 11.13 Chemoenzymatic transformations of natural lecithins into 2-lysophospholipids.

by hydrolysis of natural lecithin. GPC is acylated with fatty acid anhydride and DMAP, and the resulting PC with the desired acyl chains on hydrolysis with PLA_2 gives the desired 2-LPC (Figure 11.13).

Phospholipase D (PLD, phosphatidylcholine phosphatidohydrolase) is isolated from cabbage and other plant material and is widely used in the hydrolysis of natural PC to phosphatidic acid (PA) and choline. PA is an important second messenger in mammalian signal transduction pathways (English, 1996). In the synthesis of LPs, PLD can be used on a diacylphosphatidylcholine to obtain PL with the desired polar head group or the PA. In a subsequent step, the PL is subjected to the action of the right PLA_1, PLA_2, or lipase to get the desired 1-LPs or 2-LPC. Fluorogenic analogues of LPC were also synthesized using similar procedure, starting from commercially available GPC and PLD, and evaluated as substrates for enzymatic assayas (Rose and Prestwich, 2006).

LPs are also the substrates for PLD (Figure 11.14). In the absence of alcohol, a crude preparation of cabbage PLD catalyzes the hydrolysis of LPC to LPA. The hydrolysis can happen in two ways: either by direct hydrolysis in analogy with the formation of PA from PC, or by the formation of the intermediate cyclic LPA via an intramolecular phosphodiester bond and successive ring opening (Long et al., 1967; Virto et al., 2000). In the presence of glycerol, 2-Lysophosphatidylglycerol is formed.

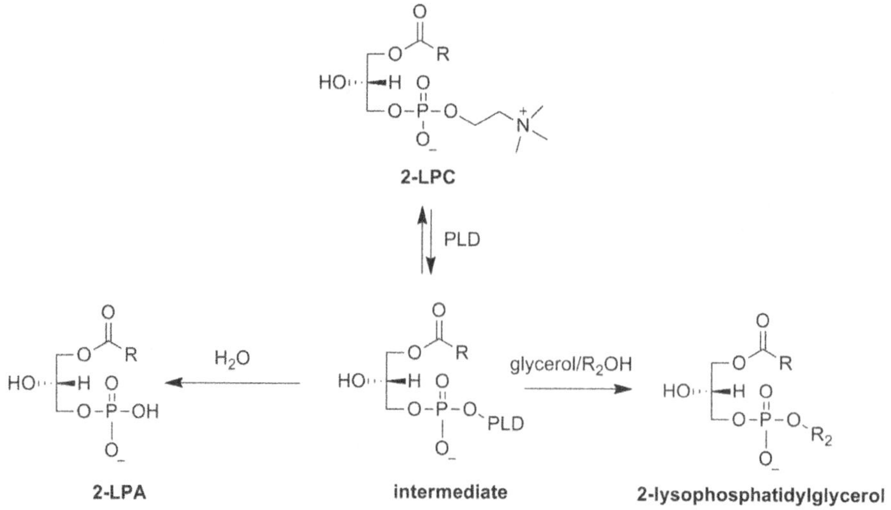

Figure 11.14 PLD-catalyzed hydrolysis and transphosphatidylation of 2-LPC with PLD.

S1P is an LP mediator of cellular processes and is generated from sphingolipids, which are essential plasma membrane lipids concentrated in liquid-ordered domains (Rivera et al., 2008). It can be synthesized following the activation of enzymatic cascade. Sphingomyelin is converted into ceramide by sphingomyelinase, ceramide into sphingosine by ceramidase, and sphingosine into S1P by sphingosine kinase (Figure 11.15) (Pyne and Pyne, 2000).

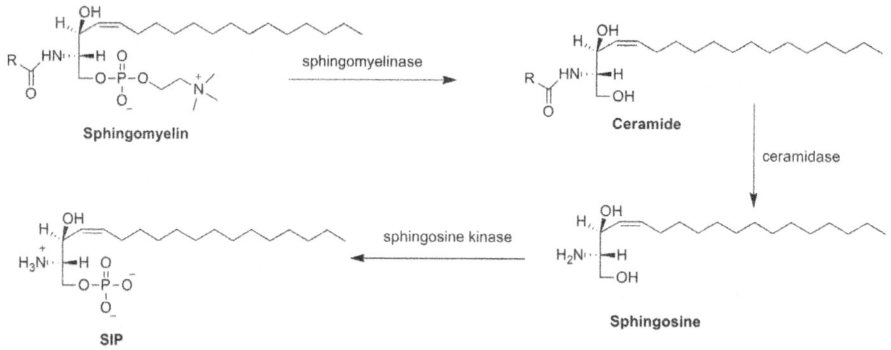

Figure 11.15 Enzymatic synthesis of sphingosine 1-phosphate (SIP) from sphingomyelin.

Lysophospholipids Containing DHA (DHA-LPs)

Lysophospholipids containing docosahexaenoic acid (DHA, 22:6 omega-3) esterified on the *sn*-1 or *sn*-2 position of the glycerol backbone, from now on simply called DHA-LPs, are getting considerable interest because they are recognized to promote the biological effects of DHA, an important dietary fatty acid. DHA is a well-known dietary fatty acid of the omega-3 series (Dyerberg and Bang, 1979). DHA is recognized as a key element for the growth and the functional maintenance of the brain and retina and plays a preventive role in cardiovascular and inflammatory diseases and in some cancers (Horrocks and Yeo, 1999). Use of DHA as a therapeutic agent in the treatment of depression, schizophrenia, or bipolar disorder is also reported in the literature (Bourre, 2005). In addition, DHA-LPs possess similar physicochemical properties to other glycerophospholipids and other LPs and, therefore, they can be used in food products as an additive with a function of an emulsifier, texturing agent, or antioxidant (D'Arrigo and Servi, 2010).

To meet the growing demand of DHA-LPs, several researchers synthesized DHA-LPs either by chemical synthesis or enzymatic synthesis. The major emphasis in this synthesis is to prevent DHA-oxidation during synthesis, and special attention must be paid to the stability of the final products.

Chemical Synthesis of DHA-LPs

There are not many methods available for the synthesis of DHA-LPs. Only one synthetic route is described in the literature for the production of 1-lyso-2-DHA-PLs starting from ether phospholipids (Ether-PLs). Ether-PLs are phospholipids characterized by the presence of an ether bond instead of the ester bond at the *sn*-1 position on the glycerol backbone. Tokumura et al. (2002a) reported the synthesis of 1-lyso-2-DHA-phosphatidic acid from lecithin extracted from beef heart. The beef heart lecithin was a mixture of mainly phosphatidylcholines (PCs) and a small amount of ether-PCs. Both alkaline and acidic hydrolysis steps were used in this synthesis (Figure 11.16). PCs from beef heart were first treated with alkali to get alkali-stable lyso-PCs having a *sn*-1-O-alkyl or alkenyl group, which is isolated and purified. Ether-PCs having a DHA at the *sn*-2 position were further prepared by treatment of the alkenyl- and alkyl-lyso-PCs with DHA anhydrides. The PCs thus obtained were hydrolyzed with phospholipase D (PLD) to generate PAs. The final step consists of an acidic hydrolysis of the *sn*-1 ether link by dispersion of the PA in the mixture of methanol, chloroform, and hydrochloric acid at room temperature under vigorous stirring, and to provide 1-lyso-2-DHA-PAs.

Recently, the same procedure with slight modifications was used for the synthesis of 1-lyso-2-DHA-PC and 1-lyso-2-DHA-PE (Hung et al., 2011a,b). The pure

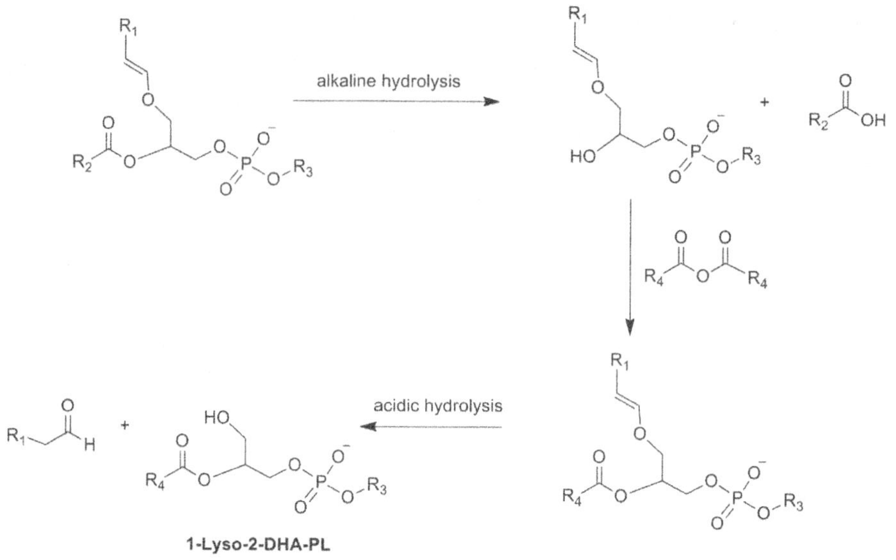

Figure 11.16 Chemical synthesis of 1-lyso-2-DHA-PL from ether PL. R_1, R_2: saturated, monosaturated or polyunsaturated carbon chain; R_3: choline, inositol, ethanolamine, serine, or glycerol; R_4: DHA carbon chain.

1-octadecenyl-2-DHA-PC and 1-octadecenyl-2-DHA-PE, respectively, were used as starting materials and the PLD-catalyzed removal of either the choline or ethanolamine head group was omitted.

Enzymatic Synthesis of DHA-LPs

As we previously described, the enzymatic synthesis of LPs has many advantages over chemical synthesis because of the selectivity and specificity of enzymes, less exposure of the chemical reagents, and less or no contamination of the trace amount of toxic chemicals used in the synthetic process. Moreover, enzymatic reactions usually occur under mild conditions, such as lower temperatures, compared to chemical reactions.

Synthesis of DHA-LPs Using Phospholipases

We described above that enzymatic synthesis of LPs is mainly performed using phospholipases (Figure 11.11), a family of enzymes that catalyze PLs hydrolysis. In the synthesis of 1-DHA-2-LPs, the enzyme PLA_2 is used to remove the *sn*-2-located DHA acyl group from the homogeneous PCs containing two DHA groups (di-

DHA-PCs) (Huang et al., 2010; Tokumura et al., 2002a). The di-DHA-PCs of high purity and PLA2 from bee venom are commercialy available. The hydrolysis was done in an alkaline buffer; either Tris-HCl buffer (pH 8.9) or sodium tetraborate buffer (pH 9.0) at a suitable temperature. The calcium chloride was added to the buffer to meet the calcium-dependence of bee venom PLA_2 activity. Tokumura and coworkers used 1-DHA-2-LPs to produce 1-DHA-2-LPA by using an additional enzymatic step with PLD in order to remove the choline group. A moderate yield of around 40% was obtained for the two enzymatic steps from di-DHA-PC to 1-DHA-2-LPA. However, not many data are available concerning use of PLD in the area of synthesis of DHA-rich LPs.

Tokumura et al. (2002a) produced DHA-LPs utilizing either PLD activity or lyso-PLD activity displayed by the *Streptomyces chromofuscus* enzyme (Friedman et al., 1996). On the one hand, lyso-PLD activity allowed the formation of 1-DHA-2-LPA by the release of choline moiety from formerly produced 1-DHA-2-LPC. On the other hand, PLD activity was used for producing 1-plasmenyl (or plasmanyl)-2-DHA-PA from previously synthesized PC counterparts prior to the acidic hydrolysis of the *sn*-1 ether bond to produce 1-lyso-2-DHA-PA.

Other than PLD hydrolytic activity, the PLD unique transphosphatidyltion capability could also be used in the synthesis of DHA-containing lysophosphatidylethanolamine (LPE) and lysophosphatidylserine (LPS). Nojima et al. (1994) reported the synthesis of eicosapentaenoic acid (EPA, 20:5 omega-3) containing LPE and LPS. In this process, 1-EPA-2-acyl-PC was first prepared by lipase-catalyzed acidolysis of soy PC with free EPA. In the second step, PLD-catalyzed transphosphatidylation generates 1-EPA-2-acyl-PE or 1-EPA-2-acyl-PS with ethanolamine and L-serine, respectively, as nucleophilic acceptors. Finally, 1-EPA-2- LPE and 1-EPA-2- LPS were obtained by PLA_2 catalyzed hydrolysis. Similar procedures could be used for the synthesis of DHA-containing LPE and LPS by replacing EPA with DHA.

Synthesis of DHA-LPs Using Lipases

Lipases are the most widely used class of enzymes in organic synthesis and are defined as enzymes catalyzing the hydrolysis of water-insoluble long-chain triacylglycerols. The most often described pathway is lipase-catalyzed partial hydrolysis of PLs rich in DHA. In nature, DHA is exclusively located at the *sn*-2 position on the glycerol backbone of both triacylglycerols and PLs (Chen and Li, 2007; Devos et al., 2006; Farkas et al., 2000). Most lipases display 1, 3-regioselectivity regarding the position of released fatty acyl groups in triacylglycerols and 1-regioselectivity in PLs (Kapoor and Gupta, 2012). They are, therefore, efficient tools for the production of 1-lyso-2-DHA-PLs through partial hydrolysis of DHA-rich PLs. However, the non-enzymatic migration of the remaining DHA group from the *sn*-2 to the *sn*-1 position can occur and is

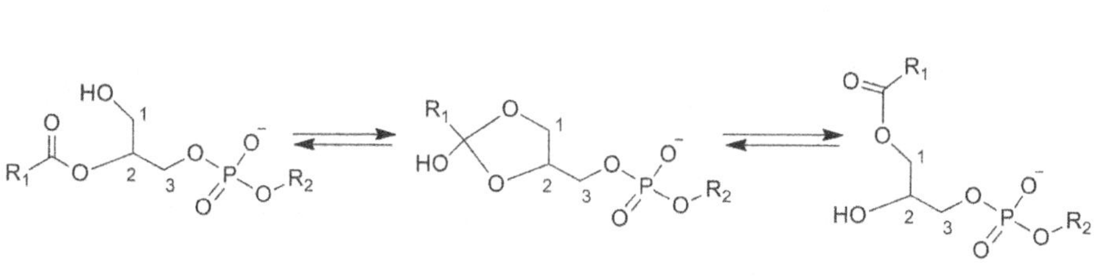

Figure 11.17 Mechanism of acyl migration from *sn*-2 to *sn*-1 position.

common during synthesis. The optimization of the reaction condition is necessary to minimize the migration and production of byproducts. It is reported that acyl migration from the *sn*-2 position to the *sn*-1 position occurs spontaneously, whereas migration from the *sn*-1 to the *sn*-2 occurs to a lesser extent because 2-acyl compounds are thermodynamically more unstable (Li et al., 2010; Poisson et al., 2009; Xu, 2000). This migration is initiated by a nucleophilic attack of the hydroxyl group on the carbonyl carbon leading to a cyclic ortho ester intermediate (Figure 11.17).

The DHA-LPs and more precisely 1-lyso-2-DHA-PLs are important and are of biological interest and industrial applications, such as food additives or as ingredient in therapeutic agents. In a recent review the DHA oxidation, acyl migration during DHA-LPs synthesis, LPs analysis, and quantification are described in detail (Pencreac'h et al. 2013). However, it is important to understand different pathways for chemical and enzymatic synthesis of DHA-LPs. Production of DHA-LPs also needs more studies. Moreover, the parameters affecting the stability of DHA-LPs such as acyl migration and oxidation should be better understood to improve the monitoring of their production.

Alkyl-Lysophospholipids (ALPs)

Synthetic ether-linked analogues of phosphatidylcholine and lysophosphatidylcholine, collectively named as antitumor lipids (ATLs), were synthesized in the late 1960s but have attracted more interest since the finding that the ether lipid 1-*O*-octadecyl-2-*O*-methyl-glycero-3-phosphocholine (ET-18-OCH_3, Edelfosine), a synthetic analogue of 2-LPC, induces a selective apoptotic response in cancer cells while sparing normal cells. The synthetic ATLs are divided in two major subtypes: (1) alkyl ether phospholipids (AEPs), referred to as antitumor ether lipids (AELs) or alkyl-lysophospholipids (ALPs), containing ether bonds in the glycerol backbone of the phospholipids, for example edelfosine (ET-18-OCH_3); and (2) alkylphosphocholines (APCs), lacking the glycerol backbone and formed by esterification of long-chain alcohol into a phosphobase, for example hexadecylphosphocholine (HPC, Miltefosine) (Gajate and Mollinedo, 2002). In addition to edelfosine and miltefosine, several other ALP and APC analogues have been synthesized that show antitumor activities, including 1-hexadecylthio-2-methoxymethyl-glycero-3-phosphocholine (BM41.440, Ilmofosine) (Herrmann et al., 1987a,b,c, 1990), the cyclic analogue SRI 62-834 (Houlihan et al., 1995), octadecyl-(N,N-dimethyl-piperidinio-4-yl)-phosphate (D-21266, Perifosine) (Crul et al., 2002; Patel et al., 2002; Ruiter et al., 2003), and erucylphosphocholine (Jendrosske and Handrick, 2003; Jendrossek et al., 1999, 2001, 2003) (Figure 11.18).

It is well established that a number of synthetic ether phospholipids, collectively referred to as ALPs, possess antineoplastic activities in vitro and in vivo (Andressen, 1988; Berdel et al., 1985). ALPs are synthetic analogues of LPC that are synthesized by replacing the acyl group with an alkyl group within the LPC molecule, and these molecules are very stable in contrast to LPC. The important features of ALPs and APCs are their low metabolism rates in vitro and in vivo, and they do not target the DNA but act at the cell membrane. In 1997, the findings of selective apoptosis in cancer cells treated with edelfosine, ET-18-OCH_3, created more interest in this compound as a promising and selective anticancer drug (Mollinedo et al., 1997) and, in general, created interest in all ALP compounds.

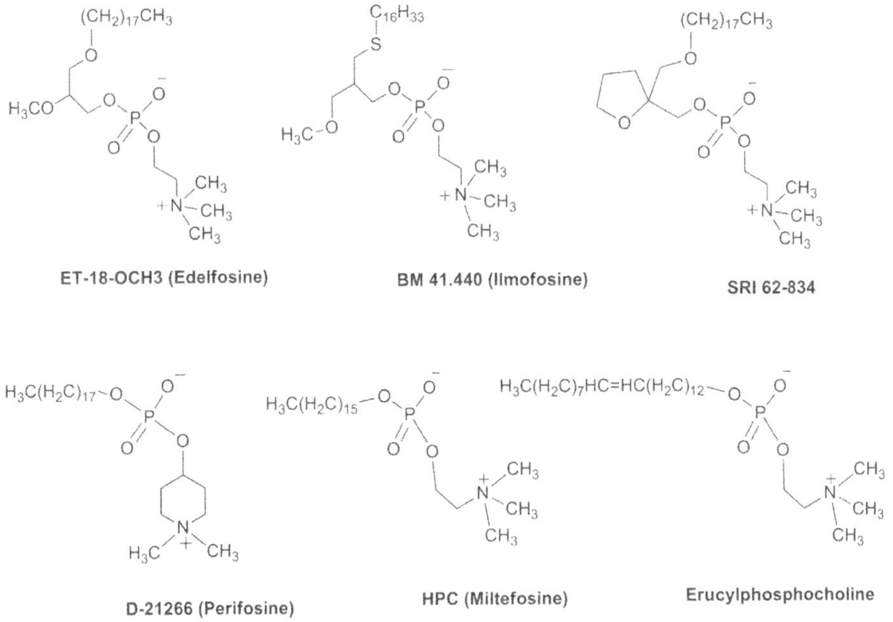

Figure 11.18 Chemical structures of synthetic ether lipids.

Although ET-18-OCH_3 has in vivo antitumor activity, a major dose-limiting toxicity is hemolysis (Ahmad et al., 1997). A liposomal formulation of ET-18-OCH_3 named ELL-12 was previously characterized (Mayhew et al., 1997; Perkins et al., 1997) and compared with free ET-18-OCH_3 for growth inhibitory and hemolytic effects (Peters et al., 1997). It is reported that ELL-12 shows a decrease in growth inhibition between 1.5- and 3.0-fold, and the inhibitory effects are not strongly dependent on liposome composition. In contrast, hemolysis was highly dependent on liposome composition. ET-18-OCH_3 and ELL-12, an optimal liposomal ET-18-OCH_3 formulation, inhibited growth in the micromolar range in drug-sensitive and drug-resistant cells. It was concluded that ELL-12 differs from ET-18-OCH_3 in that the specific in vitro growth inhibitory effects are retained, whereas nonspecific membrane-perturbing effects are reduced. ET-18-OCH_3 acts as a direct activator of cellular signaling, in particular activating the apoptotic mechanism of the cell through intracellular Fas/CD95 activation, and therefore it can constitute a new class of synthetic drugs targeting apoptosis. The clinical phase I trial on ELL-12 indicated that the drug is well-tolerated, and no myelosuppression was reported (Bittman et al., 2001). Previous phase I clinical studies with orally given ET-18-OCH_3 have also shown low toxicity and no myelosuppression. The hematological or systemic side effects, such as myelosuppression and nephro-, neuro-, or hepatotoxicity, have been rarely observed in the clinical trials with ET-18-OCH_3, even after prolonged therapy (Gajate and Mollinedo, 2002). This apparent lack of toxicity differentiates ET-18-OCH_3 from other anticancer agents.

The interest in ET-18-OCH_3 lies in its capacity to resensitize cancer cells to apoptosis. Recent evidence indicates that ET-18-OCH_3 acts through intracellular activation of the Fas/CD95 death receptor, involving a raft-mediated process (Gajate and Mollinedo, 2001; Gajate et al., 2000; Mollinedo et al., 1997; van der Luit et al., 2002). This represents a novel and unique mechanism of action in cancer chemotherapy. The mechanisms of the antitumor effect of ET-18-OCH_3 were reviewed by Mollinedo et al. (2004). Molecular understanding of how ET-18-OCH_3 enters the tumor cell and triggers Fas/CD95 activation from inside the cell opens a new subject for future research. Understanding of these processes will open a new channel to target tumor cells in cancer chemotherapy. In addition, the mechanism of action of ET-18-OCH_3 could also be of use for the treatment of other diseases, such as autoimmune diseases

ALPs were initially developed to be used as antitumor agents. They are also shown to be highly effective in the treatment of visceral leishmaniasis, a disease caused by the species making up the protozoan complex *Leishmania donovani, L. infantum,* and *L. chagasi* that can cause 90% mortality when untreated. The antiproliferative action of three ALPs, edelfosine (ET-18-OCH_3), miltefosine (Hexadecylphosphocholine), and ilmofosine (BM 14.440), have been studied on the promastigotes and amastigotes of *L. donovani*. ALP molecules have demonstrated efficacy against leishmaniasis (Croft et al., 1987, 1996). Miltefosine was used for oral treatment of visceral leishmaniasis in India after a successful clinical trial (Sundar et al., 2000) and for the treatment of cutaneous leishmaniasis (Soto et al., 2001). However, drug resistance has emerged as a major problem in the treatment of leishmaniasis (Perez-Victoria et al., 2001). The molecular mechanisms involved in the resistance to ALPs are of considerable interest in the area of antiparasite chemotherapy. The widespread use of ALPs in the treatment of visceral and cutaneous leishmaniasis could induce resistance. Therefore, understanding how resistance arises could lead to strategies for new and more effective antiprotozoal drugs. Further investigation to fully unravel its mechanism of action will lead to improvements in leishmania treatment.

Biological Significance of LPs

Lysophosphatidic acid (LPA) and S1P (Figure 11.1) are two distinct biologically active LP molecules that exert a wide range of cellular effects through specific G protein-coupled receptors (GPCRs) and provide invaluable insight into the biological activities of these molecules in vivo. Activated platelets are a major source of LPA and S1P; however, these LPs are produced by different cell types such as mature neurons, Schwann cells, adipocytes, and fibroblasts, which have been implicated in the production of LPA (Fukushima et al., 2000; Pages et al., 2001a,b; Sano et al., 2002; Weiner et al., 2001; Yatomi et al., 2000). Similarly, in response to external stimuli, hematopoietic cells such as peripheral mononuclear cells, erythrocytes, and neutrophils can contribute to the level of S1P in the blood (Yang et al., 1999). Taken together, these multiple sources contribute to LPA and S1P levels in biological fluids including serum, saliva, and follicular fluid. Ovarian cancer cells also produce S1P and LPA, which may contribute to tumor cell progression and metastasis (Eder et al., 2000; Yatomi et al., 2001). LPA and S1P regulate the development and functions of different organ systems such as the cardiovascular (Karliner, 2002), nervous (Sengupta et al., 2004), immune (Ren et al., 2006), and reproductive (Murph and Mills, 2007) systems.

Sphingosine 1-Phosphate (S1P)

S1P is produced by sphingosine kinase (SphK)-mediated phosphorylation of sphingosine, a lipid synthesized through the metabolic conversion of ceramide. Two sphingosine kinase isoforms, SphK1 and SphK2, are known to exist in mice (Kohama et al., 1998; Liu et al., 2000). It is suggested that the deletion of SphK1 reduces serum and plasma levels of S1P to 50% of those observed in wild type mice. This indicates that SphK2 or unidentified sphingosine kinases are responsible for the remaining SphK catalytic activity (Allende et al., 2004). SphK activity has been described in cytosolic, membrane, and extracellular compartments (Ancellin et al., 2002; Olivera et al., 1999). The S1P and the kinases are critical regulators of numerous fundamental biological processes important for human health and illness. They are potent signaling molecules and regulate both cell growth and survival, induce growth arrest and apoptosis (Hannun and Obeid, 2008), and are known as powerful inflammatory mediators.

The inflammatory lipid S1P is involved in the malignant progression of breast cancer, but only SphK1 (and not SphK2) increases the generation of S1P that is released from MCF-7 cells (the human breast carcinoma cells). The estradiol (E_2) and epidermal growth factor (EGF) enhance SphK1 activity, but only E_2 induces S1P release. SphK1 is overexpressed in human breast cancer, is involved in breast cancer progression, and is associated with tumor angiogenesis. Studies have shown that SphK1 is linked to radiation and chemotherapy resistance as well as to the resistance of MCF-7 cells to Tamoxifen (Nolvadex®), a breast cancer drug, which interferes with the activity of estrogen (Shida et al., 2008a; Sukocheva et al., 2009). The estrogen can promote the growth of breast cancer cells because the estrogen receptors are overexpressed in majority of breast cancers. This suggests that SphK1/S1P might be an excellent target to overcome antiestrogen resistance in human breast cancer pathophysiology, and this may be an effective therapy like any other chemotherapeutic drugs (Thurlimann et al., 2005).

SphK1 has become a cancer research hotspot and has recently been considered a bonafide oncogene (Vadas et al., 2008). Many studies have shown that SphK1 is overexpressed in various cancers, and the meta-analysis recently performed established the relationship between SphK1 expression and various cancers. There is a possibility that SphK1 might be used as a cancer biomarker (Zhang et al., 2014). S1P produced inside cancer cells is exported and exerts its extracellular functions by binding to its specific receptors in an autocrine, paracrine, and endocrine manner, which is known as inside-out signaling. S1P is also known to exert its intracellular functions in the inflammatory process. Growing interests in the role of S1P in breast cancer progression including angiogenesis and lymphangiogenesis, and these have been recently reviewed (Aoyagi et al., 2012).

Lysophosphatidic Acid (LPA)

LPA is a simple lipid molecule made up of a glycerol backbone with a hydroxyl group, a phosphate group, and a long saturated or unsaturated fatty acid chain. Despite the simplicity in structure, LPA is a member of a family of lipid mediators and second messengers. In contrast to S1P, which is mostly synthesized intracellularly by sphingosine kinases and exported across the plasma membrane, LPA is produced extracellularly from more complex PLs by specific phospholipases. In particular, the major route of LPA production occurs through the hydrolysis of extracellular LPC by a secreted lysophospholipase D (lysoPLD) called autotaxin (ATX); this enzyme is present in the circulation and accounts for LPA production in plasma.The secreted ATX can interact with target cells allowing the production and delivery of LPA close to its receptors.

Serum LPA levels are regulated by the secreted lysophospholipases PLA_1 and PLA_2 and lysophospholipase D(lysoPLD). PLA_1 and PLA_2 are also thought to be responsible for the de novo generation of LPC (Figure 11.1), a substrate for lysoPLD, which can cleave LPC to produce LPA (Aoki et al., 2002; Sano et al., 2002). The recent identification of lysoPLD as autotaxin (ATX) has been helpful for understanding LPA pathways. ATX was first identified as a protein in a melanoma cell culture medium that stimulated cancer cell motility (Murata et al., 1994; Stracke et al., 1992). ATX is important for LPA activity because it can access pools of LPC that exist in high concentrations in many body fluids and allow the synthesis of LPA and subsequent activation of cellular responses through LPA receptors. ATX promotes the proliferation of several cancer cell lines. The increase in proliferation is enhanced in the presence of LPC, suggesting that LPA activity is important for this effect (Hama et al., 2004; Umezu-Goto et al., 2002). In vivo, ATX is expressed in adipocytes and is up-regulated in genetically obese, diabetic mice (Ferry et al., 2003). These results, together with the fact that LPA is present in the extracellular fluid of adipose tissue in vivo and released by adipocytes in vitro, indicate that an LPA-dependent paracrine control of adipose tissue is mediated by ATX (Valet et al., 1998). The available data suggest that many LPA-mediated effects are paralleled by ATX, particularly effects mediated by the receptor LPA_1, and the ATX activity creates a microenvironment for LPA activity (Hama et al., 2004). It is interesting to note that S1P is also generated by the hydrolysis of sphingosylphosphorylcholine (SPC) by ATX (Clair et al., 2003) (Figure 11.1). However, this occurs at a lower efficiency than the generation of LPA from LPC, and therefore, its biological significance is not very clear.

In fact LPA, like S1P, is a bioactive lipid mediator of diverse cellular processes fundamental for cancer progression and is involved in proliferation, apototosis, migration, and cancer cell invasion (Mills and Moolenaar, 2003). The identification and cloning of GPCRs and their high affinity for LPA and S1P have enabled a mechanistic understanding of their diverse roles in biological processes (Moolenaar et al., 2004), such as development and function of numerous organ systems like the cardiovascular, nervous, immune, and reproductive systems. In general, LPA is involved in various biological processes including brain development and embryo implantation, pathophysiological conditions including neuropathic pain and pulmonary and renal fibrosis (Okudaira et al., 2010), as well as biological activities on blood cells (platelets, monocytes) and on cells of the vessel wall (endothelial cells, smooth muscle cells, macrophages) (Schober and Siess, 2012). LPA has also been shown to act as an important intermediate in the transmembrane signal transduction processes as a platelet activating factor and in the stimulation of cell proliferation (Moolenaar, 1995). Rivera and Chun (2006) critically reviewed the identification and expression of the LPs receptors, receptors signaling properties and function, and have provided new insight into our understanding of the role of LPs in biology systems and diseases.

Lysophosphatidylcholine (LPC)

2-Lysophosphatidylcholine (2-LPC), from now on called LPC, is the most abundant LP in nature. It is a minor component on the cell membranes and an intermediate in the PL metabolism in membranes during PL turnover (Stafford and Dennis, 1988). LPC is a water soluble, amphiphilic molecule, and it aggregates in micelles. The micelle concentration depends on the fatty acid chain length in the molecule. As previously described, LPA is produced from LPC by lysoPLD (named Autotaxin ATX) activity, and thus converts LPC to LPA (Nakanaga et al., 2010). Many studies suggest that LPA is relevant to the pathogenesis of ovarian carcinoma, where LPC was used as substrate to determine lysoPLD activity in ovarian carcinoma studies (Tokumura et al., 2002b; Xie and Meier, 2004). LPC in high concentration has been found in the plasma of ovarian cancer patients, suggesting that LPA produces via a lysoPLD activity (Shuying Liu et al., 2009). LPC is believed to be bioactive mediator in wound healing, inflammation, and atherogenesis. It has been shown to be involved in atherosclerosis and inflammatory diseases by altering different functions in a number of cell types.

Many other cell types are also LPC targets; for example, LPC yields biological effects on putative inflammatory cells such as endothelial cells (ECs), monocytes, macrophages, neutrophils, and T-lymphocytes (Matsumoto et al., 2007). LPC stimulates the proliferation and the migration of smooth muscle cells (SMCs), which are responsible for the pathogenesis of atherosclerosis and of restenosis after angioplasty. LPC is originally known to induce neuropathic pain, such as behavioral allodynia, thermal hyperalgesia, and demyelination (Inoue et al., 2004). These effects of LPC require ATX to produce LPA, the product of ATX-mediated LPC hydrolysis (Inoue et al., 2008).

Lysophospholipid G Protein-Coupled Receptors

The many biological responses reported for the LPs that include LPA and S1P are attributed to signaling through specific G protein-coupled receptors (GPCRs). The LPA and S1P have generated interest for their extracellular signaling properties. The majority of the responses documented for extracellular LPs are attributable to the activation of specific, seven-transmembrane domain GPCRs. Currently there are nine distinct LP receptors that have been identified, four of which mediate effects of LPA and five that mediate effects of S1P, and it is expected that the total number of receptors may be larger. These nine receptors have been identified in mammals and recently reviewed by Anliker and Chun (2004); the review highlighted major features of LPA and S1P GPCRs. The four known LPA receptors that mediate LPA signaling in a wide variety of tissues and cell types are designated as LPA_1, LPA_2, LPA_3, and LPA_4. The numbering system reflects their initial order in the literature. The LPA_1 receptor was the first LP receptor identified and is the best characterized LPA receptor. Similarly, the five identified S1P receptors are designated as $S1P_1$, $S1P_2$, $S1P_3$, $S1P_4$, and $S1P_5$. These nine receptors are expressed in a large number of tissues and cell types, allowing a variety of cellular responses to LP signaling, such as cell adhesion, cell motility, cytoskeletal changes, proliferation, angiogenesis, process retraction, and cell survival.

In addition to these receptors, a large number of orphan receptors have been provisionally identified as LP receptors, but the data that exist on their identity are conflicting. In particular, some putative receptors for sphingosylphosphorylcholine (SPC) and lysophosphatidylcholine (LPC) may be proton sensors (Xu, 2002), and they are unrelated to LP signaling (Ludwig et al., 2003). These and other orphan/putative LP receptors have also been reviewed (Ishii et al., 2004). The biochemistry and metabolism of LPs are also covered in many reviews (Ishii et al., 2004; Kluk and Hla, 2002; Meyer zu Heringdrof et al., 2002; Mills and Moolenaar, 2003; Okajima, 2002; Osborne and Stainier, 2003; Pages et al., 2001a; Siehler and Manning, 2002; Spiegel and Milstein, 2003; Takuwa, 2002; Xie et al., 2002; Yatomi et al., 2001). LPs signaling through GPCRs have major influences on multiple organ systems; major systems influenced by LPs include the developing and adult cardiovascular system, reproductive system, immune system, and nervous system. An increased understanding of the physiological and pathophysiological effects of LPs is the major growth area in this field.

Lysophospholipid Acyltransferase (LPATs)

Cellular membranes contain several classes of glycerophospholipids (GPLs), which have numerous structural and functional roles in cells. The GPLs are an important component of biological membranes and constituents of serum lipoproteins and the pulmonary surfactant. They play an important role as precursors of lipid mediators such as platelet-activating factor (PAF) and eicosanoids (Ishii and Shimizu, 2000; Shimizu, 2009). Cellular membranes in each tissue contain a distinct composition of various GPLs (Schlame et al., 2000; Yamashita et al., 1997). The acyl groups of GPLs are highly diverse, depending on the polar head group, and are distributed in an asymmetric manner. The saturated and monounsaturated fatty acids are usually esterified at the *sn*-1 position, whereas polyunsaturated fatty acids are esterified at the *sn*-2 position, and several other distributions are also reported (Lands, 2000; Shimizu, 2009; Shindou and Shimizu, 2009; Yamashita et al., 1997). GPL was first synthesized from glycerol-3-phosphate by the de novo pathway (Kennedy pathway) using acyl-CoAs as donors (Kennedy and Weiss, 1956), and subsequently Lands reported the maturation of phospholipids in the remodeling pathway known as Lands's cycle (Lands, 1958) to produce membrane diversity. Rapid turnover of *sn*-2 acyl moiety of GPL is attributed to the concerted and coordinated actions of phospholipase A_2s (PLA_2s) and lysophospholipid acyltransferases (LPATs) (Lands, 2000; Shimizu, 2009; Shindou and Shimizu, 2009; Waku and Nakazawa, 1972). Many species of GPL differ in the phosphoryl head groups, fatty acids chain lengths, and degree of saturation; therefore, many LPATs should exist. In recent years, many LPATs functioning in the remodeling pathway have been identified, resulting in the advancement in the field of LPATs (Shimizu, 2009; Shindou and Shimizu, 2009). Membrane diversity is important for membrane fluidity and curvature and is produced by concerted and overlapping reactions of multiple LPATs that recognize both the polar head group of lyso-GPL and various acyl-CoAs in the remodling pathway. The findings on the cloning and characterization of LPATs, which contribute to membrane asymmetry and diversity, were recently reviewed (Shindou et al., 2009). Identification of additional LPATs may contribute to further elucidation of membrane diversity and asymmetry.

Physiological Significance of LPs

The LP mediator family has attracted much attention for its role in the prevention, maintenance, and treatment of human chronic diseases. Natural foods carry potentially active components that prevent injury by harmful chemicals and bacteria and regulate multiple functions in the digestive tract. The food-derived lipids are also capable of affecting gastrointestinal functions. In addition, the digestive tract releases bioactive lipids together with polypeptide hormones into the lumen. Thus, gastrointestinal immunity and inflammation are known to be affected by both dietary lipids and lipid metabolism in the digestive tract (Turner, 2009). In addition to fatty acid–derived mediators such as eicosanoids, the complex lipid mediators possessing one long hydrophobic chain in the digestive system have recently attracted much attention. Modulation of gastrointestinal wound repair and inflammation by mucous phospholipids was reported (Sturm and Dignass, 2002). In addition, LPA (1- or 2-acyl-*sn*-glycerol-3-phosphate) (Figure 11.19), an important member of the phospholipid mediator family that also includes platelet-activating factor (1-O-alkyl-2-acetyl-*sn*-glycero-3-phosphocholine), has attracted much attention as a novel modulator of intestinal wound healing (Tokumura, 1995). To date, at least six GPCRs specific for LPA have been well characterized (Choi et al., 2010; Noguchi et al., 2009; Okudaira et al., 2010). GPR35, an orphan GPCR, has been shown to be expressed in intestinal lumen and postulated to participate in the development of gastric cancer (Okmura et al., 2004) and to respond more specifically to polyunsaturated LPAs than kynurenic acid (Oka et al., 2010).

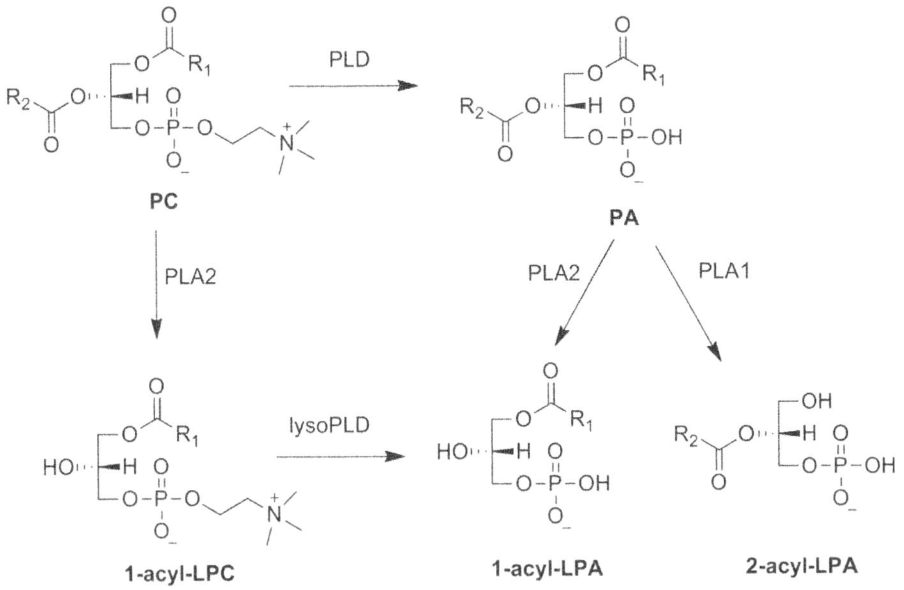

Figure 11.19 Possible metabolic pathways of LPA production in mammalian digestive tracts. PC: phosphatidylcholine; LPC: lysophosphatidylcholine; PA: phosphatidic acid; LPA: lysophosphatidic acid; PLD: phospholipase D; lysoPLD: lysophospholipase D.

The physiological and pathophysiological roles of LP mediators, especially LPA and LPC derived from foods or released from the mucosal layers of the gastrointestinal wall, are well described in a recent review (Tokumura, 2011). These LPs coming from foods, produced in the lumen of the lower digestive tract, or released from the digestive mucosa, are capable of acting directly on the lumen surface for maintenance of the mucosal structural integrity and functional homeostasis. It is suggested that LPs such as LPA and LPC in the lumen of the lower digestive tract have important physiological effects, although most of the effects of LPA are opposite of those of LPC. The LPA receptors expressed on the lumen surface of the lower digestive tract are distinct from those in the basolateral surface, leading to the different types of LPA signals to the gastrointestinal epithelium. More work on this aspect is necessary for better understanding of physiological and pathophysiological effects of LP mediators that act on the lumen side of the digestive system and for the development of functional foods rich in LPs. However, special care may be necessary because such LP-rich food supplements have the potential to aggravate gastrointestinal cancer.

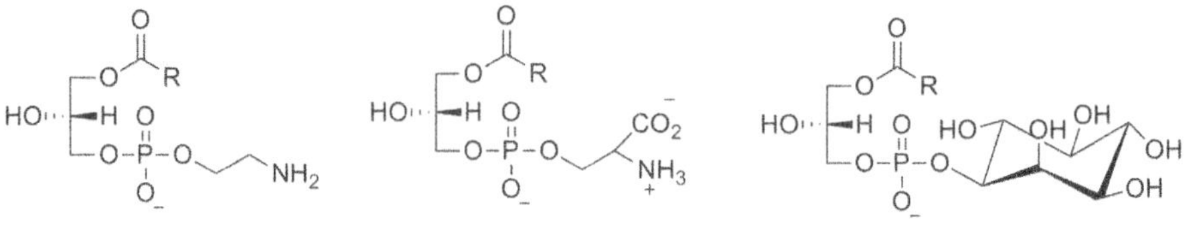

LPE LPI LPS

Figure 11.20 Lysophospholipid mediator family. All structures shown are 1-acyl isomers.

The presence and function of other LPs, such as lysophosphatidylinositol (LPI), lysophosphatidylserine (LPS), and lysophosphatidylethanolamine (LPE) (Figure 11.20), in the digestive lumen are not well studied. Further study of the mechanisms of LP production and degradation in the gastrointestinal tract and its movement in the digestive tract are important for better understanding of foods in our daily diet.

These studies will help to understand the food preference in the daily diet depending on the pathological states, such as suffering from cancer or gastrointestinal ulcer. The systemic analysis of the concentrations of LP and its precursor, diradylglycerophospholipid, in daily diets will indicate what types of foods are beneficial for human health.

Conclusion

Lysophospholipids (LPs) are bioactive compounds that regulate biological functions and disease processes. LPs and their receptors are attractive targets for developing drugs for the treatment of life-threatening diseases. The availability of synthetic LPs is important for biological studies. However, only a few synthetic methods for LPs are available because of the difficulties associated with multistep synthesis, starting from enantiomerically pure glycerol derivatives through tedious protection-deprotection steps and the possible acyl- and phosphoryl-group migration during the synthetic process, resulting in regioisomeric mixtures of products. The synthesis of enantiomerically pure LPs is a challenge because it requires a chiral synthon as a starting material and extensive use of protective groups and activating agents. The use of GPC derived from the natural PLs is the best strategy for producing product with the natural absolute configuration. Enzymatic synthesis of LPs has many advantages compared to chemical methods. Many different pathways for chemical and enzymatic synthesis of LPs have been described in literature, but only few of them have been checked for DHA-LP synthesis. It is important to study the parameters affecting the stability of DHA-LPs, such as acyl migration and oxidation, during the production process. Furthermore, process research for DHA-LP production from a few hundred micrograms to industrial scale should be performed.

Alkyl-lysophospholipids (ALPs) represent a family of promising anticancer drugs that induce apoptosis in a variety of tumor cells. Initially, ALPs were developed as antitumor agents and were also proven to be highly effective in the treatment of visceral leishmaniasis. The selectivity of ET-18-OCH_3 (edelfosine) for cancer cells, as well as its accumulation in tumor tissues, makes this molecule extremely interesting for cancer therapy.

References

Adlercreutz, D.; Budde, H.; Wehtje, E. Synthesis of Phosphatidylcholine with Defined Fatty Acid in the *sn-1* Position by Lipase-catalyzed Esterification and Transesterification Reaction. *Biotechnol. Bioeng.* **2002,** *78*, 403–411.

Adlercreutz, P.; Lyberg, A. M.; Adlercreutz, D. Enzymatic Fatty Acid Exchange in Glycerophospholipids. *Eur. J. Lipid Sci. Technol.* **2003,** *105*, 638–645.

Ahmad, M. U. unpublished work, **2004**.

Ahmad, I.; Filep, J. J.; Franklin, J. C.; Janoff, A.; Masters, G. R.; Pattassery, J.; Peters, A.; Schupsky, J. J.; Zha, Y.; Mayhew, E. Enhanced Therapeutic Effects of Liposome Associated 1-*O*-Octadecyl-2-*O*-methyl-*sn*-glycerol-3-phosphocholine. *Cancer Res.* **1997,** *57*, 1915–1921.

Allende, M. L.; Sasaki, T.; Kawai, H.; Olivera, A.; Mi, Y.; van Echten-Deckert, G.; Hajdu, R.; Rosenbach, M.; Keohane, C. A.; Mandala, S.; et al. Mice Deficient in Sphingosine Kinase 1 Are Rendered Lymphopenic by FTY720. *J. Biol. Chem.* **2004,** *279*, 52487–52492.

Ancellin, N.; Colmont, C.; Su, J.; Li, Q.; Mittereder, N.; Chae, S. S.; Stefansson, S.; Liau, G.; Hla, T. Extracellular Export of Sphingosine Kinase-1 Enzyme. Sphingosine 1-Phosphate Generation and the Induction of Angiogenic Vascular Maturation. *J. Biol. Chem.* **2002,** *277*, 6667–6675.

Andressen, R. Ether Lipids in the Therapy of Cancer. *Prog. Biochem. Pharmacol.* **1988,** *22*, 118–131.

Anliker, B.; Chun, J. Lysophospholipid G Protein-coupled Receptors. *J. Biol. Chem.* **2004,** *279*, 20555–20558.

Aoi, N. Soy Lysolecithin. *Yukagaku* **1990,** *39*, 10–15.

Aoki, J.; Taira, A.; Takanezawa, Y.; Kishi, Y.; Hama, K.; Kishimoto, T.; Mizuno, K.; Saku, K.; Taguchi, R.; Arai, H. Serum Lysophosphatidic Acid Is Produced through Diverse Phospholipase Pathways. *J. Biol. Chem.* **2002,** *277*, 48737–48744.

Aoyagi, T.; Nagahashi, M.; Yamada, A.; Takabe, K. The Role of Sphingosine 1-phosphate in Breast Cancer Tumor-induced Lymphangiogensis. *Lymphatic Res. Biology* **2012,** *10*, 97–106.

Berdel, W. E.; Andresen, R.; Munder, P. G. Synthetic Alkyl-phospholipid Analogues; A New Class of Antitumor Agents. In *Phospholipids and Cellular Regulation;* Kuo, J. F., Ed.; CRC Press: Boca Raton, FL, 1985; Vol. 2, pp 41–73.

Bibak, N.; Hajdu, J. A New Approach to the Synthesis of Lysophosphatidylcholines and Related Derivatives. *Tetrahedron Lett.* **2003,** *44*, 5875–5877.

Birgbauer, E.; Chun, J. N. New Developments in the Biological Functions of Lysophospholipids. *Cell. Mol. Life Sci.* **2006,** *63*, 2695–2701.

Bittman, R.; Perkins, W. R.; Swenson, C. E. TLC ELL-2: Liposomal ET-18-OCH3. *Drugs of the Future* **2001,** *26,* 1052.

Bourre, J. M. Dietary Omega-3 Fatty Acids and Psychiatry: Moods, Behavior, Stress, Depression, Dementia and Aging. *J. Nutr. Health Aging* **2005,** *9,* 31–38.

Chada, J. S. Preparation of Crystalline L-alfa-glycerophosphorylcholine-cadmium Chloride Adduct from Commercial Egg Lecithin. *Chem. Phys. Lipids* **1970,** *4,* 104–108.

Chen, S.; Li, K. W. Mass Spectrometric Identification of Molecular Species of Phosphatidylcholine and Lysophosphatidylcholine Extracted from Shark Liver. *J. Agric. Food Chem.* **2007,** *55,* 9670–9677.

Choi, J. W.; Herr, D. R.; Noguchi, K.; Yung, Y. C.; Lee, C. W.; Mutoh, T.; Lin, M. E.; Teo, S. T.; Park, K. E.; Mosley, A. N.; et al. LPA Receptors: Subtypes and Biological Actions. *Annu. Rev. Pharmacol, Toxicol.* **2010,** *50,* 157–186.

Chun, J. Lysophospholipids in the Nervous System. *Prostaglandins Other Lipid Mediat.* **2005,** *77,* 46–51.

Chun, J.; Rosen, H. Lysophospholipid Receptors as Potential Drug Targets in Tissue Transplantation and Autoimmune Diseases. *Curr. Pharm. Des.* **2006,** *12,* 161–171.

Clair, T.; Aoki, J.; Koh, E.; Bandle, R. W.; Nam, S. W.; Ptaszynska, M. M.; Mills, G. B.; Schiffmann, E.; Liotta, L. A.; Stracke, M. L. Autotaxin Hydrolyzes Sphingosylphosphorylcholine to Produce the Regulator of Migration, Sphingosine 1-Phosphate. *Cancer Res.* **2003,** *63,* 5446–5453.

Croft, S. L.; Neal, R. A.; Pendergast, W.; Chan, J. H. The Activity of Alkyl Phosphorylcholines and Related Derivatives against *Leishmania donovani. Biochem. Pharmacol.* **1987,** *36,* 2633–2636.

Croft, S. L.; Snowdon, D.; Yardley, V. The Activities of Four Anticancer Alkyllysophospholipids against Lesmania donovani, Trypanosoma cruzi and Trypanosoma brucei. *J. Antimicrob. Chemotherapy* **1996,** *38,* 1041–1047.

Crul, M.; Rosing, H.; de Klerk, G. J.; Dubbelman, R.; Traiser, M.; Reichert, S.; Knebel, N. G.; Schellens, J. H.; Beijinen, J. H.; ten Bokkel Huinink, W. W. Phase I and Pharmacological Study of Daily Administration of Perifosine (D-21266) in Patients with Advanced Solid Tumors *Eur. J. Cancer* **2002,** *38,* 1615–1621.

D'Arrigo, P.; Fasoli, E.; Pedrocchi-Fantoni, G.; Rossi, C.; Saraceno,C.; Servi, S.; Tessaro, D. A Practical Selective Synthesis of Mixed Short/Long Chains Glycerophosphocholines. *Chem. Phys. Lipids* **2007,** *147,* 113–118.

D'Arrigo, P.; Scotti, M. Lysophospholipids: Synthesis and Biological Aspects. *Curr. Org. Chem.* **2013,** *17,* 812–830.

D'Arrigo, P.; Servi, S. Using Phospholipases for Phospholipid Modification. *Trends Biotech.* **1997,** *15,* 90–96.

D'Arrigo, P.; Servi, S. Synthesis of Lysophospholipids. *Molecules* **2010,** *15,* 1354–1377.

Dennis, E. A.; Brown, H. A.; Deems, R. A.; Glass, C. K.; Merrill, A. H.; Murphy, R. C.; Raetz, C. R. H.; Shaw, W.; Subramaniam, S.; Russell, D. W.; et al. *Functional Lipodomics.* CRC Press: Boca Raton, FL, 2006; pp 1–15.

Devos, M.; Poisson, L.; Ergan, F.; Pencreac'h, G. Enzymatic Hydrolysis of Phospholipids from *Isochrysis galbana* for Docosahexaenoic Acid Enrichment. *Enzyme Microb. Technol.* **2006,** *39,* 548–554.

Drobnik, W.; Liebisch, G.; Audebert, F. X.; Frohlich, D.; Gluck, T.; Vogel, P.; Rothe, G.; Schmitz, G. Plasma Ceramide and Lysophosphastidylcholine Inversely Correlate with Mortality in Sepsis Patients. *J. Lipid Res.* **2003,** *44,* 754–761.

Dyerberg, J.; Bang, H. Haemostatic Function and Platelet Polyunsaturated Fatty Acids in Eskimos. *Lancet* **1979,** *314,* 433–435.

Eder, A. M.; Sasagawa, T.; Mao, M.; Aoki, J.; Mills, G. B. Constitutive and Lysophosphatidic Acid (LPA)-induced LPA Production: Role of Phospholipase D and Phospholipase A2. *Clin. Cancer Res.* **2000,** *6,* 2482–2491.

English, D. Phosphatidic Acid: A Lipid Messenger Involved in Intracellular and Extracellular Signaling. *Cell Signal* **1996,** *8,* 341–347.

Farkas, T.; Kitajka, K.; Fodor, E.; Csengeri, I.; Lahdes, E.;Yeo, Y. K.; Krasznai, Z.; Halver, J. E. Docosahexaenoic Acid-containing Phospholipid Molecular Species in Brains of Vertebrates. *Proc. Natl. Acad. Sci. USA* **2000,** *97,* 6362–6366.

Fasoli, E.; Arnone, A.; Caligiuri, A.; D'Arrigo, P.; de Ferra, L.; Servi, S. Tin-mediated Synthesis of Lysophospholipids. *Org. Biomol. Chem.* **2006,** *4,* 2974–2978.

Ferry, G.; Tellier, E.; Try, A.; Gres, S.; Naime, I.; Simon, F.; Rodriguez, M.; Boucher, J.; Tack, I.; Gesta, S.; et al. Autotaxin Is Released from Adipocytes, Catalyzes Lysophosphatidic Acid Synthesis, and Activates Preadipocyte Proliferation. Upregulated Expression with Adipocyte Differentiation and Obesity. *J. Biol. Chem.* **2003,** *278,* 18162–18169.

Friedman, P.; Haimovitz, R.; Markman, O.; Roberts, M. F.; Shinitzky, M. Conversion of Lysophospholipids to Cyclic Lysophisphatidic Acid by Phospholipase D. *J. Biol. Chem.* **1996,** *271,* 953–957.

Fujita, S.; Suzuki, K. Surface Activity of the Lipid Products Hydrolyzed with Lipase and Phospholipase A2. *J. Amer. Oil Chem. Soc.* **1990,** *67,* 1008–1014.

Fukushima, N.; Weiner, J. A.; Chun, J. Lysophosphatidic Acid (LPA) Is a Novel Extracellular Regulator of Cortical Neuroblast Morphology. *Dev. Biol.* **2000**, *228,* 6–18.

Gajate, C.; Mollinedo, F. The Antitumor Ether Lipid ET-18-OCH3 Induces Apoptosis through Translocation and Capping of Fas/CD95 into Membrane Rafts in Human Leukemic Cells. *Blood* **2001,** *98,* 3860–3863.

Gajate, C.; Fonteriz, R. I.; Cabaner, C.; Alvarez-Noves, G.; Alvarez-Rodriguez, Y.; Modolell, M.; Mollinedo, F. Intracellular Triggering of Fas, Independently of Fash, as a new Mechanism of Antitumor Ether Lipid Induced Apoptosis. *Intl. J. Cancer* **2000,** *85,* 674–682.

Gajate, C.; Molliendo, F. Biological Activities, Mechanism of Action and Biomedical Prospect of the Antitumor Ether Phospholipid ET-18-OCH3 (edelfosine), a Proapoptotic Agent in Tumor Cells. *Curr. Drug Metab.* **2002,** *3,* 491–525.

Gardell, S. E.; Dubin, A. E.; Chun, J. Emerging Medicinal Roles for Lysophospholipid Signaling. *Trends Mol. Med.* **2006**, *12,* 65–75.

Ham, J. J.; Rhee, J. S. Effect of Salt Hydrate Pairs for Water Activity Control on Lipase-catalyzed Synthesis of Lysophospholipids in a Solvent-free System. *Enzyme Microb. Technol.* **1998,** *22,* 158–164.

Hama, K.; Aoki, J.; Fukaya, M.; Kishi, Y.; Suzuki, R.; Ohta, H.; Yamori, T.; Watanabe, M.; Chun, J.; Arai, H. Lysophophatidic Acid and Autotaxin Stimulate Cell Motility of Neoplastic and Non-neoplastic Cells through LPA1. *J. Biol. Chem.* **2004,** *279,* 17634–17639.

Han, J. J.; Rhee, J. S. Effect of Salt Hydrate Pairs for Water Activity Control on Lipase-catalyzed Synthesis of Lysophospholipids in a Solvent-free Systems. *Enzyme Microb. Technol.* **1998,** *22,* 158–164.

Hannun, Y. A.; Obeid, L. M. Principles of Bioactive Lipid Signaling: Lessons from Sphingolipids. *Nat. Rev. Mol. Cell. Biol.* **2008,** *9,* 139–150.

Hass, M. J.; Scott, K.; Janssen, G. Enzymatic Phosphatidylcholine Hydrolysis in Organic Solvents: An Examination of Selected Commercially Available Lipases. *J. Amer. Oil Chem. Soc.* **1994,** *71,* 483–490.

Herrmann, D. B.; Besenfelder, E.; Bicker, U.; Pahlke, W.; Bohm, E. Pharmacokinetics of the Thioether Phospholipid Analogue BM 41.440 in Rats. *Lipids* **1987a,** *22,* 952–954.

Herrmann, D. B.; Bicker, U.; Pahlke, W. BM 41.440: A New Antineoplastic, Antimetastatic, and Immune-stimulating Drug. *Cancer Detect Prev. Suppl.* **1987b,** *1,* 361–371.

Herrmann, D. B.; Neumann, H. A.; Berdel, W. E.; Heim, M. E.; Fromm, M.; Boerner, D.; Bicker, U. Phase I Trial of the Thioether Phospholipid Analogue BM 41.440 in Cancer Patients. *Lipids* **1987c,** *22,* 962–966.

Herrmann, D. B.; Pahlke, W.; Opitz, H. G.; Bicker, U. *In Vivo* Antitumor Activity of Ilmfosine. *Cancer Treat Rev.* **1990,** *17,* 247–252.

Horrocks, L. A.; Yeo, Y. K. Health Benefits of Docosahexaenoic Acid (DHA). *Pharmacol. Res.* **1999,** *40,* 211–225.

Houlihan, W. J.; Lohmeyer, M.; Workman, P.; Cheon, S. H. Phospholipid Antitumor Agents. *Med Res Rev.* **1995,** *15,* 157–223.

Huang, L. S.; Hung, N. D.; Sok, D.-E.; Kim, M. R. Lysophosphatidylcholine Containing Docosahexaenoic Acid at the *sn*-1 Position Is Anti-inflammatory. *Lipids* **2010,** *45,* 225–236.

Hung, N. D.; Kim, M. R.; Sok, D.-E. Oral Administration of 2-Docosahexaenoyl Lysophosphatidylcholine Displayed Anti-inflammatory Effects on Zymosan A-induced Peritonitis. *Inflammation* **2011a,** *34,* 147–160.

Hung, N. D.; Kim, M. R.; Sok, D.-E. 2-Polyunsaturated Acyl Lysophosphatidylethanolamine Attenuates Inflammatory Response in Zymosan A-induced Peritonitis in Mice. *Lipids* **2011b,** *46,* 893–906.

Inoue, M.; Rashid, M. H.; Fujita, R.; Contos, J. J.; Chun, J.; Ueda, H. Initiation of Neuropathic Pain Requires Lysophosphatidic Acid Receptor Signaling. *Nat. Med.* **2004,** *10,* 712–718.

Inoue, M.; Xie, W.; Matsushita, Y.; Chun, J.; Aoki, J.; Ueda, H. Lysophosphatidylcholine Induces Neuropathic Pain through an Action of Autotaxin to Generate Lysopphosphatidic Acid. *Neuroscience* **2008,** *152,* 296–298.

Ishii, I.; Fukushima, N.; Ye, X.; Chun, J. Lysophospholipid Receptors: Signaling and Biology. *Annu. Rev. Biochem.* **2004,** *73,* 321–354.

Ishii, S.; Shimizu, T. Platelet-activating Factor (PAF) Receptor and Genetically Engineered PAF Receptor Mutant Mice. *Prog. Lipid Res.* **2000,** *39,* 41–82.

Jendrossek, V.; Handrick, R. Membrane Targeted Anticancer Drugs: Potent Inducers of Apoptosis and Putative Radiosensitisers. *Cur. Med. Chem. Anti-Canc. Agents* **2003,** *3,* 343–353.

Jendrossek, V.; Erdlenbruch, B.; Hunold, A.; Kugler, W.; Eibl, H.; Lakomek, M. Erucylphosphocholine, a Novel Antineoplastic Ether Lipid, Blocks Growth and Induces Apoptosis Brain Tumor Cell Lines in Vitro. *Intl. J. Oncol.* **1999,** *14,* 15–37.

Jendrossek, V.; Kugler, W.; Erdlenbruch, B.; Eibl, H.; Lang, F.; Lakomek, M. Erucylphosphocholine-induced Apoptosis in Chemoresistant Glioblastoma Cell Lines: Involvement of Caspase Activation and Mitochondrial Activation. *Anticancer Res.* **2001,** *21,* 3389–3396.

Jendrossek, V.; Muller, I.; Eibl, H.; Belka, C. Intracellular Mediators of Erucylphosphocholine-induced Apoptosis. *Oncogene* **2003,** *22,* 2621–2632.

Kapoor, M.; Gupta, M. N. Lipase Promiscuity and Its Biochemical Applications. *Process Biochem.* **2012,** *47,* 555–569.

Karliner, J. S. Lysophospholipids and the Cardiovascular System. *Biochim. Biophys. Acta* **2002,** *1582,* 216–221.

Karliner, J. S. Mechanisms of Cardioprotection by Lysophospholipids. *J. Cell. Biochem.* **2004,** *92,* 1095–1103.

Kennedy, E. P.; Weiss, S. B. The Function of Cytidine Coenzymes in the Biosynthesis of Phospholipids. *J. Biol. Chem.* **1956,** *222,* 193–214.

Kim, J.; Kim, B. G. Lipase-catalyzed Synthesis of Lysophosphatidylcholine. *Ann. NY Acad. Sci.* **1998,** *864,* 341–344.

Kim, J.; Kim, B. G. Lipase-catalyzed Synthesis of Lysophosphatidylcholine Using Organic Cosolvent for *in Situ* Water Activity Control. *J. Amer. Oil Chem. Soc.* **2000,** *77,* 791–797.

Kim, J.-K.; Kim, M.-K.; Chung, G.-H.; Choi, C.-S.; Rhee, J.-S. Production of Lysophospholipid Using Extracellular Phospholipase A_1 from *Serratia* sp. MK1. *J. Microbiol. Biotechnol.* **1997,** *7,* 258–261.

Kim, K. S.; Sengupta, S.; Berk, M.; Kwak, Y. G.; Escobar, P. F.; Belinson, J.; Mok, S. C.; Xu, Y. Hypoxia Enhances Lysophosphatidic Acid Responsiveness in Ovarian Cancer Cells and Lysophosphatidic Acid Induces Ovarian Tumor Metastasis *in Vivo*. *Cancer Res.* **2006,** *66,* 7983–7990.

Kim, Y. A.; Park, M. S.; Kim, Y. H.; Han, S. Y. Synthesis of 1-Lyso-2-palmitoyl-*rac*-glycero-3-phosphocholine and Its Regioisomers and Structural Elucidation by NMR Spectroscopy and FAB Tandem Mass Spectrometery. *Tetrahedron* **2003,** *59,* 2921–2928.

Kohama, T.; Olivera, A.; Edsall, L.; Nagiec, M. M.; Dickson, R.; Spiegel, S. Molecular Cloning and Functional Characterization of Murine Sphingosine Kinase. *J. Biol. Chem.* **1998,** *273,* 23722–23728.

Kluk, M. J.; Hla, T. Signaling of Sphingosine-1-phosphate via the SIP/EDG-family of G-protein-coupled Receptors. *Biochim. Biophys. Acta.* **2002,** *1582,* 72–80.

Lands, W. E. Metabolism of Glycerolipides: A Comparison of Lecithin and Triglycerides Synthesis. *J. Biol. Chem.* **1958,** *231,* 883–888.

Lands, W. E. Stories about Acyl Chains. *Biochim. Biophys. Acta* **2000,** *1483,* 1–14.

Li, W.; Du, W.; Li, Q.; Sun, T.; Liu, D. Study on Acyl Migration Kinetics of Partial Glycerides: Dependence on Temperature and Water Activity. *J. Mol. Catal. B: Enzymatic* **2010,** *63,* 17–22.

Lindberg, J.; Ekeroth, J.; Konradsson, P. Efficient Synthesis of Phospholipids from Glycidyl Phosphates. *J. Org. Chem.* **2002,** *67,* 194–199.

Liu, H.; Sugiura, M.; Nava, V. E.; Edsall, L. C.; Kono, K.; Poulton, S.; Milstien, S.; Kohama,T.; Spiegel, S. Molecular Cloning and Functional Characterization of a Novel Mammalian Sphingosine Kinase Type 2 Isoform. *J. Biol. Chem.* **2000,** *275,* 19513–19520.

Long, C.; Odavic, R.; Sargent, E. J. The Action of the Cabbage-leaf Phospholipase D upon Lysolecithin. *Biochem. J.* **1967,** *102,* 221–229.

Ludwig, M. G.; Vanek, M.; Guerini, D.; Gasser, J. A.; Jones, C. E.; Junker, U.; Hofstetter, H.; Wolf, R. M.; Seuwen, K. Proton-sensing G-protein-coupled Receptors. *Nature* **2003,** *425,* 94–98.

Matsumoto, T.; Kobayashi, T.; Kamata, K. Role of Lysophosphatidylcholine (LPC) in Atherosclerosis. *Curr. Med. Chem.* **2007,** *14,* 3209–3220.

Mayhew, E.; Ahmad, I.; Bhatia, S.; Dause, R.; Filep, J.; Janoff, A. S.; Kaisheva, E.; Perkins, W. R.; Zha, Y.; Franklin, J. C. Stability of Association of 1-*O*-octadecyl-2-*O*-methyl-*sn*-glycero-3-phosphocholine with Liposome Is Composition Dependent. *Biochim. Biophys. Acta* **1997,** *1329,* 139–148.

Mazur, A. W.; Hiler, G. D. H.; Lee, S. S. C.; Amstrong, M. P.; Wendel, J. D. Regioselective and Stereoselective Enzymatic Esterification of Glycerol and its Derivatives. *Chem. Phys. Lipids* **1991**, *60,*189-199.

Meyer zu Heringdrof, D.; Himmel, H. M.; Jakobs, K. H. Sphingosylphosphorylcholine-biological Functions and Mechanism of Action. *Biochim. Biophys. Acta* **2002,** *1682,* 178–189.

Mills, G. B.; Moolenaar, W. H. The Emerging Role of Lysophosphatidic Acid in Cancer. *Nat. Rev. Cancer* **2003,** *3,* 582–591.

Mollinedo, F.; Fernandez-Luna, J. L.; Gajate, C.; Martin-Martin, B.; Benito, A.; Martinez-Damau, R.; Modolell, M. Selective Induction of Apoptosis in Cancer Cells by the Ether Lipid ET-18-OCH3 (Edelfosine): Molecular Structure Requirements, Cellular Uptake, and Protection of Bcl-2 and Bcl-x(L). *Cancer Res.* **1997,** *57,* 1320–1328.

Mollinedo, F.; Gajate, C.; Martin-Santamaria, S.; Gago, F. ET-18-OCH_3 (Edelfosine): A Selective Antitumor Lipid Targeting Apoptosis through Intracellular Activation of Fas/CD95 Death Receptor. *Curr. Med. Chem.* **2004,** *11,* 3163–3184.

Moolenaar, W. H. Lysophosphatidic Acid, a Multifunctional Phospholipid Messenger. *J. Biol. Chem.* **1995,** *270,* 12949–12952.

Moolenaar, W. H. Development of our Current Understanding of Bioactive Lysophospholipids. *Ann. N.Y. Acad. Sci.* **2000**, *905,* 1–10.

Moolenaar, W. H.; van Meeteren, L. A.; Giepmans, B. N. The Ins and Outs of Lysophosphatidic Acid Signaling. *Bioassays* **2004,** *26,* 870–881.

Murata, J.; Lee, H. Y.; Clair, T.; Krutzsch, H. C.; Arestad, A. A.; Sobel, M. E.; Liotta, L. A.; Stracke, M. L. cDNA Cloning of the Human Tumor Motility-stimulating Protein, Autotaxin, Reveals a Homology with Phosphodisterases. *J. Biol. Chem.* **1994,** *269,* 30479–30484.

Murph, M.; Mills, G. B. Targeting the Lipids LPA and S1P and Their Signaling Pathways to Inhibit Tumor Progression. *Expert Rev. Mol. Med.* **2007,** *9,* 1–18.

Mustranta, A.; Forsell, P.; Aura, A. M.; Suortti, T.; Poutanen, K. Modification of Phospholipids with Lipases and Phospholipases. *Biocatalysis* **1994,** *9,* 181–194.

Mustranta, A.; Forssel, P.; Poutanen, K. Comparision of Lipases and Phospholipases in the Hydrolysis of Phospholipids. *Process Biochem.* **1995,** *30,* 393–401.

Nakanaga, K.; Hama, K.; Aoki, J. Autotaxin—An LPA Producing Enzyme with Diverse Functions. *J. Biochem.* **2010,** *148,* 13–24.

Noguchi, K.; Herr, D.; Mutoh, T.; Chun, J. Lysophosphatidic Acid (LPA) and Its Receptors. *Curr. Opin. Pharmacol.* **2009,** *9,* 15–23.

Nojima, M.; Hosokawa, M.; Takahashi, K.; Hatano, M. Effect of EPA and DHA Containing Glycerophospholipid Molecular Species on the Fluidity of Erythrocyte Cell Membrane. *Fisheries Sci.* **1994,** *60,* 729–734.

Oka, S.; Ota, R.; Shima, M.; Yamashita, A.; Sugiura, T. GPR35 Is a Novel Lysophosphatidic Acid Receptor. *Biochem. Biophys. Res. Commun.* **2010,** *395,* 232–237.

Okajima, F. Plasma Lipoproteins Behave as Carriers of Extracellular Sphingosine-1-phosphate: Is This an Atherogenic Mediator or an Anti-atherogenic Mediator? *Biochim. Biophys. Acta* **2002,** *1582,* 132–137.

Okudaira, S.; Yukiura, H.; Aoki, J. Biological Roles of Lysophosphatidic Acid Signaling through Its Production by Autotaxin. *Biochimie* **2010,** *92,* 698–702.

Okumura, S.; Baba, H.; Kumada, T.; Nanmoku, K.; Nakajima, H.; Nakane, Y.; Hioki, K.; Ikenaka, K. Cloning of a G-protein-coupled Receptor That Shows an Activity to Transform NIH3T3 Cells and Is Expressed in Gastric Cancer Cells. *Cancer Sci.* **2004,** *95,* 131–135.

Olivera, A.; Kohama, T.; Edsall, L.; Nava, V.; Cuvillier, O.; Poulton, S.; Spiegel, S. Sphingosine Kinase Expression Increases Intracellular Sphingosine 1-Phosphate and Promotes Cell Growth and Survival. *J. Cell Biol.* **1999,** *147,* 545–558.

Osborne, N.; Stainier, D. Y. Lipid Receptors in Cardiovascular Development. *Annu. Rev. Physiol.* **2003,** *65,* 23–43.

Pages, C.; Simon, M. F.; Valet, P.; Saulnier-Blache, J. S. Lysophosphatidic Acid Synthesis and Release. *Prostaglandins Other Lipid Mediat.* **2001a,** *64,* 1–10.

Pages, C.; Daviaud, D.; An, S.; Krief, S.; Lafontan, M.; Valet, P.; Saulnier-Blache, J. S. Endothelial Differentiation Gene-2 Receptor Is Involved in Lysophosphatidic Acid-dependent Control of 3T3F442A Preadipocyte Proliferation and Spreading. *J. Biol. Chem.* **2001b,** *276,* 11599–11605.

Parrill, A. L. Lysophospholipid Interactions with Protein Targets. *Biochim. Biophys. Acta* **2008,** *1781,* 540–546.

Patel, V.; Lahusen, T.; Sausville, E. A.; Gutkind, J. S.; Senderowicz, A. M. Perifosine, a Novel Alkylphospholipid, Induces $p21^{WAF1}$ Expression in Squamous Carcinoma Cells through a p53-Independent Pathway, Leading to Loss in Cyclin-dependent Kinase Activity and Cell Cycle Arrest. *Cancer Res.* **2002,** *62,* 1401–1409.

Pencreac'h, G.; Ergan, F.; Poisson, L. DHA-lysophospholipid Production. *Curr. Org. Chem.* **2013,** *17,* 797–801.

Perez-Victoria, J. M.; Perez-Victoria, F. J.; Parodi-Talice, A.; Jimenez, I. A.; Ravelo, A. G.; Castanys S.; Gamaro, F. Alkyl-lysophospholipid Resistance in Multidrug-resistance *Leismania tropica* and Chemosensitization by a Novel P-glycoprotein Like Transporter Modulator. *Antimicrob. Agents Chemother.* **2001,** *45,* 2468–2474.

Perkins, W. R.; Dause, R. B.; Li, X.; Franklin, J. C.; Carbral-Lily, D. J.; Dank, E. H.; Mayhew, E.; Janoff, A. S. Combination of Antitumor Ether Lipid with Lipids of Complementary Molecular Shape Reduces its Transfer to RBCs. *Biochim. Biophys. Acta* **1997,** *88,* 61–68.

Peters, A. C.; Ahmad, I.; Janoff, A. S.; Pushkareva, M. Y.; Mayhew, E. Growth Inhibitory Effects of Liposome-associated 1-*O*-Octadecyl-2-*O*-methyl-*sn*-glycero-3-phosphocholine. *Lipids* **1997,** *32,* 1045–1054.

Poisson, L.; Devos, M.; Godet, S.; Ergan, F.; Pencreac'h, G. Acyl Migration during Deacylation of Phospholipids Rich in Docosahexaenoic Acid (DHA): An Enzymatic Approach for Evidence and Study. *Biotechnol. Lett.* **2009,** *31*, 743–749.

Pyne, S.; Pyne, N. J. Sphingosine 1-Phosphate Signaling in Mammalian Cells. *Bichem. J.* **2000,** *349*, 385–402.

Raynal, P.; Montangner, A.; Dance, M.; Yart, A. Lysophospholipids and Cancer: Current Status and Prospectives. *Pathol. Biol.* **2005,** *53*, 57–62.

Reblova, Z.; Pokorny, J. Effect of Lecithin on the Stabilization of Foods. In *Phospholipids: Characterization, Metabolism, and Novel Biological Applications;* Ceve, G., Paltauf, F., Eds; AOCS Press: Urbana, IL, 1995; pp 378–383.

Ren, J.; Xiao, Y. J.; Singh, L. S.; Zhao, Z.; Feng, L.; Rose, T. M.; Prestwich, G. D.; Xu, Y. Lysophosphatidic Acid Is Constitutively Produced by Human Peritoneal Mesothelial Cells and Enhances Adhesion, Migration, and Invasion of Ovarian Cancer Cells. *Cancer Res.* **2006,** *66*, 3006–3014.

Rivera, J.; Proia, R. L.; Oliveira, A. The Alliance of Sphingosine 1-Phosphate and Its Receptors in Immunity. *Nat. Rev. Immunology* **2008,** *8*, 753–763.

Rivera, R.; Chun, J. Biological Effects of Lysophospholipids. *Rev. Physiol. Biochem. Pharmacol.* **2006,** *160*, 25–46.

Rose, M. T.; Prestwich, G. D. Synthesis and Evaluation of Fluorogenic Substrates for Phospholipise D and Phospholipase C. *Org. Lett.* **2006,** *8*, 2575–2578.

Rossetto, R.; Bibak, N.; Hajdu, J. A New Approach to the Synthesis of Lysophospholipids: Preparation of Lysophosphatidic Acid and Lysophosphatidylcholine from p-Nitrophenylglycerate. *Tetrahedron Lett.* **2004,** *45*, 7371–7373.

Rossetto, R.; Bibak, N.; DeOcampo, R.; Shah, T.; Grabriellian, A.; Hajdu, J. A New Synthesis of Lysophosphatidylcholines and Related Derivatives. Use of *p*-Toluensulfonate for Hydroxyl Group Protection. *J. Org. Chem.* **2007,** *72*, 1691–1698.

Ruiter, G. A.; Zerp, S. F.; Bartelink, H.; Van Blitterswijk, W. J.; Verhejj, M. Anticancer Alkyllysophospholipid Inhibit the Phosphatidyl-inositol 3-Kinase-Akt/PKB Survival Pathway. *Anticancer Drugs* **2003,** *14*, 167–173.

Sano, T.; Baker, D.; Virag, T.; Wada, A.; Yatomi, Y.; Kobayashi, T.; Igarashi, Y.; Tigyi, G. Multiple Mechanisms Linked to Platelet Activation Result in Lysophosphatidic Acid and Sphingosine 1-Phosphate Generation in Blood. *J. Biol. Chem.* **2002,** *277*, 21197–21206.

Sarney, D. B.; Fregapane, G.; Vulfson, E. N. Lipase—Catalyzed Synthesis of Lysophospholipids in a Continuous Bioreactor. *J. Amer. Oil Chem. Soc.* **1994,** *71*, 93–96.

Schlame, M.; Rua, D.; Greenberg, M. L. The Biosynthesis and Functional Role of Cardiolipin. *Prog. Lipid Res.* **2000,** *39*, 257–288.

Schober, A.; Siess, W. Lysophosphatidic Acid in Atherosclerotic Diseases. *Br. J. Pharmacol.* **2012,** *167*, 465–482.

Sengupta, S.; Wang, Z.; Tipps, R.; Xu, Y. Biology of LPA in Health and Disease. *Semin. Cell Dev. Biol.* **2004,** *15*, 503–512.

Servi, S. Phospholipases as Synthetic Catalysts. *Top. Curr. Chem.* **1999,** *200*, 127–158.

Shida, D.; Takaba, K.; Kapitonov, D.; Milstien, S.; Spiegel, S. Targeting SphK1 as a New Strategy against Cancer. *Curr. Drug Targets* **2008a,** *9*, 662–673.

Shida, D.; Fang, X.; Kordula, T.; Takabe, K.; Lepine, S.; Alvarez, S. E.; Milstien, S.; Spiegel, S. Cross-talk between LPA1 and Epidermal Growth Factor Receptors Mediates Up-regulation of Sphingosine Kinase 1 to Promote Gastric Cancer Cell Motility and Invasion. *Cancer Res.* **2008b,** *68*, 6569–6577.

Shimizu, T. Lipid Mediators in Health and Disease: Enzymes and Receptors as Therapeutic Targets for the Regulation of Immunity and Inflammation. *Annu. Rev. Pharmacol.* **2009,** *49*, 123–150.

Shindou, H.; Shimizu, T. Acyl-CoA: Lysophospholipid Acyltransferases. *J. Biol. Chem.* **2009,** *284*, 1–5.

Shindou, H.; Hishikawa, D.; Harayama, T.; Yuki, K.; Shimizu, T. Recent Progress on Acyl CoA: Lysophospholipid Acyltransferase Research. *J. Lipid Res.* **2009,** *50*, S46–S51.

Shuying Liu, S.; Murphy, M.; Panupinthu, N.; Mills, G. B. ATX-LPA Receptor Axis in Inflammation and Cancer. *Cell Cycle* **2009,** *8*, 3695–3701.

Siehler, S.; Manning, D. R. Pathways of Transduction Engaged by Sphingosine-1-phosphate through G-protein-coupled Receptors. *Biochim. Biophys. Acta* **2002,** *1582*, 94–99.

Soto, J.; Toledo, J.; Gutierrez, P. Treatment of American Cutaneous Leishmaniasis with Miltefosine, an Oral Agent. *Clin. Infect. Dis.* **2001,** *33*, 57–61.

Spiegel, S.; Milstien, S. Sphingosine-1-phosphate: An Enigmatic Signaling Lipid. *Natl. Rev. Mol. Cell. Biol.* **2003,** *4*, 397–407.

Spiegel, S.; Olivera, A.; Zhang, H.; Thompson, E. W.; Su, Y.; Berger, A. Sphingsone-1-Phosphate, a Novel Second Messenger Involved in Cell Growth Regulation and Signal Transduction, Affects Growth and Invasiveness of Human Breast Cancer Cells. *Breast Cancer Res. Treat.* **1994,** *31*, 337–348.

Stafford, R. E.; Dennis, E. A. Lysophospholipids as Biosurfactants. *Colloids Surfaces* **1988,** *30*, 47–64.

Stracke, M. L.; Krutzsch, H. C.; Unsworth, E. J.; Arestad, A.; Cioce, V.; Schiffmann, E.; Liotta, L. A. Identification, Purification, and Partial Sequence Analysis of Autotaxin, a Novel Motility-stimulating Protein. *J. Biol. Chem.* **1992,** *267*, 2524–2529.

Sturm, A.; Dignass, A. U. Modulation of Gastrointestinal Wound Repair and Inflammation by Phospholipids. *Biochim. Biophys. Acta* **2002,** *1582*, 282–288.

Sukocheva, O.; Wang, L.; Verrier, E.; Vadas, M. A.; Xia, P. Restoring Endocrine Response in Breast Cancer Cells by Inhabitation of the Sphingosine Kinase-1 Signaling Pathway. *Endocrinology* **2009,** *150*, 4484–4492.

Sundar, S.; Makharia, A.; More, D. K. Short Course of Oral Miltefosine for Treatment of Visceral Leishmaniasis. *Clin. Infect. Dis.* **2000,** *31*, 1110–1113.

Sutphen, R.; Xu, Y.; Wilbanks, G. D.; Fiorica, J.; Grendys, E. C.; LaPolla, J. P. J.; Arango, H.; Hoffman, M. S.; Martino, M.; Wakeley, K.; et al. Lysophospholipids Are Potential Biomarkers of Ovarian Cancer. *Cancer Epidemiol. Biomarkers Prev.* **2004,** *13*, 1185–1191.

Takuwa, Y. Subtype-specific Differential Regulation of Rho Family G Proteins and Cell Migration by the Edg Family Sphingosine-1-phosphate Receptors. *Biochim. Biophys. Acta* **2002,** *1582*, 112–120.

Thurlimann, B.; Keshaviah, A.; Coates, A. S.; Mouridsen, H.; Mouridsen, H.; Mauriac, L.; Frobes, J. F.; Paridaens, R.; Castiglione-Gertsch, M.; Gelber, R. D.; Rabaglio, M.;

Smith, I.; Wardley, A.; Price, K. N.; Goldhirsch, A. A. Comparision of Letrozole and Tamoxifen in Postmenopausal Women with Early Breast Cancer. *N. Engl. J. Med.* **2005**, *353*, 2747–2757.

Tigyi, G.; Parrill, A. L. Molecular Mechanisms of Lysophosphatidic Acid Action. *Prog. Lipid Res.* **2003,** *42*, 498–526.

Tokumura, A. A Family of Phospholipid Autocoids: Occurrence, Metabolism and Bioactions. *Prog. Lipid Res.* **1995,** *34*, 151–184.

Tokumura, A. Physiological and Pathophysiological Roles of Lysophosphatidic Acids Produced by Secretory Lysophospholipase D in Body Fluids. *Biochim. Biophys. Acta* **2002,** *1582,* 188–225.

Tokumura, A. Physiological Significance of Lysophospholipids That Act on the Lumen Side of Mammalian Lower Digestive Tracts. *J. Health Sci.* **2011,** *57*, 115–128.

Tokumura, A.; Sinomiya, J.; Kishimoto, S.; Tanaka, T.; Kogure, K.; Sugiura, T.; Satouchi, K.; Waku, K.; Fukuzawa, K. Human Platelets Respond Differently to Lysophosphatidic Acids Having a Highly Unsaturated Fatty Acyl Group and Alkyl Ether-linked Lysophosohatidic Acids. *Biochem. J.* **2002a,** *365*, 617–628.

Tokumura, A.; Tominaga, K.; Yasuda, K.; Kanzaki, H.; Kogure, K.; Fukuzawa, K. Lack of Significant Differences in the Corrected Activity of Lysophospholipase D, Producer of Phospholipid Mediator Lysophosphatidic Acid, in Incubated Serum from Women with and without Ovarian Tumors. *Cancer* **2002b,** *94*, 141–151.

Torkhovskaya, T. I.; Ipatova, O. M.; Zakharova, T. S.; Kochetova, M. M.; Khalilov, T. S. Lysophospholipids Receptors in Cell Signaling. *Biochemistry* (Moscow) **2007,** *72*, 125–131.

Turner, J. R. Intestinal Mucosal Barrier Function in Health and Disease. *Nat. Rev. Immunol.* **2009,** *9*, 799–809.

Ulbrich-Hofman, R. Phospholipases Used in Lipid Transformation. In *Enzymes in Lipid Modification*; Bornscheuer, U. T., Ed.; Wiley-VCH: Weinheim, Germany, 2000; pp 219–262.

Umezu-Goto, M.; Kishi, Y.; Taira, A.; Hama, K.; Dohmae, N.; Takio, K.; Yamori, T.; Mills, G. B.; Inoue, K.; Aoki, J.; Arai, H. Autotaxin Has Lysophospholipase D Activity Leading to Tumor Cell Growth and Motility by Lysophosphatidic Acid Production. *J. Cell Biol.* **2002,** *158*, 227–233.

Vadas, M.; Xia, P.; McCaughan, G.; Gamble, J. The Role of Sphingosine Kinase 1 in Cancer: Oncogene or Non-oncogene Addiction. *Biochimica et Biophysica Acta* **2008,** *1781,* 442–447.

Valet, P.; Pages, C.; Jenneton, O.; Daviaud, D.; Barbe, P.; Record,M.; Saulnier-Blache, J. S.; Lafontan, M. Alpha2-adrenergic Receptor-mediated Release of Lysophophatidic Acid by Adipocytes. A Pancreatic Signal for Preadipocyte Growth. *J. Clin. Invest.* **1998,** *101*, 1431–1438.

Van der Luit, A. H.; Budde, M.; Ruurs, P.; Verheij, M.; van Blitterswijk, W. J. Alkyl-lysophospholipid Accumulates in Lipid Rafts and Induces Apoptosis via Raft-dependent Endocytosis and Inhibition of Phosphatidylcholine Synthesis. *J. Biol. Chem.* **2002,** *277*, 39541–39547.

Vaz, S.; Persson, M.; Svensson, I.; Adlercreutz, P. Lipase-catalyzed Fatty Acid Exchange in Digalactosyldiacylglycerol. Improvement of Yield Due to the Addition of Phenylboronic Acid. *Biocatal. Biotransfor.* **2000,** *18*, 1–12.

Virto, C.; Sevensson, I.; Adlercreutz, P. Enzymatic Synthesis of Lysophosphatidic Acid and Phosphatidic Acid. *Enzyme Microb. Technol.* **1999,** *24*, 651–658.

Virto, C.; Sevensson, I.; Adlercreutz, P. Hydrolytic and Transphosphatidylation Activities of Phospholipase D from Savoy Cabbage towards Lysophosphatidylcholine. *Chem. Phys. Lipids* **2000,** *106*, 41–51.

Waku, K.; Nakazawa, Y. Acyltransferae Activity to 1-Acyl-, 1-o-alkenyl-, and 1-o-alkyl-glycero-3-phosphorylcholine in Ehrlich Ascites Tumor Cells. *J. Biochem.* **1972,** *72,* 495–497.

Weiner, J. A.; Fukushima, N.; Contos, J. J.; Scherer, S. S.; Chun, J. Regulation of Schwann Cell Morphology and Adhesion by Receptor-mediated Lysophosphatidic Acid Signaling. *J. Neurosci.* **2001,** *21*, 7069–7078.

Xia, J.; Hui, Y. Z. The Stereospecific Synthesis of Mixed-acid Phospholipids with Polyunsaturated Fatty Acid from D-mannitol. *Tetrahedron Assymetry* **1997,** *8*, 451–458.

Xie, Y.; Meier, K. E. Lysophospholipase D and its Role in LPA Production. *Cellular Signaling* **2004,** *16*, 975–981.

Xie, Y.; Gibbs, T. C.; Meier, K. E. Lysophosphatidic Acid as an Autocrine and Paracrine Mediator. *Biochim. Biophys. Acta* **2002,** *1582*, 270–281.

Xu, X. Enzymatic Production of Structured Lipids: Process Reactions and Acyl Migration. *Inform* **2000,** *11*, 1121–1131.

Xu, Y. Sphingosylphosphorylcholine and Lysophosphatidylcholine: G Protein-coupled Receptors and Receptor-mediated Signal Transduction. *Biochim. Biophys. Acta* **2002,** *1582*, 81–88.

Xu, Y.; Shen. Z.; Wiper, D. W.; Wu, M.; Morton, R. E.; Elson, P.; Kennedy, A. W.; Belinson, J.; Markman, M.; Casey, G. Lysophosphatidic Acid as a Potential Biomarker for Ovarian and Other Gynecological Cancers. *J. Amer. Med. Assoc.* **1998,** *280*, 719–723.

Yamashita, A.; Sugiura, T.; Waku, K. Acyltransferases and Transacylases Involved in Fatty Acid Remodeling of Phospholipids and Metabolism of Bioactive Lipids in Mammalian Cells. *J. Biochem.* **1997,** *122*, 1–16.

Yang, L.; Yatomi, Y.; Miura, Y.; Satoh, K.; Ozaki, Y. Metabolism and Functional Effects of Sphingolipids in Blood Cells. *Br. J. Haematol.* **1999,** *107*, 282–293.

Yatomi, Y.; Ohmori, T.; Rile, G.; Kazama, F.; Okamoto, H.; Sano, T.; Satoh, K.; Kume, S.; Tigyi, G.; Igarashi, Y.; Ozaki, Y. Sphingosine 1-Phosphate as a Major Bioactive Lysophospholipid That Is Released from Platelets and Interacts with Endothelial Cells. *Blood*, **2000,** *96*, 3431–3438.

Yatomi, Y.; Ozaki, Y.; Ohmori, T.; Igarashi, Y. Sphingosine 1-Phosphate: Synthesis and Release. *Prostaglandins* **2001,** *64*, 107–122.

Zhang, Y.; Wang, Y.; Wan, Z.; Liu, S.; Cao, Y.; Zeng, Z. Sphingosine Kinase1 and Cancer: A Systemic Review and Meta-analysis. *PLOS One* **2014,** *9*, 1–12.

Zhao, Z.; Xiao, Y.; Elson, P.; Tan, H.; Plummer, S. J.; Berk, M.; Aung, P. P.; Lavery, I. C.; Achkar, J. P.; Li, L.; et al. Plasma Lysophosphatidylcholine Levels: Potential Biomarkers for Colorectal Cancer. *J. Clin. Oncol.* **2007,** *25*, 2696–2701.

12

NMR of Polar Lipids

Bernd W.K. Diehl ▪ *Spectral Service AG, Cologne, Germany*

Principles in NMR Spectroscopy

Short History

Nuclear magnetic resonance (NMR) spectroscopy has been developed to be the most powerful analytical method. It allows the visualization of single atoms and molecules in various media in solution as well as in solid state. NMR is nondestructive and gives a molar response that allows structure elucidation and quantification simultaneously. Magnetic interactions between NMR active nuclei along covalent bindings result in spin-spin ^{n}J-couplings. Through-space interactions can be detected using the Nuclear Overhauser Effect (NOE). Both interactions enable three-dimensional structure elucidation.

The steady progress of NMR spectroscopy can clearly be seen in the list of Nobel Prize winners. In 1944, the first Nobel Prize in physics was awarded to Isidor Issac Rabi for the development of a resonance method that enabled recording the magnetic properties of atomic nuclei. Felix Bloch and Edward Mills Purcell received the Nobel Prize in 1952 for the first practical NMR experiments that both independently operated in 1945 at different places. By then NMR spectroscopy had become more than a physical experiment. Through the discovery of the "chemical shift," the method became a tool for chemists in structure elucidation. The first useful NMR spectrometers were continuous wave (CW) instruments using permanent or electromagnets. Superconductor magnets were the next generation of NMR equipment since the 1970. However, only when Richard R. Ernst in the mid-1960s developed the basics of the Fourier transformation (FT) method was the foundation of the modern NMR spectroscopy methods was laid. Because NMR spectroscopy was by then a domain of physicians, Ernst was the first chemist on the list of Nobel Prize winners in 1991. One decade later, Kurt Wüthrich was the second honored chemist. He received the Nobel Prize in 2002 for the elucidation of three-dimensional structures of macromolecules. The NMR technique has become an important tool in other scientific fields, especially in medicine. It is not surprising that only one year later in 2003, the NMR technique was honored again, and the Nobel Prizes were allocated to Paul C. Lauterbach and Peter Mansfield for their research in magnetic resonance imaging. Rightly, the NMR community expects further prizes in one of the widespread application areas of NMR spectroscopy in the future (Holzgrabe et al., 1999).

Introduction

Since the development of the high-resolution NMR spectrometer in the 1950s, NMR spectra have been a major tool for the study of both newly synthesized and natural products isolated from plants, bacteria, and so forth. In the 1980s, a second revolution occurred. The introduction of reliable superconducting magnets combined with newly developed, highly sophisticated pulse techniques, and the associated Fourier transformation, provided the chemist with a method suitable for determining the three-dimensional structure of very large molecules (e.g., biomacromolecules such as neutral and polar lipids in complex matrices like lecithin).

Because drugs in clinical use are synthetic or natural products, NMR spectroscopy has been mainly used for the elucidation and confirmation of structures. For the last decade, cryogenic probe technology has opened a wide door for routine analysis due to an increase of the signal to noise ratio by a factor of 5. NMR methods now have been introduced to quantitative analysis in any composition of lipids.

Multinuclei Spectroscopy and Heteronuclear Coupling

Organic compounds are composed basically of the elements hydrogen, carbon, phosphorus, nitrogen, and oxygen. In addition, there are the halogens—fluorine, chlorine, bromine, and iodine—and sometimes metal atoms. Each of these elements has an isotopic nucleus that can be detected by the NMR experiment. The low natural abundance of ^{15}N and ^{17}O in nature prevents NMR from being routinely applied to these elements without the use of labeled substances, but ^{1}H, ^{13}C, ^{19}F, and ^{31}P NMR spectroscopy are daily routine work. Many instruments are equipped with broadband (BB) or quattro nucleus probes (QNPs) for sequential NMR analysis of ^{1}H, ^{13}C, ^{31}P, and ^{19}F (or ^{15}N), without the hardware having to be changed. Modern NMR spectrometers are available up to field strengths of 18.8 Tesla or a proton resonance frequency of 800 MHz. Routine analyses are made at proton frequencies between 300 and 600 MHz.

The spin-spin coupling is not restricted to protons (homonuclear couplings). All NMR-active isotopes show spin-spin couplings among each other. The type and the intensity depend on the natural abundance and the spin quantum number. A single nucleus with $I = ½$ couples another signal to a duplet. The 100% natural abundance of ^{31}P and ^{19}F allows the handling of its coupling in equivalence to an additional proton. Each phosphorous or fluorine causes a duplication of the homonuclear multiplets. All other important atoms in organic molecules have lower amounts of NMR active nuclei, especially carbon, silicon, and nitrogen.

The majority of the inactive nuclei (e.g., ^{12}C and ^{28}Si) show no magnetic interaction, or, as in case of ^{14}N, the coupling mostly is not observable due to its high quadrupole moment. Exceptions are highly symmetrical ammonium compounds as shown in the ^{13}C NMR signal of glycerol phosphocholine (GPC), which shows two couplings with ^{14}N (1:1:1 $^{2}J_{C,N}$ triplets due to I = 1) and four with ^{31}P ($^{2}J_{C,P}$ and $^{3}J_{C,P}$ doublets due to I = ½) (see Figure 12.1).

The principles of homo- and heteronuclear couplings can be transferred to all NMR spectra of lipids (especially phospholipids [PL]) and are the basis for several two-dimensional NMR techniques.

Principles of Quantitative NMR

The NMR experiment makes possible the direct observation of atoms. The relative area (integral value) of an NMR signal is strictly linearly proportional to the amount of atoms in the probe volume. The signals are a measure of molar ratios of molecules, independent of the molecular weight. There are no response factors such as those in UV detection caused by varying extinctions dependent on molecular structures; nonlinear calibration curves such as those found with light-scattering detectors are unknown to NMR spectroscopy. The NMR experiment in principle counts spins; the unit is mol or mol/L. Analytical methods that directly detect SI units as their measurement variable are so-called primary methods.

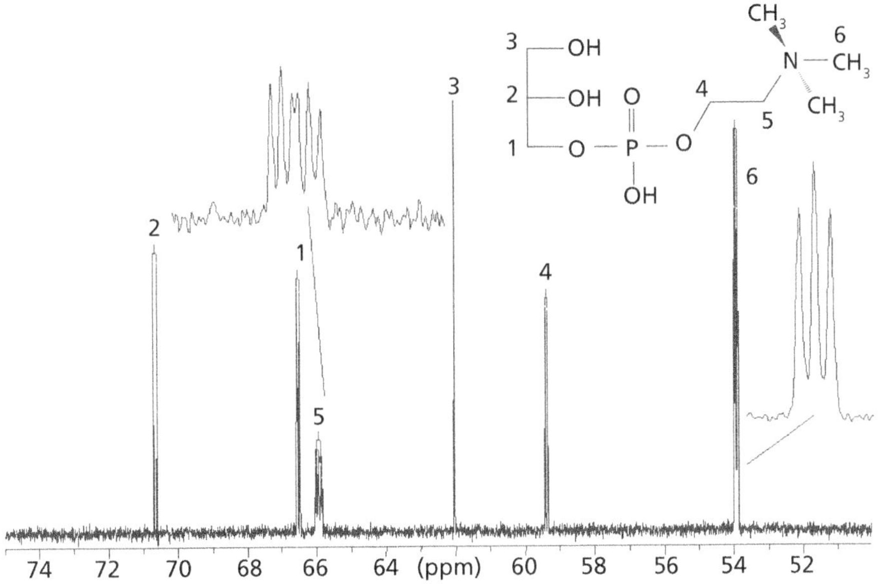

Figure 12.1 ^{13}C NMR spectrum of glycerophosphocholine (GPC) in D_2O containing ^{31}P and ^{14}N couplings.

Other existing primary methods are, for example, weighing or titration; the units there are gram or MOL, respectively. The instrument for weighing is a balance, which must not be validated. However, the physics behind the measurement in primary methods must be calibrated.

The calibration of a balance is done by defined balance weights mostly certified by government organizations. The calibration of a quantitative NMR measurement is done using a defined chemical substance with certificates from, for example, the National Institute of Standards (NIST) or Federal Institute for Materials and Testing (Bundesanstalt für Materialforschung und prüfung, BAM) (Holzgrabe et al., 2008).

Regardless of the absolute information, some uncertainties must be considered, when a primary method is used in practice. A balance must not be used under water, in a vacuum, at irregular deep or high temperatures, or during vibrancies or electrostatic irregular conditions. The correct site and use of a balance as well as the correct site and use of a NMR spectrometer are based on the compliance with good laboratory practice.

When the site and the external conditions are fulfilled, the qNMR is only influenced by the following items:

1. The volume of the NMR tube inside the magnetic coil for the analysis of the test item must be identical with that of the reference standard.
2. The temperature and magnetic susceptibility should be in the same narrow range for the analysis of the test item and the reference standard.
3. The spins must be totally relaxed after each transient.
4. The test item and the reference standard must be fully dissolved in the solvent.
5. The signal and noise must be in a proper ratio.
6. The magnetic field must be homogeneous for both measurements.
7. The matching and tuning of the probe must be in the same order.

The first two points can easily be overcome by using the internal standard method. In this case, which is the most popular qNMR approach, the test item and the reference standard have to be exactly weighted and dissolved in one solution. The uncertainty of sample probe volume, temperature, magnetic susceptibility, even field homogeneity and the matching and tuning of the probe do not influence the result, and, therefore, can be neglected. However, the internal standard method needs some other requirements to be fulfilled; these four golden rules must be kept:

1. No chemical reaction must occur between the solvent or the test item and the internal reference standard.
2. At least one signal of the test item and the internal standard must not interfere (more than one is better) with each other or with the solvent or system signals (water, impurities, etc.). In other words, selectivity must be given.

3. The signal and noise must be in a proper ratio.
4. The spins must be totally relaxed after each transient.

Fulfilling these four principles, qNMR is a valid method by physical and mathematical principles.

NMR analysis is linear between LOQ and the upper level, which represents the solubility of the test item in most cases. A linear regression without intercept must be provided. If an intercept is observed, this is a clear violation of the previous four rules.

NMR is a nondestructive method. The sample and the detector system don't have contact, and a contamination of the probe is impossible. Due to the use of an internal standard, the reanalysis of a sample (comparable to a multiple injection in chromatography) must show identical results; however, a second analysis of the same sample after a certain time can document that no chemical degradation or reaction has occurred. This is a recommended robustness test; results from physics and mathematics are valid by nature but not the chemistry!

Classical reproducibility tests by independent multiple sample preparation are used in analytical chemistry to validate the whole system that includes the sample preparation, the balance, the capability of the operator, and the data evaluation. Results of several validation studies and round-robin tests are shown later in this chapter.

Any NMR analysis of a system of lipids is a multicomponent analysis, even for purified components like triacylglycerol (TAG) or single isolated PLs, due to the natural individual fatty acid (FA) composition. Because NMR is an absolute molar method, it is possible that within one analysis each chemical compound in a mixture can be quantified as long as the four principles are fulfilled. This is valid for the intramolecular FA distribution as well as for mixtures of lipid classes. In the case in which the whole method, including sample preparation for one compound, was shown to be valid, this fact can be applied to any other compound in the mixture. No reference standard is necessary because NMR does not discriminate spins from different molecules. Specific byproducts and solvents can be quantified in one step without any additional validation or even calibration.

NMR Spectroscopy of Complex Lecithin of Several Sources

The most powerful method in PL analysis is ^{31}P NMR spectroscopy (Dennis and Plückthun, 1984; Diehl, 1998; Holzgrabe et al., 2008; Kühl, 2008; Meneses, 1988). The 100% natural amount of ^{31}P NMR in nature, the selectivity, and the sensitivity make it the reference method of the International Lecithin and Phospholipid Society (ILPS). ^{31}P NMR is an ab initio method, so only artificial standards (e.g., triphenylphosphate [TPP] or distearoylphosphatidylglycerol [DSPG]) are used to quantify all PLs from complex mixtures of any origin.

Many high performance liquid chromatography (HPLC) methods for PLs have been developed, but chromatographic resolution and dynamics of detection are not always satisfactory. For each source of PLs, special standards are needed due to the different distributions of fatty acids. These standards are expensive and in some cases are not available. Another problem is represented by the analysis of PLs in complex matrices. In many cases, separation is impossible or very difficult, often due to the surface activity, which is desired in the application of PLs but which complicates the analysis of these compounds. Therefore, a method is needed that is selective in the detection of PLs in order to avoid a separation from the matrix. The ^{31}P NMR spectroscopy of PLs meets these requirements. The ILPS has chosen the ^{31}P NMR method as the reference method. It has been tested worldwide by round-robin tests in comparison to various HPLC and thin-layer chromatography (TLC) methods.

The ^{31}P NMR spectroscopy differentiates between the various PLs and has high dynamics in quantification. Only phosphorus-containing substances are detected by this method; the analysis is not disturbed by other nonphosphorus components. The resonance frequency of phosphorus depends on the chemical environment within the molecule. PLs with different chemical structures are therefore recorded at distinct frequencies. Frequency can be measured very precisely, with even small differences in the chemical structure of PLs are easily detected. Each PL is represented by a single signal. Separation of the various PLs is not necessary because the PLs are characterized by their different resonance frequencies (Diehl, 2001).

Different Lecithin Sources

For historical reasons, the investigation of soy lecithin and modified soy lecithin (such as hydrolyzed, hydrogenated, acetylated, and fractionated) become the first interesting items of NMR analysis.

Typical cases are chemically or enzymatically treated lecithin (hydrolyzed, acetylated, or hydrogenated) as well as other lecithin from animal sources (egg, milk, or fish). The universality of the ^{31}P NMR approach can be demonstrated by comparison of hydrolyzed lecithin from soybean. ^{31}P NMR distinguishes between different types of lysolecithin, so 1-lysophosphatidylcholine (1-LPC) and 2-lysophosphatidylcholine (2-LPC) show different signals. This enables us to see what type of phospholipase (A1 or A2) was used. Figure 12.2 shows a ^{31}P NMR spectrum of soy lecithin, and Figure 12.3 shows a hydrolyzed type.

The selective quantification of the phosphatidylcholine (PC) degradation products 1-LPC, 2-LPC, and GPC makes ^{31}P NMR spectroscopy the ideal method for

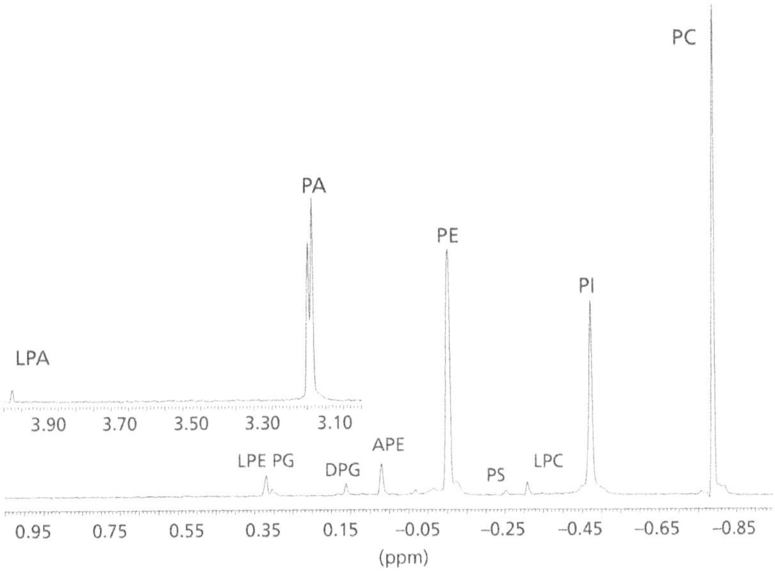

Figure 12.2 ^{31}P NMR of soybean lecithin. (L)PA: (lyso)phosphatidylic acid; PG: diacyl-phosphatiylglycerol; (L)PE: (lyso)phosphatidylethanolamine; APE: acylphosphatidylethanolamine; PS: phosphatidylserine; PI: phosphatidylinositol; (L)PC: (lyso)phosphatidylcholine.

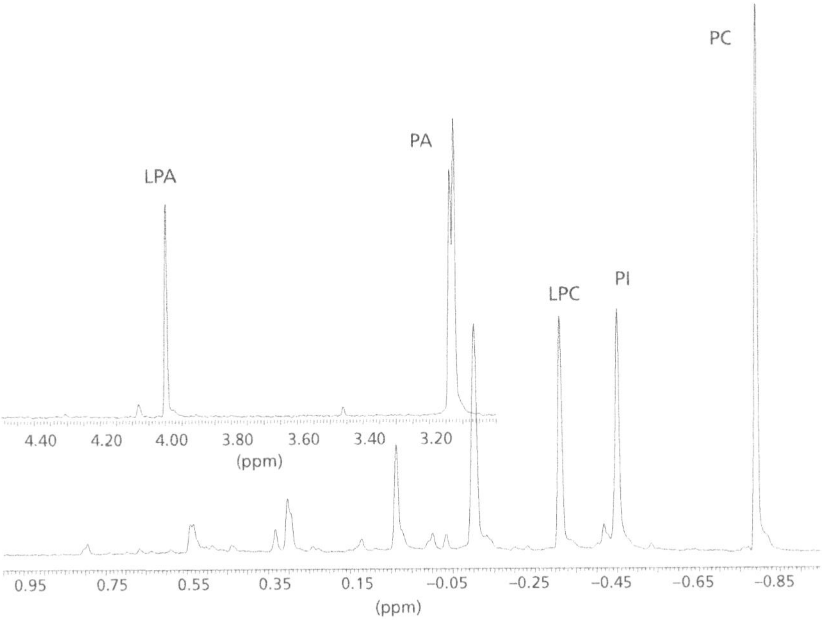

Figure 12.3 ^{31}P NMR of hydrolyzed soybean lecithin.

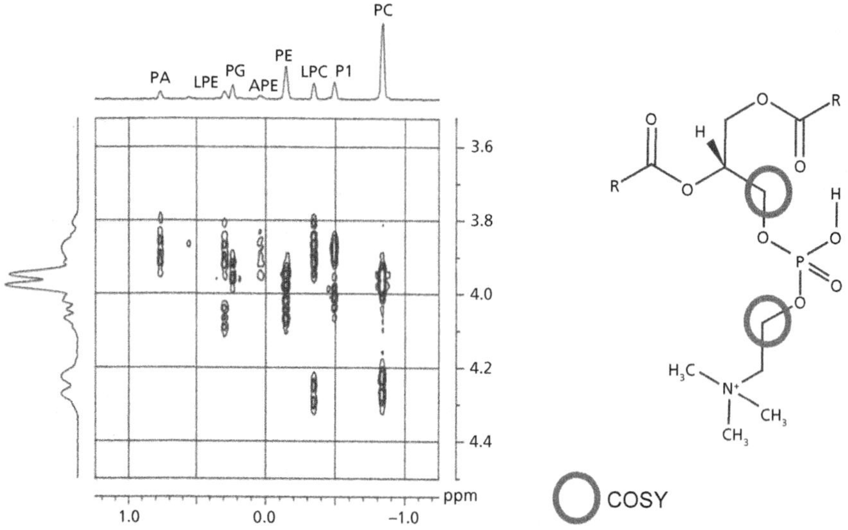

Figure 12.4 ^{31}P,H-COSY spectrum of native soybean lecithin.

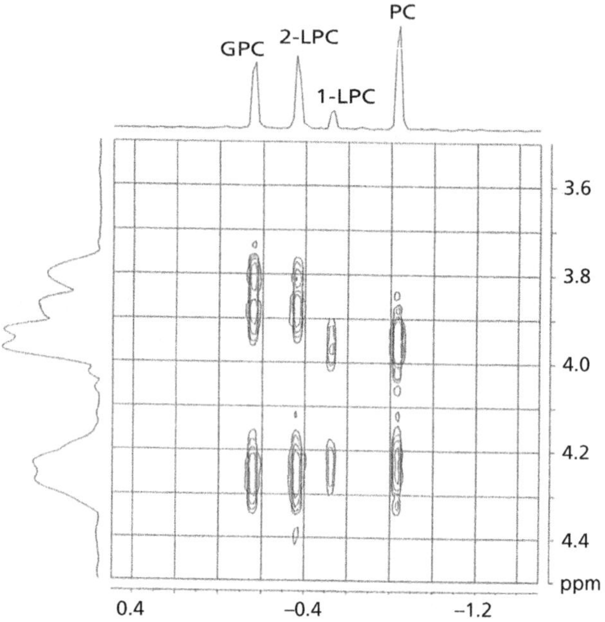

Figure 12.5 ^{31}P,H-COSY spectrum of PC and its degradation products, 1-LPC, 2-LPC, and glycerophosphatidylcholine (GPC).

testing the stability of liposome-containing formulations in pharmaceutical products. The amount of the different PLs can be calculated from their integral areas.

P,H-COSY and H,P-TOCSY

Two-dimensional techniques, especially the heteronuclear methods P,H-COSY (P-detected P,H correlation spectroscopy) and H,P-TOXY (H-detected H,P total correlation spectroscopy), enable the signal assignment of complex mixtures (Braun et al., 1998). A typical P,H-COSY of native soybean lecithin is given in Figure 12.4. The cross peaks in the second dimension indicate the alkyl-ester groups connected to the phosphorous, the CH_2 group *sn*-1 position of the glycerol, and the first esterified group of the polar head group. The chemical shift of the glycerol cross peaks distinguishes between PL and its 1- or 2-lysoderivatives. Even double LPCs like GPC are separated by ^{31}P NMR chemical shift and the characteristic cross peaks in the P,H-COSY (see Figure 12.5).

The proton-detected type of H,P-TOCSY (see Figure 12.6) enables the detection of the whole proton coupling system of the phosphor ester groups, as well as the COSY cross peaks, the signals of the *sn*-2 methine carbons, and the corresponding methylene protons of the polar head groups.

Glycolipids, Sphingolipids

Lecithin from animal sources such as dairy or egg represents a specific mixture of neutral lipids, polar lipids, sterols, proteins, and water. The polar lipids can be distinguished between PLs and glycolipids where sphingolipids play a specific role. The phospholipid profile is monitored by ^{31}P NMR spectroscopy; the FA composition can be detected by ^{1}H NMR and ^{13}C NMR (Diehl and Ockels, 1995a).

Beside sterols, glycolipids can also be analyzed by ^{13}C NMR spectroscopy. Significant signals of carbohydrates can be found in the region between $\delta = 95$ and 110 ppm. Figure 12.7 shows ^{13}C NMR spectra of respective fractions from soy lecithin. In some instances, with high concentrations of glycolipids, a two-dimensional heteronuclear multiple-quantum correlation spectroscopy (HMQC) spectrum is also useful (cf. Figure 12.8, p. 402). Therefore, as in the case of soy lecithin, the identity of sitosterylglycosid can be determined directly from the two-dimensional cross peaks.

Even more complex glycolipids, such as ceramides, cerebrosides, gangliosides, and spingosine, also give characteristic signals in ^{1}H NMR spectra of complex mixtures (e.g., milk) (see Figure 12.9, p. 403). However, a direct quantification is not readily possible due to the small amount and possible signal interferences. It is helpful to fractionate the lipids on silica gel. Therefore, the nonpolar triglyceride fats, such as glycerides, elute with hexane. Further on, fractions of chloroform, chloroform/

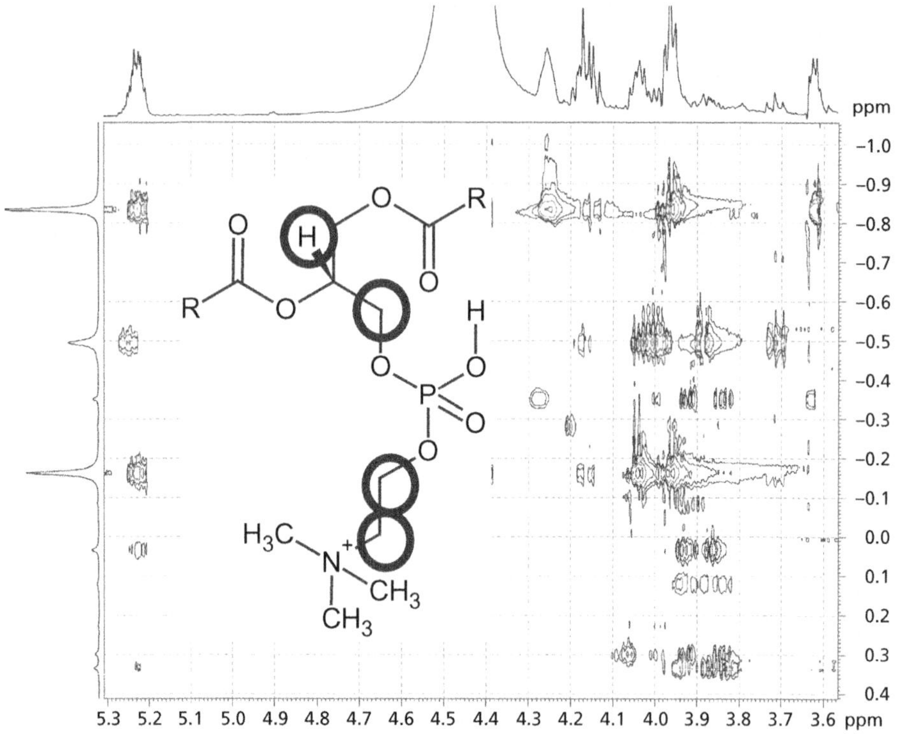

Figure 12.6 $^1H,^{31}P$-TOCSY spectrum of native sunflower lecithin.

acetone (1:1), and chloroform/methanol (1:1) can be obtained, resulting in an elementary preparative separation into neutral lipids, glycolipids, and PLs.

Dairy Products and Milk

The analysis of milk plays a special role here. A simple ^1H NMR spectrum is taken within a few minutes, after 0.5 ml of milk is mixed with 0.3 ml of D_2O and has been filled in the NMR tube. The dominant water signal is eliminated by the pulse sequence water suppression using presaturation (PRESAT) (Figure 12.10, p. 403). The signals of the phospholipids PC and sphingomyelin (SPH) can directly be seen at $\delta = 3.2$ ppm.

After separation of peptides and carbohydrates, the lipid and lecithin fraction remain. The FA composition can be worked out from a ^1H NMR spectrum. Furthermore, the phospholipids PC and SPH can be quantified (cf. Figure 12.11, p. 404).

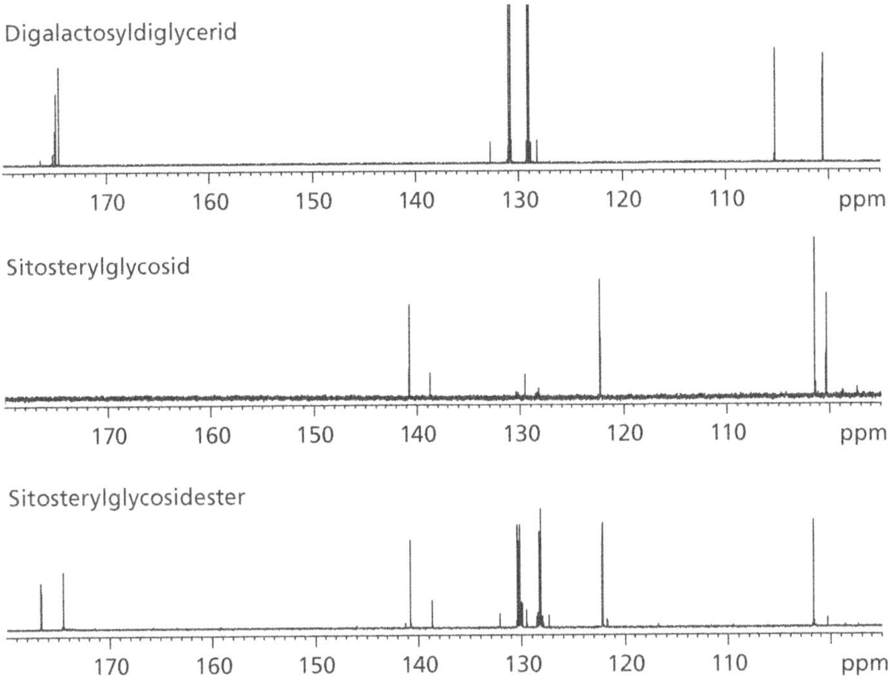

Figure 12.7 ^{13}C NMR spectra of glycolipids from soybean.

Milk contains three specific FAs that are not present in vegetable oils: conjugated linoleic acid (CLA), butyric acid, and caproleic acid, a FA with a terminal double bond. All three FAs are important factors in the assessment of milk fat or milk fat mixtures. The characteristic signals of CLA that represent conjugated double bonds in FA can be seen in Figure 12.12, p. 404. Of course, a quantitative analysis of these three FAs in detail is possible.

CLA and caproleic acid can be determined by using the signal of the double-bond proton. In the ^{1}H NMR spectrum of milk fat from a yogurt, the signal of CH group of triglycerides (TG) was normalized to 100 units. From the three signals of CLA or caproleic acid, molar concentrations of 1.2% and 0.7%, respectively, may be determined. As a byproduct, a relative content of 2.1% 1,2-diglycerides can be obtained.

In the ^{1}H NMR spectrum, butyric acid shows the signal of the terminal methyl group as a triplet at δ = 0.93 ppm (Figure 12.13, right, p. 405). But also in the ^{13}C NMR spectrum, butyric acid is well-determined. In the area of the carbonyl groups, even its region-selective position in *sn*-1–3 can be determined (Figure 12.13, left, p. 405).

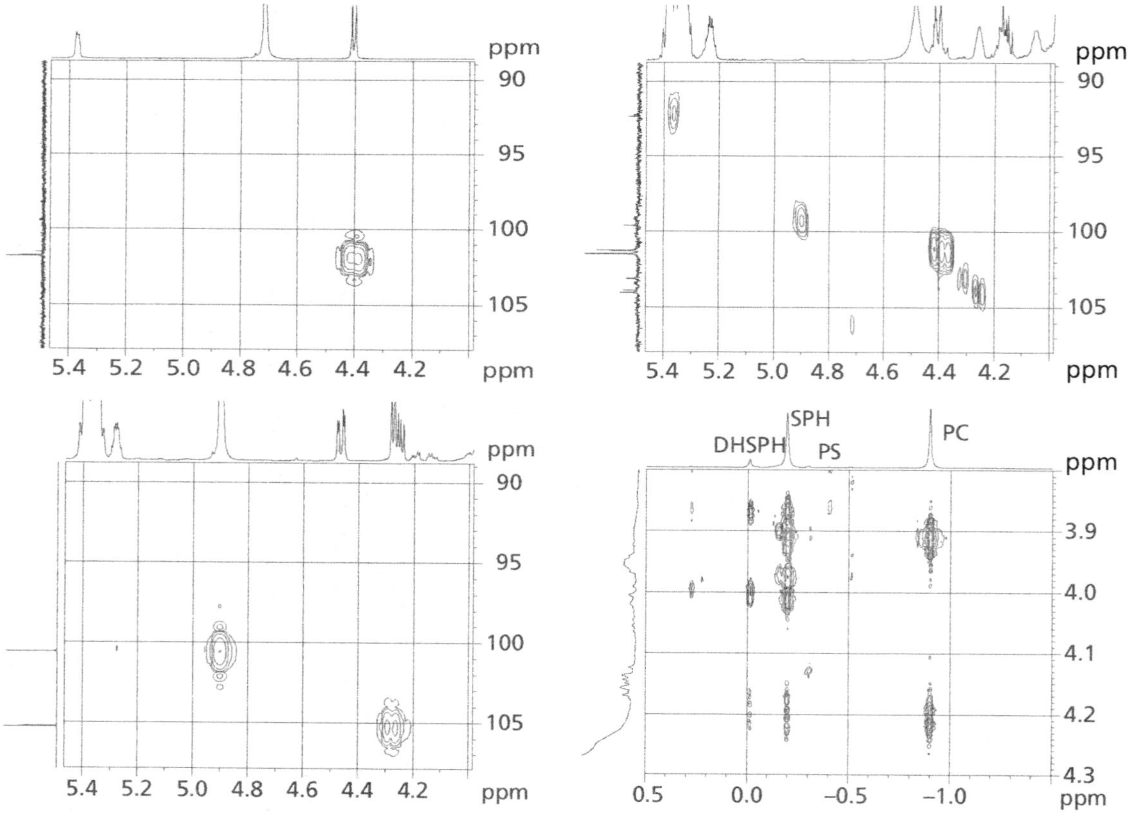

Figure 12.8 HMQC spectra of soy lecithin (top right), sitosteryl glycoside (top left), digalactosyl glyceride (bottom left), and a P,H-COSY of polar milk lipids (bottom right).

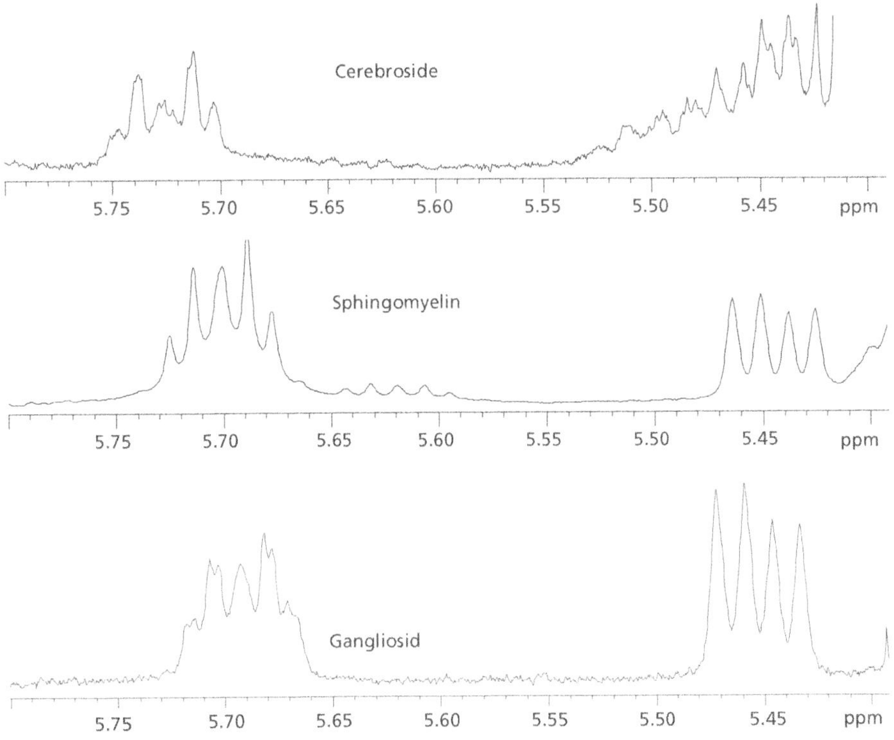

Figure 12.9 ^{1}H NMR spectra detail of double bonds from sphingo lipids.

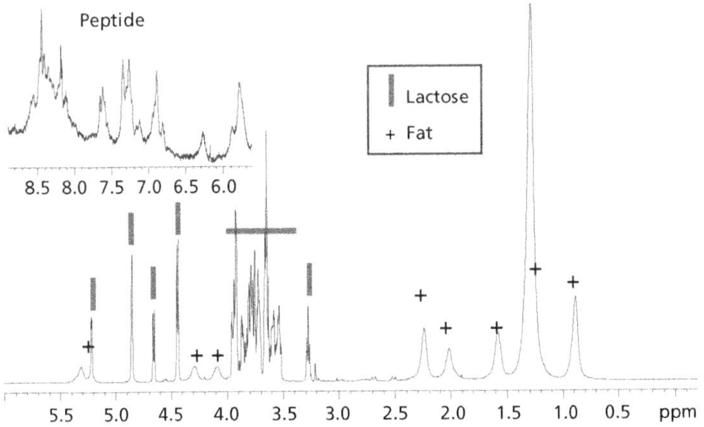

Figure 12.10 ^{1}H NMR spectrum of dairy milk and water suppression.

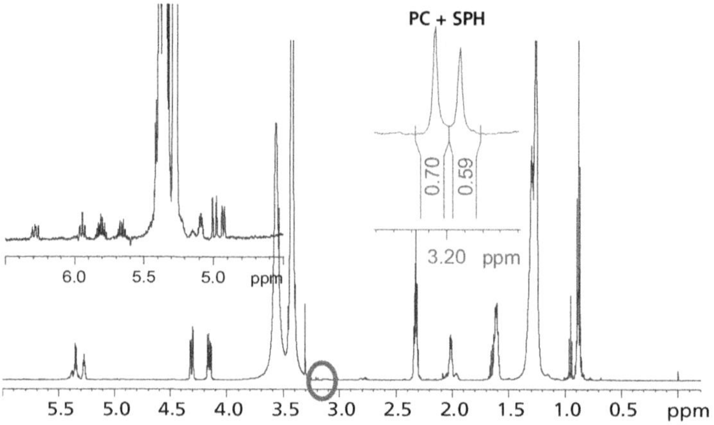

Figure 12.11 ^{1}H NMR spectrum of the lipid fraction of dairy milk.

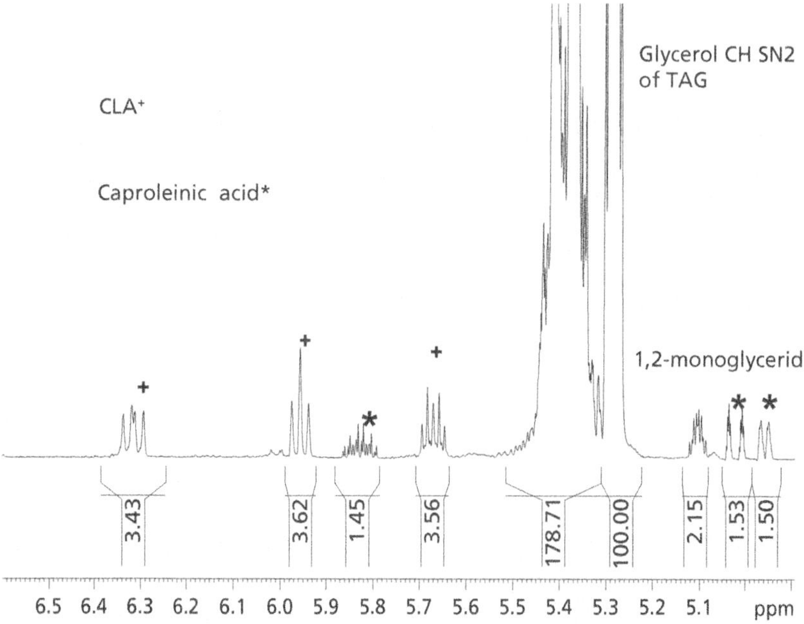

Figure 12.12 Double bonds in a ^{1}H NMR spectrum of milk fat.

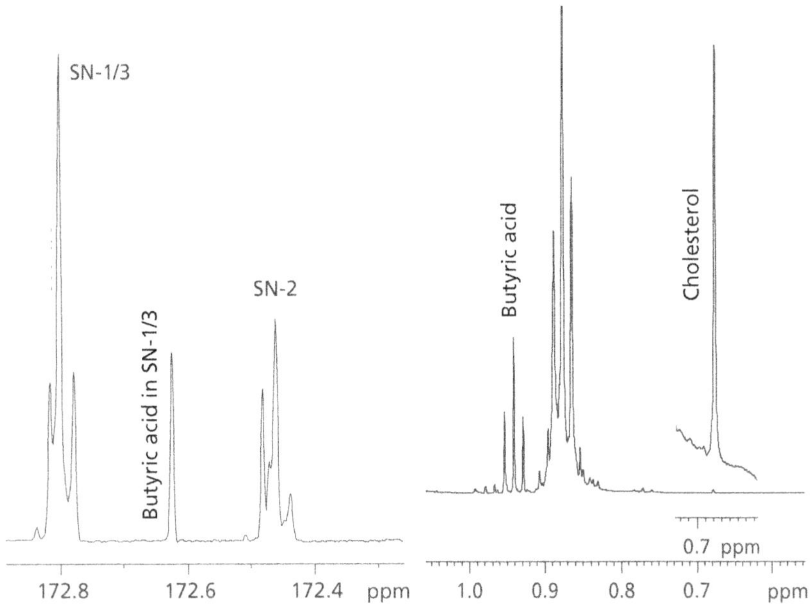

Figure 12.13 Milk fat, ^{13}C NMR spectrum detail of carbonyl atoms (left) and ^{1}H NMR spectrum detail of methyl groups (right).

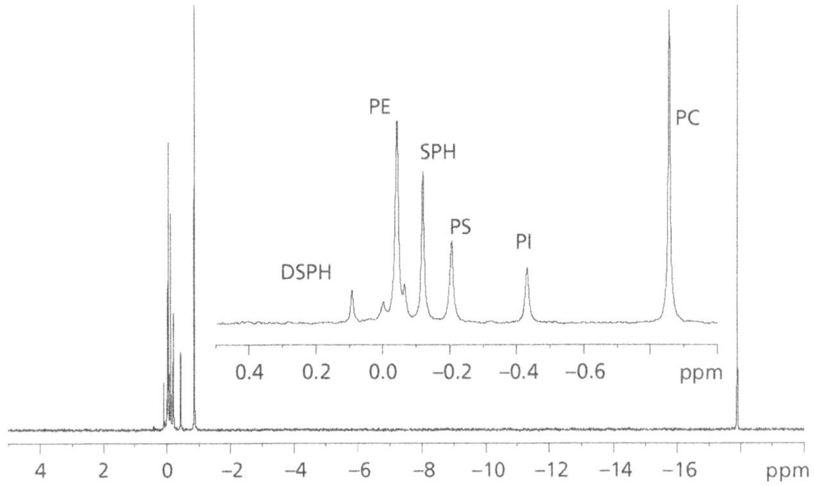

Figure 12.14 ^{31}P NMR spectrum of milk lecithin, internal standard TPP at $\delta = -17.8$ ppm.

Figure 12.14 shows the ^{31}P NMR spectrum used for the assay of milk lecithin. The internal standard is TPP at δ = −17.8 ppm. The analysis of PLs by ^{31}P NMR is reported separately.

Other Sources

NMR also is a modern tool in monitoring human plasma studies (Tukiainen et al., 2008; Tynkkynen, 2012). The combination of ^{1}H and ^{31}P NMR spectroscopy enables the quantification of PLs in blood plasma (see Figure 12.15) and red blood cells, as well as the amounts of triacylglycerol and total cholesterol. Total cholesterol can be detected by ^{1}H NMR using the methyl resonance at δ = 0.65 ppm, and the distinction of cholesterol and cholesterol esters easily can be performed using the methyl resonances at δ = 1.0 ppm (see Figure 12.16).

In ^{1}H NMR spectra the glycerol methine protons of triacylglycerol at δ = 5.27 and PLs at δ = 5.22 ppm are also well separated. The integral values directly result in the molar ratio of both important compounds. A detection of PC and other choline species as sphingomyelin (SPH) can be detected simultaneously at δ = 3.2 ppm (see Figure 12.17). The method was used in combination with the classical gas chromatographic (GC) analysis.

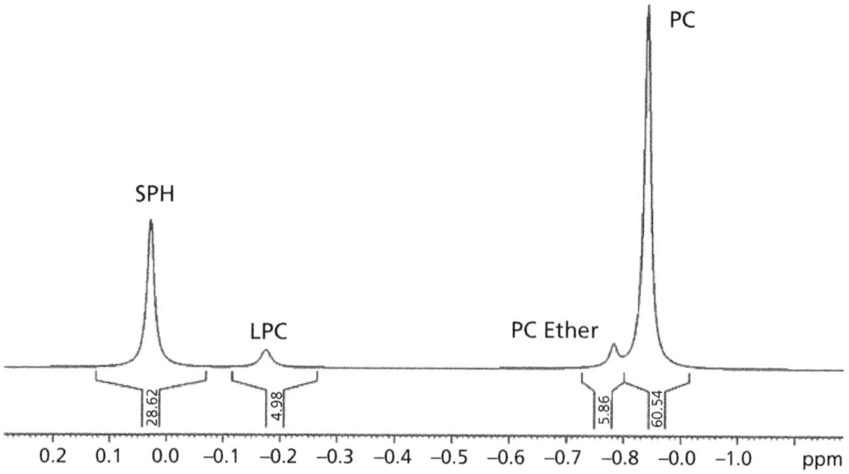

Figure 12.15 ^{31}P NMR of human blood plasma.

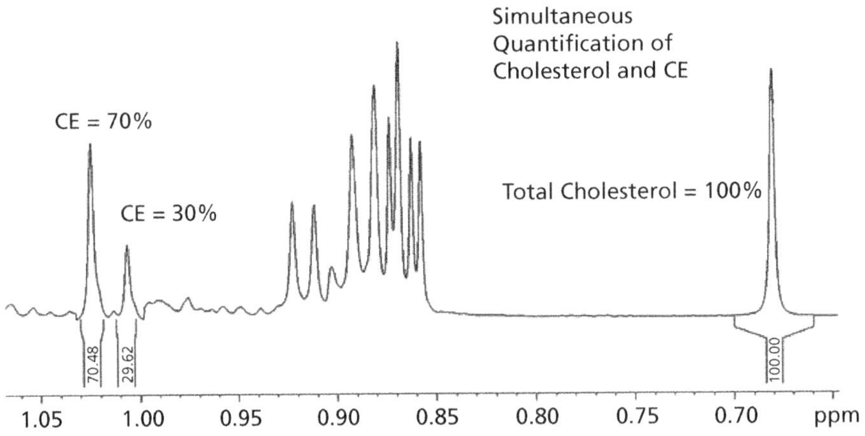

Figure 12.16 ^{1}H NMR of human blood plasma, details.

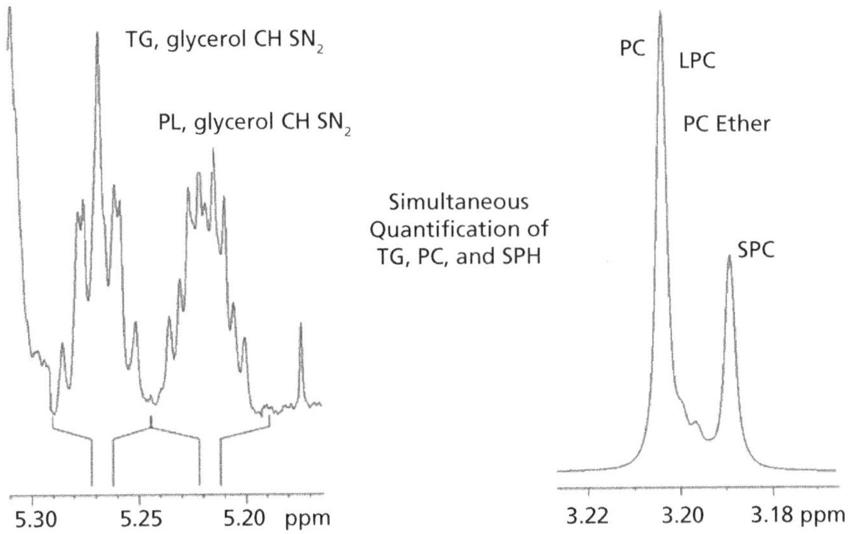

Figure 12.17 ^{1}H NMR of human blood plasma, cholesterol region.

Ammonium Phosphatides

The ^{31}P NMR also enables the analysis of ammonium phosphatide E 422—a reaction product of phosphoric acid and vegetable oil—which is used as an artificial emulsifiers, especially in the chocolate industry. The overview ^{31}P NMR spectrum (see Figure 12.18) shows several functional groups of mono- and di-esters of the phosphate, as well as cyclic LPCs and pyrophosphate species.

The list of possible applications of qNMR is long and cannot be shown in detail in this chapter. Therefore, some specific items will be discussed in more detail; the analysis of enzymatically prepared phosphatidylserine (PS) and its formulations as well as the complex marine-sourced PLs of krill oil will be shown in the section "Marine Polar Lipids."

Phoshotidylserin, Enzymatic Reactions, and Stability Tests

PS is a natural compound in many lecithin types from animal and plant sources. PS is a major compound in animal brain, but due to problems with Bovine Spongiforme Enzephalopathie (BSE), the use of PS of animal sources was unwarranted. Alterna-

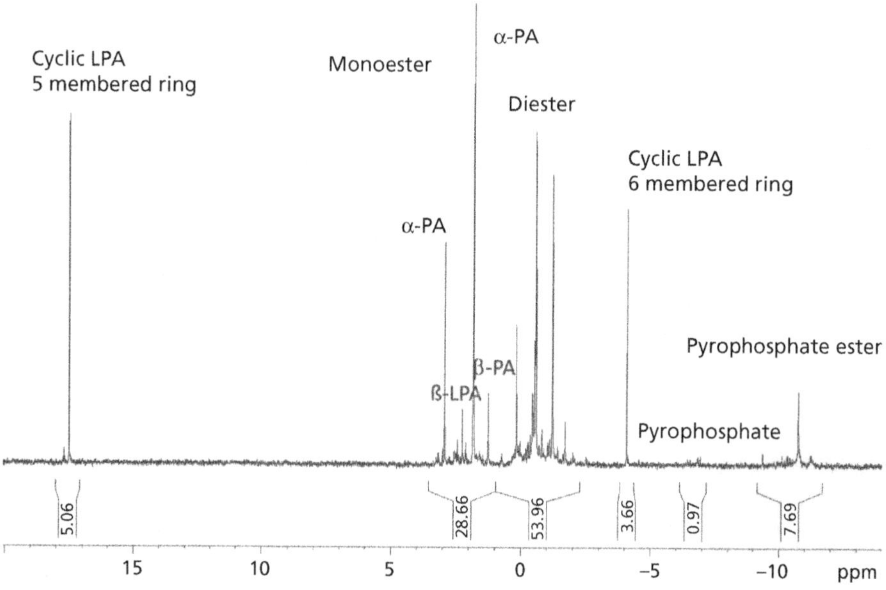

Figure 12.18 ^{13}P NMR of ammonium phosphatide E422.

tively, the enzymatic reaction of vegetable lecithin (mostly soy) was used as a basis to produce PS. ^{31}P NMR (see Figure 12.19) became the tool to optimize the synthesis, to detect and quantify the byproducts of the enzymatic reaction, and to monitor the storage stability.

Typical byproducts are produced by the enzymatic reactions with water (for example, hydrolysis can lead to phosphatidic acid [PA] formation) as well as reactions with alcohols.

From some experiments it is proposed that PS is thermo-unstable. We have done a short test series of a PS/PC mixture in chloroform/methanol and water. The mixture was filled in several tubes and sealed by screw caps. All samples were put on a heater that had a temperature of 155 °C. After different reaction times, the mixtures were analyzed by ^{31}P NMR.

The result documents the instability of PS. Although PC only shows small hydrolysis to LPC, the PS was almost totally hydrolyzed to PA. The sum of molar amounts of PC and LPC or PS and PA, and the corresponding lyso-types are constant within the limits of detection.

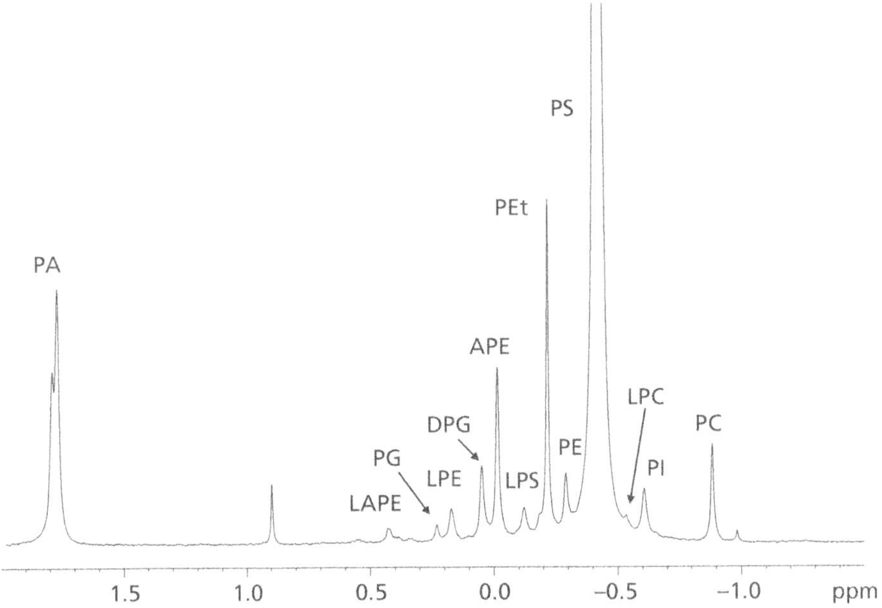

Figure 12.19 ^{31}P NMR spectrum of a PS raw material prepared from soybean lecithin with phospholipase reaction.

Marine Polar Lipids: A Combination of ^{31}P, ^{1}H, and ^{13}C NMR Spectroscopy

Krill oil is extracted from *Euphausia superba*, a small shrimp-like krill species harvested from the wild in the Southern Ocean around Antarctica. The lipid content and profile of krill oil may vary depending on season, species, age, and factors such as storage and processing conditions. But overall, krill oil has a unique composition of lipid classes. Similar to other seafood, the essential FAs, eicosapentaenoic acid (EPA) and docosahexaenoic acid (DHA), are provided in krill oil. The predominant saturated FA in krill oil is palmitic acid, and omega-3 EPA and DHA account for the highest value of polyunsaturated fatty acids (PUFA). Of note are also several other lipophilic substances that can be found in krill oil, such as cholesterol and cholesterol esters (approximately 1.5%) and phytinic esters (metabolites of chlorophyll, 1–2%). Due to the fact that krill's primary diet is based on algae, phytanic acid is metabolized in the TAG part of the lipids; however, phytanic acid is not present in PLs. Astaxanthin esters (strong antioxidants) are found in the form of a mix of di- and mono-esters (approximately 100 mg/kg) as well as homarine (N-methyl picolinic acid). The differences that can be used to distinguish krill oil from other oils include the fact that most of the FAs in fish oil are incorporated into TAG, whereas almost all omega-3 FAs of krill oil are incorporated into PL.

Different methods can be used to analyze and fingerprint krill oil in relation to other oils and to take advantage of the previously mentioned differences. The methods are based on chromatographic, mass spectrometric, or other spectroscopic procedures. Of particular interest for the analysis of krill oil PL and FA are ^{31}P, ^{1}H, and ^{13}C NMR spectroscopy. NMR gives the relative amount and also the distribution of FAs in the different positions of the glycerol backbone in PL and TAG molecules (Medina and Sacchi, 1994; Schiller and Arnold, 2002; Wollenberg, 1990). Moreover, other compounds like cholesterol and free FA can be detected by NMR. The procedure is followed by the U.S. pharmacopeia (USP) convention standards (USP 29 S.2710–2715).

The signal assignment for analysis by ^{31}P NMR was tested by performing two-dimensional ^{31}P NMR spectra and by standard addition of PL from other sources. Thereafter, ^{31}P NMR spectra of krill oil revealed a complex mixture of PL species (Figure 12.20).

Ether instead of ester in the *sn*-2 position accounts for more than 95% of PC, phosphatidylethanolamine (PE), and N-acyl phosphatidyl ethanolamine (APE) (i.e., 10% ether PC of the total sum of PC; 10% ether 2-LPC of the total sum of 2-LPC; 30% ether PE of the total sum of PE; 45% ether APE of the total sum of APE; 30% ether LPE of the total sum of LPE). The signal assignment can be done by two-dimensional P,H-COSY (see Figure 12.21). The relative amount of ether lipids is

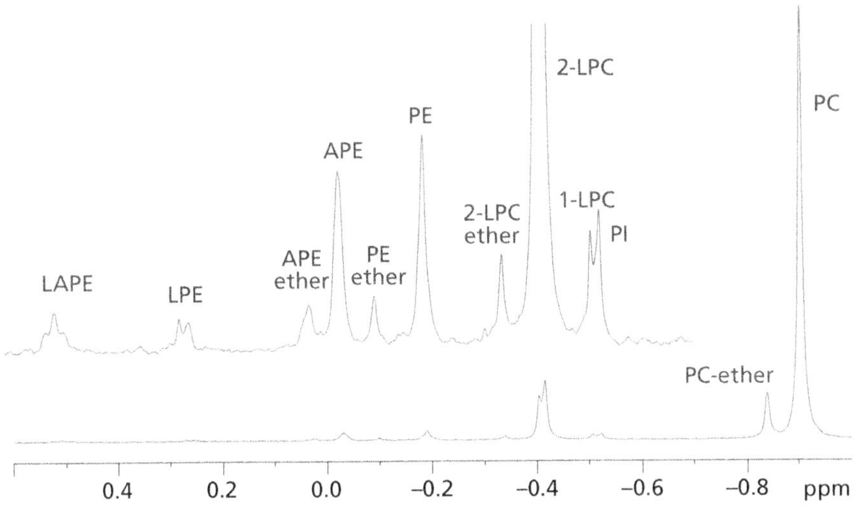

Figure 12.20 ^{31}P NMR of krill oil taken from an USP reference standard.

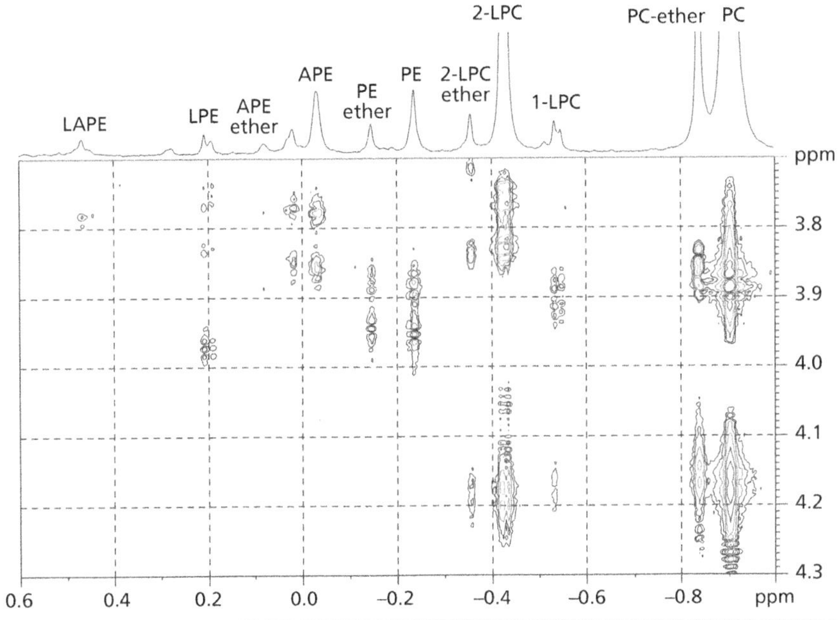

Figure 12.21 P,H-COSY NMR for signal assignment taken from an USP reference standard.

characteristic for the krill sources. In principle, ^{31}P NMR can distinguish between the different types of PL ether and plasmalogens, which are common for PLs from animal sources, or even between different organs within a group of animals depending on its possible function for the metabolism. Figure 12.22 shows the comparison of PC and PE ether ^{31}P NMR signals of bovine heart and lung surfactant.

^{1}H NMR spectroscopy of krill oil is presented in Figure 12.23. PLs and triacylglycerides can be detected separately at the glycerol *sn*-2 proton between $\delta = 5.2$ and 5.3 ppm. The corresponding spectrum of the isolated neutral lipids (mostly TAG), cholesterol, and phytyl esters is presented in Figure 12.24.

A typical and specific minor lipid class found in krill oil consists of phytyl wax esters, which appear at $\delta = 4.6$ ppm as a doublet (see Figure 12.24). Krill uses chlorophyll-containing algae as a primary source of nutrition; therefore, these phytyl wax esters are highly specific and not detected in normal fish oil. The detection of the total amounts of omega-3 FAs is possible by observing the terminal methyl groups of the lipids in ^{1}H NMR spectra (see Figure 12.25).

The asymmetric distribution of omega-3 PUFAs in PL and TAG molecules is demonstrated with the ^{13}C NMR spectra in both isolated PC (see Figure 12.26, p. 416) and TAG fractions (see Figure 12.27, p. 416) with a predominance of omega-3 PUFAs in the PC. Addition of fish oil would increase the omega-3 PUFA content in

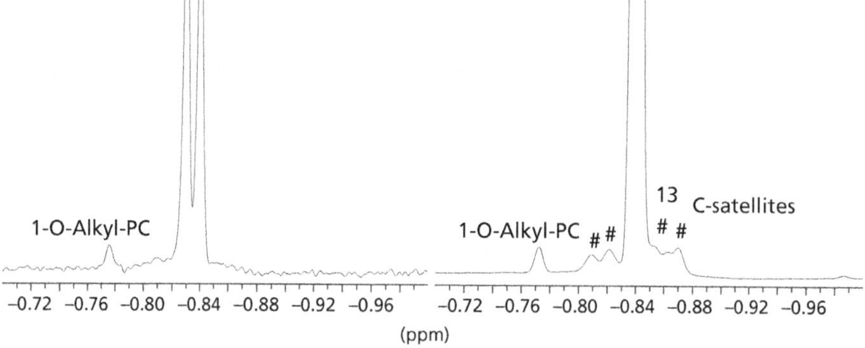

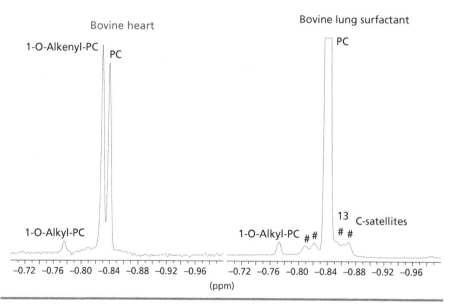

Figure 12.22 Phospholipids and their ether derivatives from different organs of bovine origin.

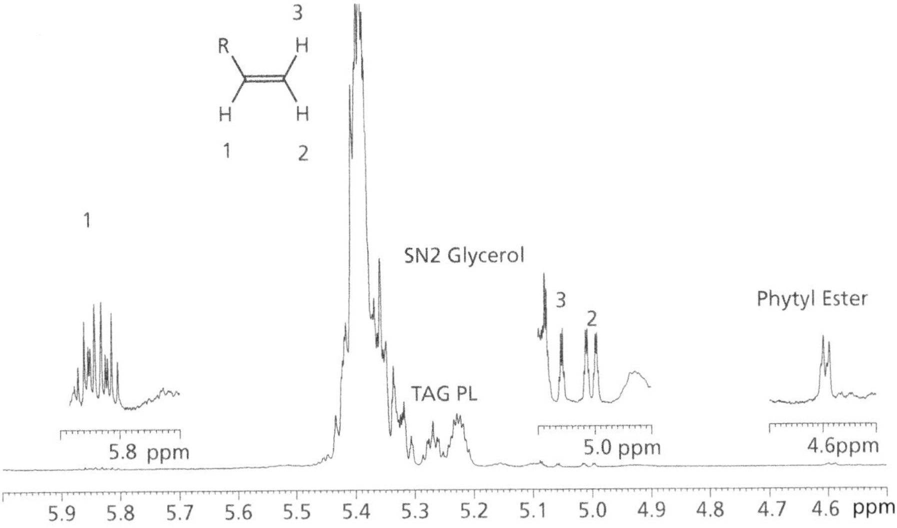

Figure 12.23 ¹H NMR spectrum details of the double-bond region of total krill oil.

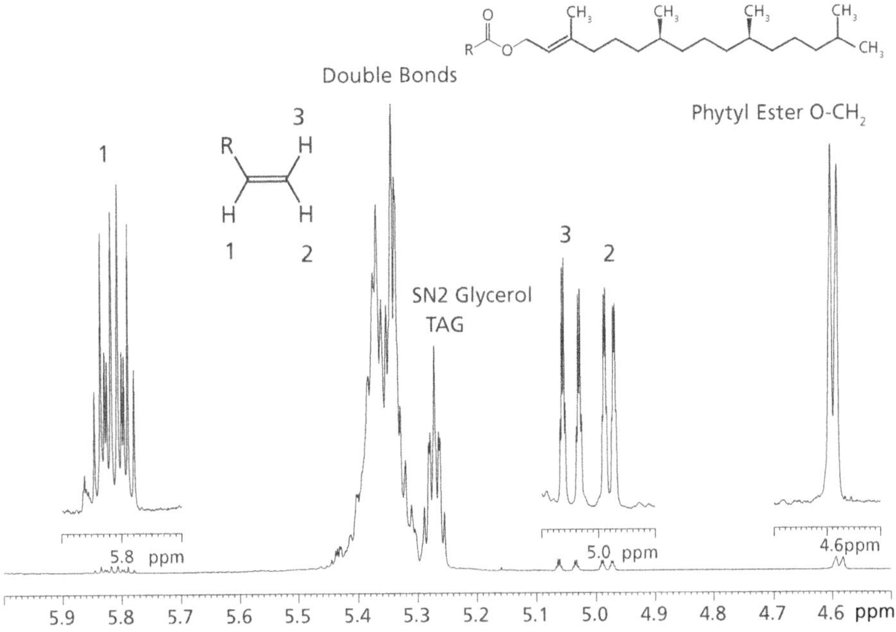

Figure 12.24 ¹H NMR spectrum details of the double-bond region of the neutral lipid fraction of krill oil.

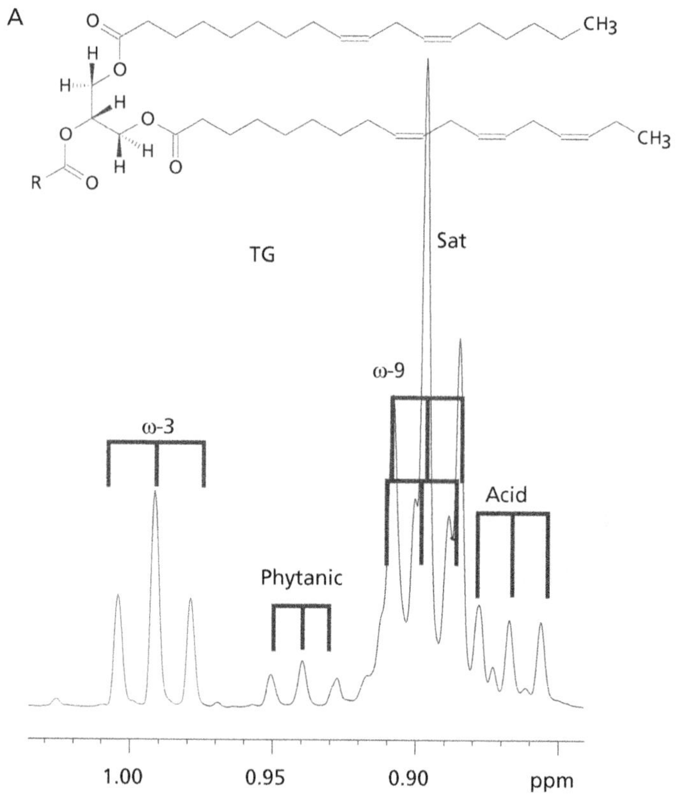

Figure 12.25 Comparison of ^{1}H NMR spectra from krill TAG (A) and PC (B) demonstrating the asymmetric fatty acid distribution in neutral and polar lipids.

the TG fraction, an adulteration that can easily be detected. The ^{13}C NMR spectra of krill oil show low amounts of DHA and EPA in TG, which is specifically found at the signal of *sn*-2 glycerol.

The comparison of the ^{13}C NMR carbonyl regions of PC and TAG from krill demonstrates the ability of the method not only to quantify a FA profile of lipids but also the regioselective asymmetric distribution.

The simultaneous detection of PUFAs in PC and TAG is possible by ^{13}C NMR spectroscopy based on the chemical shift of the carbonyl atom by the position of double bond (Diehl, 1995; Diehl and Ockels, 1995b; Ng, 1985).The analysis of whole lipids by ^{13}C NMR shows the typical asymmetric FA pattern for PC (four doublets at

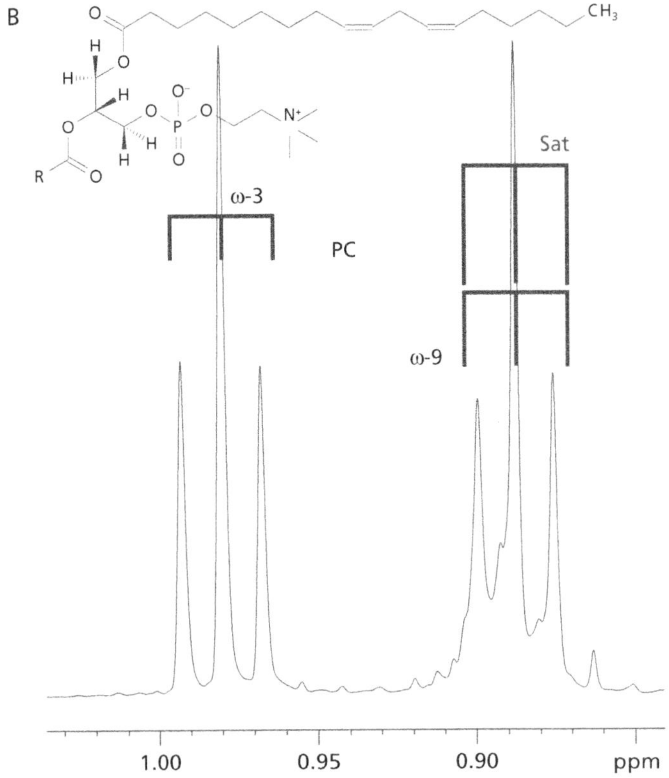

Figure 12.25 *Continued*

δ = 71 ppm) and PC ether (three doublets at δ = 72.5 ppm, a singlet at δ = 72 ppm; (see Figure 12.28) in the glycerol backbone. The amounts of DHA and EPA can be directly calculated from the multiplet at δ = 71 ppm.

^{13}C NMR spectra of the whole krill oil samples also clearly separate the free FAs from esterified FAs. The composition of the FA classes is mainly affected by processing conditions. Smaller amounts of significant compounds were water-soluble homarine and its degradation product methyl pyridinium chloride, as well as inosine (all characteristic for crustaceans), proline as the major free amino acid, and ethanol as a byproduct from processing (see Figure 12.29).

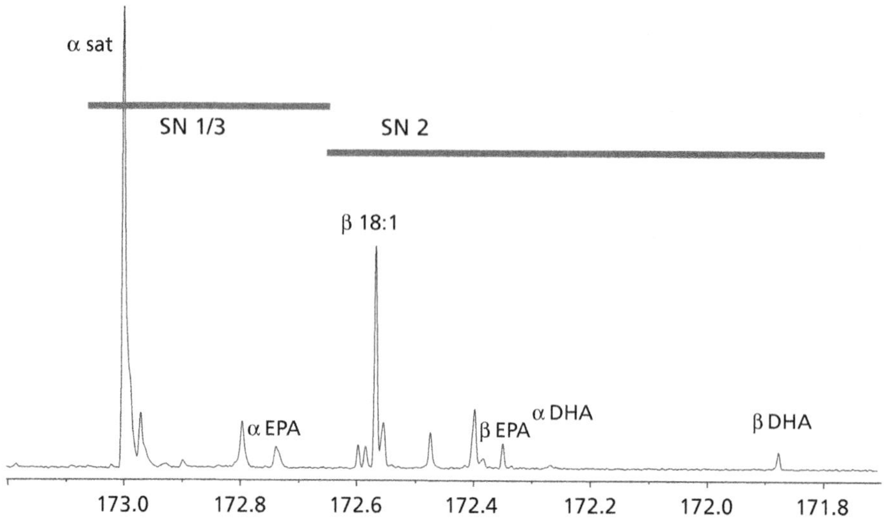

Figure 12.26 ^{13}C NMR spectrum of krill oil, carbonyl C atoms of isolated PC.

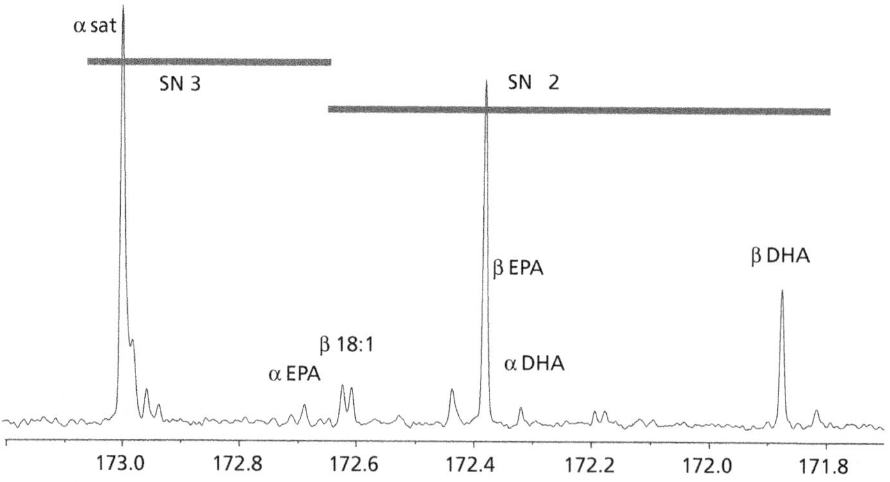

Figure 12.27 ^{13}C NMR spectrum of krill oil, carbonyl C atoms of isolated TAG.

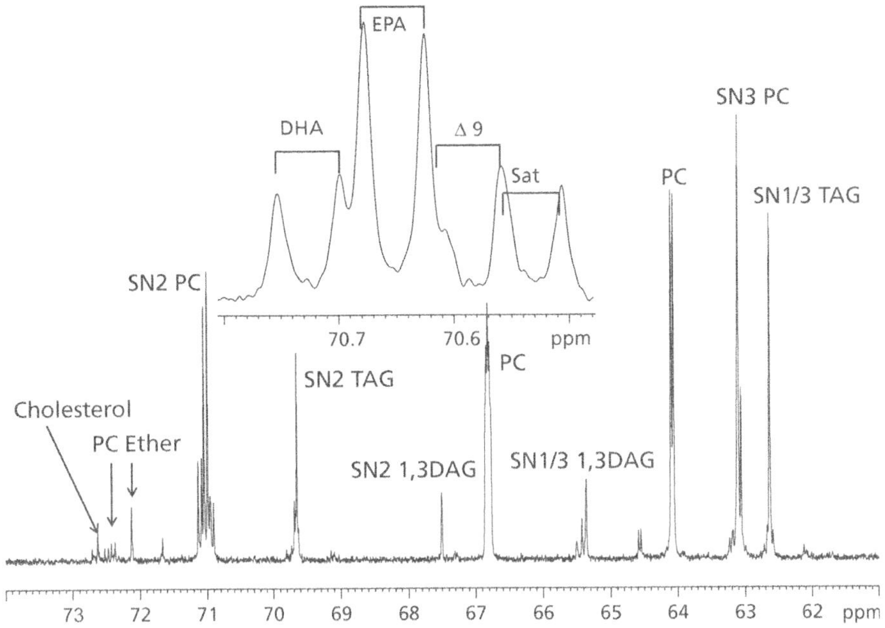

Figure 12.28 ^{13}C NMR spectrum of krill oil, detail of glycerol C atoms.

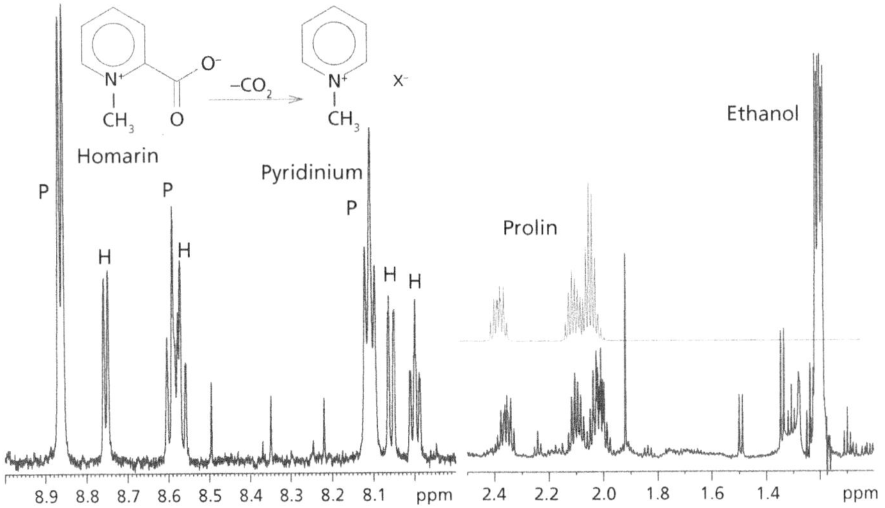

Figure 12.29 ^{1}H NMR spectrum details of the water-soluble part of krill oil.

FA Composition and Stereochemistry

In ultra-high-resolved ^{31}P NMR spectra, the phosphorous signals show splitting depending on the FA distribution of the PL. In soybean lecithin, normally 20% of all FAs are saturated. The distribution of these FAs is not random, whether in different PLs or within the glycerol backbone of a single PL.

Nearly 100% of the saturated acids—mostly palmitinic acid—can be found in the *sn*-1 position, whereas the unsaturated FAs are distributed between the *sn*-1 and *sn*-2 positions. ^{31}P NMR can distinguish between the two types of double unsaturated (U-U) and mixed saturated/unsaturated (S-U) fatty acids. The ^{31}P NMR of phosphatidic acid (PA) shows two separate signals: one is for saturated-unsaturated (S-U) FA composition and the second is accounts for double unsaturated (U-U) at *sn*-1 and *sn*-2 positions (40% S-U and 60% U-U) (Figure 12.30). No double saturated (S-S) types were detected.

Synthetic distearoylphosphatidylcholine (DSPC) of course shows only one signal in the ^{31}P NMR spectrum. The natural phosphatidylglycerol (PG) from soybean lecithin, in contrast, shows a splitting of the two dominant types with two unsaturated FAs and the mixture of S-U FAs causes a third signal downfield (see Figure 12.31). In principle, all PLs are chiral molecules, but some (e.g., PG) have diastereomeric configuration.

Pure enantiomeric molecules only show a single ^{31}P NMR signal. A splitting into two signals indicates a diastereomeric mixture. To find out which chiral center is racemic, a combination of ^{31}P NMR and enzymatic reactions can be used. The phospholipase A2 reaction determines the stereochemistry at the glyceride part of the molecule.

The percentage of hydrolyzation indicates the D/L ratio and can be observed by classical analytical procedures and ^{31}P NMR. Complete hydrolyzation (100%) occurs only when the glyceride has its natural L configuration. To ensure either the lipase

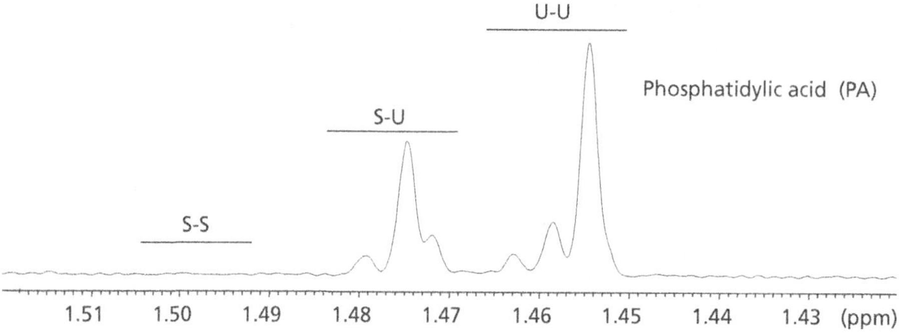

Figure 12.30 High resolution ^{31}P NMR spectra of soy PA.

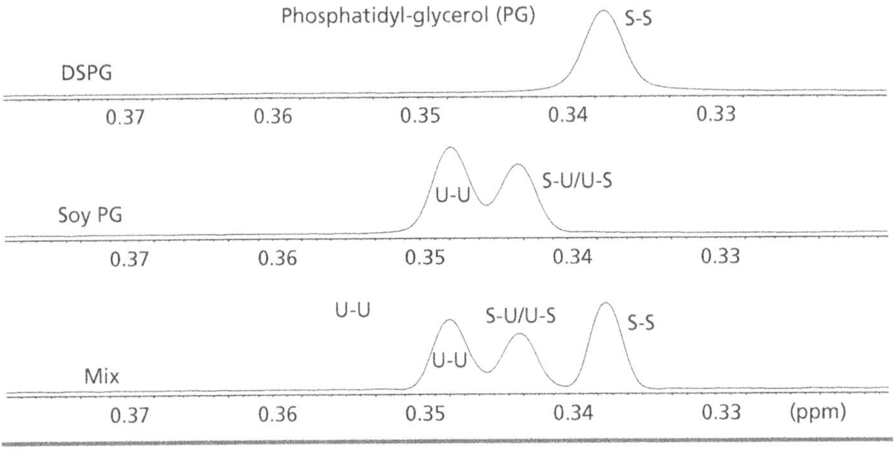

Figure 12.31 High resolution ^{31}P NMR spectra of synthetic DSPG (top), PG with natural fatty acid distribution from soybean (middle), and the mixture of both (bottom).

reaction is specific or not, DSPC was mixed with DSPG. For DSPC, no hydrolyzation and no splitting in ultra-high-resolved ^{31}P NMR was found. On the contrary for the semisynthetic DSPG, a 1:1 diastereomeric ^{31}P NMR splitting and, consequently, 100% hydrolyzation was observed in the ^{31}P NMR spectrum.

The possible combinations of enzymatic and NMR results of DSPG are given in Figure 12.32.

Validation and Round-Robin Tests

The internal standard method had been fully validated concerning selectivity, recovery, reproducibility, instrument precision, and robustness. A lyophilized drug formulation based on egg yolk PC was used in 1998 at 300 MHz (Gunstone, 1993).

To determine the recovery the standard-addition method was used, showing additionally the linearity of the method. A defined amount of egg yolk PC has been added. The mean value of a number of independent measurements fits linear regression with intercept and the regression coefficient is near 1, demonstrating that the recovery is almost 100%. Table 12.A shows the excellent correspondence of the values. To determine the instrument precision, the same sample was measured six times in succession (see Table 12.B).

To determine the intermediate precision on two different days, three independent samples prepared by two different persons were measured. Both data groups are compared in Table 12.C.

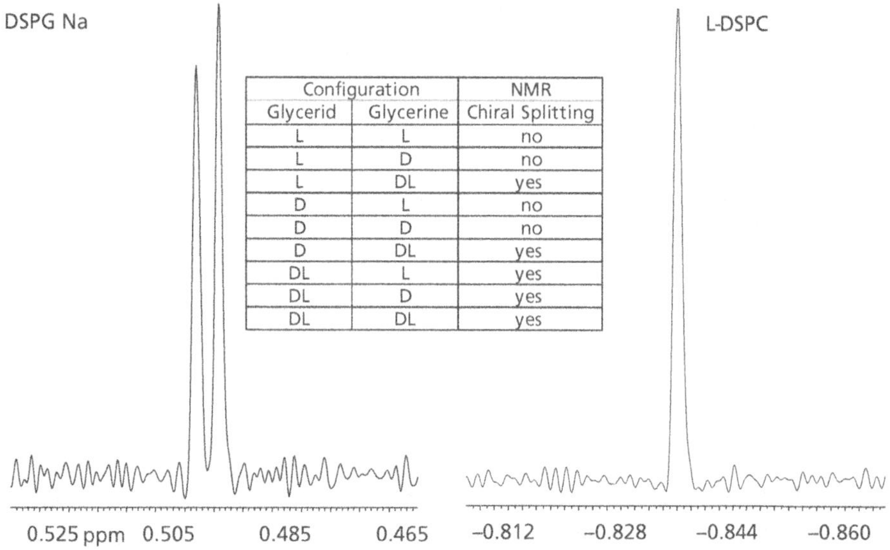

Figure 12.32 ^{31}P NMR spectra of D,L-DSPG (left) and L-DSPG (right) after enzymatic treatment and the possible combinations of enzymatic and NMR results.

Table 12.A Linearity and Recovery Test

	R2	Intercept	Content[a]
PC	0.99867	33.278	33.248
SPH	0.93160	0.167	0.162

[a]Mean value of seven independent measurements.

Table 12.B Reproducibility Test by Six Independent Sample Preparations

Percentage by Weight	PC	SPH	Sum
B1	33.6	0.18	33.8
B2	32.8	0.17	33.0
B3	33.5	0.17	33.7
B4	33.4	0.16	33.5
B5	33.3	0.16	33.4
B6	33.5	0.16	33.7
Mean	33.4	0.16	33.5
StdDev	0.27	0.005	0.26

Table 12.C Intermediate Precision of Six Independent Measurements on Two Different Days

Day 1	PC	SPH	Sum	Day 2	PC	SPH	Sum
Mean	33.4	0.16	33.6	Mean	33.1	0.161	33.3
StdDev	0.24	0.003	0.246	StdDev	0.205	0.007	0.20

Table 12.D Robustness Test by Variation of the pH Value

pH Value	PC	SPH	Sum	pH Value	PC	SPH	Sum
9.5	33.2	0.15	33.4	7.0	33.6	0.15	33.7
9.0	33.4	0.15	33.5	6.5	32.8	0.16	33.0
8.5	33.1	0.17	33.3	Mean	33.3	0.16	33.4
8.0	33.3	0.16	33.4	StdDev	0.23	0.008	0.22
7.5	33.3	0.16	33.5				

Table 12.E Robustness Test by Variation of the Solvent Composition

	CDCl3	MeOH	Cs-EDTA	PC	SPH
B1	1	1100	1	33.6	0.16
K	0.9	1.1	1	33.2	0.15
L	0.8	1.2	1	33.1	0.16
M	0.7	1.3	1	33.0	0.17
N	0.6	1.4	1	33.2	0.16
Mean				33.2	0.16
StdDev				0.207	0.007

Robustness in terms of sample preparation and stability of the prepared samples was performed by variation of the pH value (see Table 12.D). Within the given limits, variation of the pH value has no influence on the results. The recommended pH value is 8.5; values above 9.5 and below 6.5 lead to hydrolysis of the PLs. In addition, the line shape of the NMR signals becomes broad, changing to more acidic conditions, and the chemical shift shows larger variations.

According to the standard operation procedure for sample preparation, equal amounts (by volume) of $CDCl_3$, methanol, and Cs complex of ethylenediaminetetraacetic acid (Cs-EDTA) solution are used. A series of measurements with changed ratios of the solvent components shows the low influence of inaccuracies in handling the solvents (see Table 12.E).

There is no significant influence on the results changing the composition of the solvent mixture within certain limits. Further changes can lead to line broadening and/or to reduced extraction yield. In routine work, 256 scans are used. The number of scans has an influence on the signal-to-noise (S/N) ratio. In general, the S/N ratio is a square root function of the number of scans. To show the effect, the same sample was measured eight times, every time doubling the number of scans (see Table 12.F) except in the last case due to lack of time. The last measurement was stopped after 6500 pulses (5.5 hours).

The amount of the sample was varied from 60 to 140 mg and the amount of the internal standard was between 20 and 40 mg, taking care that even extreme ratios are covered. The mean value and the standard deviation prove that there is no significant influence on the results (see Table 12.G).

The sample changer holds up to 120 samples. With a 20 minute measuring time (256 scans), it can take up to 40 hours until the last sample is measured. It should be checked whether the time between sample preparation and measurement has an

Table 12.F Robustness Test by Variation of the Number of Scans

NS	PC	SPH	Sum	NS	PC	SPH	Sum
64	94.674	0.450	95.124	2048	94.764	0.469	95.233
128	94.831	0.446	95.277	4096	94.721	0.444	95.165
256	94.852	0.453	95.305	6484	94.550	0.468	95.018
512	94.904	0.465	95.369	Mean	94.748	0.454	95.203
1024	94.691	0.441	95.132	StdDev	0.107	0.010	0.107

Table 12.G Robustness Test by Variation of the Ration of Sample and Internal Standard TPP

	Sample	Standard	PC	SPH
E	60.54	40.11	33.138	0.161
F	83.49	35.31	33.544	0.170
G	90.79	35.38	33.546	0.172
D	101.51	32.65	33.472	0.161
H	111.41	25.41	33.685	0.174
I	120.61	25.46	33.405	0.165
J	141.16	20.81	33.201	0.167
Mean			33.427	0.167
StdDev			0.196	0.005

influence on the result. The last measurement was made more than 120 hours after sample preparation. Table 12.H shows the measurement of the same sample on five successive days.

There is no trend; the results scatter similar to all other measured series. Samples are found to be stable for at least 120 hours after preparation. Some more tests showed that there are signals caused by LPC near the limit of detection for the first time after 3 weeks.

All measurements have been evaluated (integrated) with the dedicated software DIS NMR as well as with WIN NMR under Windows 3.11, which is provided by the spectrometer manufacturer. DIS NMR was used in automatic mode, whereas spectra were integrated manually with WIN NMR. It was found that DIS NMR produced significantly lower values than WIN NMR. With small random tests, the difference is covered up by the scattering, but with larger random tests (>30), the statistical tests show a significant deviation. Table 12.I shows a comparison of integration procedures.

The automatic integration led to erroneous results, especially in cases with more than 512 pulses. The integration limits are influenced by the S/N ratio; with a higher S/N ratio, too much of the slope of the NMR signal (Lorentz curve) is added to the integration, leading to increased values. After the correction of integration limits, the results are statistically valid. In routine work 256 pulses are used, so manual as well as automatic integration lead to correct results.

Table 12.H Robustness Test by Variation of the Time between Preparation and Measurement

	PC	SPH	Sum		PC	SPH	Sum
Day 1	33.1	0.16	33.3	Day 5	33.2	0.15	33.3
Day 2	33.3	0.15	33.5	Mean	33.2	0.16	33.3
Day 3	33.1	0.16	33.3	StdDev	0.080	0.007	0.080
Day 4	33.2	0.17	33.4				

Table 12.I Comparison of the Mean Values Using Different NMR Processing Software

	PC	SPH
DIS NMR	33.18 ± 0.20	0.154 ± 0.024
WIN NMR	33.27 ± 0.22	0.162 ± 0.007

The quantitation of PLs was validated as described previously. Method precision (reproducibility) and instrument precision show that main components have a standard deviation less than 1% and components near the limit of quantitation have a standard deviation below 5%. In the case of the used liposome lyophilisate preparation, the extraction recovery of the PLs is 100%.

This analytical method is robust. It is almost insensitive to variation in sample preparation and parameters of measurement. Prepared samples can be stored up to 5 days at room temperature (however, typically it does not exceed 24 h) until they are measured without negative influence on the results. Manual integration should be preferred; control of integration limits is recommended using automatic integration.

Besides the main component PC, the degradation products LPC and PLs from egg yolk like SPH can be observed and quantified down to 0.01%.

Between 1992 and 2000, the ILPS started some trials to establish HPLC methods for PL analysis. The recommended detector is a light scattering system (LSD). For native soy lecithin this method is useful, but in cases of more complex lecithin samples it failed.

Several other validation studies were performed, for example, for soy lecithin and krill lecithin (ILPS, 2013). A sample of krill oil was analyzed by 10 NMR laboratories worldwide using the USP monograph method. The challenge was the quantification of PC including the ether PC signals between –0.8 and –1 ppm and the sum of the other krill PLs between $\delta = 0.5$ and –0.5 ppm.

Ten laboratories delivered the data within 2 months after sample preparation. The samples were stable over this long time period and confirm the robustness of the sample preparation and analysis by the USP method. The standard deviation for the PC and PC ether is 0.5. This is an excellent value. The total PLs show a standard deviation of 1% and the summarized small signals a value of 3%. The different magnetic field strength (400 to 600 MHz) and the probes (room temperature and cryo) do not affect a significant difference even if the S/N ratio is important for a proper integration of all small signals (see Table 12.J).

Oxidation

Generally, the level of primary oxidation products in oils and fats is indicated by the peroxide value according to the official Wheeler method (POV) (DIN EN ISO 3960, 2010). As alternative oxidation level indices, some authors propose the NMR-determined ratio between the olefinic and aliphatic protons and the ratio between the aliphatic and diallylmethylene protons in fatty acids. A good correlation between these ratios and the conventional POV method has been found (Diehl and Ockels, 1995a; Saito and Nakamura, 1990; Saito and Udagawa, 1992; Wanasundara and Shahidi, 1993). However, relatively high levels of lipid oxidation were observed before

Table 12.J Integral Values of ^{31}P NMR of Krill Oil Obtained by 10 Different Laboratories

Lab No.	1	2	3	4	5	6	7
Int 1	225.01	225.88	222.38	225.35	225.35	224.19	224.33
Int 2	93.37	92.49	97.95	100.24	100.24	96.25	100.58
Sum	318.38	319.37	320.33	325.59	325.59	320.44	324.91

Lab No.	8	9	10	Mean	Dev	Dev %
Int 1	225.61	223.57	225.76	224.8	1.3	0.6
Int 2	96.75	97.4	101.39	97.7	3.1	3.1
Sum	322.36	320.97	327.15	322.5	3.1	1.0

any detectable changes occurred in the NMR spectrum. Another possibility is the direct quantification of all hydroperoxide proton signals by ^{1}H NMR (Abou-Gharbia et al., 1997).

It was observed that a degradation caused by hydrolyzation or oxidation of PLs, for example, PC, can occur in drug formulations. Through the use of the official titration methods for lipids, the detection and quantification of PL oxidation products is not successful. In complex liposomal formulations these analyses are even more complicated. Therefore, there is a need for a new method. Because ^{1}H NMR spectroscopy analysis is equivalent for the detection of POV in edible oils, this technique was used for monitoring the oxidative stability of liposomal formulations with PC. The combination of ^{1}H, ^{13}C, and ^{31}P NMR analyses well-documented the primary oxidation step by formation of peroxides and the secondary step by formation of aldehydes. Even the following polymerization steps could be evaluated using these NMR methods.

During the autoxidation process of lipids, hydroperoxides are formed as primary products that are easily decomposed to secondary products such as aldehydes, ketones, alcohols, and acids (Skiera et al., 2012a). The classical method of POV determination is a titration procedure that is not selective to peroxides but also reacts with several other components.

The POV value in edible oils is strongly disturbed by the amount of natural or added phenols, for example, oleoropein and other tyrosoles in olive oil or tocopherols in soybean oil. These compounds decrease the POV values. Especially in olive oil, the POV values determined by the official Wheeler method are lower than the direct observed peroxides by ^{1}H NMR spectroscopy. In some cases (e.g., black cumin oil), the presence of high amounts of quinones produces too high POV. In polar lipids as PLs these disturbances are more intense. The classical POV determination often failed totally. In pharmaceutical formulations, besides the active ingredients, antioxidants are often added, which make a precise POV determination impossible. This applies to solid formulations or pastes and even to liquid and liposomal types of pharmaceuticals.

The detection of aldehydes as a class of secondary oxidation products is traditionally performed by the anisidine value (AV), a nonspecific photometric method that uses the UV absorption of Schiff's base reaction products of aldehydes with anisidine. These reactions between the different aldehydes and the amino group are not specific. NMR tests have shown that hexanal gives only low amounts of the Schiff's base, whereas the major products resulted from the condensation in the ortho position of anisidine (Frankel et al., 1990).

The yield of the Schiff's base of hexenal and hexadienal is higher but not complete. Only aldehydes without α-acidic protons like benzaldehyde react completely with anisidine, forming the Schiff's base. This fact reflects that the response factors for hexanal, hexenal, and hexadienal are 0.5, 6, and 30, respectively.

The use of classical AV is, therefore, totally unfeasible. However, a correlation between classical AV analysis and ^{1}H NMR shows good comparability of both primary methods (Doum, 2013). The NMR method shows absolute values of total aldehydes in any matrix, whereas the AV titration needs response factors for different aldehyde species.

^{1}H NMR spectroscopy enables simultaneous detection of primary (hydroperoxides) and secondary oxidation products (aldehydes) of all lipid classes. Systematic studies dealing with the characterization of primary oxidation products in edible oils have been published. Lipid oxidation starts with hydroperoxides and ends with polymerization. Aldehydes and acids are secondary intermediates or end products.

NMR spectroscopy is a tool that is sensitive to all steps of the oxidation process, and due to its molar response, quantitative analysis is possible. In addition, because all processes can be observed within a FA that has a defined structure, an external calibration is not necessary. NMR spectra of FAs and the oxidation products deliver self-calibrated data. Figure 12.33 shows the reaction scheme that may explain the theory.

For the internal calibration, one can use the methylene group at C2 position (δ = 2.3 ppm) or the terminal methyl groups (0.9 ppm). A reasonable value for a calibration is to set up the integral value of the signal of C2 protons to 2000 units or the methyl to 3000. Now each calculated integral value of a hydroperoxide or an aldehyde proton represents 1 per mill molar amount, independently of whether we analyze a TAG, a PL, a glycolipid, or even free FAs. The type of FA also doesn't affect the calibration; the method is useful for low unsaturated lipids such as milk and lipids of marine origin. The detection of aldehydes is also absolute and without any response factor. Aldehyde values detected by NMR are comparable between any kinds

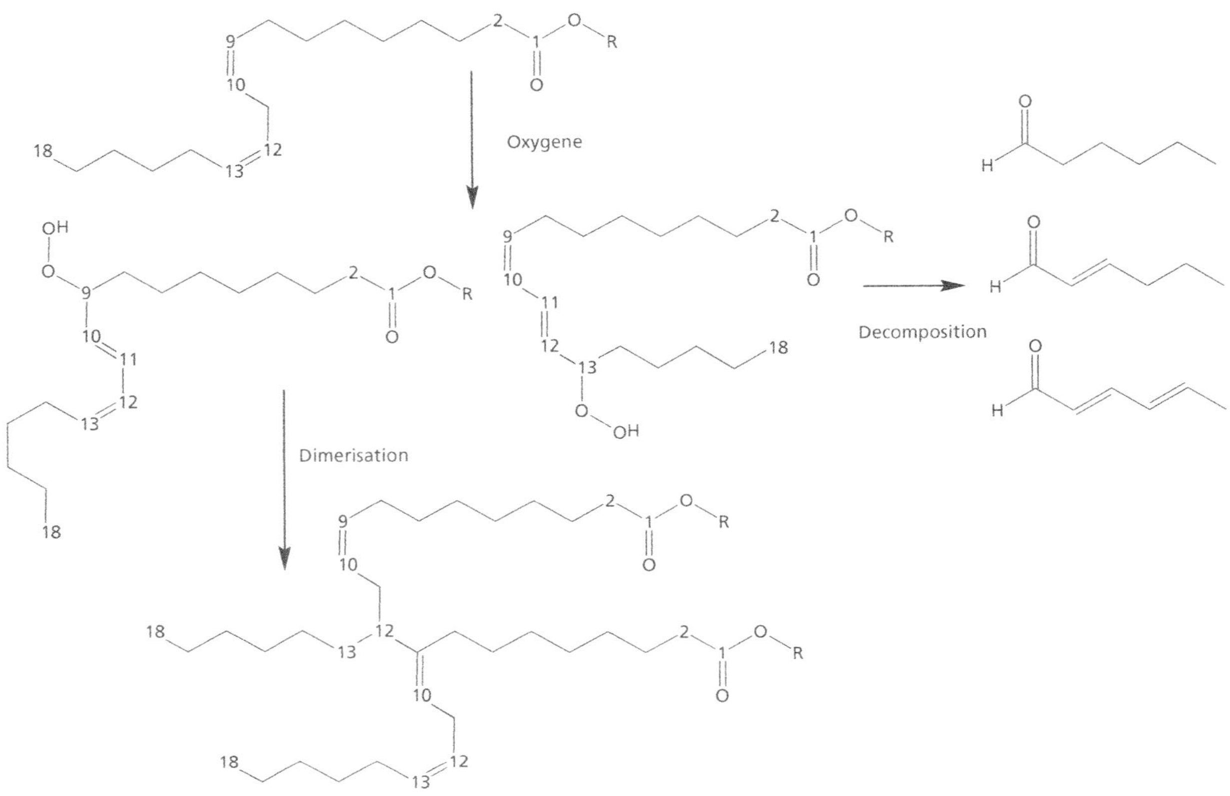

Figure 12.33 Oxidation pathway.

of vegetable or animal lipid. Normally the degree of oxidation is within a range between ppm and 0.5%. Therefore, this relation is not really disturbed.

Besides the increasing signals of hydroperoxides and aldehydes, the oxidative degradation of lipids can be observed by the subsequent decreasing signals of methylene groups in allylic and double allylic positions, as well as the disappearance of the triplet belonging to δ-3 methyl groups, which are well separated from others at δ = 0.95 ppm. In any case, oxidation produces *trans* and *trans* conjugated FAs that normally are only minor components of oils (e.g., CLA in milk and meat). The protons of these conjugated double bonds appear separated from the signals of natural *cis* homoconjugated bonds at a deeper field between δ = 5.5 and 7 ppm. The number of different types of hydroperoxides, aldehydes, and conjugated double bonds depends on the degree of unsaturation of a FA. Classical vegetable oils only produce a manageable number of degradation products; most of the oxidized species can be obtained as single signals. The higher the degree of unsaturation, the faster and more complex is the number of possible degradation products. In marine oils, a single assignment of hundreds of hydroperoxides or aldehyde signals is not possible anymore. However, the total integral value of the responding areas gives suitable and quantitative information about the molar amount of the targets.

One of the biggest problems in the judgment of oxidative degradation is the fact that the total amount of hydroperoxides or aldehydes is not necessarily sufficient. The values of hydroperoxides decrease during the storage at higher temperatures (e.g., at 40 °C) over a longer period, whereas the amount of aldehydes increases. After further degradation even of the aldehydes, the long-time stored drug formulation or oil seems to be stable because the forming of the tertiary degradation products, oligomers, and polymers is hard to obtain. We could show that even these degradations can be observed by ^{1}H NMR and ^{31}P NMR and at least semiquantitatively estimated.

Three different approaches for testing the oxidative state in PLs are now shown. First, the oxidative degradation of purified PC from soybean is discussed. This is an often-used starting raw material for drug formulations in solid form, in pasty form mixed with TAGs, and in liposomal formulations. The quality controls of an enriched PC fraction of soybean origin (raw material for drug excipients) using the ^{1}H NMR spectroscopy will be discussed in detail.

The different signals represent several chemical groups of the glycerol backbone and the choline polar head group as well as the different segments of the FAs (see Figure 12.34). In addition, several additives such as α-, β-, γ- and δ-tocopherol can be observed using their phenolic OH protons (see Figure 12.35). The solvent residue of ethanol (triplet at δ =1.2 ppm and quadruplet at δ = 3.6 ppm [see Figure 12.36]) can be obtained and quantified, as can the hydroperoxides.

The most common solvent in NMR analysis of lipids is $CDCl_3$. As a result of proton transfer effects that are caused by this solvent and impurities, hydroperoxide

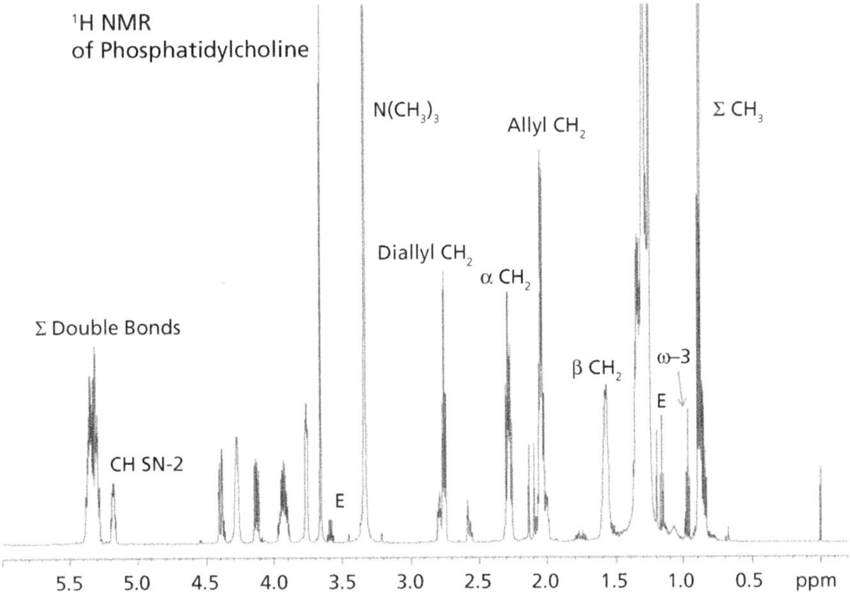

Figure 12.34 ¹H NMR spectrum of PC from soybean, signal annotation.

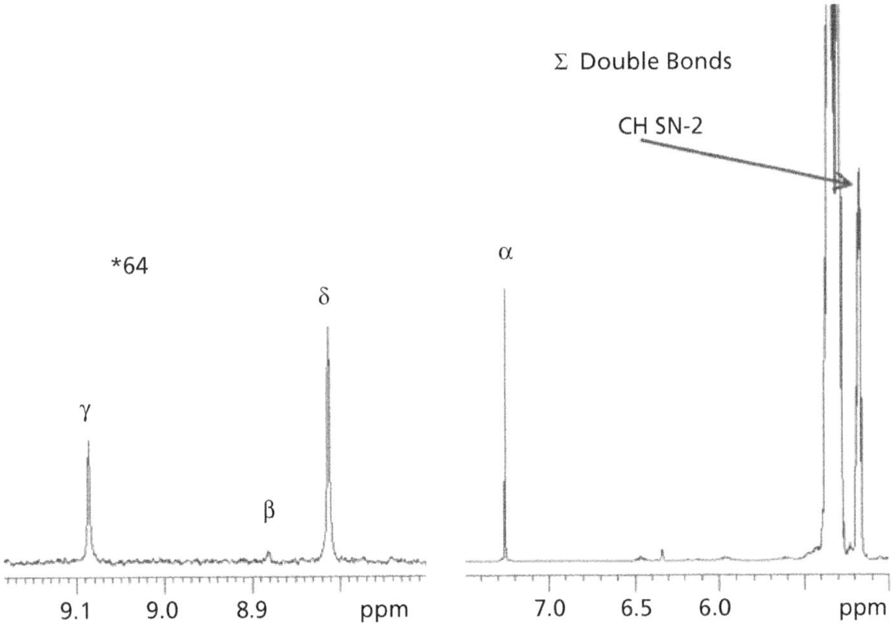

Figure 12.35 Phenolic protons of a tocopherol mixture in edible oil.

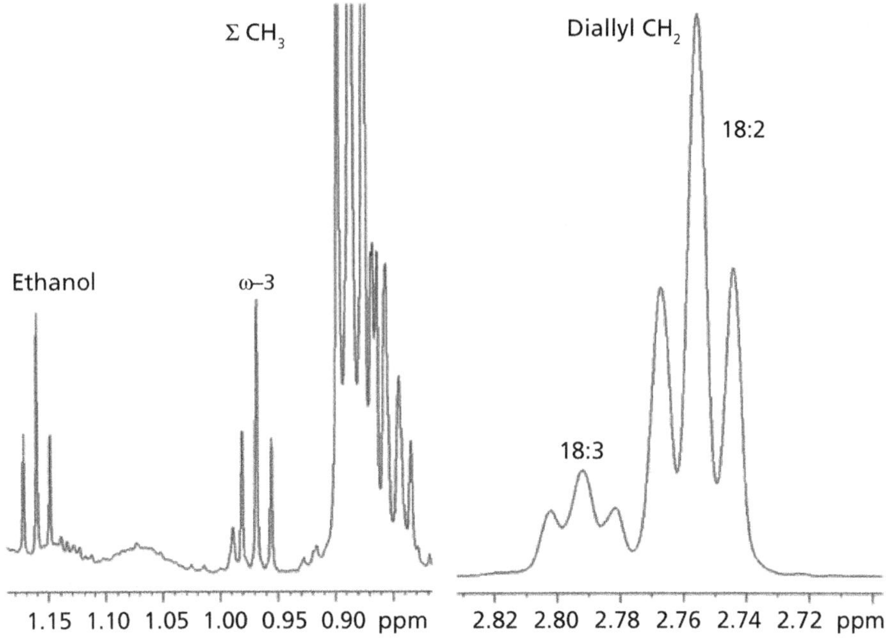

Figure 12.36 Details of a lipid ^{1}H NMR spectrum with methyl groups (left) and double allylic methylene groups (right).

proton signals appear as broad peaks with varying chemical shifts in the region of δ = 8 to 9 ppm (Skiera et al., 2012b). To overcome these difficulties, one can utilize $CDCl_3$ with 20% of deuterated DMSO as a solvent. Under these conditions, the hydroperoxide proton signals are shifted to δ = 10–11 ppm even in PL solutions (Figure 12.37).

Higher amounts of free FAs or other acidic protons in combination with high water contents disturb the analysis of hydroperoxides in complex polar lipids. In these cases, a special sample treatment can help, as can the use of ^{13}C NMR. Methine carbons wearing the hydroperoxide appear at about 90 ppm. However, longer measuring times and a very sensitive high-field cryo system is recommended to limit the measuring time to a maximum of 1 hour.

A solid formulation was monitored by NMR during storage at different temperatures between –18 °C and 40 °C. The hydrolyzation in solid formulations only plays a marginal role; degradation mostly is caused by oxidation due to the higher surface and the lack of humidity. The analysis of the ^{31}P NMR data shows two different effects. The line shape of the PC signal at δ = –0.76 ppm changes, and additional signals are observable as the shoulders of the main peak (Figure 12.38). In addition, the

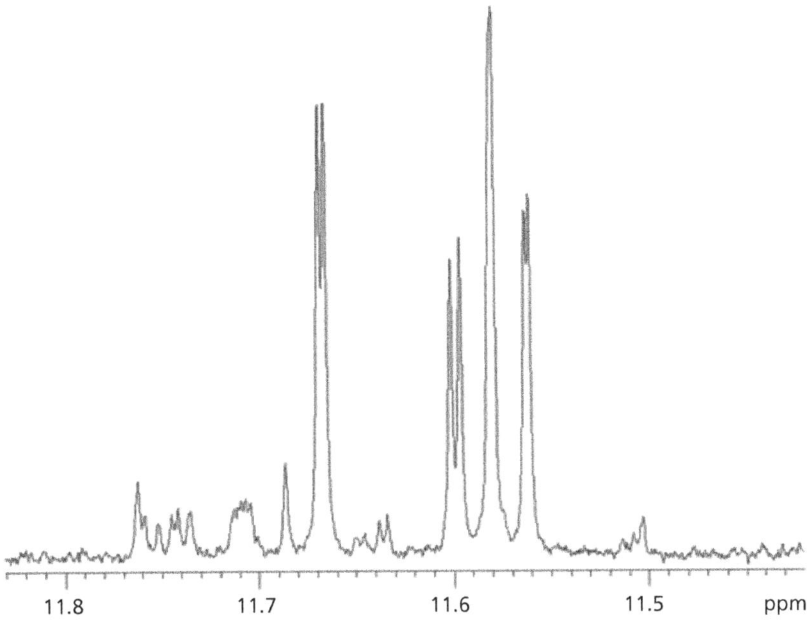

Figure 12.37 Hydroperoxide signals of PC from soybean.

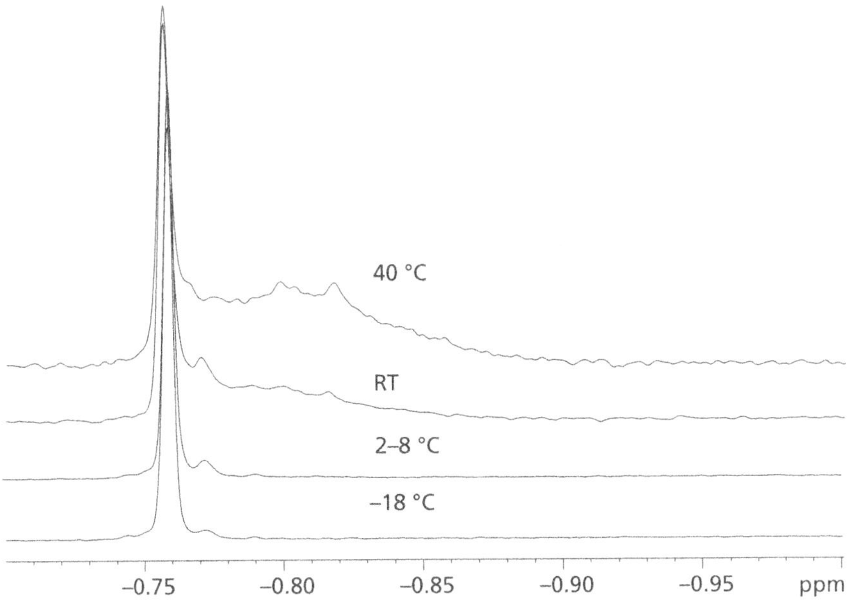

Figure 12.38 ^{31}P NMR of PC in solid drug formulation during storage at different temperatures.

detected total amount of PC at 40 °C decreases nearly to half of the amount detected at −18 °C if only the narrow signal of intact PC is used for quantification. The reason is that the higher oxidation products of PC further react to oligo- and polymers. These molecules show different additional signals at a higher field combined with a line broadening. By using the whole integral area between 0 and −1 ppm, the total PC content didn't change. The monitoring of lipid oxidative degradation of PC and PE in fish oil was tested by ^{31}P NMR line-shape analysis. The polymer degradation process was checked through the use of model substances of DHA-PC and DHA-PE (Hämäläinen and Kamal-Eldin, 2005).

The kinetics of FA degradation also can be well-observed by ^{1}H NMR spectroscopy. In the downfield region of the ^{1}H NMR spectra, the signals of hydroperoxide protons appear at $\delta = 11–11.5$ ppm, whereas the aldehyde protons are observable between $\delta = 9.5$ and 9.8 ppm (Figure 12.39). Each group represents several types of oxidation products. However, the total integral area can be used for quantitative issues.

The storage at −18 °C caused no degradation; after storage at 2–8 °C hydroperoxides could be detected as primary oxidation products. At room temperature, the formation of aldehydes as secondary oxidation products can be observed, and at least at 40 °C the hydroperoxides had chemically reacted and major amounts of the PC molecules became polymer. The details of NMR spectra enable the detection of the total degradation of the double bonds, and even the differences in individual kinetics of linoleic and linolenic acid degradation.

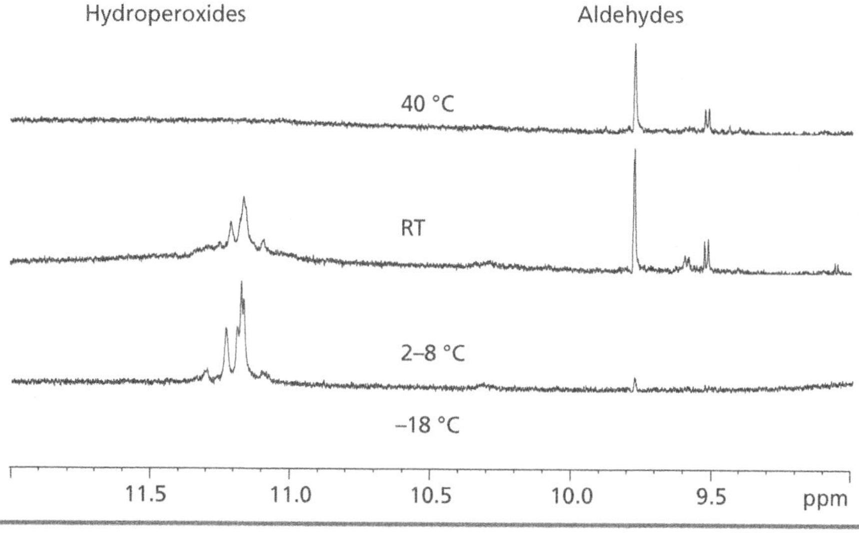

Figure 12.39 ^{1}H NMR spectra of hydroperoxides extracted from solid PC formulations at different storage temperatures.

A liquid 1% liposomal formulation was monitored by NMR during storage at different temperatures between –18 °C and 40 °C. The hydrolyzation in liquid formulations plays a major role; however, degradation by oxidation can be observed in parallel. During storage, the signals of PC decrease and those of lyso-PC increase. The analysis of the ^{31}P NMR signal line shape shows the same effect as mentioned for the solid material. The PC signal at $\delta = -0.84$ ppm shows an increasing signal observable as high field shoulder. To prove the shoulder signal as a PC derivative, H,P-TOCSY spectrum was performed, and the corresponding cross peaks clearly show the different PC and LPC species (Figure 12.40). In addition, the ^{1}H NMR spectrum shows increasing signals of aldehydes (Figure 12.41). The ^{31}P NMR spectrum of a PE reference standard from egg indicates the polymerization according the broad shoulder, too (Figure 12.42). To prove the assumption of oligomeric structures, ^{31}P DOSY NMR spectra (diffusion-ordered spectroscopy [Henna et al., 2012]) were acquired. The larger a molecule, the slower is the diffusion in the NMR tube. The experiments clearly show the different diffusion coefficient for the remaining monomeric PE and LPE as a narrow spot and the tailing region of the oligomeric species (Figure 12.43).

The third degradation test is done using an encapsulated (hard gelatin) mixture of saturated TAG and soy PC. In this case, the hydrolyzation dominates the degradation process and can be easily monitored by ^{31}P NMR (Figure 12.44).

The oxidation is suppressed due to the fact that the wax-like material has no contact with the atmosphere. However, the amount of oxidative markers can depend on the storage temperature and time. The initial peroxide amount corresponds to the amount of the soy PC raw material. No increase of the hydroperoxides was obtained

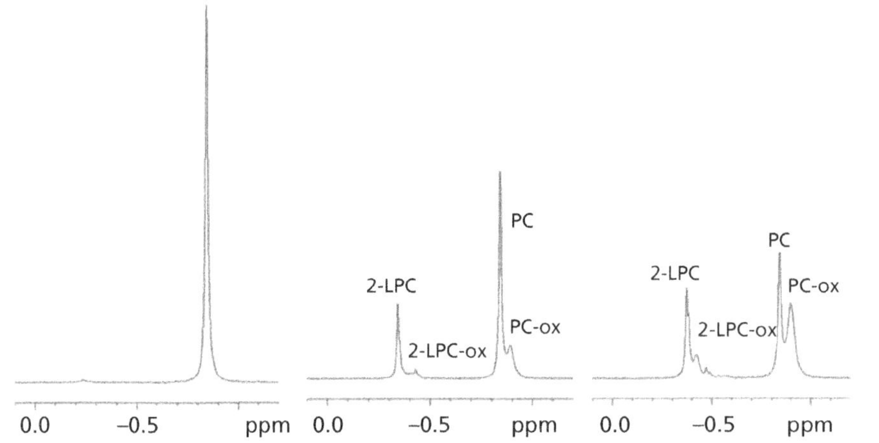

Figure 12.40 ^{31}P NMR spectra of liposome degradation start (left); 1 month at 40° C (middle); 2 months at 40 °C (right).

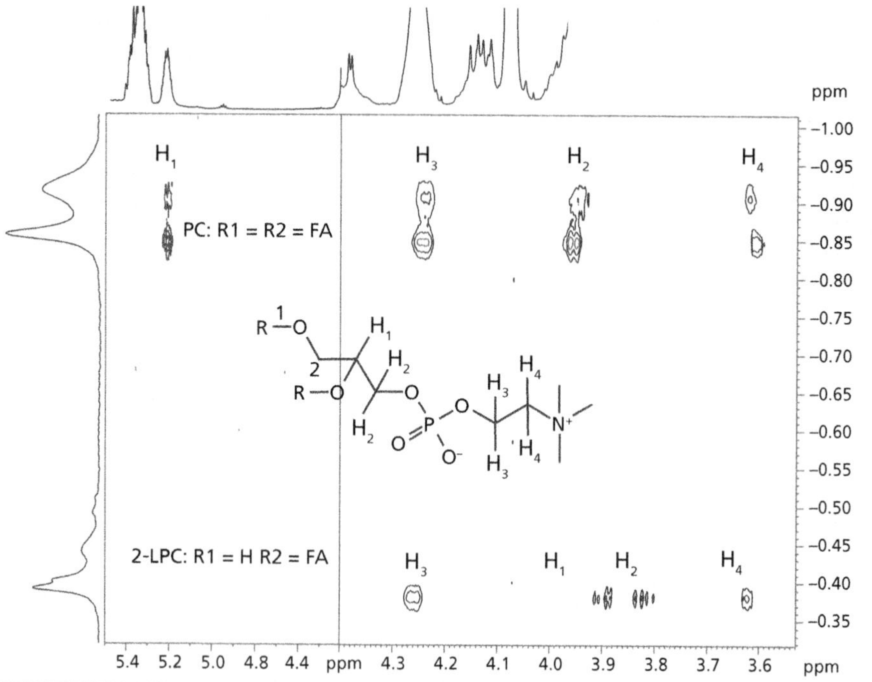

Figure 12.41 H,P-TOCSY of degraded liposomes from PC.

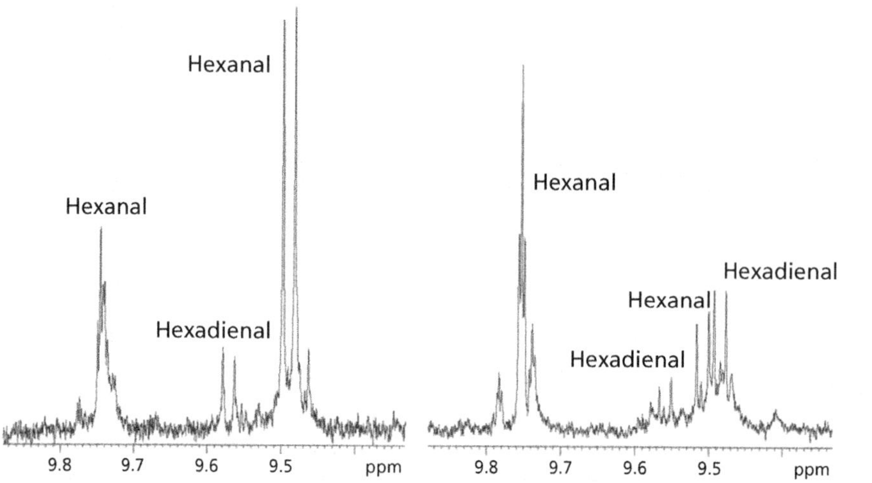

Figure 12.42 ^{1}H NMR spectrum of oxidized PE of egg yolk (left), ^{31}P-DOSY (right).

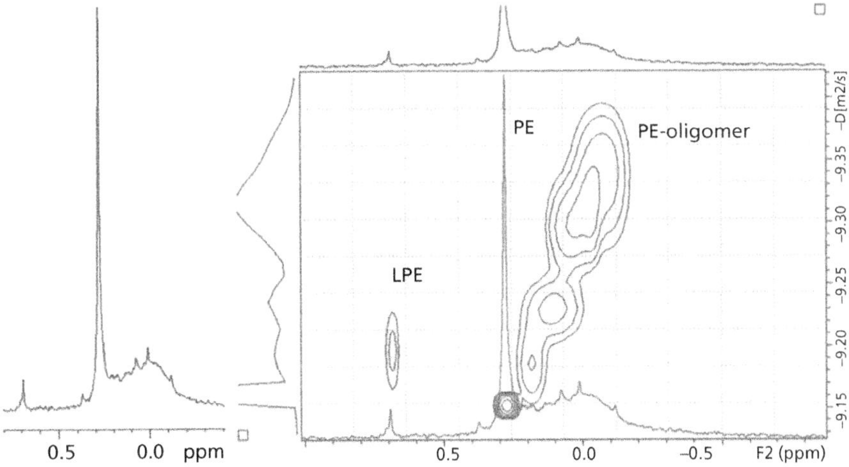

Figure 12.43 ^{31}P NMR (left) and ^{31}P NMR DOSY of oxidized PE from egg yolk (right)

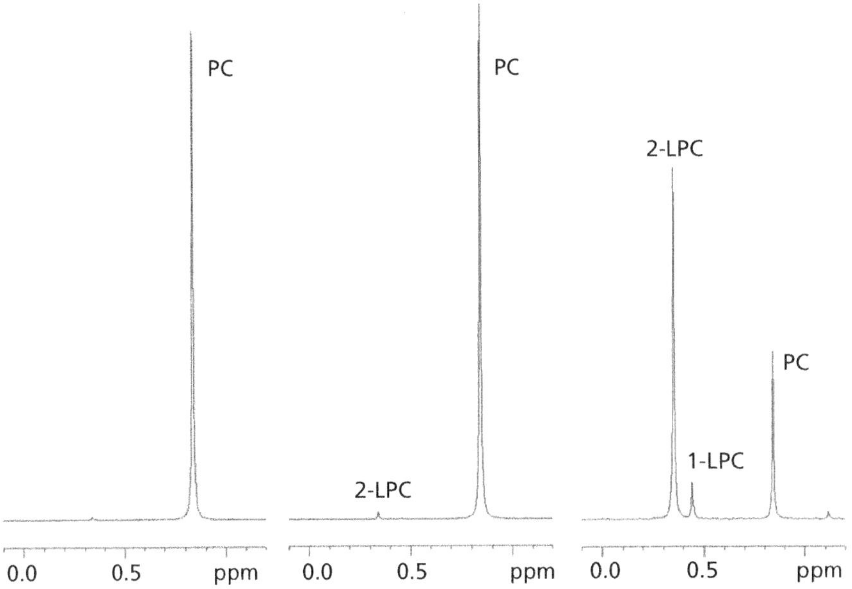

Figure 12.44 ^{31}P NMR of storage stability test on PC start (left); 4 months at 8 °C (middle); and 4 months at 40 °C (right).

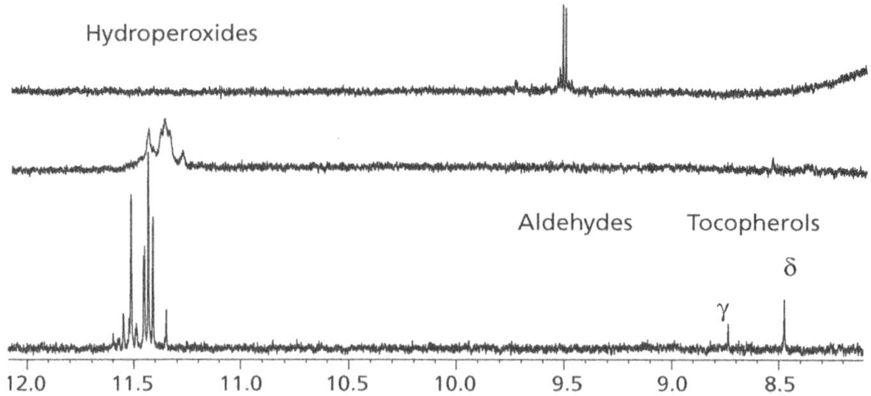

Figure 12.45 Degradation of hydroperoxides and tocopherol during storage at different temperatures, 8 °C (bottom), room temperature (middle), and 40 °C (top).

during the storage at room temperature, but these compounds appear at higher storage temperatures. Aldehydes are formed and the added antioxidant, a mixture of tocopherols, disappears (Figure 12.45). The combination of different NMR methods of only one sample preparation enables detection, kinetics, and quantification of oxidative degradation. The primary NMR method does not need any reference standard to generate absolute values of oxidative states as an alternative to common official methods like POV or AV and polymerization. Another advantage is the simultaneous detection of increasing signals of the degradation products and the decreasing signals of native double bonds and even the signals of antioxidants. So, the efficacy of antioxidants can easily be investigated in simple models. In cases of edible oils, NMR now is a routine quality check; its application is increasingly used in PL degradation monitoring.

References

Abou-Gharbia, H. A.; Shahidi, F.; Shehata, A. A. Y.; Youssef, M. M. Effects of Processing on Oxidative Stability of Sesame Oil Extracted from Intact and Dehulled Seeds. *J. Am. Oil Chem. Soc.* **1997,** *74,* 215–221.

Braun, S.; Kalinowski, H. O.; Berger, S. *150 and More Basic NMR Experiments.* Wiley-VCH: Weinheim, 1998.

Dennis, E. A.; Plückthun, A. Phosphorus-31 NMR of Phospholipids in Micelles. In *Phosphorus-31 NMR, Principles and Applications;* Gorenstein, D. G., Ed.; Academic Press: New York, 1984; pp 423–446.

Diehl, B. W. K. Fatty Acid Distribution by ^{13}C NMR Spectroscopy. *Fat Sci. Tech.* **1995,** *97,* 115–118.
Diehl, B. W. K. Multinuclear High Resolution NMR Spectroscopy. In *Lipid Analysis of Oils and Fats;* Hamilton, R., Ed.; Chapman & Hall: London, 1998; pp 87–137.
Diehl, B. W. K. High Resolution NMR Spectroscopy. *Eur. J. Lipid Sci. Technol.* **2001,** *103,* 830–834.
Diehl, B. W. K.; Ockels, W. Fatty Acid Distribution by ^{13}C NMR-Spectroscopy. *Fat Sci. Tech.* **1995a,** *97* (3), 115–118.
Diehl, B. W. K.; Ockels, W. Quantitative Analysis of Phospholipids. *Proceedings of the 6th International Colloquium Phospholipids, Characterization, Metabolism and Novel Biological Applications,* Cevc, G., Paltauf, F., Eds.; AOCS Press: Urbana, 1995b, pp. 29–32.
Diehl, B. W. K.; Herling, H.; Ernst, H.; Riedl, I. ^{13}C NMR Analysis of the Positional Distribution of Fatty Acids in Plant Glycolipids. *Chem. Phys. Lipids* **1995,** *77,* 147–153.
DIN EN ISO 3960, Animal and Vegetable Fats and Oils—Determination of Peroxide Value—Iodometric (Visual) Endpoint Determination (ISO 3960:2007, corrected version 2009-05-15); German version EN ISO 3960, 2010.
Doum, V. *Measurements of the Aldehyde Amounts in Lipids by ^{19}F NMR Spectroscopy after Derivatisation,* University of Applied Science, Niederrhein, 2013.
Frankel, E. N.; Neff, W. E.; Miyashita, K. Autoxidation of Polyunsaturated Triacylglycerols. II. Trilinolenoylglycerol. *Lipids* **1990,** *25,* 40–47.
Gilard, V.; Trefi, S.; Balayssac, S.; Delsuc, M.-A.; Gostan, T.; Malet-Martino, M.; Martino, R.; Prigent Y.; Taulelle, F.; DOSY NMR for Drug Analysis. In *NMR Spectroscopy in Pharmaceutical Analysis;* Holzgrabe, U., Wawer, I., Diehl, B. W. K., Eds.; Elsevier: Amsterdam, Holland, 2008.
Gunstone, F. D. High Resolution ^{13}C NMR-spectroscopy of Lipids. In *Advances in Lipid Methodology Two;* Christie, W. W., Ed.; The Oily Press: Dundee, 1993; pp 1–68.
Hämäläinen, T. I.; Kamal-Eldin, A. Analysis of Lipid Oxidation Products by NMR Spectroscopy. In *Analysis of Lipid Oxidation;* Kamal-Eldin, A., Pokorny, J., Eds.; AOCS Press: Urbana, IL; 2005; pp 70–126.
Henna, L.; Baron, C.; Diehl, B. W. K.; Jacobssen C. Oxidative Stability of Emulsions Prepared from Purified Marine Phospholipids and the Role of α-tocopherol. *J. Agr. Food Chem.* **2012,** *60,* 12388–12396.
Holzgrabe, U.; Wawer, I.; Diehl, B. *NMR Spectroscopy in Drug Development and Analysis;* Wiley VCH: Weinheim, Germany, 1999.
Holzgrabe, U.; Wawer, I.; Diehl, B. *NMR Spectroscopy in Pharmaceutical Analysis;* Elsevier: Amsterdam, Holland, 2008.
Kühl, O. *Phosphorus-31 NMR Spectroscopy.* Springer: Berlin, 2008.
Medina, I.; Sacchi, R. Acyl Stereospecific Analysis of Tuna Phospholipids via High Resolution ^{13}C NMR. *Chem. Phys. Lipids* **1994,** *70,* 53–61.
Meneses, P.; Glonek, T. High Resolution 31P NMR of Extracted Phospholipids. *J. Lipid Res.* **1988,** *29,* 679–689.
Ng, S. Analysis of Positional Distribution of Fatty Acids in Palm Oil by ^{13}C NMR Spectroscopy. *Lipids* **1985,** *20,* 778–782.

Saito, H.; Nakamura, K. Application of the NMR Method to Evaluate the Oxidative Deterioration of Crude and Stored Fish Oils. *Agric. Biol. Chem.* **1990,** *54,* 533–534.

Saito, H.; Udagawa, M. Assessment of Oxidative Deterioration of Salted Dried Fish by Nuclear Magnetic Resonance. *J. Am. Oil Chem. Soc.* **1992,** *69,* 1157–1159.

Schiller, J.; Arnold, K. Application of High Resolution ^{31}P NMR Spectroscopy to the Characterization of the Phospholipid Composition of Tissues and Body Fluids—A Methodological Review. *Med. Sci. Monit.* **2002,** *8,* MT205–MT222.

Skiera, C.; Steliopoulos, P.; Kuballa, T.; Holzgrabe, U.; Diehl, B. ^{1}H NMR Spectroscopy as a New Tool in the Assessment of the Oxidative State in Edible Oils. *J. Am. Oil Chem. Soc.* **2012a,** *89* (8), 1383–1391.

Skiera, C.; Steliopoulos, P.; Kuballa, T.; Holzgrabe, U.; Diehl, B. W. K.^{1}H NMR Approach as an Alternative to the Classical p-Anisidine Value Method. *Eur. Food Res. Technol.* **2012b,** *235* (6), 1101–1105.

Tukiainen, T.; Tynkkynen, T.; Mäkinen, V.-P.; Jylänki, P.; Kangas, A.; Hokkanen, J.; Vehtari, A.; Gröhn, O.; Hallikainen, M.; Soininen, H.; et al. A Multi-metabolite Analysis of Serum by ^{1}H NMR Spectroscopy: Early Systemic Signs of Alzheimer's Disease. *Biochem. Biophys. Res. Com.* **2008,** *375,* 356–361.

Tynkkynen, T. *^{1}H NMR Analysis of Serum Lipids.* University of Eastern Finland, 2012.

Wanasundara, U. N.; Shahidi, F. Application of NMR Spectroscopy to Assess Oxidative Stability of Canola and Soybean Oils. *J. Food Lipids* **1993,** *1,* 15–24.

Wollenberg, K. F. Quantitative High Resolution 13C Nuclear Magnetic Resonance of the Olefinic and Carbonyl Carbons of Edible Vegetable Oils. *J. Am. Oil Chem. Soc.* **1990,** *67* (8), 487–494.

13

Polar Lipid Profiling by Supercritical Fluid Chromatography/ Mass Spectrometry Method

Takayuki Yamada, Yumiko Nagasawa, Kaori Taguchi, Eiichiro Fukusaki, and Takeshi Bamba ▪ *Department of Biotechnology, Graduate School of Engineering, Osaka University, Osaka, Japan*

Introduction

Significance of Polar Lipid Profiling

Polar lipids such as phospholipids and sphingolipids are abundant in plant oils (Cert et al., 2000) and are the major component of biological membranes (van Meer et al., 2008). These lipids also play important physiological roles in cellular signaling (Wymann and Schneiter, 2008). High-throughput simultaneous analyses for all types of lipids, including polar lipids, are favorable to elucidate the function of individual l ipid molecular species or to distinguish among phenotypes. The research field aimed at comprehensive lipid profiling is called *lipidomics.* Because of the diversity of chemical structures and the wide range of polarities in lipids, a powerful separation technique is required for comprehensive lipid profiling. Supercritical fluid chromatography/mass spectrometry (SFC/MS) is the ideal tool for lipid profiling due to the features of the mobile phase. This chapter introduces the benefits of SFC/MS and focuses on analytical methods for comprehensive polar lipid profiling using SFC/MS.

Characteristics of Supercritical Fluid Chromatography

A supercritical fluid (SF) is a substance whose temperature and pressure are above the critical point (Figure 13.1). The density and solvation power of an SF approach those of a liquid, but the viscosity is similar to that of a gas. In addition, the diffusion coefficient of an SF is in between that of a gas and a liquid (Table 13.A). Because of these properties, an SF is a suitable mobile phase in chromatography. The gas-like diffusion coef-

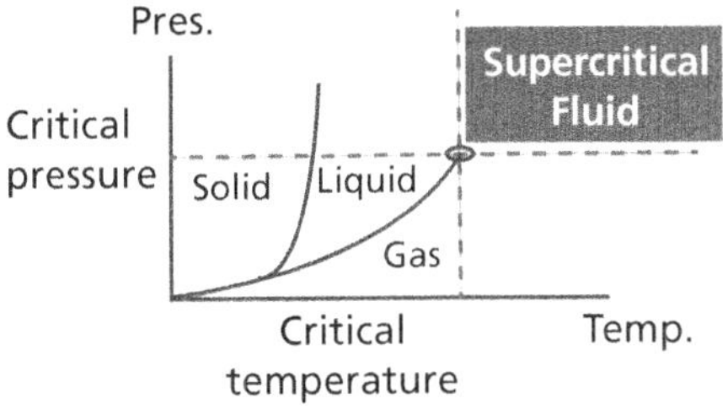

Figure 13.1 Phase diagram.

Table 13.A Physical Properties in Each Phase

Properties	Gas	Supercritical Fluid	Liquid
Density [$kg \cdot m^{-3}$]	0.6–2	300–900	700–1600
Diffusion coefficient [10^{-9} $m^2 \cdot s^{-1}$]	1000–4000	20–700	0.2–2
Viscosity [10^{-5} $Pa \cdot s$]	1–3	1–9	200–300

ficient results in higher resolution when compared to liquid chromatography (LC), and the low viscosity allows for a higher flow rate, which improves throughput without lowering chromatographic resolution. The solvent properties of supercritical carbon dioxide ($ScCO_2$), which is frequently used as the mobile phase in SFC, can be continuously changed by controlling the pressure and temperature. The polarity of $ScCO_2$ is similar to the polarity of *n*-hexane, and therefore, nonpolar compounds are readily soluble in $ScCO_2$, whereas polar compounds are not. In addition, $ScCO_2$ and methanol, which have very different polarities, are miscible, and therefore the polarity of the mobile phase can be changed by adding organic solvents, such as methanol, as a modifier to $ScCO_2$.

Because the polarity of the mobile phase can be increased by the addition of modifiers such as methanol, ethanol, acetonitrile, and 2-propanol, polar lipids are also soluble in the SFC mobile phase. This is a key advantage of SFC. The addition of a modifier has a greater effect on solvation power than changes to the column oven temperature or pressure. Therefore, the separation can typically be adjusted by changing the composition of the $ScCO_2$ and modifier, and both isocratic and gradient elution methods can be utilized. In other words, the solvation power can be controlled and can be changed continually over a wide range of polarities. Consequently, SFC is a powerful separation tool, not only for nonpolar lipids but also for a diverse variety of polar lipids.

SFC can be coupled with MS as well as with general ionization, atmospheric pressure chemical ionization (APCI), or electrospray ionization (ESI). Moreover, desolvation from the SFC mobile phase can be carried out more rapidly compared with that from water-rich mobile phases used in LC. When $ScCO_2$ is sprayed on the ionization unit, it quickly changes from the liquid phase to the gas phase by the release of pressure. It is reported that the sensitivity of detection for several compounds increased 10 times or more when using SFC/ESI/tandem mass spectrometry (MS/MS) as compared with that in the case of LC/ESI/MS/MS (Grand-Guillaume Perrenoud et al., 2014). Commonly used types of MS (e.g., quadrupole, time of flight, or ion trap) can be coupled with SFC, and the MS mode can be selected depending on the purpose of the experiment.

Polar Lipid Profiling Using Mass Spectrometry

MS has been frequently used for lipid detection in lipidomics studies. High selectivity and sensitivity are especially required in lipidomics. Using MS, the masses of molecules are determined. These data are used in order to obtain qualitative and quantitative information regarding the targeted compounds. Polar lipids have a polar head group and several fatty acid side chains connected to a backbone such as glycerol or sphingosine (Figure 13.2). Because of the variety of structures present in the fatty acid side chains and the polar head groups, there are enormous numbers of molecular species in polar lipids. In particular, there are nearly 100 fatty acid molecular species because of the combination of the number of carbon atoms (carbon chain length) and double bonds (degree of unsaturation). Furthermore, polar lipid molecular species have a large number of isomers. It is impossible to identify compounds by measuring only the mass of the precursor ions (Table 13.B) produced by ionizing the lipid molecules. Detailed identification or discrimination of polar lipids can be performed using fragment analysis using MS/MS (Table 13.C).

Analytical Aspects

Mass Spectrometric Techniques for Polar Lipid Profiling

There are several types of MS modes based on MS/MS analysis used for polar lipid profiling (Figure 13.3) (Han et al., 2012). These MS modes are used depending on the purpose of the experiment.

For screening of unknown samples, nontargeted analysis is required using full scanning and product ion scanning. For precise identification of individual lipid molecular species, high mass accuracy and high-resolution mass spectrometers such as time-of-flight (TOF) mass spectrometers (Kim et al., 2011) and Orbitrap Fourier

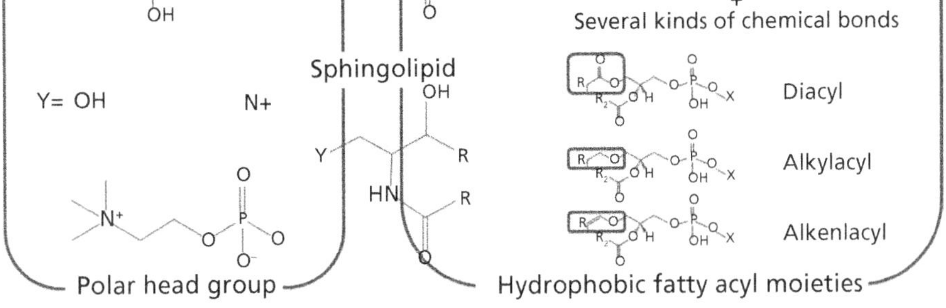

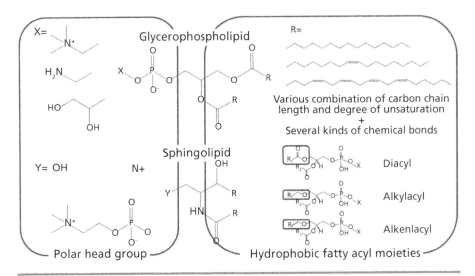

Figure 13.2 Chemical structures of polar lipids.

Table 13.B Precursor Ions Detected in Each Analysis by Flow Injection

Lipid Class	Positive	Negative	Lipid Class	Positive	Negative
PC	$[M + H]^+$	$[M + HCOO]^-$	PI	$[M + NH_4]^+$	$[M - H]^-$
LPC	$[M + H]^+$	$[M + HCOO]^-$	LPI	$[M + NH_4]^+$	$[M - H]^-$
PE	$[M + H]^+$	$[M - H]^-$	PG	$[M + NH_4]^+$	$[M - H]^-$
LPE	$[M + H]^+$	$[M - H]^-$	LPG	$[M + NH_4]^+$	$[M - H]^-$
Cer	$[M + H]^+$	—			

Source: Reproduced from Yamada et al. (2013) with permission.

Table 13.C Summary of Product Ions Used to Identify Lipid Molecular Species

Lipid Class	Positive	Negative
PC	PI 184.0729	NL 60.0205, PI FA
LPC	PI 184.0729	NL 60.0205, PI FA
PE	NL 141.0184	PI FA
LPE	NL 141.0184	PI FA
PI	NL 277.0368	PI 241.0119, PI FA
LPI	NL 277.0368	PI 241.0119, PI FA
PG	NL 189.0431	PI 152.9958, PI FA
LPG	NL 189.0431	PI 152.9958, PI FA
SM	PI 184.0729	NL 60.0205
Cer	PI 264.2685	—
CE	PI 369.3516	—
TAG	NL FA	—
DAG	NL FA	—
MAG	NL FA	—

Note: PI (except for lipid class): product ion; NL: neutral loss; FA: fatty acid.

Source: Reproduced from Yamada et al. (2013) with permission.

transform (Orbitrap FT) mass spectrometers (Gallart-Ayala, 2013; Taguchi and Ishikawa, 2010) are frequently used. Using full scanning, the molecular formula can be obtained and isobaric molecules can be distinguished (Figure 13.4). Using product ion scanning, polar head groups and fatty acyl moieties can be determined.

For the detection of specific lipid classes or molecular species with specific fatty acid side chains, precursor ion scanning (PIS) or neutral loss scanning (NLS) of a triple quadrupole mass spectrometer (QqQ-MS), targeting specific polar head groups or fatty acyl moieties, are used (Taguchi et al., 2005). For example, the phosphatidylcholine (PC) molecular species generates a product ion of phosphorylcholine (*m/z* 184.1) in the MS/MS analysis. Therefore, using PIS at *m/z* 184.1, all PC molecular species in the scan range are detected simultaneously, independent of their fatty acid side chains. On the other hand, the phosphatidylethanolamine (PE) molecular species generates product ions by neutral loss of 141.0 Da (phosphorylethanolamine). Therefore, using NLS at 141.0 Da, all PE molecular species in the scan range are detected simultaneously (Figure 13.5).

In order to analyze predetermined compounds, multiple reaction monitoring (MRM) or selected reaction monitoring (SRM) is used (Ikeda, 2008). In MRM mode, precursor ions whose *m/z* values coincide with that of the target compound are selected in the first quadrupole (Q1), the selected precursor ions are fragmented in the second quadrupole (Q2), and product ions whose *m/z* values coincide with those of the target compound are finally selected in the third quadrupole (Q3). In this way, MRM facilitates highly selective and highly sensitive analysis.

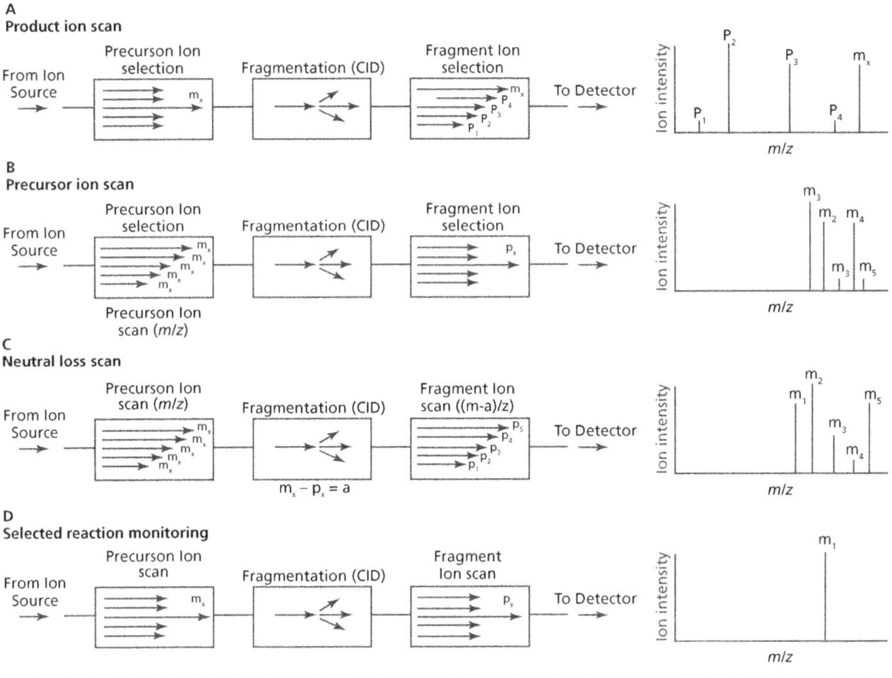

Figure 13.3 Data acquisition modes of mass spectrometry for lipid profiling (reproduced from Han et al., 2012 with permission).

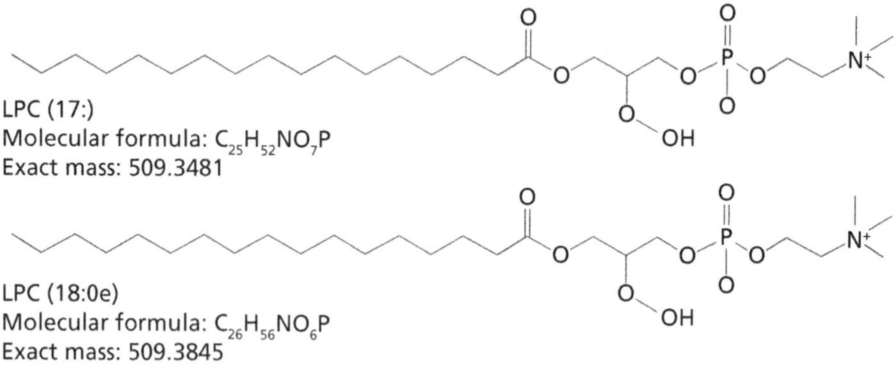

Figure 13.4 An example of isobaric lipid molecular species.

Separation Behavior of Polar Lipids in SFC

The basic configuration of SFC is similar to that of LC with some key differences (Figure 13.6). First, for SFC the pump head is chilled. CO_2 provided from a gas tank is pumped in the liquid state to obtain a stable flow. Second, a back pressure regulator (BPR) is employed in SFC. As shown in Figure 13.1, it is necessary to apply pressure above the critical point to keep the phase state as supercritical. The BPR is located after the column, and the phase state of the mobile phase inside the column is therefore maintained as supercritical or subcritical by controlling the pressure and column temperature.

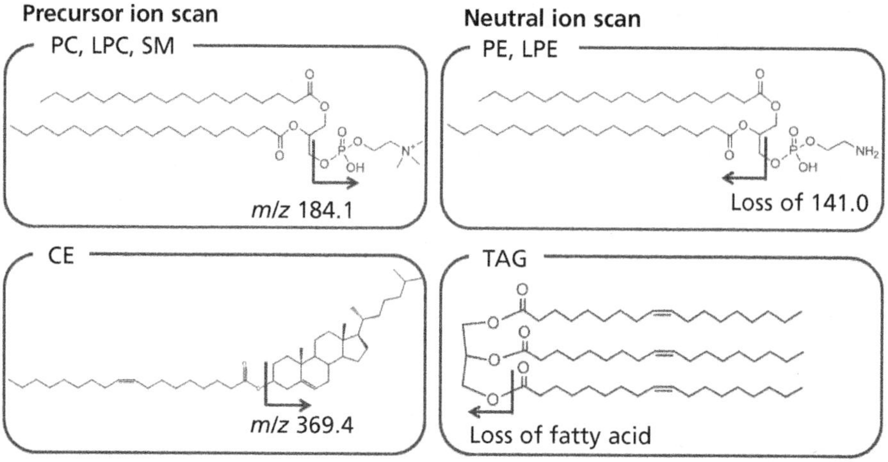

Figure 13.5 Precursor ion scanning and neutral loss scanning for lipid profiling.

SFC using a packed column has been developed for chiral separation, and it has been well accepted owing to the cost performance. However, SFC has been adapted in the light of its separation properties and has gained attention as a separation technique suitable for analytical scale separations. This subsection explains the beneficial separation behaviors of SFC for polar lipid analysis.

SFC allows for the use of dedicated columns like 2-ethylpyridine as well as the use of conventional LC columns. Notably, SFC can utilize almost all LC columns employed for both normal-phase liquid chromatography (NPLC) and reversed-phase liquid chromatography (RPLC) (Bamba, 2008). The broad selection of columns suitable for SFC enables the application of LC columns to lipid separation. The columns for NPLC, such as silica, amino, or cyano columns, allow for the separation of polar lipid molecular species based on their polar head groups. The columns for RPLC, such as octadecylsilyl (ODS, C18), C8, or phenyl columns, can separate polar lipid molecular species based on their fatty acyl moieties (Figure 13.7). In addition, C30 columns are amenable to regioisomer separation in SFC (Lee et al., 2014). The stationary phases of these columns recognize the steric differences in triacylglycerol isomeric molecular species and this separation mode would be applicable to isomeric polar lipid molecular species. The separation behaviors mentioned can be employed in SFC as well as LC, and they provide different information in the chromatogram; hence, it is important to select the proper separation mode for the required purpose in polar lipid profiling.

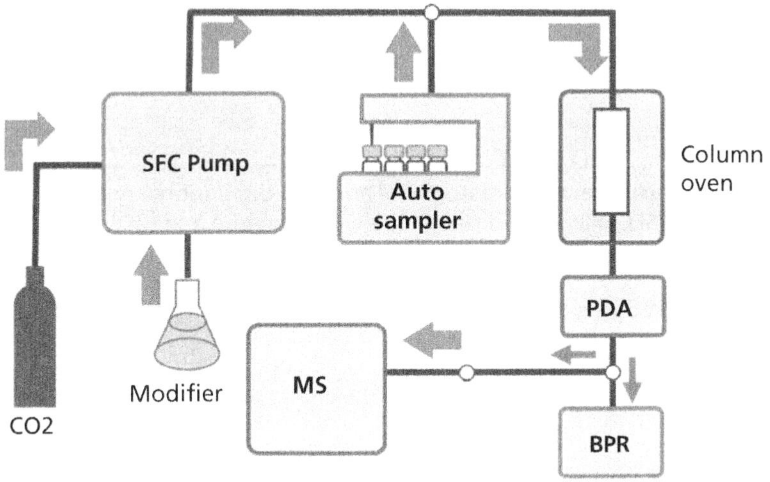

Figure 13.6 Structure of an SFC apparatus.

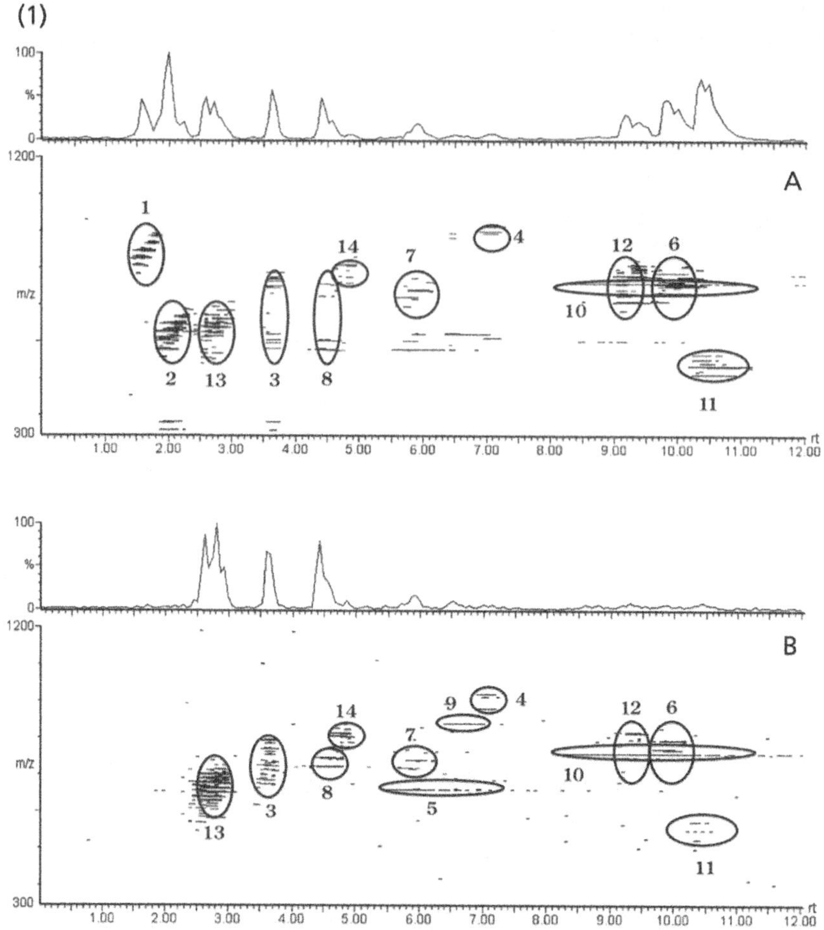

Figure 13.7 Total current chromatograms and two-dimensional maps of lipids obtained from (SFC/MS analysis using (1) a cyano column and (2) an ODS column in (A) positive ionization mode and (B) negative ionization mode. 1, TAG; 2, DAG; 3, MGDG; 4, DGDG; 5, PA; 6, PC; 7, PE; 8, PG; 9, PI; 10, PS; 11, LPC: 12, SM; 13, Cer; and 14, CB (reproduced from Bamba et al., 2008, with permission).

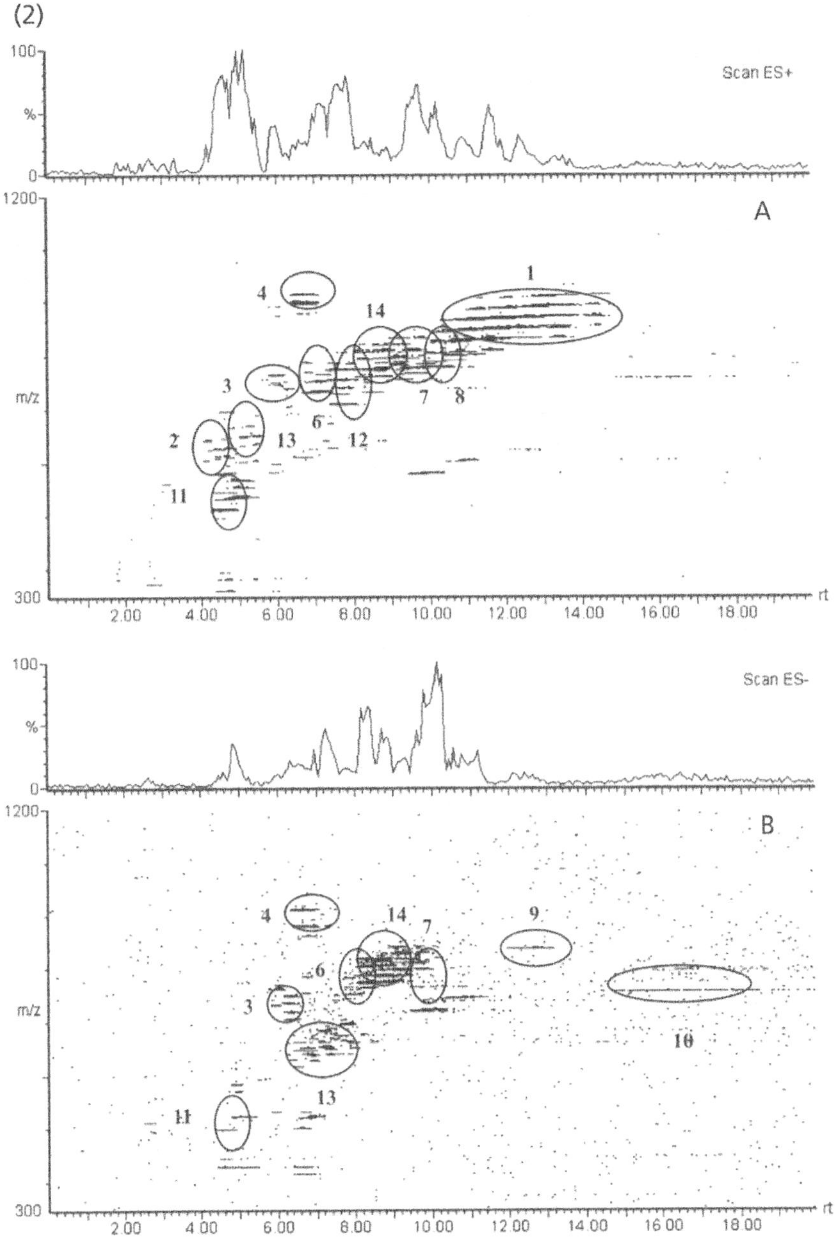

Figure 13.7 *Continued*

On the other hand, SFC-specific separation behaviors are also known in lipid profiling. An embedded polar group column separates polar lipids based on both the polar head groups and fatty acyl moieties in a single column (Figure 13.8C). This specific separation behavior is only obtained in SFC and has not been observed in LC.

Although NPLC and RPLC columns are utilized in SFC, there are other factors for the chromatographer to consider during the method development process. The first factor is the gradient condition. Typically, the gradient runs in opposite directions for NPLC and RPLC. For example, the polarity of the mobile phase changes from polar to nonpolar in RPLC but from nonpolar to polar in NPLC. However, as a rule of thumb, the gradient in SFC always runs in a single direction, from low to high composition of modifier, regardless of the column type. The second factor is the choice of column. Even columns of the same type, like ODS, have slight differences in the end-capping or the method of chemical bonding to the base particles, and these differences can have a considerable effect on the separation behavior (Yamada et al., 2013). Peak tailing is sometimes observed in the lipids composed of choline, such as PC and sphingomyeline (SM), using ODS columns without end-capping (Figure 13.8B). The strong interaction between the silanol residues and the positively charged polar head groups causes peak broadening and results in distortion of the peak shape. In other words, a column with end-capping of silanol residues is recommended if any choline-containing polar lipids are included in the target compounds.

Compared with an LC column, the 2-ethylpyridine column for SFC enables more rapid separation of oxidized phospholipid isomers (Uchikata, 2012b) (Figure 13.9). The column consisting of 2-ethylpyridine bound to silica particles has a net positive charge, whereas the oxidized lipids are negatively charged because of the oxidative modifications. Therefore, it is believed that the separation is achieved by interaction between the charges of the column and the compounds.

Several separation behaviors are observed in SFC, and it is important to take them into consideration for polar lipid separation. Generally, the first choice of column is an ODS column with end-capping because of the good peak shape obtained for polar lipids. However, an embedded polar group column may be a better option if the priority is to separate polar lipids based on their polar head groups. A C30 column or the 2-ethylpyridine column can also be chosen as an alternative option to improve the separation of structural analogs or isomers of polar lipids.

Sample Preparation for Polar Lipid Profiling

The most versatile method for the extraction of lipids is Bligh and Dyer's method (Bligh and Dyer, 1959). This extraction method is applicable to almost all types of lipids. Folch's method (Folch et al., 1957) is also frequently used. The difference between Bligh and Dyer's method and Folch's method is the use of water. Because water

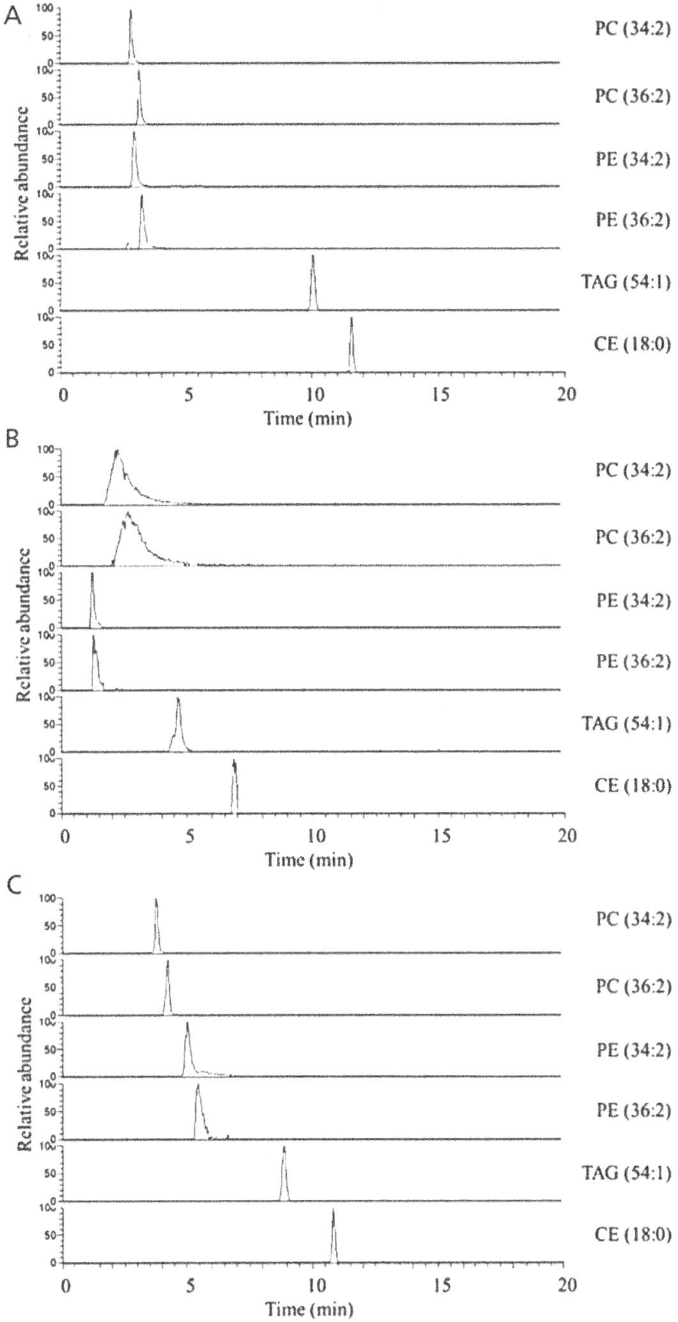

Figure 13.8 Extracted ion chromatograms of polar lipids using (A) an Inertsil ODS-4 (250 mm × 4.6 mm i.d., 5 μm; GL Sciences) column, (B) an Inertsil ODS-P (250 mm × 4.6 mm i.d., 5 μm; GL Sciences) column, and (C) an Inertsil ODS-EP (250 mm × 4.6 mm i.d., 5 μm; GL Sciences) column (reproduced from Yamada et al., 2013, with permission).

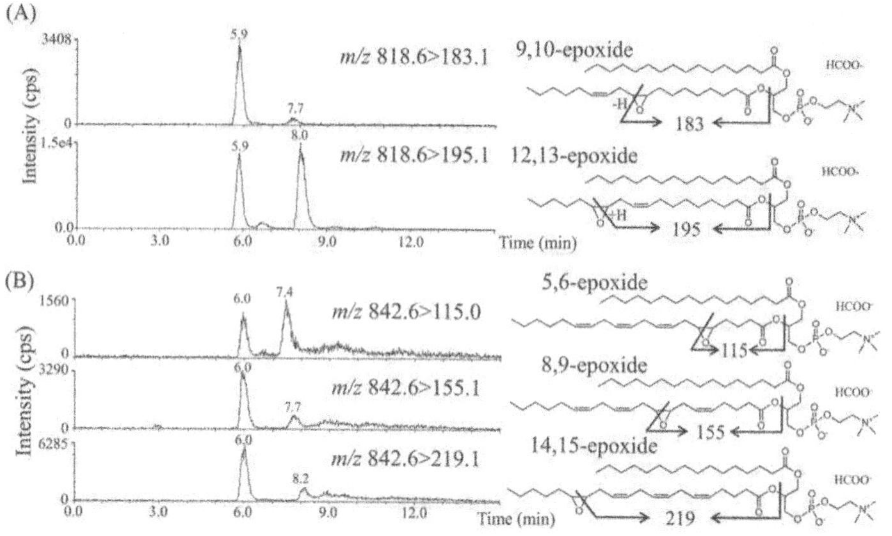

Figure 13.9 MRM chromatograms of epoxides derived from (A) PC 16:0/18:2 and (B) PC 16:0/20:4 (reproduced from Uchikata et al., 2012, with permission).

is not used in Folch's method, this method is more suitable for lipid species that are susceptible to hydrolysis.

As previously discussed, SFC is applicable to the simultaneous analysis of a diverse selection of lipids with a wide range of polarities, but peak tailing is observed for several acidic polar lipids such as phosphatidic acid (PA) and phosphatidylserine (PS) (Figure 13.10). Broad peaks lower the chromatographic separation efficiency and the detection sensitivity in MS. Because this extended elution is caused by a too-strong interaction between the polar head groups in the lipid molecules and the column stationary phase, the peak shape is improved by chemical derivatization that modifies the polar functional groups. Although other techniques for reducing peak tailing are reported (Ogiso and Suzuki, 2008; Sato et al., 2010), these techniques are complicated and the reproducibility in long-time analysis would be low. On the other hand, the derivatization-based sample preparation method is simple with high reproducibility.

In trimethylsilyl (TMS) derivatization (Lee et al., 2011), hydroxyl groups in polar lipid molecules are substituted for trimethylsiloxy groups (Figure 13.11, Table 13.D). As a result, the peak shape of the polar lipids such as PA and sphongosine-1-phosphate (So1P) is improved (Figure 13.12). In addition, the limit of detection (LOD) in MS for most of the target polar lipids is decreased except for PC and PE, which are abundant in many biological tissues.

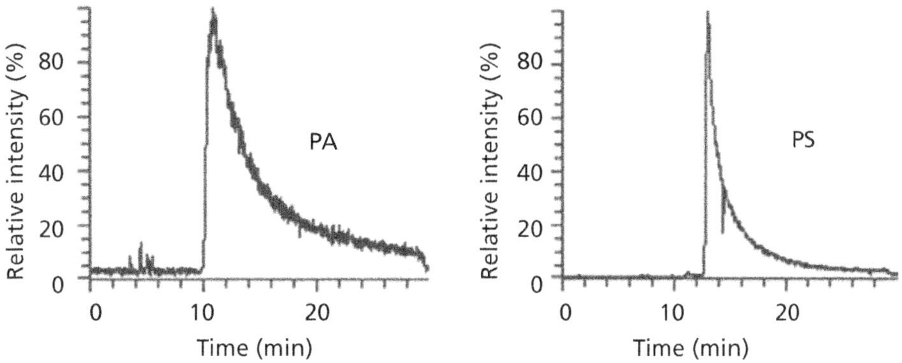

Figure 13.10 Tailings of (A) phosphatidic acid and (B) phosphatidylserine (reproduced from Ogiso et al., 2008, with permission).

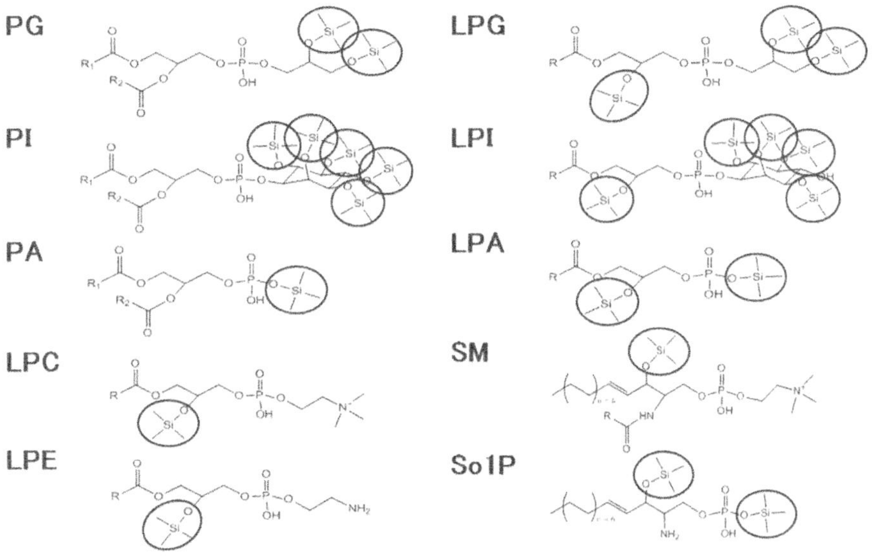

Figure 13.11 Chemical structures of silylated polar lipids.

Table 13.D MRM Transitions for Silylated Polar Lipids

Polar Lipids	Nonsilylation		TMS Silylation	
	Polarity	MRM Transitions	Polarity	MRM Transitions
PG	Positive	[M+NH4] > [M+NH4-189]	Positive	[M+H] > [M+H-316]
PI	Positive	[M+H] > [M+H-260]	Positive	[M+H] > [M+H-620]
PA	Positive	[M+NH4] > [M+NH4-115]	Positive	[M+H] > [M+H-170]
LPC	Positive	[M+H] > 184	Positive	[M+H] > 184
LPE	Positive	[M+H] > [M+H-141]	Positive	[M+H] > [M+H-141]
LPG	Negative	[M-H] > [M-H-228]	Positive	[M+H] > [M+H-316]
LPI	Positive	[M+Na] > 283	Positive	[M+H] > [M+H-548]
LPA	Negative	[M-H] > 153	Positive	[M+H] > [M+H-170]
SM	Positive	[M+H] > 184	Positive	[M+H] > 184
So1P	Positive	[M+H] > 264	Positive	[M+H] > 264

Source: Compiled from Lee at al. (2011).

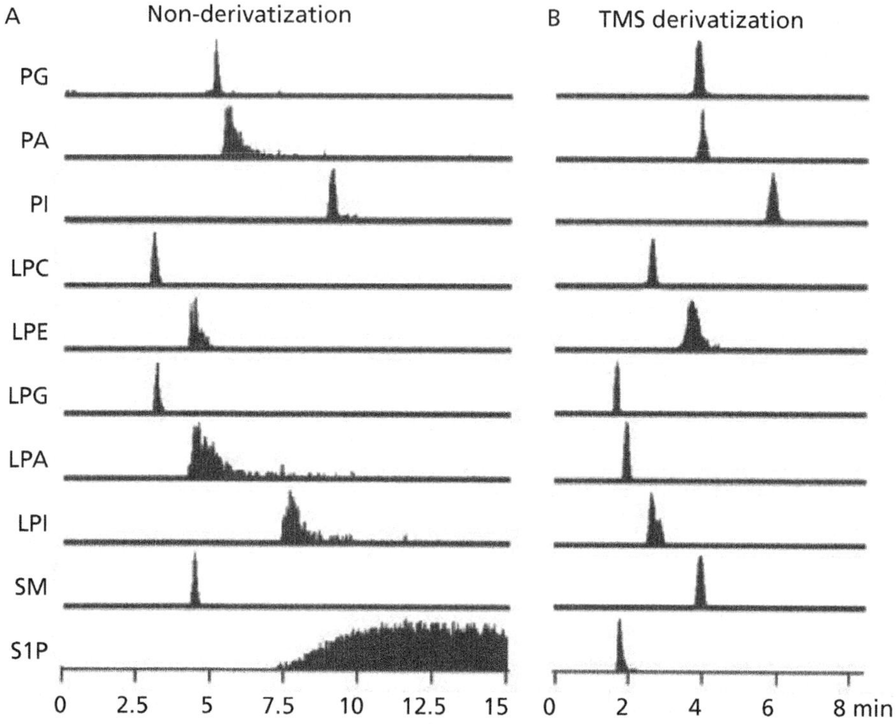

Figure 13.12 Chromatograms of (A) nonsilylated and (B) silylated polar lipids (reproduced from Lee et al., 2011, with permission)..

Although TMS derivatization is not applicable to phosphatidylserine (PS) and lysophosphatidylserine (LPS), the peak shapes of PS and LPS are improved by trimethylsilyldiazomethane (TMSD) derivatization (Lee et al., 2013) (Figure 13.13). The use of TMSD allows for methylation of the free hydroxyl groups in the phosphate moieties (Figure 13.14, Table 13.E). TMSD methylation also enhances the LOD for most of the polar lipids.

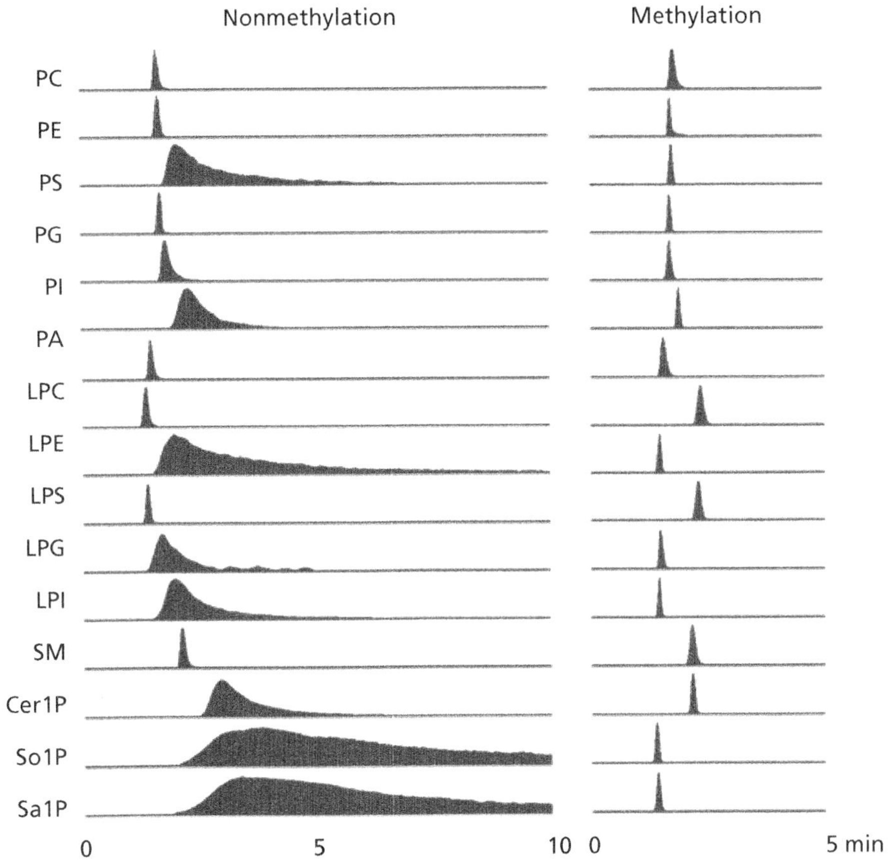

Figure 13.13 MRM chromatograms of nonmethylated and methylated polar lipids (reproduced from Lee et al., 2011, with permission).

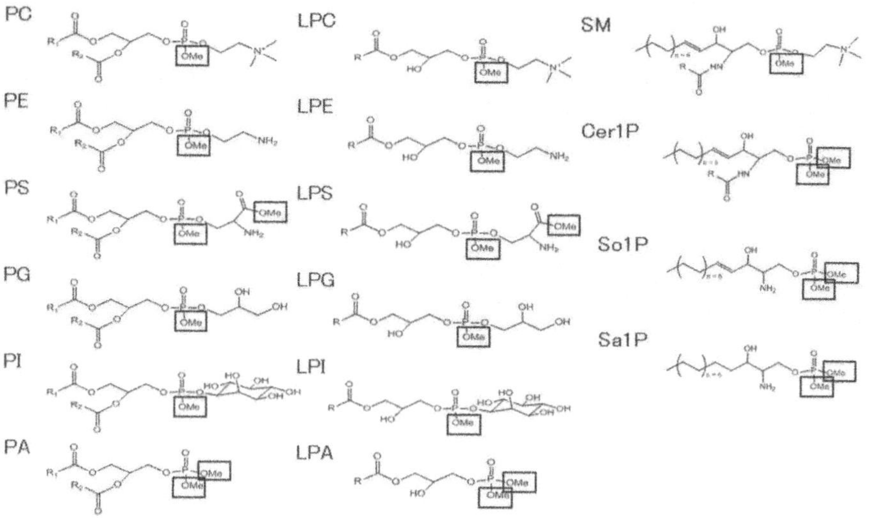

Figure 13.14 Structure of methylated polar lipids (reproduced from Lee et al., 2013, with permission).

Table 13.E MRM Transitions for Methylated Polar Lipids

Polar Lipids	Nonmethylation		Methylation	
	Polarity	**MRM Transitions**	**Polarity**	**MRM Transitions**
PC	Positive	[M+H] > 184	Positive	[M+H] > 198
PE	Positive	[M+H] > [M+H-141]	Positive	[M+H] > [M+H-155]
PS	Positive	[M+H] > [M+H-185]	Positive	[M+H] > [M+H-213]
PG	Positive	[M+NH4] > [M+NH4-189]	Positive	[M+H] > [M+H-186]
PI	Positive	[M+H] > [M+H-260]	Positive	[M+H] > [M+H-274]
PA	Positive	[M+NH4] > [M+NH4-115]	Positive	[M+H] > [M+H-126]
LPC	Positive	[M+H] > 184	Positive	[M+H] > 198
LPE	Positive	[M+H] > [M+H-141]	Positive	[M+H] > [M+H-155]
LPS	Positive	[M+H] > [M+H-185]	Positive	[M+H] > [M+H-213]
LPG	Negative	[M-H] > [M-H-228]	Positive	[M+H] > [M+H-186]
LPI	Positive	[M+Na] > 283	Positive	[M+Na] > 297
LPA	Negative	[M-H] > 153	Positive	[M+H] > [M+H-126]
SM	Positive	[M+H] > 184	Positive	[M+H] > 198
Cer1P	Positive	[M+H] > 264	Positive	[M+H] > 264
So1P	Positive	[M+H] > 264	Positive	[M+H] > 264
Sa1P	Positive	[M+H] > 266	Positive	[M+H] > 266

Source: Compiled from Lee at al. (2013).

Processing Technologies

Workflow for Data Analysis in Polar Lipid Profiling

MRM analysis can identify a target compound using the *m/z* values of precursor and product ions. It is possible to calculate the quantity of the target compound using the peak area values of a chromatogram. If several compounds have the same transition (isobaric compounds or isomers), identification can be performed based on the retention times. However, it is often the case that a compound cannot be identified using the information for only one product ion. Generally, a compound should be identified after confirming that the peaks of several chromatograms were detected at the same retention time.

Lipid profiling by precursor and neutral loss ion scanning can be performed first by selecting the *m/z* of monoisotopic ions from the mass spectra and then performing qualitative and quantitative analysis of these extracted ion chromatograms (EIC). In most cases, multiple scanning might be needed for polar lipid identification, as with MRM.

Finally, data analysis of product ion scanning of accurate mass analysis was performed by referencing exact masses of precursor ions and product ions as previously discussed.

Software Tools for Lipid Identification and Quantitation

During analysis of the MS scan data, precursor ions and product ions should be checked for all detected compounds. Therefore, analysis of large datasets is very complicated and time-consuming. In recent years, software tools for high-throughput identification and quantitation of individual lipid molecular species, such as Lipid Search and LipidBlast (Kind et al., 2013) (Figure 13.15), have been developed. Using these software programs or databases, several hundreds of lipid molecular species existing in the samples are identified and quantified in several minutes.

In addition, for QqQ-MS data, overlapping of isotopomers, as previously mentioned, should be considered. Therefore, a suitable data-processing tool has been constructed for qualitative and quantitative data analysis of PIS and NLS (Zeng et al., 2013). Moreover, software for simulating remodeling of fatty acid side chains (Kiebish et al., 2010) and metabolic pathways (Han et al., 2013) has been developed.

Applications

Lipidomics Using SFC/MS

As previously discussed, SFC is eminently suitable for simultaneous analysis of diverse lipids with a wide range of polarities, and the various amounts of lipid molecular species present in biological samples can be detected, identified, and quantified by coupling SFC with MS. When the total lipid extract obtained from plasma was

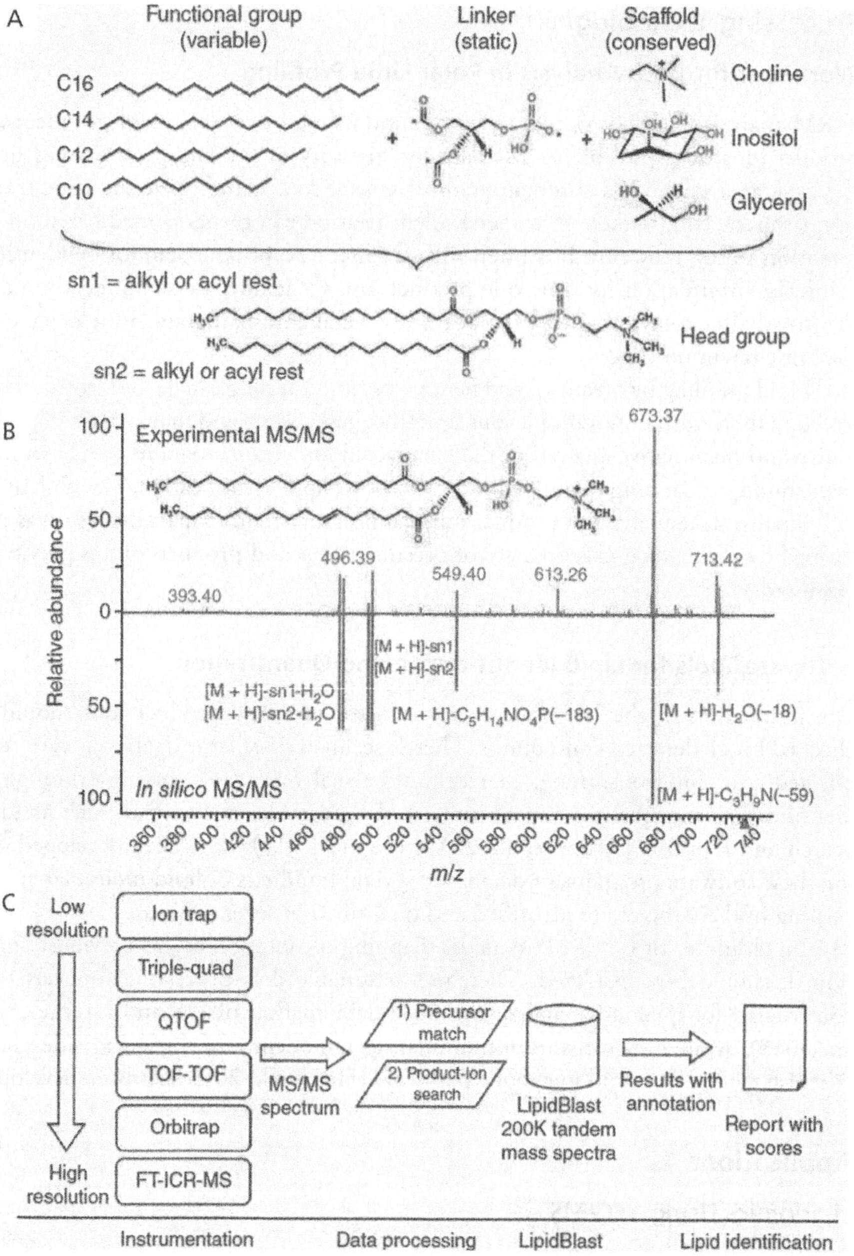

Figure 13.15 Data analysis using LipidBlast library (reproduced from Kind et al., 2013).

subjected to SFC/MS analysis and the raw MS data were entered into Lipid Search software, more than 500 lipid molecular species were detected within 20 minutes. Using this SFC/MS method, PC and PE molecular species, which are abundant in many biological samples, were successfully separated.

SFC/MS is applicable to lipid profiling of diverse samples. The lipid mixture extracted from *Catharanthus roseus* was analyzed, and polar lipids including phospholipid and glycolipid were detected. Moreover, in the lipid profiling of soybeans, PC molecular species were detected, as were triacylglycerols (Lee et al., 2012).

Online Supercritical Fluid Extraction-Supercritical Fluid Chromatography/ Mass Spectrometry (Online SFE-SFC/MS)

Supercritical fluid extraction (SFE) is applicable to hydrophobic compounds with high extraction efficiency owing to the high infusibility and low polarity of $ScCO_2$. In addition, polar compounds can also be extracted by the addition of a polar modifier to $ScCO_2$. Moreover, SFE can be directly connected to SFC. The online SFE-SFC system facilitates automation of the extraction and separation processes and enables high-throughput performance for the extraction and analysis of compounds in darkness and under anoxic conditions (Figure 13.16). Therefore, this analysis method is suitable for compounds susceptible to photolytic degradation and oxidation (Matsubara et al., 2012).

An online SFE-SFC/MS has applicability to polar lipid profiling, and a method for high-throughput polar lipid profiling has been developed for determination of polar lipids in dried blood plasma (Uchikata, 2012a). In this method, the extraction and peak separation were completed in 20 minutes (Figure 13.17), and over 130 molecular species were annotated in the blood plasma. Moreover, this method confirmed that SFE has significantly higher extraction efficiency than Bligh and Dyer's extraction method. Extraction conditions for online SFE-SFC/MS should be optimized for the separation conditions. With phospholipids, the increase in modifier in the composition enhances the extraction efficiency, but this also causes broadening of peaks (Figure 13.18). The peak shape can be improved by reducing the amount of modifier, and as a result, phospholipids can be separated from other compounds. Online SFE-SFC/MS analysis has the potential to emerge as the preferred method for the high-throughput analysis of a large number of specimens or in clinical assays.

Future Trends

Future Perspectives in Chromatographic Separation

Polar lipids consist of a polar head group and fatty acid side chains, and therefore this combined complexity results in the existence of hundreds of thousands of molecular

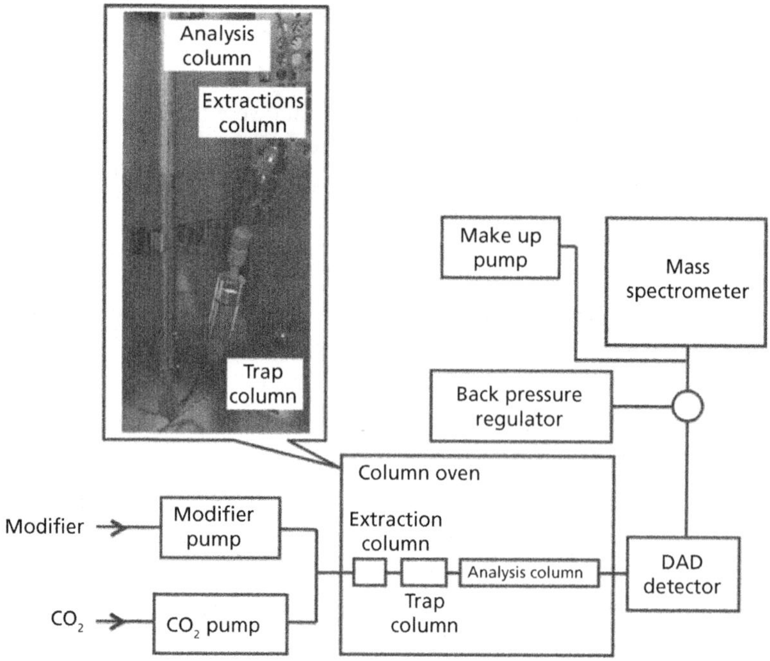

Figure 13.16 Online SFE-SFC/MS apparatus (reproduced from Matsubara et al., 2012, with permission).

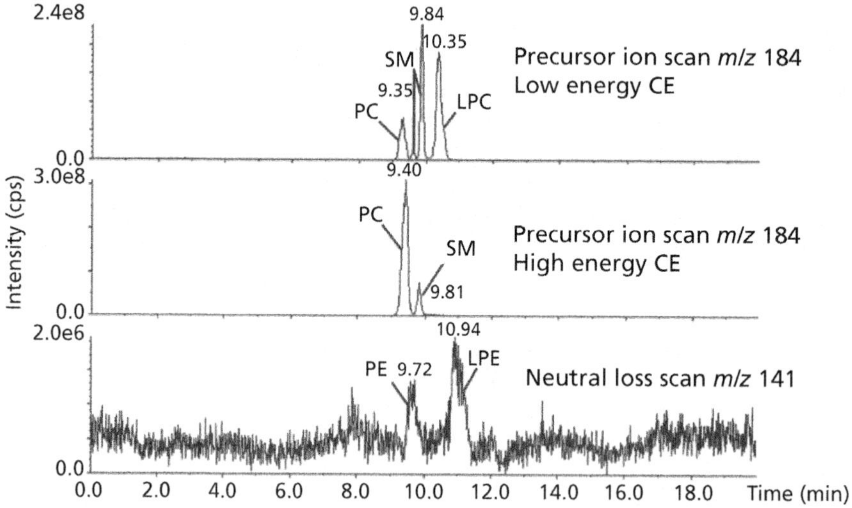

Figure 13.17 Total ion current chromatograms obtained from online SFE-SFC/MS analysis (reproduced from Uchikata et al., 2012, with permission).

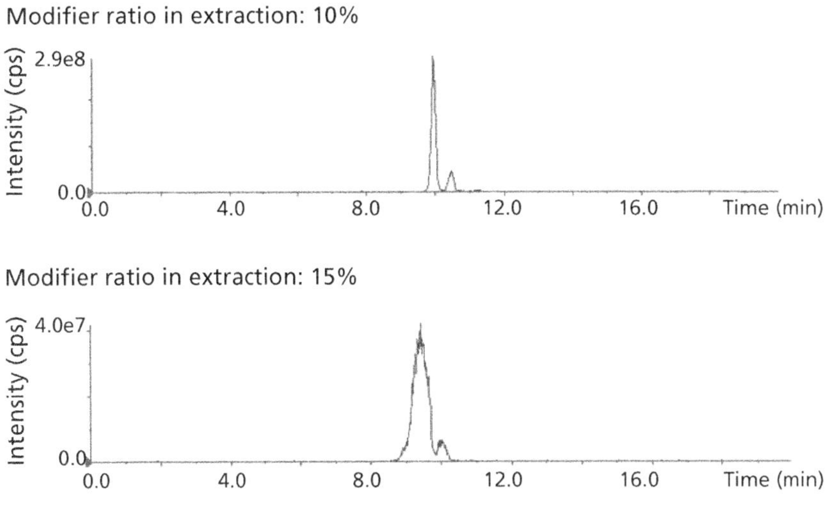

Figure 13.18 Extracted ion chromatograms of polar lipid molecular species with different modifier ratio in extraction.

species, including structural analogs and regioisomers. Information on the retention time and mass fragmentation is necessary for their accurate identification, and the identification factors are usually confirmed using a standard sample. We often encounter an identification issue if a compound whose retention time and/or fragmentation are unknown appears on the chromatogram. Because of the diversity of polar lipids, it is almost impossible to prepare standard samples of all the lipids present in biological samples. However, we can overcome this problem if we can develop a prediction model of retention time and mass fragmentation for each polar lipid molecular species. Although mass fragmentation is predictable by means of chemoinformatics, retention time has remained unpredictable until now because of the lack of sufficient work on understanding and comparing the retention behaviors of polar lipid molecular species in SFC. Therefore, we hope further investigations will improve the understanding of these mechanisms and help establish a retention time prediction model of polar lipids using SFC. This will help in identifying polar lipids more precisely and will allow us to enhance comprehensive polar lipid profiling.

Future Perspectives in Data Analysis

As previously mentioned, use of retention times obtained from SFC enables more precise identification of individual lipid molecular species. In RPLC, the retention pattern is determined by the carbon chain length and degree of unsaturation (Figure 13.19)

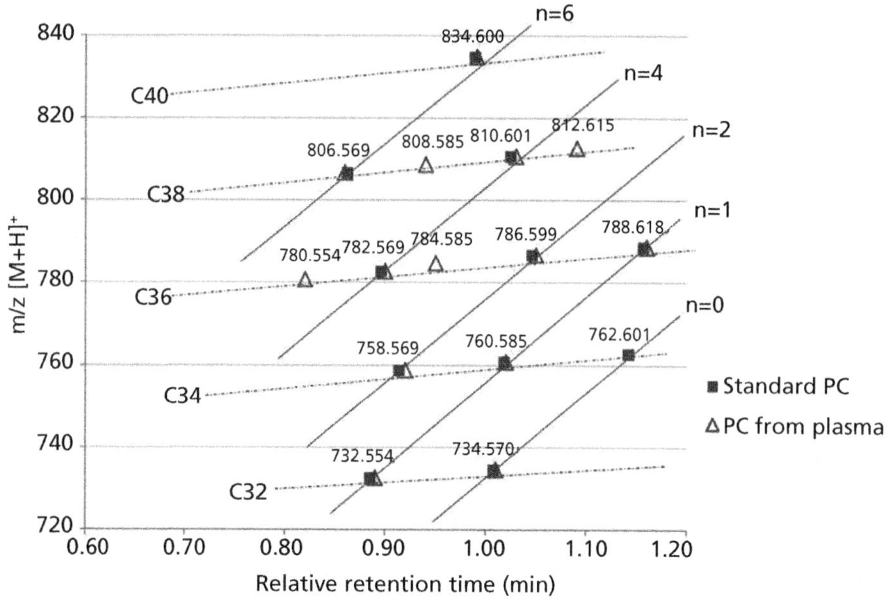

Figure 13.19 Elution patterns of lipid molecular species in reversed-phase liquid chromatography (reproduced from Choi et al., 2014, with permission).

(Choi et al., 2014). Using a reversed-phase column in SFC, the retention time of compounds in the same lipid class possessing the same carbon chain length in the fatty acid chain become shorter as the degree of unsaturation increases, whereas the retention time of compounds in the same lipid class and with the same degree of unsaturation become longer as the carbon chain length increases. Therefore, a retention time library can be constructed by analyzing the separation behavior of SFC.

In addition, an alignment method for SFC can be developed by confirming the reproducibility of the retention times, and this would aid more precise identification.

Conclusion

In this chapter, the present situation and future perspectives of polar lipid profiling using SFC/MS are summarized. SFC enables high throughput and comprehensive separation of lipids with diverse polarities. In addition, the sensitivity can be enhanced by coupling to MS. We expect that further study of this separation mechanism will make SFC a useful tool that provides the complex lipid profile of polar lipids, including the diverse structural analogs and isomers that are frequently found in biological systems.

References

Bamba, T.; Shimonishi, N.; Matsubara, A.; Hirata, K.; Nakazawa, Y.; Kobayashi, A.; et al. High Throughput and Exhaustive Analysis of Diverse Lipids by Using Supercritical Fluid Chromatography-Mass Spectrometry for Metabolomics. *J. Biosci. Bioeng.* **2008,** *105*, 460–469.

Bligh, E. G.; Dyer, W. J. A Rapid Method of Total Lipid Extraction and Purification. *Canadian Journal of Biochemistry and Physiology* **1959,** *37*, 911–917.

Cert, A.; Moreda, W.; Pérez-Camino, M. C. Chromatographic Analysis of Minor Constituents in Vegetable Oils. *J. Chromatogr. A* **2000,** *881*, 131–148.

Choi, J. M.; Kim, T.-E.; Cho, J.-Y.; Lee, H. J.; Jung, B. H. Development of Lipidomic Platform and Phosphatidylcholine Retention Time Index for Lipid Profiling of Rosuvastatin Treated Human Plasma. *J. Chromatogr. B. Analyt. Technol. Biomed. Life Sci.* **2014,** *944*, 157–165.

Folch, J.; Lees, M.; Sloane Stanley, G. H. A Simple Method for the Isolation and Purification of Total Lipids from Animal Tissues. *J. Biol. Chem.* **1957,** *226*, 497–509.

Gallart-Ayala, H.; Courant, F.; Severe, S.; Antignac, J.-P.; Morio, F.; Abadie, J.; LeBizec, B. Versatile Lipid Profiling by Liquid Chromatography-High Resolution Mass Spectrometry Using All Ion Fragmentation and Polarity Switching. Preliminary Application for Serum Samples Phenotyping Related to Canine Mammary Cancer. *Anal. Chim. Acta.* **2013,** *796*, 75–83.

Grand-Guillaume Perrenoud, A.; Veuthey, J.-L.; Guillarme, D. Coupling State-of-the-art Supercritical Fluid Chromatography and Mass Spectrometry: From Hyphenation Interface Optimization to High-sensitivity Analysis of Pharmaceutical Compounds. *J. Chromatogr. A* **2014,** *1339*, 174–184.

Han, R. H.; Wang, M.; Fang, X.; Han, X. Simulation of Triacylglycerol Ion Profiles: Bioinformatics for Interpretation of Triacylglycerol Biosynthesis. *J. Lipid Res.* **2013,** *54*, 1023–1032.

Han, X.; Yang, K.; Gross, R. W. Multi-dimensional Mass Spectrometry-based Shotgun Lipidomics and Novel Strategies for Lipidomic Analyses. *Mass Spec. Rev.* **2012,** *31*, 134–178.

Ikeda, K.; Shimizu, T.; Taguchi, R. Targeted Analysis of Ganglioside and Sulfatide Molecular Species by LC/ESI-MS/MS with Theoretically Expanded Multiple Reaction Monitoring. *J. Lipid Res.* **2008,** *49*, 2678–2689.

Kiebish, M. A.; Bell, R.; Yang, K.; Phan, T.; Zhao, Z.; Ames, W.; Seyfried, T. N.; Gross, R. W.; Chuang, J.; Han, X. Dynamic Simulation of Cardiolipin Remodeling: Greasing the Wheels for an Interpretative Approach to Lipidomics. *J. Lipid Res.* **2010,** *51*, 2153–2170.

Kim, H.-J.; Kim, J. H.; Noh, S.; Hur, H. J.; Sung, M. J.; Hwang, J.-T.; Park, J. H.; Yang, H. J.; Kim, M. S.; Kwon, D. Y.; et al. Metabolomic Analysis of Livers and Serum from High-Fat Diet Induced Obese Mice. *J. Proteome Res.* **2011,** *10*, 722–731.

Kind, T.; Liu, K.-H.; Lee, D. Y.; DeFelice, B.; Meissen, J. K.; Fiehn, O. Lipid Blast in Silico Tandem Mass Spectrometry Database for Lipid Identification. *Nat. Methods.* **2013,** *10*, 755–758.

Lee, J. W.; Yamamoto, T.; Uchikata, T.; Matsubara, A.; Fukusaki, E.; Bamba, T. Development of a Polar Lipid Profiling Method by Supercritical Fluid Chromatography/Mass Spectrometry. *J. Sep. Sci.* **2011,** *34*, 3553–3560.

Lee, J. W.; Uchikata, T.; Matsubara, A.; Nakamura, T.; Fukusaki, E.; Bamba, T. Application of Supercritical Fluid Chromatography/Mass Spectrometry to Lipid Profiling of Soybean. *J. Biosci. Bioeng.* **2012,** *113*, 262–268.

Lee, J. W.; Nishiumi, S.; Yoshida, M.; Fukusaki, E.; Bamba, T. Simultaneous Profiling of Polar Lipids by Supercritical Fluid Chromatography/Tandem Mass Spectrometry with Methylation. *J. Chromatogr. A.* **2013,** *1279*, 98–107.

Lee, J. W.; Nagai, T.; Gotoh, N.; Fukusaki, E.; Bamba, T. Profiling of Regioisomeric Triacylglycerols in Edible Oils by Supercritical Fluid Chromatography/Tandem Mass Spectrometry. *J. Chromatogr. B. Analyt. Technol. Biomed. Life Sci.* **2014,** *966*, 193–199.

Matsubara, A.; Harada, K.; Hirata, K.; Fukusaki, E.; Bamba, T. High-accuracy Analysis System for the Redox Status of Coenzyme Q10 by Online Supercritical Fluid Extraction-Supercritical Fluid Chromatography/Mass Spectrometry. *J. Chromatogr. A* **2012,** *1250*, 76–79.

Ogiso, H.; Suzuki, T.; Taguchi, R. Development of a Reverse-phase Liquid Chromatography Electrospray Ionization Mass Spectrometry Method for Lipidomics, Improving Detection of Phosphatidic Acid And Phosphatidylserine. *Anal. Biochem.* **2008,** *375*, 124–131.

Sato, Y.; Nakamura, T.; Aoshima, K.; Oda, Y. Quantitative and Wide-ranging Profiling of Phospholipids in Human Plasma by Two-dimensional Liquid Chromatography/Mass Spectrometry. *Anal. Chem.* **2010,** *82*, 9858–9864.

Taguchi, R.; Ishikawa, M. Precise and Global Identification of Phospholipid Molecular Species by an Orbitrap Mass Spectrometer and Automated Search Engine Lipid Search. *J. Chromatogr. A* **2010,** *1217*, 4229–4239.

Taguchi, R.; Houjou, T.; Nakanishi, H.; Yamazaki, T.; Ishida, M.; Imagawa, M.; Shimizu, T. Focused Lipidomics by Tandem Mass Spectrometry. *J. Chromatogr. B. Analyt. Technol. Biomed. Life Sci.* **2005,** *823*, 26–36.

Uchikata, T.; Matsubara, A.; Fukusaki, E.; Bamba, T. High-throughput Phospholipid Profiling System Based on Supercritical Fluid Extraction-Supercritical Fluid Chromatography/Mass Spectrometry for Dried Plasma Spot Analysis. *J. Chromatogr. A.* **2012a,** *1250*, 69–75.

Uchikata, T.; Matsubara, A.; Nishiumi, S.; Yoshida, M.; Fukusaki, E.; Bamba, T. Development of Oxidized Phosphatidylcholine Isomer Profiling Method Using Supercritical Fluid Chromatography/Tandem Mass Spectrometry. *J. Chromatogr A* **2012b,** *1250*, 205–211.

van Meer, G.; Voelker, D. R.; Feigenson, G. W. Membrane Lipids: Where They Are and How They Behave. *Nat. Rev. Mol. Cell Biol.* **2008,** *9*, 112–124.

Wymann, M. P.; Schneiter, R. Lipid Signaling in Disease. *Nat. Rev. Mol. Cell Biol.* **2008,** *9*, 162–176.

Yamada, T.; Uchikata, T.; Sakamoto, S.; Yokoi, Y.; Nishiumi, S.; Yoshida, M.; Fukusaki, E.; Bamba, T. Supercritical Fluid Chromatography/Orbitrap Mass Spectrometry Based Lipidomics Platform Coupled with Automated Lipid Identification Software for Accurate Lipid Profiling. *J. Chromatogr. A* **2013,** *1301*, 237–242.

Zeng, Y.-X.; Mjøs, S.; Meier, S.; Lin, C.-C.; Vadla, R. Least Squares Spectral Resolution of Liquid Chromatography-Mass Spectrometry Data of Glycerophospholipids. *J. Chromatogr. A* **2013,** *1280*, 23–34.

14

Omega-3 Phospholipids

Kangyi Zhang ■ *Institute of Agricultural Products Processing, Henan Academy of Agriculture Sciences, Henan Province, China*

Introduction

Phospholipids (PLs) are major constituents of cell membranes and play crucial roles in the biochemistry and physiology of the cell (Mead et al., 1986; Vance et al.; 1993). They have been widely used in food, pharmaceutical, and cosmetic products as highly proficient emulsifiers. Omega-3 phospholipids (n-3 PLs) are defined as PLs containing n-3 long-chain polyunsaturated fatty acids (PUFAs). It is increasingly accepted that n-3 fatty acids, especially n-3 PUFAs eicosapentaenoic acid (EPA, C20:5n-3) and docosahexaenoic acid (DHA, C22:6n-3), in high levels benefit a variety of human diseases, and their effects on neurological diseases are still emerging (Guinot et al., 2013; Mangano et al., 2013; Sedlmeier et al., 2014). Research on beneficial effects of n-3 PUFAs against atherosclerosis and thrombosis was initiated by Bang and Dyerberg in the early 1970s (Bang et al., 1971; Dyerberg et al., 1975). Later, n-3 PUFAs were shown to be crucial at cellular levels for maintaining membrane homeostasis, and for an optimal balance with n-6 fatty acids to secure a healthy body composition homeostasis, a regulated inflammatory response, a psychiatric balance and equilibrated neurology, and for maintaining membrane homeostasis (Ghasemifard et al., 2014; Zendedel et al., 2015). The positive influence of n-3 PUFAs on human health is now well established (Abedi and Sahari, 2014; van de Rest and de Groot, 2014), and there is a growing demand for n-3 PUFAs by the pharmaceutical industry in the form of PL as well as in their natural triacylglycerol (TAG) form (Carlson, 1991). Most of the clinical studies have been carried out using n-3 FAs bound to TAGs or ethyl esters (Tocher, 2015).

As will be discussed in this chapter, the therapeutic and preventive effects of n-3 PLs seem to go beyond those of n-3 neutral lipids. Well documented are their cardiovascular benefits and their positive effects on inflammation, blood pressure, blood sugar, and premenstrual symptoms, just to mention a few of the clinical documentations (Hosomi et al., 2012). This chapter provides an overview of the chemical structure and biological functions of n-3 phospholipids. Processing, physical properties, and enzymatic modifications are discussed in detail. Some clinical and nutritional properties of phospholipids are also reviewed.

Phospholipid Molecular Structure

PLs are a class of lipids that normally have two FAs esterified to a glycerol backbone, while the third position is esterified to a phosphorous group that is further linked to

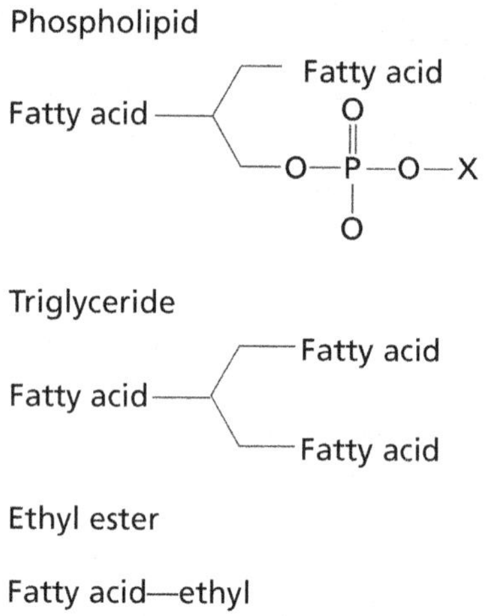

Figure 14.1 Triacylglycerol and phospholipid structures.

a headgroup, whereas TAGs consist of three FAs esterified to a glycerol backbone. In nature, n-3 FAs are mainly esterified either in PLs or TAGs and, due to a partial hydrolysis, are present in the free form. The two basic classes of PLs are glycerol phospholipids and sphingomyelins. The structures of major plant and animal PLs are presented in Figure 14.1.

The phospholipid structure is built on a glycerol backbone, linked to two fatty acids in the *sn*-1 and *sn*-2 positions and a phosphoric acid ester in the third position (Gunstone, 2008). The second free hydroxyl group of the phosphate ester (phosphatidic acid, PA) can consist of choline, ethanolamine, inositol, or serine to form phosphatidylcholine (PC), phosphatidylethanolamine (PE), phosphatidylinositol (PI), and phosphatidylserine (PS). Phosphatidylglycerol (PG) and diphosphatidylglycerol (DPG, or cardiolipin) involve a second molecule of glycerol (Figure 14.1). Sphingomyelins (SM) consist of sphingosine (a dihydroxy amine) bonded to one fatty acid at the NH_2 function and one polar head group, such as phosphocholine or phosphoethanolamine, at the CH_2OH function. The phospholipid is called an n-3 PL when there are one or two n-3 PUFAs at the *sn*-1 and *sn*-2 positions in the structure.

Sources of n-3 PLs

Food sources for n-3 PLs are very limited. The majority of n-3 PL products are made from marine organisms (Burri et al., 2012). A small amount of PLs highly enriched with n-3 fatty acids are produced from animal or plant lecithins using enzymes (Yamamoto et al., 2014). The biotechnology methods using enzymes are the most reliable for producing n-3 PLs from commercially available, less expensive animal or plant lecithins.

Marine n-3 PLs

Marine n-3 PLs are defined as PLs bounding long-chain n-3 PUFAs derived from marine organisms, which makes them different from PLs derived from vegetable sources because those do not contain long-chain n-3 PUFAs. Long-chain n-3 PUFAs are characteristic of marine oils and occur pervasively in the PLs of fish and marine species, with EPA and DHA commonly accounting for up to 50% of their fatty acid compositions (Haraldsson et al., 1993). The predominant source of marine n-3 PLs products is krill (Burri and Berge, 2013). In recent years, several alternative raw materials, such as fish eggs and meal, have been identified as attractive sources (Calder, 2014).

The distribution of the different PLs in marine resources was recently reviewed (Lu et al., 2011). According to Lu's review, the most predominant PLs in marine sources such as salmon, tuna, rainbow trout, and mackerel is PC; the second most abundant is shown to be PE. But PI, PS, lyso-PC, and sphingomyelin are found in smaller amounts.

Krill Oil

One important source of marine n-3 PLs is krill oil. Krill oil has become increasingly popular as a food supplement during the last decade, and the research on krill oil in human and animal studies is growing. The majority of the n-3 fatty acids, in particular EPA and DHA, found in krill oil are bound to phospholipids, whereas other marine oils confine these to TAGs or ethyl ester forms; the majority of these two fatty acids, EPA and DHA, are contained in the PC form (Phleger et al., 2002). The proportion of PLs in the total lipids of krill has been reported to vary between 30 and 60%, depending on krill species, age, season, and harvest time (Tou et al., 2007).The FA composition of the PLs in krill oil was recently stated in two studies (Le Grandois et al., 2009; Winther et al., 2011). A complex composition of PL species in krill oil was revealed. In most of the identified 69 different choline-containing phospholipid substances (Winther et al., 2011), 58% contained n-3 PUFA in the *sn-1* position, and 10% of the PC molecules contained n-3 PUFAs in both the *sn-1* and *sn-2* positions. The dominant prevalent n-3 PUFAs were EPA and DHA, but stearidonic acid and docosapentaenoic acid (DPA) were also detected in smaller amounts. Krill oil is extracted from the Antarctic crustacean krill (*Euphausia superba*), a shrimp-like zooplankton. Krill oil are either processed on board the harvesting vessel or kept frozen as it is until they are processed on land. The final extraction step in all cases is based on acetone extraction (Zhao et al., 2014) and ethanol extraction. Attempts to apply supercritical CO_2/ethanol as an extraction solvent failed mainly because of cost.

Fish Roe

Fish roes are fully ripe internal egg masses in the ovaries or the released external egg masses of fish and certain marine animals, such as shrimp, scallops, and sea urchins (Rao et al., 2015). Fish roe is used for human consumption and is found to be a good source of n-3 PLs (Higuchi et al., 2006; Rao et al., 2015; Shirai et al., 2006a). Roe lipids of *Wallagu attu* and *Ompok pabda* were found to contain 28% and 56% PLs, respectively (Mukhopadhyay and Ghosh, 2007). PC and PE accounted for more than 80% of the PL of roe from carps (*Cyprinus carpio* and *Carassius aurantusa*) (Lili et al., 2007). Fish roe from herring, salmon, pollock, and flying fish contain 38%–75% of their lipids in the form of PLs, for which the predominant lipid class is PC (Shirai et al., 2006b). Salmon roe contained 27.2% PL with PC as the major component (80%) (Armstrong et al., 1994). The PL fraction of Atlantic halibut (*Hippoglossus hippoglossus*) roe was 70.8%, which contained PC (62.2%) and PE (6.7%) with a high concentration of n-3 PUFA, namely EPA and DHA (Peterson et al., 1986). A high presence of these two FAs in fish roe has also been demonstrated in other studies (Higuchi et al., 2006; Lieber et al., 1994; Loy et al., 1991).

The most efficient, but probably not the most cost-effective, method for fish egg processing is drying of the raw biomass followed by ethanol extraction. The processing for fish meal PLs is completely different from that of fish egg and quite similar to vegetable oil processing. Hexane (or iso-hexane) extraction of all fish meal lipids is followed by a conventional degumming process to separate a PL concentrate from the extracted crude oil.

Fish

Fish oil, which is produced as a byproduct of the fish meal industry, was until recently the main oil incorporated into fish feed. Fish is currently the primary dietary source of EPA and DHA, in particular cold-water oily fish such as herring, tuna, salmon, sardine, anchovy, or mackerel (Strobel et al., 2012). However, the n-3 PUFA proportion and absolute content in farmed fish vary strongly depending on the fish species, the fat content of fish, and the feeding conditions in comparison to wild fish (Christiansen, 1984; Henriques et al., 2014; Kitessa et al., 2014). Because vegetable oils and oilseeds are used in aquaculture feed, the EPA and DHA to arachidonic acid and linolenic acid ratio can be as low as 1.5 for farmed salmon compared to 14.5 for wild salmon (Strobel et al., 2012). In comparison, the corresponding ratio in krill oil is 12.0 (Ulven et al., 2011).

Fish contains between 1% and 1.5% PLs and 10% and 15% TAGs (Hjaltason and Haraldsson, 2006). Depending on the kind of fish, up to one-third of the EPA and DHA content might exist in the form of PLs (Xu et al., 1996). One study has shown that in Atlantic salmon, EPA and DHA are bound to PLs and TAGs in a 40:60

ratio (Polyi and Ackman, 1992). Hence, fish represents a potential source of marine PLs, but the production of marine PLs from fish has so far been limited. Moreover, the amount of byproducts from fish is significant and thus further represents a valuable source of marine PLs. Products currently on the market differ slightly in their phospholipid and fatty acid composition depending on their source. Table 14.A gives an example (Schneider, 2013).

Structured n-3 PLs

Replacement of fatty acid residues present in a native PL by fatty acids with beneficial physiological effects can lead to tailored PLs that offer intriguing marketing opportunities for manufacturers of nutraceuticals (Vaidya and Cheema, 2014). Although PLs containing n-3 PUFAs are available in fish and marine products, refinement procedures, including laborious extraction and separation, are required for their industrial use. Modification is usually performed to obtain desired functional properties of lecithin or PLs. Thus, enzymatic modification using inexpensive plant lecithin (PC) and a rich n-3 PUFA source might be an effective method for obtaining n-3 PUFA-enriched PLs (Maligan et al., 2012).

Enzymatic modification is of interest because enzymes can be used to modify PLs in different ways (Vikbjerg et al., 2006; Yamamoto et al., 2014). Attempts to prepare PLs enriched with n-3 PUFAs from commercially available plant or animal lecithins using biotechnology methods have been carried out (Estiasih et al., 2013; Mutua and Akoh, 1993). In general, enzymes used in structured PL synthesis are a specific 1,3-lipase derived from *Rizhomucor miehei* (Haraldsson and Thorarensen, 1999; Totani and Hara, 1991) or *Thermomyces lanuginose* (Vikbjerg et al., 2005a, 2005b) and phospholipase A_1 (Cavazos-Garduno et al., 2015; García et al., 2008; Zhao, No, et al.

Table 14.A Composition of Different Marine Phospholipid Products

	Salmon Eggs	Fish Meal	Krill
Phospholipid content	35	40	40
PC	29	22	35
PE	2	6	2
PI	1	2	1
Other PLs	3	10	2
Total omega-3	35	33	24
DHA[a]	11	9	13
EPA[a]	19	20	7

[a]Based on total fatty acids.

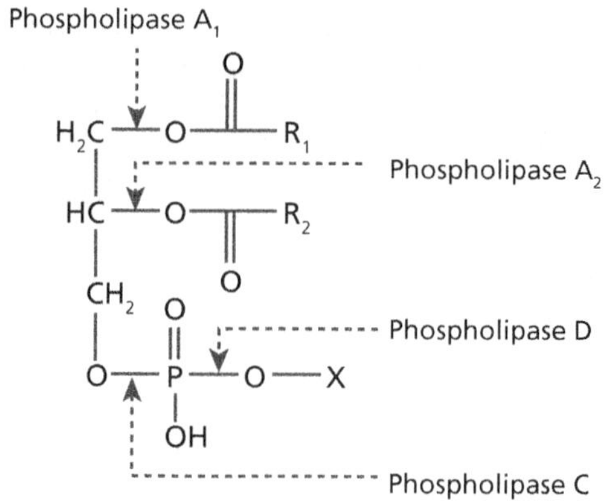

Figure 14.2 Modified position in phospholipid structure by different phospholipases.

2014) or phospholipase A_2 (Na et al., 1990), which can modify fatty acid in the *sn*-1 or *sn*-2 positions. For the modified positions in phospholipid structure by different phospholipases, see Figure 14.2.

Incorporation of n-3 fatty acid into PL structure can be controlled during reaction. Some factors that may affect the acidolysis reaction in n-3 fatty acid's incorporation into the PL structure are enzyme concentration, reaction time, and acyl donor to PL ratio. *Rhizomucor miehei* lipase (Lipozyme™) with 1,3 regiospecificity had been employed to corporate phospholipids of commercial soybean lecithin and n-3 fatty acid (Haraldsson and Thorarensen, 1999). The use of commercial soybean lecithin is more economical as a carrier of n-3 fatty acid; the price is cheaper, and the process is not too difficult. Other than traditional fish oil sources, the current and future potential sources of n-3 long-chain PUFA for n-3 PLs, specifically EPA and DHA or precursors to these key essential fatty acids (Miller et al., 2008), are:

1. Other marine sources including by-catch and marine invertebrates such as krill and copepods
2. Vegetable oils that contain biosynthetic precursors that can be used by aquatic animals to biosynthesize n-3 long-chain PUFA (Codabaccus et al., 2011)
3. Several different microbial taxa that produce single-cell oils that are rich in n-3 long-chain PUFA (Collins et al., 2014)
4. Vegetable oils derived from oilseed crops that have undergone genetic modification to contain n-3 long-chain PUFA (Ruiz-Lopez et al., 2013, 2014)

The positioning of n-3 long-chain fatty acids is also important for optimal physiological benefits. Natural n-3 PLs have the n-3 fatty acids at the *sn*-2 position. Other than, for example, in the enzyme-modified soybean PLs, however, n-3 fatty acids are randomly distributed between *sn*-1 and *sn*-2 position. But only position 2 allows an efficient transport of the fatty acid into the human body and to the target cellular structures. Of greatest importance for optimal performance and delivery into the human organism is the direct link (conjugation) of the n-3 fatty acids to the PL backbone and not just a physical blend of fish oils and vegetable PLs.

Stability of n-3 PLs

The most stable n-3 fatty acid is in the form of PLs. DHA is highly sensitive to oxidation, and toxic oxidized products such as aldehydes and hydroperoxides can be formed (Frankel, 1998; Kamal-Eldin and Yanishlieva, 2002). High amounts of oxidized products in the human body cause oxidative stress, which can induce various kinds of diseases such as diabetes, cancer, and rheumatoid arthritis (Frankel, 1998). PUFAs, EPA and DHA in particular, need to be protected, and some different strategies have been used to avoid oxidation. A common way to reduce oxidation is to apply a synthetic or natural antioxidant (Frankel, 1998). As there is a desire to reduce the amount of synthetic material in food, natural antioxidants are more attractive. One group of natural compounds that has shown antioxidant activity is PLs. Incorporation of n-3 fatty acids into the PL structure increases their oxidative stability, suggesting a more beneficial form of PLs (Maligan et al., 2012).

DHA incorporated into PLs has been found to be more resistant to oxidation than both TAGs and ethyl esters bounding DHA (Song et al., 1997). PLs are therefore highly interesting as food additives to stabilize DHA. DHA was most protected against hydroperoxide formation when it was incorporated at the *sn*-1 position of either PC or PE. In these lipids, hydroperoxide formation at carbon atoms 4, 7, 8, and 11 was completely prevented (Lyberg et al., 2005). Earlier findings have shown that the combination of tocopherol, ascorbic acid, and lecithin has a higher protective effect on PUFA than tocopherol, ascorbic acid, or lecithin alone (Yanishlieva and Marinova, 2001). To achieve the maximum protective effect, DHA should be incorporated into PC or PE (one DHA molecule per lipid molecule) and both tocopherol and ascorbic acid should be added (Lyberg et al., 2005).

All marine PL products have been shown surprisingly stable against oxidation (Burri et al., 2012). There is speculation as to whether this is due to the natural content of antioxidants (e.g., astaxanthin) co-extracted together with the other lipids and PLs from the biomass or this is a function of the PLs themselves. It seems to be both, because other nonmarine PLs, even when highly purified and devoid of antioxidants, are usually quite resistant to oxidation (Murru et al., 2013). In the case of fish meal PLs, for example, a Rancimat test showed no sign of oxidation, even at 120 °C over a time of 40 hours. This result extrapolated to shelf life at ambient temperature would give more than 1900 days of stability.

Nutritional Properties of n-3 PLs

n-3 PUFAs

The n-3 PUFAs distinguish themselves from the n-6 FAs by the position of the first double bond from the methyl end of the chain. The n-3 (DHA and EPA) and n-6 (arachiodionic acid) FAs greatly impact inflammation: n-3 reduces inflammation, and in large quantities n-6 will increase inflammation. The metabolites of n-6 cannot be converted into n-3 FAs, and humans can only poorly synthesize n-3 PUFAs from their precursor alpha-linolenic acid (ALA; C18:3n-3), which is present in plants such as flaxseed, rapeseed, soybean, walnut, perilla, hemp, and chia (Christie, 2003). The ratio between n-3 and n-6 FAs is therefore a good indication for the risk of many inflammatory diseases (Lands, 2014). If cells get damaged through oxidative stress, the fats that are attached to the phospholipid are released and go down an inflammation or anti-inflammation path (Simopoulos, 2002). If the quantity of n-6 and n-3 FAs is out of balance on a cellular level, the inflammatory processes can get out of control (Winther et al., 2011). Examples of inflammatory processes are cardiovascular disease and arthritis.

Consumption of n-3 PUFAs, particularly the long-chain FAs EPA and DHA, has been reported to have beneficial physiological effects (Janssena et al., 2015), such as reduction in the incidences of cardiovascular disease (Schuchardt and Hahn, 2013), cancer, diabetes, arthritis, and central nervous system disorders (schizophrenia, depression, Alzheimer's disease). Therefore, the development of products containing n-3 EPA and DHA are of interest for use in nutraceuticals (Kong-González et al., 2015; Runau et al., 2015; Wang et al., 2015). Despite an overwhelming amount of evidence on the beneficial effects of n-3 PUFAs on human health, some controversy remains (Chowdhury et al., 2012; Hooper et al., 2006; Marik and Varon, 2009; Rizos et al., 2012; Wang et al., 2006). This might be because published meta-analyses were based on different study inclusion criteria such as n-3 dose given, number of subjects, length of treatment, disease state of participants, comedications, dietary background (intake of fish/other n-3 products), compliance of study product intake, placebo/not placebo controlled, and so on. This led to a significant heterogeneity in conclusions. Moreover, there is a variation in the statistical strictness (p-values) used in the different analyses.

Phospholipids

PLs are better carriers for n-3 fatty acids than triglycerides due to the dual hydrophilic and hydrophobic properties. They are mainly known for their function as building blocks for cell membranes in almost all known living beings. In addition to their role in cellular structure and function, PLs play an essential part in the formation of lipoproteins, which transport lipids to tissues via the blood stream (Scholey et al., 2013). In addition, certain PL metabolites serve as vital molecules within several signaling systems.

PCs by themselves, without the added benefit of n-3 PUFAs, have been shown to alleviate senescence (Hung et al., 2001; Torres García and Durán Agüero, 2014), to modulate atherosclerotic plaques (Zhang et al., 2014), and to be beneficial for cognitive function (Scholey et al., 2013) and anti-inflammatory activities (Jung et al., 2013; Murray et al., 2015; Vicenova et al., 2014). The beneficial effects of PLs on blood and hepatic lipids have been studied in a number of animal experiments (Buang et al., 2005; Hartmann et al., 2014; Lee et al., 2014; Tőkés et al., 2015), and both cholesterol and TAG levels are affected upon treatment. In particular, egg and soybean PLs have been shown to lower hepatic TAG levels by 45% in orotic acid–induced fatty liver in rats (Buang et al., 2005) and to increase levels of high-density lipoproteins (HDL) in humans (Abbasi et al., 2015; O'Brien and Andrews, 1993). A recent study suggested that the presence of tumors is associated with PC, breast adipose tissue from tumor-bearing mice expressed higher PC than that from non-tumor bearing mice. (Margolis et al., 2015; Vance, 2013). PC was further described by Lieber et al. (1994) to defend against liver cirrhosis and fibrosis induced by alcohol fed to baboons over a period of 6.5 years. Interestingly, patients with fatty liver showed a PL profile modification with a decrease of PC and PE (Puri et al., 2007).

The majority of the researches performed with PLs did not include n-3 PUFAs containing PLs, indicating that PLs in general have beneficial effects. However, it has also been shown that n-3 fatty acids are better protected from oxidation when they are incorporated into PLs compared to TAGs. Other studies have demonstrated that PL-bound n-3 PUFAs have more potent effects on blood plasma and liver lipid levels compared to PLs without n-3 PUFAs (Dasgupta and Bhattacharyya, 2007; Shirouchi et al., 2007).

n-3 PLs

In recent years, more and more focus has been laid on the beneficial health effects of PLs in both animals and humans. Dietary fatty acids regulate several physiological functions. However, to exert their properties, fatty acids have to be present in the diet in an optimal balance. Particular attention has been focused on tissue's PUFAs n-6/n-3 ratio, which is influenced by the type and the esterified form of dietary fatty acids (Decsi and Koletzko, 2005). Dietary phospholipid carriers might facilitate the transportation of n-3 fatty acids through the body and its subsequent integration into cell membranes. DHA is an important constituent of the membrane phospholipids of neural structures, usually occupying the *sn*-2 position (Lauritzen et al., 2001; Lippmeier et al., 2009; Tam and Innis, 2006). A typical example is PE, which is especially rich in the brain and retina. Other phospholipids in which DHA is a prominent feature include PC, PI, PS, cerebrosides, and sphingomyelin.

During the last decade, several papers have described the effects of two important sources of marine PLs: krill oil and fish roe. The studies about marine PLs in humans and animals are summarized in Tables 14.B and 14.C (Burri et al., 2012).

It is well known that n-3 PLs have more bioavailability of their long chain PUFAs for cellular structures and functions compared to triglycerides and other PLs without n-3 PUFAs (Amate et al., 2001; Murru et al., 2013; Ramirez et al., 2001). Thus, it has been suggested that n-3 PLs are a peculiar delivery form of n-3 fatty acids for animals and humans. Dietary EPA and DHA, when esterified to PLs, are more efficiently incorporated into tissue PLs and seem to possess peculiar properties through specific mechanism(s) of action, such as the capacity to affect endocannabinoid biosynthesis at much lower doses than EPA and DHA in triglyceride form, probably because of the higher incorporation into tissue PLs (Krill Oil Monograph, 2010; Lemaitre-Delaunay et al., 1999). The presence of long-chain and low-melting PUFA is believed to add mobility and fluidity to cell membranes, and thus, they adjust membrane function and integrity properly in lower ambient temperatures (Haraldsson and Thorarensen, 1999). In addition, fatty acids from PLs are more easily absorbed in the body than are the corresponding TAGs or ethyl esters (Galli et al., 1992). There is higher bioavailability of DHA for incorporation into erythrocytes in human adults when DHA is provided in PL form rather than in TAG (Lemaitre-Delaunay et al., 1999). PLs were about 2.1 times more effective than TAG as substrates for accretion of brain arachidonic acid in the development of neonatal primate brains (Wijendran et al., 2002).

Digestion, Absorption, and Bioavailability of n-3 PLs

In PC, two fatty acid chains are attached to a glycerol group that is connected to choline over a phosphate group; this structure enables the formation of phospholipid bilayers. In the TAGs form, however, three fatty acid chains are attached to a glycerol group. Dietary TAGs have physiological functions in the body that differ from those of PLs, being used principally as an energy source or to store energy in fat tissues. In addition, there is a difference in the digestion and absorption of PLs and TAGs in the small intestine. N-3 TAGs are insoluble in water and their absorption depends on bile salts and enzymes in the small intestine (Arnesjo et al., 1969).

In contrast to n-3 TAGs, n-3 PLs follow simpler digestion and distribution routes in the human body. PLs are not dependent on bile for digestion (Arnesjo et al., 1969). The solubility of PLs in water has the advantage of decreasing the digestive discomfort sometimes seen with n-3 provided in TAG form. PC and lysophospholipids can spontaneously form micelles via the intestinal wall and be conveyed in an aqueous environment. PLs can be absorbed without digestion in their intact form,

Table 14.B Overview of Clinical Studies with Marine n-3 PLs

Area of Study	Main Findings	Treatment	Population Characteristics	References
Obesity	Changed endocannabinoid levels	KO	Normal to obese	Banni et al. (2011)
Inflammation	Reduced arthritic symptoms	KO	Arthritis	Deutsch et al. (2007)
	Reduced oxidative damage	KO	Athletes	Skarpańska-Stejnborn et al. (2010)
Cardiovascular	Improved blood lipids	KO	Dyslipidemia	Bunea et al. (2004)
PMS	Reduced dysmenorrhea	KO	Women	Sampalis et al. 2003
Brain	Improved word recall	n-3 PS[a]	Memory complains	Richter et al. (2010)
Liver	Improved lipid parameters	Roe[c]	Chronic liver disease	Hayashi et al. (1999)
Eye	Improved attention	n-3 PLs[b]	ADHD children	Vaisman et al. (2008)
Bioavailability	Increased n-3 FA blood levels	KO	Healthy	Maki et al. (2009); Schuchardt et al. (2011); Ulven et al. (2011)

[a]The n-3 PS was synthesized from krill.

[b]The n-3 PLs were isolated from KO.

[c]Salmon roe.

Notes: FA: fatty acid; KO: krill oil; PLs: phospholipids; PMS: premenstrual syndrome; PS: phosphatidylserine.

Table 14.C Overview of Clinical Studies with Marine n-3 PLs

Area of Study	Main Findings	Treatment	Animal Model	References
Obesity	Improved metabolic profile	n-3 PLs[b]	High fat diet (m)	Rossmeisl et al. (2012)
	Decreased body weight	KO	High fat diet (r)	Ferramosca et al. (2012)
	Decreased hepatic steatosis	KO	High fat diet (m)	Tandy et al. (2009)
	Decreased hepatic and heart lipids	KO	Genetic obesity (r)	Batetta et al. (2009)
	Reduced abdominal fat	Roe[c]	High fat diet (m)	Moriya et al. (2007)
	Reduced endocannabinoid biosynthesis	KO	High fat diet (m)	Piscitelli et al. (2011)
Inflammation	Reduced oxidative stress	KO	Ulcerative colitis (r)	Ierna et al. (2010)
	Reduced arthritis scores	KO	Arthritis (m)	Ierna et al. (2010)
	Increased hepatic β-oxidation	Roe[c], KO	TNFα overexpression (m)	Bjorndal et al. (2012); Grimstad et al. (2012); Vigerust et al. (2013)
Cardiovascular	Improved blood lipids and adiponectin	Roe[a]	Healthy (m)	Higuchi et al. (2006, 2008)
	Attenuated heart remodeling	KO	Heart failure (r)	Fosshaug et al. (2011)
Bone	Did not improve bone mass/architecture	KO	Growing females (r)	Lukas et al. (2011)
Brain	Improved memory function	n-3 PLs[d]	Healthy (r)	Gamoh (2011)
	Improved learning and memory	n-3 PS[e]	Healthy (r)	Lee et al. (2010)
	Improved learning capacity	Roe[a]	Healthy (m)	Shirai et al. (2006)
	Increased DHA level in brain	KO	Genetic obesity (r)	di Marzo et al. (2010)
Other	Beneficial hepatic gene regulation	KO	Healthy (m)	Burri et al. (2011)
	Decreased hepatic lipogenesis	KO	Healthy (r)	Ferramosca et al. (2011)

[a]Salted herring roe

[b]Extracted from herring meal.

[c]Herring roe.

[d]Isolated from KO.

[e]Synthesized from KO.

Notes: (m): mice; (r): rats; KO: krill oil; PLs: phospholipids; PS: phosphatidylserine.

or as lysophosphatidylcholine (LPC) after digestion by enzymes in the small intestine (Tall et al., 1983; Zierenberg and Grundy, 1982). Their simpler digestion process in the small intestine before absorption is one of the factors indicating that n-3 PLs offer greater bioavailability in the human body than n-3 TAGs. In the blood, lysophospholipids and free fatty acids will be attached to proteins (e.g., albumin), which act as transporting molecules (Schuchardt et al., 2011). This step will enable efficient transportation of PLs and fatty acids to various organs, including the brain. Lipoprotein assemblages act as vehicles for fatty acid transport in blood. PLs located along the surface of lipoproteins play important roles in fatty acid transport in blood. Human lipoproteins include chylomicrons, HDL, low-density lipoproteins (LDL), and very low-density lipoproteins (VLDL) (Murru et al., 2013).

A study on a group of preterm infants fed with different types of formula has shown that the absorption of DHA was higher in those infants receiving the PUFA PLs formula than in infants receiving breast milk or the PUFA TAG formula (Schuchardt et al., 2011). EPA and DHA from krill oil, mainly bound to PLs, were absorbed at least as efficiently as EPA and DHA from fish oil, which is in the TAG form (Maki et al., 2009). Schuchardt et al. (2011) compared the uptake of three EPA/DHA formulations derived from fish oil (reesterified TAG [rTAG], ethyl ester, and krill oil [mainly PL]) and showed that the bioavailability of n-3 PUFAs may vary according to their esterified form. The highest incorporation of EPA/DHA into plasma PLs was obtained by krill oil, followed by fish oil, rTAG, and then ethyl ester. The availability of LPC and lysophosphatidylethanolamine (LPE) has been well documented in platelets (Bakken and Farstad, 1992; McKean et al., 1982; Neufeld et al., 1983). Also, a study on the incorporation of labeled DHA into platelet and red cell PLs showed that in platelets, [^{13}C] DHA accumulated in both PC and PE, although this was a little faster in PC (Lands, 1960). In contrast to platelets, [^{13}C]LPC could be detected in erythrocytes when [^{13}C] DHA started rising in PC (Lemaitre-Delaunay et al., 1999). This is in agreement with the hypothesis that DHA is preferentially esterified to LPC in erythrocytes (Brossard et al., 1997) and is subsequently reacylated into PC (Tamura et al., 1985). Interestingly, DHA levels in erythrocytes may be considered as an index of essential fatty acids in the brain, and LPC might be a preferential vehicle of DHA to the brain, as suggested in young rats (Innis, 1992; Thies et al., 1994).

Biological Activities of n-3 PLs

In mice fed with a high-fat diet, n-3 PUFA in PLs has more efficient bioavailability than TAGs, such as reducing hepatic steatosis, low-grade inflammation in white adipose tissue (Cansell, 2010; Cansell et al., 2003, 2006, 2009; Rossmeisl et al., 2012), glycemia, and blood lipid levels (Piscitelli et al., 2011; Tandy et al., 2009). Moreover, the down-regulation of lipogenic genes was stronger in the PL-treated mice, and reduced plasma insulin and adipocyte hypertrophy was observed only with the PL form (Rossmeisl et al., 2012). Numerous studies in humans linked DHA levels in blood plasma to brain-related disorders such as Alzheimer's disease, and several different protective roles of DHA in the brain have been suggested (Cole et al., 2010; Schaefer et al., 2006). Finally, in humans it has been demonstrated that PLs are a more efficient delivery form of DHA to platelets and erythrocytes than TAGs (Lemaitre-Delaunay et al., 1999).

Supplementation of n-3 PUFA as PLs exerts stronger biological effects compared with the TAG form because (1) various PL species can also act as ligands for nuclear receptors involved in the transcriptional regulation of cholesterol metabolism and steroidogenesis (Chakravarthy et al., 2009; Li et al., 2005), and (2) the PL form has been shown to augment the bioavailability of DHA and EPA in both rodents (Cansell et al., 2009; Tso et al., 1984) and humans (Carnielli et al., 1998; Deutsch, 2007). Results from several epidemiologic studies (Hibbeln, 1998; Noaghiul and Hibbeln, 2003) suggest that dietary consumption of n-3 PUFAs, EPA and DHA in particular, affects neuropsychiatric disorders, presumably because of their structural and neurochemical involvement in pathophysiological processes (Hibbeln and Salem, 1995; Hibbeln et al., 2006; Young and Conquer, 2005).

In Vaisman' study (Vaisman et al., 2008), ingestion of n-3 PUFA PLs increased the scores for a continuous performance task compared with those n-3 TAGs with about 250 mg/d EPA/DHA. This finding indicates that n-3 PUFAs incorporated in PC act on brain function more efficiently than those incorporated in TAGs. These results differ from previous reports in which even larger amounts of DHA (Voigt et al., 2001) or EPA+DHA (Hirayama et al., 2004) supplementation in attention-deficit hyperactivity disorder children for 2–4 months were shown to result in elevated plasma PL DHA concentrations and have no effect on continuous performance test scores. EPA-enriched phospholipid (EPA-PL) supplementation was efficacious in alleviating insulin resistance and suppressing body fat accumulation and hepatic steatosis by modulating the secretion of inflammatory cytokines and adipocytokines, suppressing sterol regulatory element-binding protein-1c mediated lipogenesis, and enhancing of fatty acid β-oxidation. These results demonstrate that EPA-PL is a novel beneficial food component for the prevention and improvement of metabolic disorders (Murru et al., 2013). Although dietary manipulation of n-3 PUFAs in the brain is complicated by the high concentrations in this organ, supplementation of DHA-enriched PLs, such as the bovine brain cortex PS, in animal models was shown to attenuate neuronal effects of aging (Zanotti et al., 1989) and to affect behavior as well (Castilho et al., 2004; Drago et al., 1981).

Experiments and clinical studies demonstrate that n-3 PUFA have shown to reduce cardiovascular risk (Mozaffarian and Wu, 2011; Simopoulos, 2008). Yet, effects of n-3 PUFA on cardiovascular diseases (CVD) and total and coronary heart disease mortality remain controversial. Mozaffarian et al. (2013) investigated the influence of plasma PL n-3 PUFA on CVD or mortality. The result showed that circulating n-3 PUFA levels are linked to lower total mortality among generally later in healthy adult's life, with potentially greatest associations with cardiovascular events and especially arrhythmic cardiac death.

The mechanism behind an increased tissue uptake of n-3 PUFAs in PLs is currently unknown, and more detailed mechanistic researches need to be done in order to fully explain the differences in tissue accumulations between the PL and TAG forms.

Endocannabinoid System Affecting of n-3 PLs

Endocannabinoids are bioactive mediators produced by metabolism of phospholipids (di Marzo, 2008). The composition of dietary fatty acids can affect energy homeostasis via changes of endocannabinoid system. The two major arachidonic acid containing endocannabinoid are arachidonoyl ethanolamide (AEA, also known as anandamide), and 2-arachidonoylglycerol (2-AG). AEA is formed by a pair of reactions involving conversion of PE to N-acylphosphatidylethanolamine (N-acyl PE) followed by the action of phospholipase D. 2-AG is formed as a result of the sequential actions of a diacylglycerol lipase and phospholipase C. Cannabinoid receptor type 1 (CB1) receptors are thought to be one of the most widely expressed G protein-coupled receptors in the brain. Cannabinoid receptor type 2 (CB2) receptors are mainly expressed on T cells of the immune system, on macrophages and B cells, and in hematopoietic cells (Console-Bram et al, 2012). AEA and 2-AG act via the CB1 and CB2 receptors (di Marzo, 2008). In the classical eicosanoid system, dietary n-3 fatty acids increase the levels of endocannabinoids with either EPA or DHA. These include docosahexaenoyl ethanolamide and eicosapentaenoyl ethanolamide (Calder, 2014). Dietary PUFAs modulate fatty acid composition of adipocyte PLs that act as endocannabinoid precursors, and then n-3 PUFAs have their beneficial effects in abdominal obesity, insulin resistance, and dyslipidemia by CB1-mediated lipogenic actions of endocannabinoids in adipocytes (Matias et al., 2007, 2008a, 2008b). Therefore, dietary fatty acids, by modulating arachidonic acid levels in tissue PLs, influence endocannabinoid biosynthesis and thereby down-regulate an overactive endocannabinoid system (Alvheim et al., 2012).

Dietary DHA and EPA in the form of PLs are superior to TAGs with respect to the reversal of adipocyte hypertrophy, hepatic steatosis, and low-grade inflammation and the preservation of glucose homeostasis (Rossmeisl et al., 2012; Visioli et al., 2000). The higher efficacy of n-3 PUFAs administered as PLs has been associated with their better bioavailability and with a relatively strong suppression of the levels of major endocannabinoids in white plasma and adipose tissue, suggesting that modulation of the endocannabinoid system activity contributed to greater efficacy of n-3 PUFA in PLs form when compared to that of in TAG form (Rossmeisl et al., 2012). The increase in obesity prevalence in the United States may be associated with the increased intake of linoleic acid, the precursor of arachidonic acid, and accordingly the precursor of endocannabinoids (Alvheim et al., 2012). On the other hand, addition of EPA and DHA to the diet resulted in a decrease of endocannabinoid levels in the liver and also in hypothalamus of mice fed with experimental diets with high linoleic acid content (Alvheim et al., 2012). Interestingly, the decrease of 2-AG was correlated to the plasma PL n-6/n-3 PUFA ratio, which was probably caused by the replacement of the 2-AG precursor arachidonic acid with n-3 PUFAs, as observed in obese Zucker rats (Malnoë et al., 1994).

Both the amount of dietary n-3 PUFAs and the dietary form, such as fish oil or krill oil, may influence EPA and DHA incorporation into brain lipids and consequently either the biosynthesis of arachidonic acid or its bound to PLs via its partial replacement with EPA and DHA (Liu et al., 2014; Schmitz and Ecker, 2008). Thus, even though most of the nutritional effects of n-3 PUFAs may occur through the modulation of the levels of PL-derived metabolites, such as oxygenated eicosanoids (Calder, 2006) and endocannabinoid concentrations (Banni and di Marzo, 2010), it is not clear whether and how modulation of the biosynthesis of these bioactive compounds may influence brain activities. The human body is very selective in which PLs it uses. For example, DHA is the main fatty acid in brain; our brains mainly use DHA-PLs because DHA is an electrically charged fat molecule that is very important for our nervous systems. Retinas need molecules that have even more charge to convert light rays to brain signals, so eye phospholipids contain two DHA lipid molecules whenever possible (Lauritzen et al., 2001). No wonder n-3 PUFAs play an important role in brain health, including focus, memory, alertness and acuity.

Nutritional Activity of n-3 PLs

Compared with TAGs, dietary n-3 PUFA PLs from marine fish exerts superior metabolic effects in the context of high-fat diet-induced obesity in mice (Rossmeisl et al., 2012). By multiple mechanisms of action, dietary PLs of marine origin might thus substantially improve treatment and prevention strategies for obesity-associated metabolic disorders. In a recent study (Rossmeisl et al., 2014), n-3 PLs supplemented to a high-fat diet in mice efficiently prevented weight gain, dyslipidemia, and insulin resistance, while reducing hepatic steatosis, which is associated with integrated inhibition of FA and cholesterol biosynthesis in the liver. These complex effects are unique to n-3 PLs; they were absent in response to soybean-derived PLs that do not contain DHA or EPA. Thus, n-3 PLs, either alone or in combination with antidiabetic drugs, could be of therapeutic value in obese subjects with nonalcoholic fatty liver disease.

Dietary EPA and DHA in PLs may reestablish the physiological endocannabinoid, up-regulated with dyslipidaemia, insulin resistance, visceral obesity, and atherogenic inflammation (di Marzo and Després, 2009), through a decrease of the n-6/n-3 PUFA ratio and the endocannabinoid precursors. It is clearly emerging that dietary

EPA-PLs and DHA-PLs affect endocannabinoid biosynthesis at much lower doses than EPA and DHA in TAG form, and positively modify several parameters of the metabolic syndrome.

Structured n-3 PLs also exert their biological activity similar to marine n-3 PLs. PLs can decrease the plasma cholesterol by influencing absorption of cholesterol in the intestine. N-3 fatty acid and soybean lecithin in structured PLs give synergic hypercholesterolemia effects. N-3 fatty acid in structured PLs may improve lipid profiles by lowering total cholesterol, total triglycerides, and LDL levels and by increasing HDL levels. Structured phospholipids can also reduce the occurrence of thickening of the aorta intima-media layer (Maligan et al., 2012).

N-3 PLs in the lipoproteins can influence the distribution of lipoproteins in the body and, hence, the availability of fatty acids. Increased transport to and utilization of n-3 by various tissues has been demonstrated when delivered as n-3 PLs. For example, elevated n-3 fatty acid concentrations of in target organs, such as the brain and liver, are observed when delivered as n-3 PLs. Thus, dietary intake of n-3 PLs appears to play a beneficial role in the distribution of fatty acids to various body tissues.

Future Perspectives

Currently, commercial PLs resulted from vegetable sources are mainly processed from soybeans and are widely used as functional ingredients in food, cosmetics, and pharmaceutical industries. PLs are produced from marine sources such as krill, roe, and fish byproducts, or nonmarine animal sources such as milk and egg. The increased accessibility of n-3 PLs over the recent years turns up new possibilities for the use of PLs both as a superior nutritional source of n-3 PUFAs, and for use in the functional food, cosmetic, and pharmaceutical industries.

PLs perform an increasingly vital role in food manufacturing and are currently applied to a wide range of food products such as dairy products, instant drinks, baked goods, margarine, and chocolate. The most widely used application of PLs in foods is as emulsifiers, both in water-in-oil and oil-in-water emulsions. However, modern diets have an insufficiency of intake in PLs due to purified raw materials and increased use of refined oils; therefore, the supplementation of n-3 PLs may serve three important actions within the functional food proportion: (1) supplementation of n-3 FAs, (2) emulsifying properties, and (3) beneficial nutritional effects of the PLs themselves.

The second significant application of PLs is within the cosmetics area, where their applications include specific ingredients and emollients for hair care, skin care, make-up, and decorative products. The emulsifying properties of PLs are used in skin moisturizing products, where PLs play two important roles: (1) they provide beneficial properties for the epidermis, and (2) they are used as carriers or vehicles in dermatologic delivery systems.

The pharmaceutical applications of n-3 PLs include two main fields: (1) nutritional use, and (2) drug delivery. In the drug delivery respect, PLs are able to form liposomes after mixing into an aqueous media. Liposomes have been investigated as drug carriers over the past few decades, and several different types have been exploited.

Although investigation on n-3 PUFAs has made important progress in different fields, there are still some issues, such as efficacy, according to the dietary form and putative mechanisms of action, which should be better characterized. EPA and DHA esterified to PLs are major n-3 PUFAs present in our diets. We have been exposed to these PUFAs throughout our evolution, maximizing the ability to fully exploit the nutritional properties of EPA and DHA. In future studies, whether EPA and DHA in PL form are also more effective than in TAG or ethyl esters should be expanded in ameliorating some pathological conditions where n-3 PUFAs seem to exert beneficial activities such as cancer and psychiatric disorders.

Interest in the production of structured n-3 PLs containing n-3 specific fatty acid residues has also increased significantly. After decades of studies on production of long chain n-3 fatty acid from oil-seed crop, the prospects are now a reality, including most recently the difficult to achieve yet nutritionally important EPA and DHA. Long chain n-3 fatty acid obtained from transgenic oilseed may in the future supply the most economically alternative source of n-3 PUFA for aquaculture and a range of other uses. It is estimated that the availability and cost of oils from transgenic plants would be similar to those of currently available commercial oilseed crops such as rapeseed and soybean. Further research and development in this area have the potential for significant social, commercial, health, and environmental benefits.

References

Abbasi, A.; Dallinga-Thie, G. M.; Dullaart, R. P. Phospholipid Transfer Protein Activity and Incident Type 2 Diabetes Mellitus. *Clin. Chim. Acta* **2015,** *15* (439), 38–41.

Abedi, E.; Sahari, M. A. Long-chain Polyunsaturated Fatty Acid Sources and Evaluation of Their Nutritional and Functional Properties. *Food Sci. Nutr.* **2014,** *2* (5), 443–463.

Alvheim, A. R.; Malde, M. K.; Osei-Hyiaman, D.; Lin, Y. H.; Pawlosky, R. J.; Madsen, L.; Kristiansen, K.; Frøyland, L.; Hibbeln, J. R. Dietary Linoleic Acid Elevates Endogenous 2-AG and Anandamide and Induces Obesity. *Obesity* **2012,** *20* (10), 1984–1994.

Alvheim, A. R.; Torstensen, B. E.; Lin, Y. H.; Lillefosse, H. H.; Lock, E. J.; Madsen, L.; Hibbeln, J. R.; Malde, M. K. Dietary Linoleic Acid Elevates Endogenous 2-Arachidonoylglycerol and Anandamide in Atlantic Salmon (Salmo salar L.) and Mice, and Induces Weight Gain and Inflammation in Mice. *Br. J. Nutr.* **2013,** *109* (8), 1508–1517.

Amate, L.; Gil, A.; Ramírez, M. Feeding Infant Piglets Formula with Long-chain Polyunsaturated Fatty Acids as Triacylglycerols or Phospholipids Influences the Distribution of These Fatty Acids in Plasma Lipoprotein Fractions. *J. Nutr.* **2001,** *131,* 1250–1255.

Armstrong, S. G.; Wyllie, S. G.; Leach, D. N. Effects of Season and Location of Catch on the Fatty Acid Composition of Some Australian Fish Species. *Food Chem.* **1994,** *51* (3), 295–305.

Arnesjo, B.; Nilsson, A.; Barrowman, J.; Borgstrom, B. Intestinal Digestion and Absorption of Cholesterol and Lecithin in the Human. Intubation Studies with a Fat-soluble Reference Substance. *Scand. J. Gastroenterol.* **1969,** *4,* 653–665.

Bakken, A. M.; Farstad, M. The Activities of Acyl-CoA: 1-Acyl-lysophospholipid Acyltransferase (S) in Human Platelets. *Biochem. J.* **1992,** *288* (3), 763–770.

Bang, H. O.; Dyerberg, J.; Nielsen, A. B. Plasma Lipid and lipoprotein Pattern in Greenlandic West-coast Eskimos. *Lancet* **1971,** *1,* 1143–1145.

Banni, S.; Carta, G.; Murru, E.; Cordeddu, L.; Giordano, E.; Sirigu, A. R.; Berge, K.; Vik, H.; Maki, K. C.; di Marzo, V.; et al. Krill Oil Significantly Decreases 2-Arachidonoylglycerol Plasma Levels in Obese Subjects. *Nutr. Metab. (Lond)* **2011,** *8,* 7.

Banni, S.; di Marzo, V. Effect of Dietary Fat on Endocannabinoids and Related Mediators: Consequences on Energy Homeostasis, Inflammation and Mood. *Mol. Nutr. Food Res.* **2010,** *54* (1), 82–92.

Batetta, B.; Griinari, M.; Carta, G.; Murru, E.; Ligresti, A.; Cordeddu, L.; Giordano, E.; Sanna, F.; Bisogno, T.; Uda, S.; et al. Endocannabinoids May Mediate the Ability of (n-3) Fatty Acids to Reduce Ectopic Fat and Inflammatory Mediators in Obese Zucker rats. *J. Nutr.* **2009,** *139,* 1495–501.

Bjorndal, B.; Burri, L.; Wergedahl, H.; Svardal, A.; Bohov, P.; Berge, R. K. Dietary Supplementation of Herring Roe and Milt Enhances Hepatic Fatty Acid Catabolism in Female Mice Transgenic for hTNFalpha. *Eur. J. Nutr.* **2012,** *51,* 741–753.

Brossard, N.; Croset, M.; Normand, S.; Pousin, J.; Lecerf, J.; Laville, M.; Tayot, J. L.; Lagarde, M. Human Plasma Albumin Transports [13C]Docosahexaenoic Acid in Two Lipid Forms to Blood Cells. *J. Lipid Res.* **1997,** *38* (8), 1571–1582.

Buang, Y.; Wang, Y. M.; Cha, J. Y.; Nagao, K.; Yanagita, T. Dietary Phosphatidylcholine Alleviates Fatty Liver Induced by Orotic Acid. *Nutrition* **2005,** *21,* 867–873.

Bunea, R.; El Farrah, K.; Deutsch, L. Evaluation of the Effects of Neptune Krill Oil on the Clinical Course of Hyperlipidemia. *Altern. Med. Rev.* **2004,** *9,* 420–428.

Burri, L.; Berge, K.; Wibrand, K.; Berge, R. K.; Barger, J. L. Differential Effects of Krill Oil and Fish Oil on the Hepatic Transcriptome in Mice. *Front. Genet.* **2011,** *2,* 1–8.

Burri, L.; Berge, K. Recent Findings on Cardiovascular and Mental Health Effects of Krill Oil and Omega-3 Phospholipids. In *Omega-6/3 Fatty Acids: Functions, Sustainability Strategies and Perspectives;* De Meester, F., Watson, R. R., Zibadi, S., Eds.; Humana Press Inc.: New York, 2013.

Burri, L.; Hoem, N.; Banni, S.; Berge, K. Marine Omega-3 Phospholipids: Metabolism and Biological Activities. *Int. J. Mol. Sci.* **2012,** *13,* 15401–15419.

Calder, P. C. n-3 Polyunsaturated Fatty Acids, Inflammation, and Inflammatory Diseases. *Am. J. Clin. Nutrition* **2006,** *83,* 1505S–1519S.

Calder, P. C. Marine Omega-3 Fatty Acids and Inflammatory Processes: Effects, Mechanisms and Clinical Relevance. *Biochim. Biophys. Acta* **2015,** *1851* (4), 469–484.

Cansell, M. Marine Phospholipids as Dietary Carriers of Long-chain Polyunsaturated Fatty Acids. *Lipid Technol.* **2010,** *22,* 223–226.

Cansell, M.; Nacka, F.; Combe, N. Marine Lipid-based Liposomes Increase *in Vivo* FA Bioavailability. *Lipids* **2003,** *38,* 551–559.

Cansell, M.; Moussaoui, N.; Petit, A. P.; Denizot, A.; Combe, N. Feeding Rats with Liposomes or Fish Oil Differently Affects Their Lipid Metabolism. *Eur. J. Lipid Sci. Technol.* **2006,** *108,* 459–467.

Cansell, M. S.; Battin, A.; Degrace, P.; Gresti, J.; Clouet, P.; Combe, N. Early Dissimilar Fates of Liver Eicosapentaenoic Acid in Rats Fed Liposomes or Fish Oil and Gene Expression Related to Lipid Metabolism. *Lipids* **2009,** *44* (3), 237–247.

Carlson, S. E. Are n-3 Polyunsaturated Fatty Acids Essential for Growth and Development? In *Health Effects of Dietary Fatty Acid;* Nelson, G. J., Ed.; Urbana, IL: AOCS Press, 1991; pp 42–49.

Carnielli, V. P.; Verlato, G.; Pederzini, F.; Luijendijk, I.; Boerlage, A.; Pedrotti, D.; Sauer, P. J. Intestinal Absorption of Long-chain Polyunsaturated Fatty Acids in Preterm Infants Fed Breastmilk or Formula. *Am. J. Clin. Nutr.* **1998,** *67* (1), 97–103.

Castilho, J. C.; Perry, J. C.; Andreatini, R.; Vital, M. A. B. F. Phosphatidylserine: An Antidepressive or a Cognitive Enhancer? *Prog. Neuro-Psychoph.* **2004,** *28* (4), 731–738.

Cavazos-Garduno, A.; Ochoa Flores, A. A.; Serrano-Nino, J. C.; Martínez-Sanchez, C. E.; Beristain, C. I.; García, H. S. Preparation of Betulinic Acid Nanoemulsions Stabilized by ω-3 Enriched Phosphatidylcholine. *Ultrason. Sonochem.* **2015,** *24,* 204–213.

Chakravarthy, M. V.; Lodhi, I. J.; Yin, L.; Malapaka, R. R.; Xu, H. E.; Turk, J.; Semenkovich, C. F. Identification of a Physiologically Relevant Endogenous Ligand for PPARα|ζ in Liver. *Cell* **2009,** *138* (3), 476–488.

Chowdhury, R.; Stevens, S.; Gorman, D.; Pan, A.; Warnakula, S.; Chowdhury, S.; Ward, H.; Johnson, L.; Crowe, F.; Hu, F. B.; et al. Association between Fish Consumption, Long Chain Omega 3 Fatty Acids, and Risk of Cerebrovascular Disease: Systematic Review and Meta-analysis. *BMJ* **2012,** *345,* e6698.

Christiansen, J. A. Changes in Phospholipid Classes and Fatty Acids and Fatty Acid Desaturation and Incorporation into Phospholipids during Temperature Acclimation of Green Sunfish *Lepomis cyanellus* R. *Physiol. Biochem. Zool.* **1984,** *57* (4), 481–492.

Christie, W. W. Lipid Analysis. In *Isolation, Separation, Identification and Structural Analysis of Lipids,* 3rd ed; The Oily Press, PJ Barnes and Associates: Bridgewater, UK, 2003; p 416.

Codabaccus, B. M.; Bridle, A. R.; Nichols, P. D.; Carter, C. G. An Extended Feeding History with a Stearidonic Acid Enriched Diet from Parr to Smolt Increases n-3 Long-chain Polyunsaturated Fatty Acids Biosynthesis in White Muscle and Liver of Atlantic Salmon (Salmo salar L.). *Aquaculture* **2011,** *322–323,* 65–73.

Cole, G. M.; Ma, Q. L.; Frautschy, S. A. Dietary Fatty Acids and the Aging Brain. *Nutr. Rev.* **2010,** *68,* S102–S111.

Collins, M. L.; Lynch, B.; Barfield, W.; Bull, A.; Ryan, A. S.; Astwood, J. D. Genetic and Acute Toxicological Evaluation of an Algal Oil Containing Eicosapentaenoic Acid (EPA) and Palmitoleic Acid. *Food. Chem. Toxicol.* **2014,** *72,* 162–168.

Console-Bram, L.; Marcu, J.; Abood, M. E. Cannabinoid Receptors: Nomenclature and Pharmacological Principles. *Prog. Neuropsychopharmacol. Biol. Psychiatry* **2012,** *38* (1), 4–15.

Dasgupta, S.; Bhattacharyya, D. K. Dietary Effect of Eicosapentaenoic Acid (EPA) Containing Soyphospholipid. *J. Oleo Sci.* **2007,** *56,* 563–568.

Decsi, T.; Koletzko, B. N-3 Fatty Acids and Pregnancy Outcomes. *Curr. Opin. Clin. Nutr. Metab. Care* **2005,** *8* (2), 161–166.

Deutsch, L. Evaluation of the Effect of Neptune Krill Oil on Chronic Inflammation and Arthritic Symptoms. *J. Am. Coll. Nutr.* **2007,** *26* (1), 39–48.

di Marzo, V. The Endocannabinoid System in Obesity and Type 2 Diabetes. *Diabetogia* **2008,** *51,* 1356–1367.

di Marzo, V.; Griinari, M.; Carta, G.; Murru, E.; Ligresti, A.; Cordeddu, L.; Giordano, E.; Bisogno, T.; Collu, M.; Batetta, B.; et al. Dietary Krill Oil Increases Docosahexaenoic Acid and Reduces 2-Arachidonoylglycerol but not *N*-acylethanolamine Levels in the Brain of Obese Zucker Rats. *Int. Dairy J.* **2010,** *20,* 231–235.

di Marzo, V.; Després, J. CB1 Antagonists for Obesity: What Lessons Have We Learned from Rimonabant? *Nat. Rev. Endocrinol.* **2009,** *5* (11), 633–638.

Drago, F.; Canonico, P. L.; Scapagnini, U. Behavioral Effects of Phosphatidylserine in Aged Rats. *Neurobiol. Aging* **1981,** *2* (3), 209–213.

Dyerberg, J.; Bang, H. O.; Hjorne, N. Fatty Acid Composition of the Plasma Lipids in Greenland Eskimos. *Am. J. Clin. Nutr.* **1975,** *28,* 958–966.

Estiasih, T.; Ahmadi, K.; Ginting, E.; Albab, A. U. Optimization of High EPA Structured Phospholipids Synthesis from ω-3 Fatty Acid Enriched Oil and Soy Lecithin. *J. Food Sci. Eng.* **2013,** *3,* 25–32.

Ferramosca, A.; Conte, A.; Burri, L.; Berge, K.; De Nuccio, F.; Giudetti, A. M.; Zara, V. A Krill Oil Supplemented Diet Suppresses Hepatic Steatosis in High-fat Fed Rats. *PLoS One* **2012,** *7,* e38797.

Ferramosca, A.; Conte, L.; Zara, V. A Krill Oil Supplemented Diet Reduces the Activities of the Mitochondrial Tricarboxylate Carrier and of the Cytosolic Lipogenic Enzymes in Rats. *J. Anim. Physiol. Anim. Nutr.* **2011,** 1–12.

Fosshaug, L. E.; Berge, R. K.; Beitnes, J. O.; Berge, K.; Vik, H.; Aukrust, P.; Gullestad, L.; Vinge, L. E.; Oie, E. Krill Oil Attenuates Left Ventricular Dilatation after Myocardial Infarction in Rats. *Lipids Health Dis.* **2011,** *10,* 245.

Frankel, E. *Lipid Oxidation.* The Oily Press: Dundee, Scotland, 1998.

Galli, C.; Sirtori, C. R.; Mosconi, C.; Medini, L.; Gianfranceschi, G.; Vaccarino, V.; Scolastico, C. Prolonged Retention of Doubly Labeled Phosphatidylcholine in Human Plasma and Erythrocytes after Oral Administration. *Lipids* **1992,** *27* (12), 1005–1012.

Gamoh, S. Krill-derived Phospholipids Rich in n-3 Fatty Acid Improve Spatial Memory in Adult Rats. *J. Agric. Sci.* **2011,** *3,* 3–12.

García, H. S.; Kim, I. W.; López-Hernandez, A.; Hill, G. C. Enrichment of Lecithin with n-3 Fatty Acids by Acidolysis Using Immobilized Phospholipase A_1. *Grasas Aceites* **2008,** *59,* 368–374.

Ghasemifard, S.; Turchini, G. M.; Sinclair, A. J. Omega-3 Long Chain Fatty Acid "Bioavailability": A Review of Evidence and Methodological Considerations. *Prog. Lipid Res.* **2014,** *56,* 92–108.

Grimstad, T.; Bjorndal, B.; Cacabelos, D.; Aasprong, O. G.; Janssen, E. A.; Omdal, R.; Svardal, A.; Hausken, T.; Bohov, P.; Portero-Otin, M.; et al. Dietary Supplementation of Krill

Oil Attenuates Inflammation and Oxidative Stress in Experimental Ulcerative Colitis in Rats. Scand. *J. Gastroenterol.* **2012,** *47,* 49–58.

Guinot, D.; Monroig, O.; Hontoria, F.; Amat, F.; Varó, I.; Navarro, J. C. Enriched On-Grown Artemia Metanauplii Actively Metabolise Highly Unsaturated Fatty Acid-rich Phospholipids. *Aquaculture* **2013,** *412–413,* 173–178.

Gunstone, F. D., Ed. Chemical Structure and Biological Function. In *Phospholipid Technology and Applications;* Elsevier: London, 2008; pp 1–16.

Haraldsson, G. G.; Thorarensen, A. Preparation of Phospholipids Highly Enriched with n-3 Polyunsaturated Fatty Acids by Lipase. *J. Am. Oil Chem. Soc.* **1999,** *76* (10), 1143–1149.

Haraldsson, G. G.; Kristinsson, B.; Gudbjarnason, S. The Fatty Acid Composition of Various Lipid Classes in Several Species of Fish Caught in Icelandic Waters. *INFORM* **1993,** *4,* 535.

Hartmann, P.; Fet, N.; Garab, D.; Szabó, A.; Kaszaki, J.; Srinivasan, P. K.; Tolba, R. H.; Boros, M. L-α-Glycerylphosphorylcholine Reduces the Microcirculatory Dysfunction and Nicotinamide Adenine Dinucleotide Phosphate-oxidase Type 4 Induction after Partial Hepatic Ischemia in Rats. *J. Surg. Res.* **2014,** *189* (1), 32–40.

Hayashi, H.; Tanaka, Y.; Hibino, H.; Umeda, Y.; Kawamitsu, H.; Fujimoto, H.; Amakawa, T. Beneficial Effect of Salmon Roe Phosphatidylcholine in Chronic Liver Disease. *Curr. Med. Res. Opin.* **1999,** *15,* 177–184.

Henriques, J.; Dick, J. R.; Tocher, D. R.; Bell, J. G. Nutritional Quality of Salmon Products Available from Major Retailers in the UK: Content and Composition of n-3 Long-chain PUFA. *Br. J. Nutr.* **2014,** *112* (6), 964–975.

Hibbeln, J. R. Fish Consumption and Major Depression. *The Lancet* **1998,** *351* (9110), 1213.

Hibbeln, J. R.; Salem, N., Jr. Dietary Polyunsaturated Fatty Acids and Depression: When Cholesterol Does Not Satisfy. *Am. J. Clin. Nutr.* **1995,** *62* (1), 1–9.

Hibbeln, J. R.; Ferguson, T. A.; Blasbalg, T. L. Omega-3 Fatty Acid Deficiencies in Neurodevelopment, Aggression and Autonomic Dysregulation: Opportunities for Intervention. *Int. Rev. Psychiatr.* **2006,** *18* (2), 107–118.

Higuchi, T.; Shirai, N.; Suzuki, H. Effects of Dietary Herring Roe Lipids on Plasma Lipid, Glucose, Insulin, and Adiponectin Concentrations in Mice. *J. Agric. Food Chem.* **2006,** *54,* 3750–3755.

Higuchi, T.; Shirai, N.; Suzuki, H. Effects of Herring Roe on Plasma Lipid, Glucose, Insulin and Adiponectin Levels, and Hepatic Lipid Contents in Mice. *J. Nutr. Sci. Vitaminol.* (Tokyo) **2008,** *54,* 230–236.

Hirayama, S.; Hamazaki, T.; Terasawa, K. Effect of Docosahexaenoic Acid-containing Food Administration on Symptoms of Attention-Deficit/Hyperactivity Disorder—A Placebo-controlled Double-blind Study. *Eur. J. Clin. Nutr.* **2004,** *58* (3), 467–473.

Hjaltason, B.; Haraldsson, G. G. Fish Oils and Lipids from Marine Sources. In *Modifying Lipids for Use in Food*; Gunstone, F. D., Ed.; Woodhead Publishing: Cambridge, UK, 2006; pp 57–79.

Hooper, L.; Thompson, R. L.; Harrison, R. A.; Summerbell, C. D.; Ness, A. R.; Moore, H. J.; Worthington, H. V.; Durrington, P. N.; Higgins, J. P.; Capps, N. E.; et al. Risks and Benefits of Omega 3 Fats for Mortality, Cardiovascular Disease, and Cancer: Systematic Review. *BMJ* **2006,** *332,* 752–760.

Hosomi, R.; Fukunaga, K.; Fukao, M.; Yoshida, M.; Arai, H.; Kanda, S.; Nishiyama, T.; Kanada, T. Combination Effect of Phospholipids and n-3 Polyunsaturated Fatty Acids on Rat Cholesterol Metabolism. *Food Sci. Biotechnol.* **2012,** *21* (5), 1335–1342.

Hung, M. C.; Shibasaki, K.; Yoshida, R.; Sato, M.; Imaizumi, K. Learning Behaviour and Cerebral Protein Kinase C, Antioxidant Status, Lipid Composition in Senescence-Accelerated Mouse: Influence of a Phosphatidylcholine-Vitamin B12 Diet. *Br. J. Nutr.* **2001,** *86,* 163–171.

Ierna, M.; Kerr, A.; Scales, H.; Berge, K.; Griinari, M. Supplementation of Diet with Krill Oil Protects Against Experimental Rheumatoid Arthritis. *BMC Musculoskeletal Disord.* **2010,** *11,* 136.

Innis, S. M. Plasma and Red Blood Cell Fatty Acid Values as Indexes of Essential Fatty Acids in the Developing Organs of Infants Fed with Milk or Formulas. *J. Pediatr.* **1992,** *120* (4), 78–86.

Janssen, C. I.; Zerbi, V.; Mutsaers, M. P.; de Jong, B. S.; Wiesmann, M.; Arnoldussen, I. A.; Geenen, B.; Heerschap, A.; Muskiet, F. A.; Jouni, Z. E.; et al. Impact of Dietary n-3 Polyunsaturated Fatty Acids on Cognition, Motor Skills and Hippocampal Neurogenesis in Developing C57BL/6J Mice. *J. Nutr. Biochem.* **2015,** *26,* 24–35.

Jung, Y. Y.; Nam, Y.; Park, Y. S.; Lee, H. S.; Hong, S. A.; Kim, B. K.; Park, E. S.; Chung, Y. H.; Jeong, J. H. Protective Effect of Phosphatidylcholine on Lipopolysaccharide-induced Acute Inflammation in Multiple Organ Injury. *Korean J. Physiol. Pharmacol.* **2013,** *17* (3), 209–216.

Kamal-Eldin, A.; Yanishlieva, N. V. n-3 Fatty Acids for Human Nutrition: Stability Considerations. *Eur. J. Lipid Sci. Technol.* **2002,** *104,* 825–836.

Kitessa, S. M.; Abeywardena, M.; Wijesundera, C.; Nichols, P. D. DHA-Containing Oilseed: A Timely Solution for the Sustainability Issues Surrounding Fish Oil Sources of the Health-benefitting Long-chain Omega-3 Oils. *Nutrients* **2014,** *6,* 2035–2058.

Kong-González, M.; Pérez-Cortéz, J. G.; Hernández-Girón, C.; Macías-Morales, N.; Flores-Aldana, M. Polyunsaturated Fatty Acids for Multiple Sclerosis Treatment. *Medwave* **2015,** *15* (1), e6062.

Krill Oil Monograph. *Altern. Med. Rev.* **2010,** *15,* 84–86.

Lands, B. Historical Perspectives on the Impact of n-3 and n-6 Nutrients on Health. *Prog. Lipid Res.* **2014,** *55c,* 17–29.

Lands, W. E. Metabolism of Glycerolipids. 2. The Enzymatic Acylation of Lysolecithin. *J. Biol. Chem.* **1960,** *235,* 2233–2237.

Lauritzen, L.; Hansen, H. S.; Jørgensen, M. H.; Michaelsen, K. F. The Essentiality of Long Chain n-3 Fatty Acids in Relation to Development and Function of the Brain and Retina. *Prog. Lipid Res.* **2001,** *40* (1–2), 1–94.

Lee, B.; Sur, B. J.; Han, J. J.; Shim, I.; Her, S.; Lee, H. J.; Hahm, D. H. Krill Phosphatidylserine Improves Learning and Memory in Morris Water Maze in Aged Rats. *Prog. Neuropsychopharmacol. Biol. Psychiatry* **2010,** *34,* 1085–1093.

Lee, H. S.; Nam, Y.; Chung, Y. H.; Kim, H. R.; Park, E. S.; Chung, S. J.; Kim, J. H.; Sohn, U. D.; Kim, H. C.; Oh, K. W.; Jeong, J. H. Beneficial Effects of Phosphatidylcholine on High-fat Diet-induced Obesity, Hyperlipidemia and Fatty Liver in Mice. *Life Sci.* **2014,** *118* (1), 7–14.

Le Grandois, J.; Marchioni, E.; Zhao, M.; Giuffrida, F.; Ennahar, S.; Bindler, F. Investigation of Natural Phosphatidylcholine Sources: Separation and Identification by Liquid Chromatography-electrospray Ionization-tandem Mass Spectrometry (LC-ESI-MS2) of Molecular Species. *J. Agric. Food Chem.* **2009,** *57,* 6014–6020.

Lemaitre-Delaunay, D.; Pachiaudi, C.; Laville, M.; Pousin, J.; Armstrong, M.; Lagarde, M. Blood Compartmental Metabolism of Docosahexaenoic Acid (DHA) in Humans after Ingestion of a Single Dose of [(13)C]DHA in Phosphatidylcholine. *J. Lipid Res.* **1999,** *40* (10), 1867–1874.

Li, Y.; Choi, M.; Cavey, G.; Daugherty, J.; Suino, K.; Kovach, A.; Bingham, N. C.; Kliewer, S. A.; Xu, H. E. Crystallographic Identification and Functional Characterization of Phospholipids as Ligands for the Orphan Nuclear Receptor Steroidogenic Factor-1. *Mol. Cell* **2005,** *17* (4), 491–502.

Lieber, C. S.; Robins, S. J.; Li, J.; DeCarli, L. M.; Mak, K. M.; Fasulo, J. M.; Leo, M. A. Phosphatidylcholine Protects against Fibrosis and Cirrhosis in the Baboon. *Gastroenterology* **1994,** *106,* 152–159.

Lili, L.; Shunsheng, C.; Yinghong, Q.; Zhuoping, D.; Wenhui, W. Lipids and Fatty Acids Composition of Roe from *Cyprinus carpio* (Carp) and *Carassius auratus* (Crucian Carp). *Acta Nutrimenta Sinica* **2007,** *29* (4), 415–416.

Lippmeier, J. C.; Crawford, K. S.; Owen, C. B.; Rivas, A. A.; Metz, J. G.; Apt, K. E. Characterization of Both Polyunsaturated Fatty Acid Biosynthetic Pathways in Schizochytrium sp. *Lipids* **2009,** *44,* 621–630.

Liu, L.; Bartke, N.; Van Daele, H.; Lawrence, P.; Qin, X.; Park, H. G.; Kothapalli, K.; Windust, A.; Bindels, J.; Wang, Z.; Brenna, J. T. Higher Efficacy of Dietary DHA Provided as a Phospholipid Than as a Triglyceride for Brain DHA Accretion in Neonatal Piglets. *J. Lipid Res.* **2014,** *55* (3), 531–539.

Liu, W.; Xue, Y.; Liu, C.; Lou, Q.; Wang, J.; Yanagita, T.; Xue, C.; Wang, Y. Eicosapentaenoic Acid-enriched Phospholipid Ameliorates Insulin Resistance and Lipid Metabolism in Diet-induced-obese Mice. *Lipids Health Dis.* **2013,** *12,* 109.

Loy, R.; Heyer, D.; Williams, C. L.; Meck, W. H. Choline-induced Spatial Memory Facilitation Correlates with Altered Distribution and Morphology of Septal Neurons. *Adv. Exp. Med. Biol.* **1991,** *295,* 373–382.

Lu, F. S.; Nielsen, N. S.; Timm-Heinrich, M.; Jacobsen, C. Oxidative Stability of Marine Phospholipids in the Liposomal Form and Their Applications. *Lipids* **2011,** *46,* 3–23.

Lukas, R.; Gigliotti, J. C.; Smith, B. J.; Altman, S.; Tou, J. C. Consumtion of Different Sources of Omega-3 Polyunsaturated Fatty Acids by Growing Female Rats Affects Long Bone Mass and Microarchitecture. *Bone* **2011,** *49,* 455–462.

Lyberg, A. M.; Fasoli, E.; Adlercreutz, P. Monitoring the Oxidation of Docosahexaenoic Acid in Lipids. *Lipids* **2005,** *40* (9), 969–979.

Maki, K. C.; Reeves, M. S.; Farmer, M.; Griinari, M.; Berge, K.; Vik, H.; Hubacher, R.; Rains, T. M. Krill Oil Supplementation Increases Plasma Concentrations of Eicosapentaenoic and Docosahexaenoic Acids in Overweight and Obese Men and Women. *Nutr. Res.* **2009,** *29* (9), 609–615.

Maligan, J. M.; Estiasih, T.; Kusnadi, J. Structured Phospholipids from Commercial Soybean Lecithin Containing Omega-3 Fatty Acids Reduces Atherosclerosis Risk in Male *Sprague dawley* Rats which Fed with an Atherogenic Diet. *International Scholarly and Scientific Research & Innovation* **2012,** *6* (9), 492–498.

Malnoë, A.; Henzelin, I.; Stanley, J. C. Phospholipid Fatty Acid Composition and Vitamin E Levels in the Retina of Obese (fa/fa) and Lean (FA/FA) Zucker Rats. *Biochim. Biophys. Acta* **1994,** *1212* (1), 119–124.

Mangano, K. M.; Sahni, S.; Kerstetter, J. E.; Kenny, A. M.; Hannan, M. T. Polyunsaturated Fatty Acids and Their Relation with Bone and Muscle Health in Adults. *Curr. Osteoporos. Rep.* **2013,** *11* (3), 203–212.

Margolis, M.; Perez, O., Jr.; Martinez, M. Phospholipid Makeup of the Breast Adipose Tissue Is Impacted by Obesity and Mammary Cancer in the Mouse: Results of a Pilot Study. *Biochimie* **2015,** *108,* 133–139.

Marik, P. E.; Varon, J. Omega-3 Dietary Supplements and the Risk of Cardiovascular Events: A Systematic Review. *Clin. Cardiol.* **2009,** *32,* 365–372.

Matias, I.; Gonthier, M. P.; Petrosino, S.; Docimo, L.; Capasso, R.; Hoareau, L.; Monteleone, P.; Roche, R.; Izzo, A. A.; di Marzo, V. Role and Regulation of Acylethanolamides in Energy Balance: Focus on Adipocytes and β-Cells. *Br. J. Pharmacol.* **2007,** *152* (5), 676–690.

Matias, I.; Petrosino, S.; Racioppi, A.; Capasso, R.; Izzo, A. A.; di Marzo, V. Dysregulation of Peripheral Endocannabinoid Levels in Hyperglycemia and Obesity: Effect of High Fat Diets. *Mol. Cell. Endocrinol.* **2008a,** *286* (1–2), S66–S78.

Matias, I.; Vergoni, A. V.; Petrosino, S.; Docimo, L.; Capasso, R.; Hoareau, L.; Monteleone, P.; Roche, R.; Izzo, A. A.; di Marzo, V. Regulation of Hypothalamic Endocannabinoid Levels by Neuropeptides and Hormones Involved in Food Intake and Metabolism: Insulin and Melanocortins. *Neuropharmacology* **2008b,** *54* (1), 206–212.

McKean, M. L.; Smith, J. B.; Silver, M. J. Phospholipid Biosynthesis in Human Platelets. Formation of Phosphatidylcholine from 1-Acyl Lysophosphatidylcholine by ACYL-CoA: 1-ACYL-sn-glycero-3-phosphocholine Acyltransferase. *J. Biol. Chem.* **1982,** *257* (19), 11278–11283.

Mead, J. F.; Alfin-Slater, R. B.; Howton, D. R.; Popják, G., Eds. *Lipids: Chemistry, Biochemistry and Nutrition.* Plenum Press: New York, 1986; pp 369–428.

Miller, M. R.; Nichols, P. D.; Carter, C. G. n-3 Oil Sources for Use in Aquaculture: Alternatives to the Unsustainable Harvest of Wild Fish. *Nutr. Res. Rev.* **2008,** *21,* 85–96.

Moriya, H.; Hosokawa, M.; Miyashita, K. Combination Effect of Herring Roe Lipids and Proteins on Plasma Lipids and Abdominal Fat Weight of Mouse. *J. Food Sci.* **2007,** *72,* C231–C234.

Mozaffarian, D.; Lemaitre, R. N.; King, I. B.; Song, X.; Huang, H.; Sacks, F. M.; Rimm, E. B.; Wang, M.; Siscovick, D. S. Plasma Phospholipid Long-chain ω-3 Fatty Acids and Total and Cause-specific Mortality in Older Adults: A Cohort Study. *Ann. Intern. Med.* **2013,** *158* (7), 515–525.

Mozaffarian, D.; Wu, J. H. Omega-3 Fatty Acids and Cardiovascular Disease: Effects on Risk Factors, Molecular Pathways, and Clinical Events. *J. Am. Coll. Cardiol.* **2011,** *58* (20), 2047–2067.

Mukhopadhyay, T.; Ghosh, S. Lipid Profile and Fatty Acid Composition of Two Silurid Fish Eggs. *J. Oleo Sci.* **2007,** *56* (8), 399–403.

Murray, M.; Hraiki, A.; Bebawy, M.; Pazderka, C.; Rawling, T. Anti-tumor Activities of Lipids and Lipid Analogues and Their Development as Potential Anticancer Drugs. *Pharmacol. Ther.* [Online early access]. DOI: 10.1016/j.pharmthera.2015.01.008. Published Online: Jan 17, 2015. http://www.sciencedirect.com/science/article/pii/S0163725815000091 (accessed Feb 17, 2015).

Murru, E.; Banni, S.; Carta, G. Nutritional Properties of Dietary Omega-3-enriched Phospholipids. *Biomed Res. Int.* **2013.** http://dx.doi.org/10.1155/2013/965417.

Mutua, L. N.; Akoh, C. C. Lipase-Catalyzed Modification of Phospholipids: Incorporation of n-3 Fatty Acids into Biosurfactants. *J. Am. Oil Chem. Soc.* **1993,** *70,* 125–128.

Na, A.; Eriksson, C.; Eriksson, S.-G.; Österberg, E.; Holmberg, K. Synthesis of Phosphatidylcholine with (n-3) Fatty Acids by Phospholipase A_2 in Microemulsion. *J. Am. Oil Chem. Soc.* **1990,** *67,* 766–770.

Neufeld, E. J.; Wilson, D. B.; Sprecher, H.; Majerus, P. W. High Affinity Esterification of Eicosanoid Precursor Fatty Acids by Platelets. *J. Clin. Invest.* **1983,** *72* (1), 214–220.

Noaghiul, S.; Hibbeln, J. R. Cross-national Comparisons of Seafood Consumption and Rates of Bipolar Disorders. *Am. J. Psychiatry* **2003,** *160* (12), 2222–2227.

O'Brien, B. C.; Andrews, V. G. Influence of Dietary Egg and Soybean Phospholipids and Triacylglycerols on Human Serum Lipoproteins. *Lipids* **1993,** *28,* 7–12.

Peterson, F. S.; Peterson, F. I. B.; Sargent, J. R.; Haug, T. Lipid Class and Fatty Acid Composition of Eggs from the Atlantic Halibut (Hippoglossus hippoglossus). *Aquaculture* **1986,** *52* (3), 207–211.

Phleger, C. F.; Nelson, M. M.; Mooney, B. D.; Nichols, P. D. Interannual and between Species Comparison of the Lipids, Fatty Acids and Sterols of Antarctic Krill from the US AMLR Elephant Island Survey Area. *Comp. Biochem. Physiol. B Biochem. Mol. Biol.* **2002,** *131,* 733–747.

Piscitelli, F.; Carta, G.; Bisogno, T.; Murru, E.; Cordeddu, L.; Berge, K.; Tandy, S.; Cohn, J. S.; Griinari, M.; Banni, S.; di Marzo, V. Effect of Dietary Krill Oil Supplementation on the Endocannabinoidome of Metabolically Relevant Tissues from High-fat-fed Mice. *Nutr. Metab.* **2011,** *8,* article 51.

Polvi, S. M.; Ackman, R. G. Atlantic Salmon (*Salmo salar*) Muscle Lipids and Their Response to Alternative Dietary Fatty Acid Sources. *J. Agric. Food Chem.* **1992,** *40,* 1001–1007.

Puri, P.; Baillie, R. A.; Wiest, M. M.; Mirshahi, F.; Choudhury, J.; Cheung, O.; Sargeant, C.; Contos, M. J.; Sanyal, A. J. A Lipidomic Analysis of Nonalcoholic Fatty Liver Disease. *Hepatology* **2007,** *46,* 1081–1090.

Ramirez, M.; Amate, L.; Gil, A. Absorption and Distribution of Dietary Fatty Acids from Different Sources. *Early Hum. Dev.* **2001,** *65,* 95–101.

Rao, P. G. P.; Balaswamy, K.; Jyothirmayi, T.; Karuna, M. S. L.; Prasad, R. B. N. Fish Roe Lipids: Composition and Changes during Processing and Storage. In *Processing and Impact on Active Components in Food;* Preedy, V., Ed.; Elsevier: London, 2015; pp 463–468.

Richter, Y.; Herzog, Y.; Cohen, T.; Steinhart, Y. The Effect of Phosphatidylserine-containing Omega-3 Fatty Acids on Memory Abilities in Subjects with Subjective Memory Complaints: A Pilot Study. *Clin. Interv. Aging* **2010,** *5,* 313–316.

Rizos, E. C.; Ntzani, E. E.; Bika, E.; Kostapanos, M. S.; Elisaf, M. S. Association between Omega-3 Fatty Acid Supplementation and Risk of Major Cardiovascular Disease Events: A Systematic Review and Meta-analysis. *JAMA* **2012,** *308,* 1024–1033.

Rossmeisl, M.; Jilkova, Z. M.; Kuda, O.; Jelenik, T.; Medrikova, D.; Stankova, B.; Kristinsson, B.; Haraldsson, G. G.; Svensen, H.; Stoknes, I.; et al. Metabolic Effects of n-3 PUFA as Phospholipids are Superior to Triglycerides in Mice Fed a High-fat Diet: Possible Role of Endocannabinoids. *PLoS One* **2012,** *7* (6), e38834.

Rossmeisl, M.; Medrikova, D.; van Schothorst, E. M.; Pavlisova, J.; Kuda, O.; Hensler, M.; Bardova, K.; Flachs, P.; Stankova, B.; Vecka, M.; et al. Omega-3 Phospholipids from Fish Suppress Hepatic Steatosis by Integrated Inhibition of Biosynthetic Pathways in Dietary Obese Mice. *Biochimica et Biophysical Acta* **2014,** *1841,* 267–278.

Ruiz-Lopez, N.; Haslam, R. P.; Usher, S. L.; Napier, J. A.; Sayanova, O. Reconstitution of EPA and DHA Biosynthesis in Arabidopsis: Iterative Metabolic Engineering for the Synthesis of n-3 LC-PUFAs in Transgenic Plants. *Metab. Eng.* **2013,** *17,* 30–41.

Ruiz-Lopez, N.; Haslam, R. P.; Napier, J. A.; Sayanova, O. Successful High-level Accumulation of Fish Oil Omega-3 Long-chain Polyunsaturated Fatty Acids in a Transgenic Oilseed Crop. *Plant J.* **2014,** *77,* 198–208.

Runau, F.; Arshad, A,; Isherwood, J.; Norris, L.; Howells, L.; Metcalfe, M.; Dennison, A. Potential for Proteomic Approaches in Determining Efficacy Biomarkers Following Administration of Fish Oils Rich in Omega-3 Fatty Acids: Application in Pancreatic Cancers. *Nutr. Clin. Pract.* [Online early access]. DOI:10.1177/0884533614567337. Published Online: Jan 23, 2015. http://ncp.sagepub.com/content/early/2015/01/22/0884533614567337.full.pdf+html (accessed Feb 17, 2015).

Sampalis, F.; Bunea, R.; Pelland, M. F.; Kowalski, O.; Duguet, N.; Dupuis, S. Evaluation of the Effects of Neptune Krill Oil on the Management of Premenstrual Syndrome and Dysmenorrhea. *Altern. Med. Rev.* **2003,** *8,* 171–179.

Schaefer, E. J.; Bongard, V.; Beiser, A. S.; Lamon-Fava, S.; Robins, S. J.; Au, R.; Tucker, K. L.; Kyle, D. J.; Wilson, P. W.; Wolf, P. A. Plasma Phosphatidylcholine Docosahexaenoic Acid Content and Risk of Dementia and Alzheimer Disease: The Framingham Heart Study. *Arch. Neurol.* **2006,** *63,* 1545–1550.

Schmitz, G.; Ecker, J. The Opposing Effects of n-3 and n-6 Fatty Acids. *Prog. Lipid Res.* **2008,** *47* (2), 147–155.

Schneider, M. Marine Phospholipids and Their Applications: Next-generation Omega-3 Lipids. *Nutr. Health* **2013,** 297–308.

Scholey, A. B.; Camfield, D. A.; Hughes, M. E.; Woods, W. K.; Stough, C. K.; White, D. J.; Gondalia, S. V.; Frederiksen, P. D. A Randomized Controlled Trial Investigating the Neurocognitive Effects of Lacprodan® PL-20, a Phospholipid-rich Milk Protein Concentrate, in Elderly Participants with Age-associated Memory Impairment: The Phospholipid Intervention for Cognitive Ageing Reversal (PLICAR): Study Protocol for a Randomized Controlled Trial. *Trials* **2013,** *14,* 404.

Schuchardt, J. P.; Hahn, A. Bioavailability of Long-chain Omega-3 Fatty Acids. *Prostaglandins Leukot. Essent. Fatty Acids* **2013,** *89,* 1–8.

Schuchardt, J. P.; Schneider, I.; Meyer, H.; Neubronner, J.; von Schacky, C.; Hahn, A. Incorporation of EPA and DHA into Plasma Phospholipids in Response to Different Omega-3 Fatty Acid Formulations—A Comparative Bioavailability Study of Fish Oil vs. Krill Oil. *Lipids Health Dis.* **2011,** *10,* 145.

Sedlmeier, E. M.; Brunner, S.; Much, D.; Pagel, P.; Ulbrich, S. E.; Meyer, H. D.; Amann-Gassner, U.; Hauner, H.; Bader, B. L. Human Placental Transcriptome Shows Sexually Dimorphic Gene Expression and Responsiveness to Maternal Dietary n-3 Long-chain Polyunsaturated Fatty Acid Intervention during Pregnancy. *BMC Genomics* **2014,** *15,* 941–960.

Shirai, N.; Higuchi, T.; Suzuki, H. Effect of Lipids Extracted from a Salted Herring Roe Food Product on Maze-Behavior in Mice. *J. Nutr. Sci. Vitaminol.* (Tokyo) **2006a,** *52,* 451–456.

Shirai, N.; Higuchi, T.; Suzuki, H. Analysis of Lipid Classes and the Fatty Acid Composition of the Salted Fish Roe Food Products, Ikura, Tarako, Tobiko and Kazunoko. *Food Chem.* **2006b,** *94,* 61–67.

Shirouchi, B.; Nagao, K.; Inoue, N.; Ohkubo, T.; Hibino, H.; Yanagita, T. Effect of Dietary Omega-3 Phosphatidylcholine on Obesity-related Disorders in Obese Otsuka Long-Evans Tokushima Fatty Rats. *J. Agric. Food Chem.* **2007,** *55,* 7170–7176.

Simopoulos, A. P. The Importance of the Ratio of Omega-6/Omega-3 Essential Fatty Acids. *Biomed Pharmacother.* **2002,** *56* (8), 365–379.

Simopoulos, A. P. The Importance of the Omega-6/Omega-3 Fatty Acid Ratio in Cardiovascular Disease and Other Chronic Diseases. *Exp. Biol. Med.* (Maywood) **2008,** *233* (6), 674–688.

Skarpańska-Stejnborn, A.; Pilaczyńska-Szcześniak, L.; Basta, P.; Foriasz, J.; Arlet, J. Effects of Supplementation with Neptune Krill Oil (*Euphasia Superba*) on Selected Redox Parameters and Pro-inflammatory Markers in Athletes During Exhaustive Exercise. *J. Hum. Kinet.* **2010,** *25,* 49–57.

Song, J. H.; Inoue, Y.; Miyazawa, T. Oxidative Stability of Docosahexaenoic Acid-containing Oils in the Form of Phospholipids, Triacylglycerols, and Ethyl Esters, *Biosci. Biotechnol. Biochem.* **1997,** *61,* 2085–2088.

Strobel, C.; Jahreis, G.; Kuhnt, K. Survey of n-3 and n-6 Polyunsaturated Fatty Acids in Fish and Fish Products. *Lipids Health Dis.* **2012,** *11,* 144–154.

Tall, A. R.; Blum, C. B.; Grundy, S. M. Incorporation of Radioactive Phospholipid into Subclasses of High-density Lipoproteins. *Am. J. Physiol.* **1983,** *244* (5), E513–E516.

Tam, O.; Innis, S. M. Dietary Polyunsaturated Fatty Acids in Gestation Alter Fetal Cortical Phospholipids, Fatty Acids and Phosphatidylserine Synthesis. *Dev. Neurosci.* **2006,** *28* (3), 222–229.

Tamura, A.; Tanaka, T.; Yamane, T. Quantitative Studies on Translocation and Metabolic Conversion of Lysophosphatidylcholine Incorporated into the Membrane of Intact Human Erythrocytes Fromthemedium. *J. Biochem.* **1985,** *97* (1), 353–359.

Tandy, S.; Chung, R. W. S.; Elaine, W. A. T.; Berge, A. K. K.; Griinari, M.; Cohn, J. S. Dietary Krill Oil Supplementation Reduces Hepatic Steatosis, Glycemia, and Hypercholesterolemia in High Fat-fed Mice. *J. Agric. Food Chem.* **2009,** *57* (19), 9339–9345.

Thies, F.; Pillon, C.; Moliere, P.; Lagarde, M.; Lecerf, J. Preferential Incorporation of sn-2 lysoPC DHA over Unesterified DHA in the Young Rat Brain. *Am. J. Physiol.* **1994,** *267* (5), R1273–R1279.

Tocher, D. R. Omega-3 Long-chain Polyunsaturated Fatty Acids and Aquaculture in Aquaculture Perspective. *Aquaculture* **2015.** DOI:10.1016/j.aquaculture.2015.01.010.

Tőkés, T.; Tuboly, E.; Varga, G.; Major, L.; Ghyczy, M.; Kaszaki, J.; Boros, M. Protective Effects of L-alpha-glycerylphosphorylcholine on Ischaemia Reperfusion Induced Inflammatory Reactions. *Eur. J. Nutr.* **2015,** *54* (1), 109–118.

Torres García, J.; Durán Agüero, S. Phospholipids: Properties and Health Effects. *Nutr. Hosp.* **2014,** *31* (1), 76–83.

Totani, Y.; Hara, S. Preparation of Polyunsaturated Phospholipids by Lipase-catalyzed Transesterification. *J. Am. Oil Chem. Soc.* **1991,** *68,* 848–851.

Tou, J. C.; Jaczynski, J.; Chen, Y. C. Krill for Human Consumption: Nutritional Value and Potential Health Benefits. *Nutr. Rev.* **2007,** *65,* 63–77.

Tso, P.; Drake, D. S.; Black, D. D.; Sabesin, S. M. Evidence for Separate Pathways of Chylomicron and Very Low-density Lipoprotein Assembly and Transport by Rat Small Intestine. *Am. J. Physiol.* **1984,** *247* (6), G599–G610.

Ulven, S. M.; Kirkhus, B.; Lamglait, A.; Basu, S.; Elind, E.; Haider, T.; Berge, K.; Vik, H.; Pedersen, J. I. Metabolic Effects of Krill Oil Are Essentially Similar to Those of Fish Oil but at Lower Dose of EPA and DHA, in Healthy Volunteers. *Lipids* **2011,** *46,* 37–46.

Vaidya, H.; Cheema, S. K. Sea Cucumber and Blue Mussel: New Sources of Phospholipid Enriched Omega-3 Fatty Acids with a Potential Role in 3T3-L1 Adipocyte Metabolism. *Food Funct.* **2014,** *5* (12), 3287–3295.

Vaisman, N.; Kaysar, N.; Zaruk-Adasha, Y.; Pelled, D.; Brichon, G.; Zwingelstein, G.; Bodennec, J. Correlation between Changes in Blood Fatty Acid Composition and Visual Sustained Attention Performance in Children with Inattention: Effect of Dietary n-3 Fatty Acids Containing Phospholipids. *Am. J. Clin. Nutr.* **2008,** *87,* 1170–1180.

Vance, D. E. Physiological Roles of Phosphatidylethanolamine n-Methyltransferase. *Biochim. Biophys. Acta* **2013,** *1831,* 626–632.

Vance, D. E.; Jacobson, G. R.; Saier, M. H., Jr. Lipids and Membranes. In *Biochemistry,* 3rd ed.; Zubay, G., Ed.; WCB Publishers: Dubuque, IA, 1993; pp 163–196.

van de Rest, O.; de Groot, L. C. P. G. M. The Impact of Omega-3 Fatty Acids on Quality of Life. In *Omega-3 Fatty Acids in Brain and Neurological Health;* Watson, R., De Meester, F., Eds.; Elsevier: London, 2014; pp 81–85.

Vicenova, M.; Nechvatalova, K.; Chlebova, K. Evaluation of in Vitro and in Vivo Anti-inflammatory Activity of Biologically Active Phospholipids with Anti-neoplastic Potential in Porcine Model. *BMC Complement. Altern. Med.* **2014,** *14* (1), 339.

Vigerust, N. F.; Bjorndal, B.; Bohov, P.; Brattelid, T.; Svardal, A.; Berge, R. K. Krill Oil *versus* Fish Oil in Modulation of Inflammation and Lipid Metabolism in Mice Transgenic for TNF-alpha. *Eur. J. Nutr.* **2013,** *52* (4), 1315-1325.

Vikbjerg, A. F.; Mu, H.; Xu, X. Monitoring of Monooctanoyl Phosphatidylcholine Synthesis by Enzymatic Acidolysis between Soybean Phosphatidylcholine and Caprylic Acid by Thin-layer Chromatography with a Flame Ionization Detector. *J. Agric. Food Chem.* **2005a,** *53,* 3937–3942.

Vikbjerg, A. F.; Peng, L.; Mu, H.; Xu, X. Continuous Production of Structured Phospholipids on a Packed Bed Reactor with Lipase from *Thermomyces lanuginose*. *J. Am. Oil Chem. Soc.* **2005b,** *82,* 237–242.

Vikbjerg, A. F.; Rusig, J. Y.; Jonsson, G.; Mu, H.; Xu, X. Comparative Evaluation of the Emulsifying Properties of Phosphatidylcholine after Enzymatic Acyl Modification. *J. Agric. Food Chem.* **2006,** *54,* 3310–3316.

Visioli, F.; Risè, P.; Plasmati, E.; Pazzucconi, F.; Sirtori, C. R.; Galli, C. Very Low Intakes of n-3 Fatty Acids Incorporated into Bovine Milk Reduce Plasma Triacylglycerol and Increase HDL Cholesterol B Concentrations in Healthy Subjects. *Pharmacol. Res.* **2000,** *41* (5), 571–576.

Voigt, R. G.; Llorente, A. M.; Jensen, C. L.; Fraley, J. K.; Berretta, M. C.; Heird, W. C. A Randomized, Double-blind, Placebo-controlled Trial of Docosahexaenoic Acid Supplementation in Children with Attention-deficit/Hyperactivity Disorder. *J. Pediatr.* **2001,** *139* (2), 189–196.

Wang, C.; Harris, W. S.; Chung, M.; Lichtenstein, A. H.; Balk, E. M.; Kupelnick, B.; Jordan, H. S.; Lau, J. n-3 Fatty Acids from Fish or Fish-Oil Supplements, but not Alpha-linolenic Acid, Benefit Cardiovascular Disease Outcomes in Primary- and Secondary-prevention Studies: A Systematic Review. *Am. J. Clin. Nutr.* **2006,** *84,* 5–17.

Wang, X.; Hjorth, E.; Vedin, I.; Eriksdotter, M.; Freund-Levi, Y.; Wahlund, L. O.; Cederholm, T.; Palmblad, J.; Schultzberg, M. Effects of n-3 Fatty Acid Supplementation on the Release of Pro-resolving Lipid Mediators by Blood Mononuclear Cells: The OmegAD Study. *J. Lipid Res.* [Online early access]. DOI: 10.1194/jlr.P055418. Published Online: Jan 23, 2015. http://www.jlr.org/content/early/2015/01/23/jlr.P055418.long (accessed Feb 17, 2015).

Wijendran, V.; Huang, M. C.; Diau, G. Y.; Boehm, G.; Nathanielsz, P. W.; Brenna, J. T. Efficacy of Dietary Arachidonic Acid Provided as Triglyceride or Phospholipid as Substrates for Brain Arachidonic Acid Accretion in Baboon Neonates. *Pediatr. Res.* **2002,** *51* (3), 265–272.

Winther, B.; Hoem, N.; Berge, K.; Reubsaet, L. Elucidation of Phosphatidylcholine Composition in Krill Oil Extracted from *Euphausia superba*. *Lipids* **2011,** *46,* 25–36.

Xu, R.; Hung, J. B.; German, J. B. Effects of Dietary Lipids on the Fatty Acid Composition of Triglycerides and Phospholipids in Tissues of White Sturgeon. *Aquacult. Nutr.* **1996,** *2,* 101–109.

Yamamoto, Y.; Mizuta, E.; Ito, M.; Harata, M.; Hiramoto, S.; Hara, S. Lipase-catalyzed Preparation of Phospholipids Containing n-3 Polyunsaturated Fatty Acids from Soyphospholipids. *J. Oleo Sci.* **2014,** *63* (12), 1275–1281.

Yanishlieva, N. V.; Marinova, E. M. Stabilisation of Edible Oils with Natural Antioxidants. *Eur. J. Lipid Sci. Technol.* **2001,** *103,* 752–767.

Young, G.; Conquer, J. Omega-3 Fatty Acids and Neuropsychiatric Disorders. *Reprod. Nutr. Dev.* **2005,** *45* (1), 1–28.

Zanotti, A.; Valzelli, L.; Toffano, G. Chronic Phosphatidylserine Treatment Improves Spatial Memory and Passive Avoidance in Aged Rats. *Psychopharmacology* **1989,** *99* (3), 316–321.

Zendedel, A.; Habib, P.; Dang, J.; Lammerding, L.; Hoffmann, S.; Beyer, C.; Slowik, A. Omega-3 Polyunsaturated Fatty Acids Ameliorate Neuroinflammation and Mitigate Ischemic Stroke Damage through Interactions with Astrocytes and Microglia. *J. Neuroimmunol.* **2015,** *278* (15), 200–211.

Zhang, K.; Liu, X.; Yu, Y.; Luo, T.; Wang, L.; Ge, C.; Liu, X.; Song, J.; Jiang, X.; Zhang, Y.; et al. Phospholipid Transfer Protein Destabilizes Mouse Atherosclerotic Plaque. *Arterioscler. Thromb. Vasc. Biol.* **2014,** *34* (12), 2537–2544.

Zhao, J.; Wei, S.; Liu, F.; Liu, D. Separation and Characterization of Acetone-soluble Phosphatidylcholine from Antarctic Krill (Euphausia superba) Oil. *Eur. Food Res. Technol.* **2014,** *238* (6), 1023–1028.

Zhao, T.; No, D. S.; Kim, B. H.; Garcia, H. S.; Kim, Y. Kim, I. H. Immobilized Phospholipase A_1-catalyzed Modification of Phosphatidylcholine with n-3 Polyunsaturated Fatty Acid. *Food Chem.* **2014,** *157,* 132–140.

Zierenberg, O.; Grundy, S. M. Intestinal Absorption of Polyenephosphatidylcholine in Man. *J. Lipid Res.* **1982,** *23* (8), 1136–1142.

About the Editors

Moghis U. Ahmad

Dr. Moghis U. Ahmad is Vice President of Chemical Technology and Manufacturing at Jina Pharmaceuticals Inc., Libertyville, Illinois. Earlier he held the position of vice president of the Lipid Chemistry Division at NeoPharm Inc., Waukegan, Illinois, and chemist at Sigma-Aldrich Corp., St. Louis, Missouri. Dr. Ahmad obtained Ph.D. in chemistry (1978) from AMU, Aligarh, India, and completed postdoctoral research at Texas A & M University, College Station, Texas, and Oregon State University, Corvallis, Oregon. Dr. Ahmad's proven dedication in research and development over the last 35 years has resulted in more than 50 research publications in top peer-reviewed scientific journals, over 30 patents and patent applications, and several book chapters. He edited and contributed book chapters in *Lipids in Nanotechnology* (AOCS Press, 2011), which was one of the top 10 best sellers in 2012–2013.

Dr. Ahmad is a very accomplished lipid chemist whose research is recognized both nationally and internationally. He is recognized as an authority on phospholipids and has been a leader in the American Oil Chemists' Society Phospholipid Division. He is the past chair of the AOCS Phospholipid Division and currently holds the vice president position for the International Lecithin and Phospholipids Society (ILPS). He has been an active member of the AOCS and the American Chemical Society (ACS) for more than 30 years. Dr. Ahmad is a fellow of AOCS and the Royal Society of Chemistry, London.

Xuebing Xu

Dr. Xuebing Xu is the general manager for Wilmar Global Research and Development Center. He is also an honorary professor in the Faculty of Science and Technology of Aarhus University, Denmark. His research interests include lipid technology, enzyme technology, food functionality, food nanotechnology, biofuel technology, and so forth, with particular focus on structured lipids, biocatalysis, process technology, phospholipids, ionic liquids, biodiesels, food physics, and others. Dr. Xu is also associate editor for the Journal of the American Oil Chemists' Society, *ad hoc*

reviewer for more than 150 papers, and a member of the editorial boards for a few journals. He has authored more than 200 publications and over 30 book chapters on lipid-related research.

Contributors

AOCS Press extends gratitude and appreciation to the *Polar Lipids: Biology, Chemistry, and Technology* authors who helped make this title possible.

Ateeq Ahmad
Jina Pharmaceuticals Inc.,
Libertyville, IL, USA

Imran Ahmad
Jina Pharmaceuticals Inc.,
Libertyville, IL, USA

Moghis Ahmad
Jina Pharmaceuticals Inc.,
Libertyville, IL, USA

Shoukath M. Ali
Jina Pharmaceuticals Inc.,
Libertyville, IL, USA

Yuta Atsumi
R&D Section, Biochemical Division,
Nagase ChemteX Corporation, Kyoto,
Japan

Takeshi Bamba
Department of Biotechnology,
Graduate School of Engineering, Osaka
University, Osaka, Japan

Manat Chaijan
Division of Food Technology,
Department of Agro-industry, School
of Agricultural Technology, Walailak
University, Thasala, Nakhon Si
Thammarat, Thailand

Koen Dewettinck
Department of Food Safety and Food
Quality, Ghent University, Belgium

Bernd W. K. Diehl
Spectral Service GmbH
Laboratoriumfür Auftragsanalytik,
Cologne, Germany

Eiichiro Fukusaki
Department of Biotechnology,
Graduate School of Engineering, Osaka
University, Osaka, Japan

Estefania N. Guiotto
Centro de Investigación y Desarrollo
en Criotecnología de Alimentos,
Universidad Nacional de La Plata,
La Plata, Argentina

Ruihua Guo
Wilmar (Shanghai) Biotechnology
Research and Development Center Co.,
Ltd., Shanghai, China

Ram Chandra Reddy Jala
Centre for Lipid Research, CSIR-Indian
Institute of Chemical Technology,
Hyderabad, India

Derya Kahveci
Department of Food Engineering,
Faculty of Engineering, Yeditepe
University, Istanbul, Turkey

Arnis Kuksis
University of Toronto

Mickaël Laguerre
UMR IATE, Montpellier, France

Thien Trung Le
Faculty of Food Science and Technology, Ho Chi Minh City Nong Lam University, Vietnam

G.R. List
USDA, Retired

Xiaoli Liu
R&D Center, Nagase Co., Ltd., Hyogo, Japan

Zonghui Ma
Wilmar (Shanghai) Biotechnology Research and Development Center Co., Ltd., Shanghai, China

Yumiko Nagasawa
Department of Biotechnology, Graduate School of Engineering, Osaka University, Osaka, Japan

Willem van Nieuwenhuyzen
Lecipro Consulting, The Netherlands

Worawan Panpipat
Division of Food Technology, Department of Agro-industry, School of Agricultural Technology, Walailak University, Thasala, Nakhon Si Thammarat, Thailand

Thi Thanh Que Phan
Department of Food Safety and Food Quality, Ghent University, Belgium

R.B.N. Prasad
Centre for Lipid Research, CSIR-Indian Institute of Chemical Technology, Hyderabad, India

W. Pruzanski
University of Toronto

Mabel C. Tomás
Centro de Investigación y Desarrollo en Criotecnología de Alimentos, Universidad Nacional de La Plata, La Plata, Argentina

Saifuddin Sheikh
Jina Pharmaceuticals Inc., Libertyville, IL, USA

Misa Shiihara
R&D Section, Biochemical Division, Nagase ChemteX Corporation, Kyoto, Japan

Naoki Shirasaka
R&D Section, Biochemical Division, Nagase ChemteX Corporation, Kyoto, Japan

Masatoshi Shiojiri
R&D Section, Biochemical Division, Nagase ChemteX Corporation, Kyoto, Japan

Kaori Taguchi
Department of Biotechnology, Graduate School of Engineering, Osaka University, Osaka, Japan

Naruyuki Taniwaki
R&D Section, Biochemical Division, Nagase ChemteX Corporation, Kyoto, Japan

John Van Camp
Department of Food Safety and Food Quality, Ghent University, Belgium

Pierre Villeneuve
UMR IATE, Montpellier, France

Benrong Xin
Wilmar (Shanghai) Biotechnology Research and Development Center Co., Ltd., Shanghai, China

Xuebing Xu
Wilmar (Shanghai) Biotechnology Research and Development Center Co., Ltd., Shanghai, China

Takayuki Yamada
Department of Biotechnology, Graduate School of Engineering, Osaka University, Osaka, Japan

Kangyi Zhang
Institute of Agricultural Products Processing, Henan Academy of Agriculture Sciences, Henan Province, China

Minying Zheng
Wilmar (Shanghai) Biotechnology Research and Development Center Co., Ltd., Shanghai, China

Yan Zheng
Wilmar (Shanghai) Biotechnology Research and Development Center Co., Ltd., Shanghai, China

Index

Page numbers followed by f and t indicate figures and tables, respectively.

CPSIA information can be obtained at www.ICGtesting.com
Printed in the USA
LVOW04*0957100415

433843LV00001B/1/P

9 781630 670443